Airport Competition
The European Experience

EDITED BY

PETER FORSYTH
Monash University, Australia
DAVID GILLEN
University of British Columbia, Canada
JÜRGEN MÜLLER
Berlin School of Economics
HANS-MARTIN NIEMEIER
University of Applied Sciences, Germany

ASHGATE

© Peter Forsyth, David Gillen, Jürgen Müller and Hans-Martin Niemeier 2010

All rights reserved. No part of this publication may be reproduced, stored in a retrieval system or transmitted in any form or by any means, electronic, mechanical, photocopying, recording or otherwise without the prior permission of the publisher.

Peter Forsyth, David Gillen, Jürgen Müller and Hans-Martin Niemeier have asserted their moral rights under the Copyright, Designs and Patents Act, 1988, to be identified as the editors of this work.

Published by
Ashgate Publishing Limited
Wey Court East
Union Road
Farnham
Surrey, GU9 7PT
England

Ashgate Publishing Company
Suite 420
101 Cherry Street
Burlington
VT 05401-4405
USA

www.ashgate.com

British Library Cataloguing in Publication Data
Airport competition : the European experience.
   1. Airports--Economic aspects--Europe. 2. Airports--
   Economic aspects--Europe--Case studies. 3. Aeronautics,
   Commercial--Government policy--Europe.
   I. Forsyth, P. (Peter)
   387.7'36'094-dc22

ISBN: 978-0-7546-7746-8 (hbk)
ISBN: 978-0-7546-9484-7 (ebk)

Library of Congress Cataloging-in-Publication Data
Airport competition : the European experience / edited by Peter Forsyth ... [et al.].
   p. cm.
   Includes bibliographical references and index.
   ISBN 978-0-7546-7746-8 (hardback) -- ISBN 978-0-7546-9484-7
   (e-book) 1. Airports--Europe--Case studies. 2. Airports--Management--Europe--Case studies. 3. Competition-
   -Europe--Case studies. I. Forsyth, P. (Peter)
   HE9842.A4A37 2009
   387.7'36--dc22
                                                                                            2009030234

Printed and bound in Great Britain by
MPG Books Group, UK

Contents

List of Figures		*ix*
List of Tables		*xi*
Acknowledgements		*xiii*
Editors and Contributors		*xv*

1 Introduction and Overview 1
   Peter Forsyth, David Gillen, Jürgen Müller and Hans-Martin Niemeier

PART A: HOW DO AIRPORTS COMPETE AND HOW STRONG IS COMPETITION?

2 Airport Competition and Network Access: A European Perspective 11
   Dr Peter Morrell

3 Airport Entry and Exit: A European Analysis 27
   Christiane Müller-Rostin, Hansjochen Ehmer, Ignaz Hannak, Plamena Ivanova, Hans-Martin Niemeier and Jürgen Müller

4 Airport Pricing 47
   Eric Pels and Erik T. Verhoef

5 Countervailing Power to Airport Monopolies 59
   Kenneth Button

6 Competition Between Major and Secondary Airports: Implications for Pricing, Regulation and Welfare 77
   Peter Forsyth

7 Airport Strategies to Gain Competitive Advantage 89
   Dr Anne Graham

8 An Empirical Analysis of Airport Operational Costs 103
   Eric Pels, Daniel van Vuuren, Charles Ng and Piet Rietveld

9 Competition Between Airports: Occurrence and Strategy 119
   Dr Michael Tretheway and Ian Kincaid

10 Airport Competition for Freight 137
    Dr Michael W. Tretheway and Robert J. Andriulaitis

PART B: TRAVELLER CHOICE AND AIRPORT COMPETITION

11 Modelling Air Travel Choice Behaviour 151
    Stephane Hess

12 Airport Choice Behaviour: Findings from Three Separate Studies 177
    Stephane Hess and John W. Polak

13 Improved Modelling of Competition among Airports through Flexible Form and Non-Diagonal Demand Structures Explaining Flows Registered within a New Traffic Accounting Matrix 197
    Marc Gaudry

PART C: CASE STUDIES OF AIRPORT COMPETITION

14 Competition in the German Airport Market: An Empirical Investigation 239
    Robert Malina

15 Competition among Airports and Overlapping Catchment Areas: An Application to the State of Baden-Württemberg 261
    Daniel Strobach

16 Airport Competition in Greece: Concentration and Structural Asymmetry 277
    Andreas Papatheodorou

17 The Airport Industry in a Competitive Environment: A United Kingdom Perspective 291
    David Starkie

18 The Effect of Low-Cost Carriers on Regional Airports' Revenue: Evidence from the UK 311
    Zheng Lei, Andreas Papatheodorou and Edith Szivas

PART D: POLICY ISSUES

19 Competition and the London Airports: How Effective Will It Be? 321
    Peter Forsyth and Hans-Martin Niemeier

20 Airport Alliances and Multi-Airport Companies: Implications for Airport Competition 339
    Peter Forsyth, Hans-Martin Niemeier and Hartmut Wolf

21 Airport Competing Terminals: Recent Developments at Dublin Airport 353
    Aisling Reynolds-Feighan

22 Competition, State Aids and Low-Cost Carriers: A Legal Perspective 365
    Hans Kristoferitsch

23	Subsidies and Competition: An Economic Perspective *Dan Elliott*	379
24	Competition for Airport Services – Ground Handling Services in Europe: Case Studies on Six Major European Hubs *Cornelia Templin*	393
25	Airport Competition: Market Dominance and Abuse *Peter Lewisch*	413
26	Airport Competition: A Perspective and Synthesis *Peter Forsyth*	427

Airport Competition: Some Key References   *437*
Airport Index   *439*
Index   *441*

List of Figures

2.1	City-pairs served by LCCs at Stansted accessible at similar prices	12
3.1	Airport entries and exits	36
3.2	Entry and exit of German airports	37
5.1	Price and output under a pure bilateral monopoly	62
5.2	Third-degree price discrimination between low-cost and legacy carriers	71
8.1	Long and short run cost curves	105
8.2	Cost as a function of the number of passengers	109
8.3	Cost as a function of the fraction of international passengers	110
8.4	Distribution of scale elasticities (PAX) over different airports	115
8.5	Distribution of scale elasticities (ATM) over different airports	115
8.6	Distribution of fixed effects over different airports	116
9.1	Percentage change in marketing staff per passenger at selected UK airports 1991–1997	120
9.2	Growth in world GDP, passenger traffic and cargo traffic	129
11.1	Main choice processes of an outbound air journey	155
11.2	Interactions between air travel choice dimensions	157
11.3	Downwards interactions between upper level choices and air travel choice dimensions	158
11.4	Upwards interactions between air travel choice dimensions and upper level choices	160
11.5	Structure of two-level NL model, using nesting along airport dimension	169
11.6	Structure of three-level NL model, using nesting along airport dimension and airline dimension	171
11.7	Structure of CNL model for the joint analysis of correlation along the airport, airline and access mode dimensions	172
13.1	Response asymmetry and non-linearity: Classical Linear-Logit vs Standard Box-Cox-Logit	217
13.2	Fare elasticities in intercity Box-Cox Logit model	218
13.3	Travel time elasticities in intercity Box-Cox Logit model	219
13.4	Comparing Linear and Box-Cox Logit forecasts of an ICE train scenario in Germany	222
13.5	Comparing Linear and Box-Cox Logit forecasts of intermodal train gains	222
13.6	Comparing Generalized Box-Cox and Linear Logit forecasts of intermodal train gains	227
14.1	Spatial distribution, size and ownership structure of the German airport market	242
14.2	Spatial distribution of German and neighbouring foreign airports open to public traffic with a runway of at least 1,000 metres length and ILS	248
15.1	Catchment area and competitive zones of Stuttgart Airport	273
15.2	Leadership position of airports in Baden-Württemberg	274
16.1	Evolution of the HHI, 1986–2005	282

16.2	GDP and passenger traffic in NUTS3 regions, 2000	284
16.3	GDP and passenger traffic in NUTS3 regions, 2000	285
16.4	Population and passenger traffic in NUTS3 regions, 2000	286
16.5	Population and passenger traffic in NUTS3 regions, 2000	286
17.1	Competition and catchment areas	300
17.2	Operating profit as a percentage of fixed assets v turnover (£000)	306
19.1	Price and quality	333
21.2	Passenger traffic and traffic growth rates at Cork, Shannon and Dublin airports, 1995–2006	356
21.1	Regional breakdown of traffic at the three Irish state airports, 2000 and 2006	356
24.1	Major players in the ground handling industry	395
24.2	Overview of the six major European airports in the analysis	397
24.3	Ground handling companies at six major hubs in Europe	397
24.4	Shifts in market share after deregulation (1999–2004)	400
24.5	Market split at London Heathrow 2004 (turnarounds)	402
24.6	Market split at Paris Charles de Gaulle 2004 (turnarounds)	404
24.7	Market split at Frankfurt 2004 (turnarounds)	405
24.8	Market growth of Acciona at Frankfurt	405
24.9	Market split at Amsterdam Schiphol 2004 (turnarounds)	406
24.10	Market split at Madrid Barajas 2004 (turnarounds)	407
24.11	Market split at Rome Fiumicino 2004 (turnarounds)	408
25.1	Competition between airports for fuelling stops	417
25.2	Threefold structure in aviation	420
25.3	Threefold structure and vertical disintegration in ground handling	421

List of Tables

5.1	Operating margins at European airports	65
5.2	Market share of passengers by airline at Europe's 10 largest airports (2002)	67
5.3	Market capitalizations of leading airlines ($ billions) in November 2005	67
5.4	Low-cost carriers and airports in metropolitan regions (2006)	69
8.1	Average values of four key variables	107–108
8.2	Estimation results for a number of specifications of cost functions	112–113
8.3	Average scale elasticities in different model specifications	114
9.1	Feature presented by typical primary and secondary airports	125
10.1	Snapshot of the making of a logistics giant: Deutsche Post AG	141
10.2	Top European cargo airports	142
12.1	Summary of choice data for SF-bay area case study	179
12.2	Model performance on SF-bay area data in terms of adjusted $\rho^2(0)$ measure	181
12.3	Trade-offs between flight frequency and access time (min/flight) in models for combined choice of airport, airline and access mode	183
12.4	Prediction performance on SF-bay area validation data	185
12.5	Model performance on London data	187
12.6	Model results for London data	188
12.7	Prediction performance on London validation data	189
12.8	MNL trade-offs, part 1: willingness to pay ($)	193
12.9	MNL trade-offs, part 2: willingness to accept increases in access time (min)	194
13.1	Traffic accounting matrix (TAM) representation of the physical transport network flows	200
13.2	Flow matrix for the four-city system without direct connections from Airport J to Airport B	202
13.3	Relationship between input-output (I-O) and traffic accounting matrices (TAM)	203
13.4	Market and network analysis: a three-level approach	205
13.5	Example of full price matrix used in three-mode representative utility function set	207
13.6	Flexible form, heteroskedasticity and spatial correlation in freight trade models	209
13.7	Flexible form, heteroskedasticity and spatial correlation in passenger transport models	211
13.8	Selected own share elasticities of Box-Cox Logit models used in the simulations	221
13.9	Comparing Linear with Standard and Generalized Box-Cox elasticities and values of time	224
13.10	Linear, Standard and Generalized Logit mode choice model of freight across the Pyrenees	225
14.1	Ownership structure of privatized German airports in spring 2008	240–240
14.2	Take-off distance required (TODR) for various aircraft with maximum take-off weight (MTOW) and 3/4 load capacity	246

14.3	Best substitutes and substitution coefficients for 35 German airports	252
15.1	Worldwide arrivals at selected European airports, 2004	266
15.2	Share of individual factors	268
15.3	Airports included in the study	269
15.4	Resulting scores	270
15.5	Leader and competitors	272
16.1	Results of correlation analysis	284
17.1	Selected financial and operating data for UK airports, 2005–06	293
17.2	Ownership patterns at main airports in the United Kingdom, 2007	294
17.3	UK operating bases for four non-legacy airlines, summer 2008	299
17.4	Driving times between adjacent airports (hours.minutes)	303
17.5	Financial data for the smaller UK airports, 2005–06	305
17.6	Net return (%), airports and UK private non-financial sector, 2005–06	307
18.1	Descriptive statistics	313
18.2	Impact on regional airport's non-aeronautical revenue	314
18.3	Impact on regional airport's aeronautical revenue	316
18.4	Impact on regional airports' aggregate revenue	317
19.1	Overview of regulatory milestones of BAA	323
19.2	Passenger numbers at London airports (millions)	323
19.3	Regulation X-factor	324
20.1	Horizontal integration of airlines and of airports	341
20.2	Airport companies: Stakeholdings and competition	345
21.1	Key legislative and regulatory instruments governing operation of Dublin Airport	354
21.3	Top six carriers and market shares at Dublin airport in 1996, 2000 and 2006	357
21.4	The road to Terminal 2 at Dublin Airport – key milestones	359
24.1	Third party competition per terminal	409
25.1	Threefold structure of transport markets	414
25.2	Market power in threefold structure of transport market	415

Acknowledgements

We would like to thank Achim Czerny, Karsten Fröhlich, Anne Graham, Cathal Guiomard, Stephane Hess, Kai Hüschelrath, Jörg Last, Zheng Lei, Andreas Papatheodorou, Eric Pels, David Starkie and Michael Tretheway for carefully reviewing chapters and providing valuable insights. We would like to also thank Lorra Ward for providing suggestions to improve English expressions. Furthermore, we would like to thank Harald Wiese and Fenja Fahle for preparing the index.

Editors and Contributors

Robert Andriulaitis is the Vice President, Transportation and Logistics Studies with Inter*VISTAS* Consulting Inc. He has 18 years' experience in transportation policy, cargo economics and marketing, foreign trade zone development and business logistics, in both the private and public sectors. With an MSc in Transportation and Logistics from the University of British Columbia, and experience in air, rail, trucking and intercity bus transportation issues, he brings a multi-modal perspective to transportation and logistics challenges. Prior to joining Inter*VISTAS*, he worked for the transportation policy departments of the governments of Alberta and Manitoba in Canada.

Kenneth Button is a professor at George Mason University, Virginia, where he is the Director of the Center for Aerospace Policy Research and the Center for Transportation Policy, Operations, and Logistics. He holds degrees in Economics from the universities of East Anglia, Leeds and Loughborough. He is editor of the *Journal of Air Transport Management* and of *Transportation Research Series D: Transport and Environment*. He is currently a visiting professor at the University of Bergamo and Porto University. He has written or edited over 100 books and authored over 400 articles and book chapters.

Hansjochen Ehmer, after his business administration studies at the University of Cologne, worked as a university assistant at the Institute of Transport Science of the University of Münster from 1984 to 1988. Since then he has been a member of the German Aerospace Center in Cologne. In 1996 he received his PhD in Economics with a dissertation on competition policy in regional air transport. Since 2002, he has been a Professor of Aviation Management, Economics and Business Ethics at the International University of Applied Sciences (IUAS) Bad Honnef, Bonn. He is one of the project leaders on the German Airport Performance (GAP) project.

Dan Elliott is a Director of Frontier Economics. He is a specialist in the economics of network industries, focusing particularly on transport, telecommunications and water. His work covers the full range of competition, regulatory and pricing issues arising in these sectors, and since privatization began in the UK in the 1980s he has advised many companies in proceedings with regulatory bodies and the Competition Commission.

Peter Forsyth has been Professor of Economics at Monash University, Australia since 1997. Prior to this he held posts at Australian National University and the University of New England. He holds degrees from the University of Sydney and the University of Oxford. He has specialized in the economics of transport, especially aviation, privatization and regulation, and the economics of tourism. Most recently, he has been paying particular attention to the privatization and regulation of airports, and to the use of computable general equilibrium models in evaluating the economic impacts of tourism.

Marc Gaudry, a specialist of transport demand, road safety and network accounting, is Director of Agora Jules Dupuit, a research network and website based at the Université de Montréal. He is also an associate researcher at INRETS, the French National Institute of Transport Research. His work has been recognized by over 1,700 citations in scientific documents, as well as by the Alexander von Humboldt and Ernst-Blickle research awards. He has implemented large-scale models for national ministries of transport, notably those of Canada, France, Germany and

Quebec. As a former Commissioner of the Canadian Royal Commission on National Passenger Transportation, he also acts as a scientific mediator in large-scale research or planning projects.

David Gillen graduated in 1975 from the University of Toronto with a PhD in Economics. He joined the University of British Columbia in 2005 and currently holds the positions of YVR Professor of Transportation Policy in the Sauder School of Business and is Director, Centre for Transportation Studies, University of British Columbia. He is also research economist at the Institute of Transportation Studies at the University of California, Berkeley. His current research includes pricing and auction mechanisms for roadways and runways, measuring performance of transportation infrastructure, vertical contracts in aviation and evolving network strategies and business models in airlines, airports, ports and gateways.

Anne Graham has been involved in the teaching, research and consultancy of air transport for over 20 years and has developed two key research interests. Her first area of research is airport management, economics and regulation. The third edition of her book *Managing Airports* was published in 2008. Her other research interest is the analysis of tourism and aviation demand and the relationship between the tourism and aviation industries. She has co-edited a book entitled *Aviation and Tourism: Implications for Leisure Travel*. She has written many conference papers and articles about these two research areas and is on the editorial board of the *Journal of Airport Management*.

Ignaz Hannak is enrolled at the International University of Applied Sciences Bad Honnef for Aviation Management Studies and was due to graduate in the summer of 2009. He gained aviation experience through internships at Katowice International Airport, Air Berlin and Lufthansa. He is a research assistant on the German Airport Performance Project at the IUAS Bad Honnef.

Stephane Hess is a principal research fellow in the Institute for Transport Studies at the University of Leeds, working in the area of choice modelling. Here, he has made several recent contributions to the state-of-the art in the specification, estimation and interpretation of such models, while also publishing a number of papers on the benefits of advanced structures in actual large-scale transport analyses, for example in the context of air travel behaviour research. His contributions have been recognized by the 2005 Eric Pas prize for the best PhD thesis in the area of travel behaviour modelling, as well as various other awards. He is also the editor-in-chief of the *Journal of Choice Modelling* and the organizer of the International Choice Modelling Conference.

Plamena Ivanova is a masters student in Economics and Management Science at Humboldt University, Berlin. She is a research assistant on the German Airport Performance (GAP) Project at FHW Berlin. Plamena is a co-author of several papers within the GAP project, working mainly in the field of airport charges and the development of the airport market.

Vanessa Kamp is a research assistant on the German Airport Performance project at the University of Applied Sciences, Bremen. She holds a Diploma degree in Economics, with the emphasis on regional policy and logistics in Bremen, and also received a Bachelor of Arts in Business Economics from the University of Hertfordshire in England. Research interests include the privatization and regulation of airports and the productivity and efficiency measurement of public utilities.

Ian Kincaid is Director, Economic Analysis with Inter*VISTAS*-EU Consulting Inc. Mr Kincaid has 14 years' experience in transport economics, demand forecasting, cost benefit analysis and economic impact analysis. Prior to joining Inter*VISTAS*, he was employed as a senior consultant at a London-based transport consultancy. Over his consulting career he has worked on projects for clients in Europe, North and South America, Africa and Australia. Mr Kincaid has a degree in Economics from the University of Leicester and a masters in Operational Research from the London School of Economics.

Hans Kristoferitsch studied at the University of Vienna (masters in Law, 2002, and History, 2004; doctorate in Law, 2004). His thesis 'Vom Staatenbund zum Bundesstaat', was a comparison between developments from confederacies of states to federal states, published in 2007. He also studied at Harvard University (LLM, 2005). From 2005 to 2007 he was an assistant at the European Institute of the University of Economics and Business Administration, Vienna with Professor Stefan Griller. Currently Hans is an associate with CHSH Attorneys at Law, Vienna.

Peter Lewisch gained doctorates in Law in 1985 and Economics (1987), both at the University of Vienna. He also achieved 'Habilitation' in Criminal Law (1992) and Constitutional Law (1993). In 1990 and 1992–93 he was a visiting scholar at the Center for Public Choice, Virginia. He has been an attorney at law since 1996, including academic assignment to the University of Vienna Law School. Since 2002 he has served as Professor of Law at Imadec University, Vienna, and visiting professor of EU Law at George Mason University, Virginia. He has also made several visits to Harvard and Cornell Law Schools. His main fields of interest are competition law, regulation, criminal and constitutional law, EU law, law and economics, and public choice.

Robert Malina is a lecturer in transport economics, competition policy and regulation at Münster University in Germany and works as a researcher for the Münster Institute of Transport Economics. He is also part of the faculty of Bonn University and Warsaw School of Economics, teaching Aviation Economics at both institutions. In 2005 Robert obtained his PhD from Münster University on issues of airport competition. His current research projects in aviation are concerned with the economic evaluation of catalytic effects of airport infrastructure.

Peter Morrell graduated in Economics from Cambridge University and has a masters in Air Transportation from the Massachusetts Institute of Technology. He also has a doctorate in airline capital productivity from Cranfield University. He is a former head of the Department of Air Transport at the College of Aeronautics at Cranfield University, where he has a Chair in Air Transport Economics and Finance. He is the department's Director of Research, is the European editor for the *Journal of Air Transport Management*, is on the editorial board of *Tourism Economics* and is a fellow of the Chartered Institute of Transport.

Jürgen Müller received his PhD at Stanford University. He worked as a research fellow at the Science Center Berlin and was Research Director at the German Institute for Economic Research in Berlin. Currently he holds a Chair in Economics at the Berlin School of Economics. He was the lead economist for both the Association of European Airlines study on the Cost of Air Traffic Control Delay and for the studies of the German airspace users' group on ATC delay and airport capacity constraints. He is one of the project leaders of the German Airport Performance Project and is the vice chair of the German Aviation Research Society (GARS).

Christiane Müller-Rostin is a sales account manager for Lufthansa Cargo AG. She holds a degree in Latin American Regional Studies with a major in Economics from the University of Cologne. Her masters thesis focused on market entry and deterrence strategies in the Brazilian aviation industry. After graduating in 2006, Christiane worked for a year as a research associate on the German Airport Performance project, preparing and delivering papers and presentations on airport entry and exit in Europe, coordinating project research and database compilation, and handling other project-related details. She also assisted in teaching Economics at the University of Applied Sciences Bremen.

Charles Ng works at the Economic Regulation Group, UK Civil Aviation Authority. He received his PhD in Applied Economics from the University of Minnesota, USA. His major research interests include econometric modelling of air travel demand and issues related to airline and airport competition.

Hans-Martin Niemeier is a professor of Transportation Economics and Logistics at the University of Applied Sciences Bremen. He received his PhD in Economics at the University of Hamburg and worked in the aviation section of the State Ministry of Economic Affairs of Hamburg. He is president of the German Aviation Research Society. His research focuses on airport regulation and management.

Dr Andreas Papatheodorou is Assistant Professor in Industrial Economics with emphasis on Tourism at the University of the Aegean, Greece. He is also an Honorary Research Fellow at the Nottingham University Business School, UK. Andreas holds an MPhil in Economics and a DPhil in Geography from the University of Oxford. He is also a Partner at the Air Consulting Group.

Eric Pels graduated in Economics at the Vrije Universiteit in Amsterdam in 1994. In 2000 he successfully defended his doctoral dissertation, entitled 'Airport economics and policy: efficiency, competition, and interaction with airlines'. He is an assistant professor at the Vrije Universiteit and teaches courses at undergraduate and graduate level. His research focuses on the development of transport networks, pricing transport networks and the pricing of infrastructure.

John Polak is Professor of Transport Demand and Director of the Centre for Transport Studies at Imperial College, London. Professor Polak is a mathematician by background, with over 25 years' experience in transport research and teaching, specializing in the areas of mathematical and statistical transport modelling and analysis. He is a past president of the International Association for Travel Behaviour Research. Professor Polak has been in the forefront of innovative transport model development in the UK for a number of years, and has published extensively on a number of aspects of travel behaviour and demand.

Dr Aisling Reynolds-Feighan is Associate Professor of Transport Economics in the School of Economics at University College, Dublin (UCD), and Director of the UCD Transport Policy Research Institute. She holds BA and MA degrees in Economics from UCD. Prior to joining the School of Economics at UCD, she attended the University of Illinois at Urbana-Champaign, graduating with a PhD in 1989. Her main research interests are air and road transport economics and logistics. Aisling has published widely on comparative US and European air transportation systems, networks and policies.

Piet Rietveld studied econometrics at Erasmus University, Rotterdam (cum laude degree) and received his PhD in Economics at Vrije Universiteit, Amsterdam. He worked at the International Institute of Applied Systems Analysis (Austria) and was research coordinator at Universitas Kristen Satya Wacana in Salatiga, Indonesia. Since 1990 he has been professor of Transport Economics in the Faculty of Economics, Vrije Universiteit, Amsterdam. He is also a fellow at the Tinbergen Institute. His research interests are transport and regional development, valuing quality of transport services, economics of public transport, pricing in transport, modelling land use and methods for policy analysis.

David Starkie has held positions at universities in the UK, Australia, Canada and Germany; directorships with US and UK consultancies and has advised parliamentary and government committees in the UK and Australia. A former associate member of the American Bar Association, he has been editor of the *Journal of Transport Economics and Policy* since 1997 and is the author of *Aviation Markets: Studies in Competition and Regulatory Reform* (Ashgate 2008).

Daniel Strobach has a degree in Business Administration and Economics and a PhD from the University of Hohenheim. He also studied Transportation at the Vienna University of Economics and Business Administration. Since 2009, he is working with the German Federal Competition Authority.

Dr Edith Szivas is a senior lecturer in Tourism at the University of Surrey, UK. She gained her BSc in International Trade in Budapest, Hungary, and her early career was in banking and international trade. She has an MSc in Tourism Planning and Development and a PhD from the University of Surrey. Dr Szivas has worked on tourism related projects for the United Nations World Tourism Organization, Travel and Tourism Intelligence (MINTEL), the Department of Trade and Industry, and USAID.

Cornelia Templin has been working in the Strategic Development of the Ground Services division at Fraport since 2002. She obtained her PhD on the deregulation of ground handling services in Europe from the University of Giessen in 2006. She started her career as a trainee in Aviation Management at Frankfurt Airport and studied Business Administration in Giessen and Los Angeles. In 2006, she was a visiting scholar at the University of British Columbia, Vancouver, in the Centre for Transportation Studies of the Sauder School of Business.

Dr Michael Tretheway is Executive Vice President and Chief Economist with Inter*VISTAS* Consulting Inc. in Vancouver, Canada. He is also Adjunct Professor with the Sauder School of Business at the University of British Columbia. Prior to co-founding Inter*VISTAS*, Dr Tretheway was an executive with the Vancouver International Airport Authority. He has served as an expert witness and advisor to governments, airlines, airports, ports and railways in Asia, Africa, North America and Europe. He has a PhD in Economics from the University of Wisconsin and was Associate Professor of Transportation and Logistics at the University of British Columbia for 14 years.

Erik Verhoef is a professor at Vrije Universiteit, Amsterdam. His research focuses on the efficiency and equity aspects of spatial externalities and their economic regulation, in particular in transport, urban and spatial systems. Important research themes include second-best regulation, network and spatial analysis, efficiency aspects versus equity and social acceptability, and policy evaluation. He has published various books and numerous articles.

Daniel van Vuuren obtained both his masters (Econometrics, 1997) and PhD (Economics, 2002) from the Vrije Universiteit, Amsterdam. He conducts empirical research in microeconomics. His work has been published in *Applied Economics*, the *Journal of Transport Economics and Policy*, *Labour*, and *Transportation Research A*.

Dr Zheng Lei is a lecturer in the Department of Air Transport at Cranfield University, where he teaches Quantitative and Research Methods. Previously, he was a senior lecturer at Anglia Ruskin University. Zheng has a PhD in Management from the University of Surrey, UK. His research interests include air transport and tourism management, regional economic development and research methodology.

Chapter 1
Introduction and Overview

Peter Forsyth, David Gillen, Jürgen Müller and Hans-Martin Niemeier

One of the themes in an earlier German Aviation Research Society (GARS) book, *The Economic Regulation of Airports* (Forsyth et al., 2004), was that 'strong competition between airports is not feasible' (xxviii); we argued this might change in the future. That book raised, under the heading of future challenges, the 'bigger question of whether, in more competitive circumstances, there is a continued need for formal price regulation' (xxix). In this current book we take up the issue of airport competition, with particular reference to Europe. We explore whether and where it exists, how strong it is and what policy implications it has.

Since the publication of the earlier book, developments have amplified the challenges and have put them on top of the agenda of airport policy in some jurisdictions. Three events stand out: the proposed break-up of BAA, the blocked takeover of Bratislava Airport by Vienna Airport and the prohibition of subsidies to Ryanair by Brussels South Charleroi Airport.

In 2008, the UK Competition Commission recommended that BAA be forced to divest itself of two of its three London airports, and to sell either Glasgow or Edinburgh Airport. The stated objective is to promote competition between the airports.

- On 18 October 2006 the Slovak government made a final decision to stop the sale of Bratislava Airport to neighbouring Vienna Airport because it had not been approved by the Slovenian anti-monopoly office. This case is similar to the BAA case, but it raises some additional issues. There was a concern that Bratislava Airport had the potential to compete with Vienna Airport, at least in some markets.
- On 3 February 2004 the European Commission prohibited direct aid granted by the Walloon Region and Brussels South Charleroi Airport to Ryanair. Ryanair appealed this decision and were successful in having the case overturned; a decision by the European Court of First Instance (2008) in December 2008. Nonetheless, the issue of state subsidies to low-cost carrier (LCC) airports, as well as to major airports served by full-service airlines, has remained a contentious issue. For example, Ryanair has complained to the European Commission about state aid of over €400 million per annum to Schiphol Airport and to KLM/Air France. The Charleroi decision involves a horizontal aspect of competition among airports, along with a vertical aspect of how distortions of airport competition affect competition between airlines in the downstream market.

The case for regulation of airports comes about because of a lack of competition between airports, and their consequent possession and use of market power. This lack of competition may come about for two types of reason: locational reasons and natural monopoly reasons.

The locational explanation argues that for most airports there are no close substitutes as attractive locations are limited (Forsyth, 1997). It is very difficult to obtain permission to build airports in cities with existing airports, for example, to build a competing airport next to hubs like Frankfurt or Paris airports. Such airports have a de facto monopoly reflecting planning and environmental restrictions and they have market power in the provision of aviation services which suggests a case for regulation.

The other explanation relies on economies of scale in airport provision. Thus, a monopoly is efficient as two or more airports would lead to higher average costs. If there is more than one airport, and competition is present between the airports, this competition will break down, since the airport which is able to gain more traffic will be able to enjoy lower costs and may be able to force the other out of the market. The existence and extent of scale economies at airports is disputed, though the presence of large indivisibilities, especially in the provision of runways, means that only large cities are likely to be able to have several runways and several airports. If natural monopoly is present, a profit-oriented airport will have an incentive to use its market power and increase prices. Alternatively, a non-profit-oriented airport might charge high prices and allow costs to rise. Governments might regulate airports to prevent this from occurring.

These explanations are not mutually exclusive. Whatever the explanation, few cities other than very large ones have more than one airport. This said, there may be secondary airports on a city's fringes that can compete, at least for some traffic, with the main airport. In addition, in the densely populated regions of countries such as the UK and Germany, there may be several airports that residents regard as equally accessible. Thus competition can exist, though often does not.

In this study, we find that competition between airports does exist in several regions. Thus, there is competition between the airports in the UK outside London, and this competition appears sufficiently strong to discipline their pricing. There are also many examples of monopoly airports where there is only one airport in a city and the nearest other airport is too far away to provide effective competition. In between, there are many situations where there is competition between airports for some traffic, for example between a fringe secondary airport and a major city airport for (LCC) traffic. Airports, though distant from one another, can compete in some markets, for example, for duty-free goods or as bases for LCCs. Major airports may be able to compete for hub traffic. Competition can also take place within airports, for example, between separately owned terminals.

The issue of competition between airports is of interest for several reasons:

- If it is strong, it may be feasible to dispense with direct regulation of airports. Granted that regulation can be costly in efficiency terms, even imperfect competition may be preferable to imperfect regulation.
- Even if competition is not strong enough to dispense with regulation, it may still provide a spur to improved performance.
- When several airports are being privatized, governments must assess whether the benefits from selling them as a group to one owner outweigh any gains that might be achieved through selling them separately to facilitate competition.
- With many airports now being privately owned, the issue arises of whether mergers or alliances between potentially competing airports should be permitted – the case of Schiphol and Paris Charles de Gaulle, which can compete as hubs, being a case in point.

We have organized the chapters under the following headings:

- How do airports compete and how robust is competition between them? This section provides an overview of the state of competition and explores whether the current form of competition can increase efficiency and economic welfare. Airports seem to compete in various ways, but is there evidence of strong competition? In our view the evidence is not clear and it varies from country to country and region to region. Furthermore, the

development of competition does not mean that the outcome is necessarily beneficial; more competition does not guarantee an improvement in economic welfare.
- How do travellers choose an airport and how does this affect the competitive process? This is an essential aspect of competition. In some less densely populated countries, travellers have little choice as to which airport to use. In Europe, with a well-developed land transport infrastructure, this can be different; but time and travel costs play an important role. Similarly, long-haul connecting passengers have a choice of a number of airports they can use, which can result in airports which are a long distance apart being in the same market. Imagine a world in which travel and time to access airports were reduced to minimal levels – in such a world, all airports would become close substitutes. However, even with high-speed trains, time and travel costs remain important, and congested motorways add to the complexity of the traveller choice problem.
- What is the evidence for competition in the airport industry? As the intensity of competition varies from country to country we have selected case studies from Greece, Germany and the UK.
- What issues for airport and competition policy emerge? Here we deal with the issues identified earlier of market power, through joint ownership of potentially competing airports.

How Do Airports Compete and How Strong is Competition?

In Part A of the book empirical evidence on the nature of competition is presented. Morrell provides an overview of how airports compete in Europe. He notes that airports are intermodal hubs through which passengers are transferred from work or home to their final destinations, combining different transport modes. Airports compete with one another to attract airlines and he gives many examples for the different ways how, and against whom, they compete. Morrell emphasizes that airports offer airlines incentives and even subsidies to attract new routes or to base aircraft at their airport. The EU Commission views state aid as discriminatory if it is not open to all airlines. Morrell analyses several cases and highlights their importance (see chapters by Kristoferitsch and Elliott).

Müller-Rostin et al. analyse a particular, though important, aspect of competition. In other industries, entry and exit are drivers of the competitive process and lead to welfare maximizing results in the long run. Compared to other industries, only a few airports have entered or exited the European airport sector. This suggests that the intensity of competition depends more on how the incumbents behave than on new entry.

There are two chapters that look at the issue of competition in relation to airlines. Pels and Verhoef argue that, without congestion, the ability by airports to abuse market power is limited if airlines are forced, through competition, to set fares equal to marginal cost. With congestion and with monopolistic behaviour of the airlines the airport can use its dominant position to raise charges above the competitive level.

Even large hubs are small firms compared to many airlines, which typically are part of an even larger airline alliance. Airports blamed for excessive market power claim that this fact has been overlooked. Instead of exploiting their users, they are facing substantial countervailing power from airlines. Button reviews this concept first developed by Kenneth Galbraith in 1952. While it was criticized by mainstream economists, he sees some relevance as it highlights outcomes which are either monopolistic or competitive. Applying it to the airport industry Button finds that countervailing power is hard to test. He analyses possible outcomes for traditional airlines and

LCCs and for collective action of airlines, and discusses forms of price discrimination by airports as an indication of the strength of countervailing power.

Forsyth takes up the topic of Morrell's chapter, namely increased competition between the major and the secondary airport combined with subsidies for secondary airports served by LCCs. He points out that the popular belief that 'more competition leads to improved efficiency' might not be true for airports where subsidies are present. While not denying that competition can have positive effects, he doubts that this is necessarily the case in the airport industry with indivisibilities, significant sunk costs, imperfect regulation of major airports and subsidies available to some secondary airports.

Cost conditions are recognized as an important driver of the feasibility of competition. Pels et al. analyse how airport costs vary with size across a sample of airports from around the world, using a translog cost function. They conclude that short run average operational costs decrease with output, with passenger numbers being more closely associated with falling costs than air transport movements. However, when a quality variable is introduced, the elasticity of cost with respect to output is decreased in some specifications.

The chapters by Graham and by Kincaid and Tretheway analyse strategies of airports facing competition. Graham uses Porter's 'five forces' analysis to determine the threats and opportunities airports face. Regarding new entrants and substitutes, the threats are low compared with other industries. The forces of supply and consumer power, as well as internal rivalry, differ and cannot be generalized. Airports so far have been reluctant to adopt generic competitive strategies such as cost leadership and product differentiation, except for some niche markets such as LCCs and freight. Graham observes no head-to-head competition, but a growing demand for applying competitive strategies. Kincaid and Tretheway, on the other hand, tend to be more optimistic about the competitive nature of the industry than Graham. They emphasize the change of airports from passive service providers to active promoters of their business and region. Airports have developed marketing strategies and search on their own for new service opportunities, which they actively present to airlines to develop. They show how airports can define their product and price, and promote and distribute them. They provide examples of best practice, such as differentiating their product by developing specific features. These include passenger facilitation and baggage handling; pricing airlines differently at different stages of their business relationship; creating awareness of specific services and promoting their airport through websites; and good placement in computer reservation systems.

Tretheway and Andriulaitis describe the general evolution of the air cargo market and the trend towards a more competitive environment. Air cargo has been capturing an increasing proportion of trade, partly because of a growing proportion of high-value and often low-volume goods that are being transported The addition of new passenger gateways in major areas of economic activity has helped enormously in pushing up air cargo volumes. The development of secondary airports has provided additional, more cargo-friendly options.

By analysing the whole value chain the authors show how the bargaining power of the individual players have shifted and what the likely effect is on airports. They emphasize the growing influence of freight forwarders, the emergence of specialized air cargo freighters and the combination of air freight with trucking that led to the expansion of the geographic scope of an airport's catchment area. This development has opened up new, and more competitive, options for shippers as numerous combinations of gateways/air service providers are now available to a wider range of shippers.

Traveller Choice and Airport Competition

Traveller and shipper choice lies at the heart of the airport competition issue – the strength of this competition depends on how good as substitutes travellers regard different airports to be. Hess reviews 'Modelling Air Travel Choice' and how econometricians have studied this choice by passengers. He outlines first the nature of air travel choices, which involves also the choice of airport, and then looks at how well the models have captured the complexity. Hess is highly critical, as most discrete choice models have incorporated only a subset of the choices – something which might seriously bias the results. He suggests various ways to improve the quality of the models, including the quality of the data.

Hess and Polak compare the methods and results of three studies on the way passengers choose airports in the San Francisco Bay area, with three main airports, and in Greater London with five airports. In general these studies confirm that passengers have a strong preference for their local airport. Convenient and fast access is important and only LCCs with massive fare reductions can overcome the disadvantage of less accessible airports. Even travellers visiting friends and relatives are willing to pay a premium of $28 to fly from their closest airport. Furthermore, the way business travellers value frequent flier programmes is instructive for airport strategy. These travellers are willing to pay a premium of up to $125 for flights covered by such programmes and are therefore less willing to substitute airports.

An accessible treatment of the complexities of modelling the demand for air travel, and of airport choice models, is provided by Gaudry. He presents a detailed treatment of a critical feature of demand modelling practice that has important implications for airport competition and hub stability – specifically the 'independence of irrelevant alternatives' assumption which imposes strong substitution assumptions on the models. He notes that in the modelling effort the treatment of the fastest growing component of air passenger demand, non-business trips, is particularly challenging. To model these types of trips within the current modelling framework there must be clear reference to alternatives in mode and path or airline company choice representation procedures. This arises because passenger flows are interdependent and therefore the utility derivable from the alternatives cannot be defined only by reference to own transport conditions.

Gaudry also provides a comprehensive summary of how Standard Box-Cox endogenous form specifications contribute to a much improved representation of the role of transport characteristics and performance within the current demand modelling structures. He argues that in order to include the rapidly growing leisure passenger component into the models investigating competition among airports requires the use of spatial correlation processes in Generation-Distribution models, as well as the use of Generalized Box-Cox specifications. These newer modeling approaches should be used in both mode choice as well as route choice analysis. In both cases, the strong substitution assumptions of choice modeling are avoided. Specifically, the Independence of Irrelevant Alternatives (IIA) assumption is avoided in realistic ways.

Case Studies of Airport Competition

We have selected a few case studies to analyse in depth the nature and strength of airport competition in Germany, Greece and the UK.

Germany has one of densest airport networks in Europe, and also a comparatively well-developed, high-speed motorway system. Malina develops a measure for market power taking account also of intermodal competition. He shows that most of the small airports face competition, but that the large airports such as Berlin, Frankfurt, Hamburg, Munich and Stuttgart lack good substitutes. Competition could be enhanced if, for example, the airports in the metropolitan region of Berlin were operated independently.

Strobach analyses the catchment areas of airports in southern Germany with a view to developing an indicator of competitiveness between airports for residents of different zones in the region. He develops an index based on access (time and cost) to potentially competing airports, on airport quality (including flight frequency) and parking availability. He evaluates the index for nine airports in the broader region. This index indicates which airport is likely to prove most competitive in attracting passengers from a given zone.

Papatheodorou examines the extent of concentration of passengers using the airports of Greece. Airport passenger traffic tends to be regionally based, with traffic being closely related to regional population or gross domestic product (GDP). Greek airports have some scope to gain more traffic by competing for direct flights from the main origin airports. However, while some airports could prove competitive, it is likely that they would need to make considerable financial commitments to develop their facilities.

In contrast to the more sceptical views on the strength of competition in continental Europe, Starkie argues that the UK airport industry is a competitive one; but even small airports are at least breaking even, and high fixed costs are not an obstacle for entry. This is largely due to the fact that there is an abundance of former military airfields which can be turned into airports and which, like the example of Doncaster airport, show that new or small airports can challenge incumbent airports such as Manchester effectively. As most airports are only about an hour's driving distance apart, the hinterland of airports overlap. Airports can be good substitutes, especially if they cannot price discriminate among passengers. Starkie challenges the view that takes for granted that airports are natural monopolies which automatically have to be regulated. Regulation might be necessary for some airports, but airport policy should adopt the view that 'a competitive framework is an achievable objective for a national airport policy' and this can be achieved by structuring ownership to ensure a reasonably competitive environment

Lei et al. explore the impacts of low-cost carriers on UK airports. Smaller airports often compete to attract low-cost carrier traffic, which is relatively uncommitted to any one airport. Econometric estimates suggest that while low-cost carriers do not contribute as much by way of aeronautical revenue as charter or full-service carriers, they do contribute slightly more non-aeronautical revenue than these other airlines at the smaller airports – though they contribute less than the other airlines for airports on average.

Policy Issues

With competition between airports being feasible in some but not all circumstances, and with the strength of this competition being quite variable, there are few simple and straightforward policy implications.

Forsyth and Niemeier assess the prospects for competition between the major London airports, Heathrow, Gatwick and Stansted, if the current owner were forced to divest two of its London airports. They explore the scope for competition between the airports; note that there are some aspects of competition which would emerge; and provide a spur for better performance. However the excess demand situation, along with the difficulties in expanding capacity, is likely to limit the effectiveness of competition. Thus it is likely that the problems recognized with the London airports would also need to be addressed through the reform of regulation as well as breaking up BAA.

Forsyth, Niemeier and Wolf explore the possible impacts of airport alliances and the emerging multi-airport companies on the intensity of competition. It has been suggested that airport alliances and international multi-airport companies will emerge and gain a dominant position comparable to that of the airline industry. However, they are sceptical about these trends. Furthermore, they analyse the rationale behind the decision to prohibit Vienna Airport from taking over neighbouring Bratislava Airport.

At airports where neither competition nor regulation seems to deliver results favored by the public, the issue of terminal competition has emerged. Reynolds-Feighan analyses the economics and politics of the proposal to separate the terminal ownership at Dublin airport. She argues that there might be efficiency gains through competition, but also higher coordination costs. She further points out that while the expert group recommended separate ownership on the grounds of a carefully designed tender process, the politics changed after a cabinet reshuffle. Reynolds-Feighan concludes that 'The Irish case is now added to the shortlist of examples cited by monopoly airport operators claiming that terminal competition is unworkable. However terminal competition, as proposed in the Irish circumstance, remains an untested policy option to date' (Chapter 21, p. 353).

Kristoferitsch and Elliot examine the Charleroi case and the emerging EU legal framework covering state aid for public airports. The EU Commission based its decision on the assumption that the Walloon Region acts as legislative and regulatory authority separated from its commercial activity as owner and operator of Brussels South Charleroi Airport (BSCA). The Walloon Region was charged for discriminating against other carriers and granting Ryanair lower charges. BSCA was charged with not acting as a private investor by accepting too generous conditions. The Court of First Instance ruled against the Commission stating that the Commission should have treated the Walloon Region and BSCA as a single entity. Under such circumstances, this single entity would have taken the broader economic impacts of Ryanair's services on the region into account and therefore could be seen to be acting according the private investor principle and not simply subsidizing Ryanair.

Kristoferitsch and Elliot both return to the initial decision and analyse its weaknesses from both legal and economic perspectives. Kristoferitsch points out that the framework remains fragmentary and is not legally binding on the Member States. Reductions in charges are allowed as a part of a general scheme, published in advance and open for all airlines. Kristoferitsch foresees a trend to circumvent the restrictions of the EU framework by semi privatising airports. Elliot looks at the economic aspects of subsidies received by airlines for the use of regional airports and considers the specific ways in which public subsidies could impact inter-airline or inter-airport competition. In his analysis, he focuses in particular on the Charleroi case and analyzes whether the EU Commission's position is consistent with the promotion of economic efficiency. He notes the problems of identifying the competitive benchmark and the requirement that any subsidies are not exclusive to one operator.

Ground handling is another area of airport operations that can be opened up to competition. Templin analyses the evolution of the market structure for the European ground handling services that had been deregulated as a consequence of the European service directive in 1998. Her study focuses on the six largest European airports. She raises the more general question of how the value chain between airports and airlines should be organized and how government regulation and market forces have been performing in finding an effective division of labour – both horizontally and vertically – in allowing more multi-airport service organizations to develop internationally

Lewisch looks at airport markets where, because of the absence of competition, antitrust rules would be applied. In his analysis he looks at how to apply the bottleneck concept to limit regulation to the key aspects; where the essential facilities doctrine could be used; how to deal with an abuse of a dominant market position; and how sector-specific competition law can be applied to airports. This also includes looking at issues such as vertical foreclosure and the refusal to deal.

The European Focus

In this book we have focused on airport competition in Europe, though in some cases reference has been made to other continents. Competition between airports does exist on other continents. For example, where there are several airports in large US cities, competition is feasible (though does not always occur because of common ownership, as in New York). Also in the more densely populated regions of the US, such as the north-east, travellers may have a choice of airports. Airports in Asia, such as those at Singapore, Kuala Lumpur and Bangkok, compete for hub traffic. In countries with low population densities, such as Canada and Australia, airports are separately owned, but they only have limited scope to compete because they are mostly very far apart. However, in several cases Canadian airports compete with nearby US airports for transborder and international traffic; for example Vancouver and Seattle, and Toronto and Buffalo.

However, it is probably Europe which provides the best case study of airport competition. Europe has a number of regions which are densely populated and which are served by several airports. The presence of former military airports has facilitated entry of smaller airports. There is a variety of institutional arrangements for European airports, with more privately owned airports in Europe than in other continents. Issues of competition between and within airports have attracted much more attention from competition authorities and regulators than elsewhere, and the trade-off between competition and regulation is being actively considered. In Europe, policy decisions are now being made to promote competition, something which is rare elsewhere. Airport competition is still in its early stages in Europe – one or two decades ago there was little attempt by the mainly government owned airports to compete. However airport competition is now a reality, though how extensive and intensive it will become remains to be seen.

References

Court of First Instance of the European Communities (2008), Case T 196/04 Ryanair Ltd v Commission of the European Communities, State aid – Agreements entered into by the Walloon Region and the Brussels South Charleroi airport with the airline Ryanair – Existence of an economic advantage – Application of the private investor in a market economy test, Luxembourg 17 December 2008 http://eur-lex.europa.eu/LexUriServ/LexUriServ.do?uri=CELEX:62004A0196:EN:HTML.

Forsyth, P. (1997), 'Price Regulation of Airports: Principles with Australian Applications', *Transportation Research E*, 33/4, 297–309.

Forsyth, P., Gillen, D., Knorr, A., Mayer, W., Niemeier, H.-M. and Starkie, D. (eds) (2004), *The Economic Regulation of Airports*, German Aviation Research Society Series, Aldershot, Ashgate.

PART A:
How Do Airports Compete and How Strong is Competition?

Chapter 2
Airport Competition and Network Access: A European Perspective

Dr Peter Morrell[1]

Introduction

As airports change from public services provided by central government to privately owned or commercialised entities there is a growing need to protect against possible monopolistic behaviour. This raises the question as to whether airports are indeed monopolies and need economic regulation. If there is sufficient competition between airports then there should be little need for regulation.

The Director of Economic Regulation for the UK Civil Aviation Authority (CAA) was asked whether he thought that, in the absence of economic regulation the UK airports would raise their charges to airlines. He replied that 'in the Heathrow case that is probably true and it may be true for Gatwick, but it might be less true for some elements of Stansted and Manchester (House of Commons Transport Committee, 2006).

What then are the features of Heathrow and perhaps Gatwick that make competition with services at other airports difficult? Alternatively, how do Stansted and Manchester airports compete more effectively with other airports?

This chapter attempts to answer these two questions in the EU context. This will be done first by discussing what an airport does, how it charges for its services and which of its functions the regulator is most concerned with. This will require the consideration of the airport as an inter-modal hub. Next, airport competition will be considered by examining the perception of airport operators and how this might fit in with reality.

A brief review of the present European situation with regard to governments as owners and regulators of airports will then be followed by examples of government support for airports and government-owned airport support for airlines, concluding with some examples of airport and regional incentives given to low cost airlines.

The Airport as an Intermodal Hub

Airports provide themselves, or through concessions, aeronautical infrastructure (runways, taxiways, aprons), passenger and cargo terminals, ground handling services and surface access facilities, including car parks. They also provide various commercial outlets such as shops, premium lounges and banks.

---

1   This chapter is based partly on a study undertaken in 2002 for the European Commission (DG TREN) by a team headed by the author (ATG et al., 2002), and partly on the author's own analysis and views.

The above airport services are provided in the context of the door-to-door transport network, whether for passengers or freight. Air services will always be 'consumed' in conjunction with one or more sectors provided by other transport modes:

- air to car/taxi/truck and vice versa;
- air to bus/rail and vice versa;
- air to air and vice versa;
- surface to surface also possible.

Competition should be considered in terms of door-to-door service, time and price, rather than just on an airport to airport basis. What distinguishes London Heathrow Airport from Stansted Airport is the large number of destinations served non-stop: many of these are long-haul and of relatively high frequency. The only alternative is an additional connecting flight to a European hub, which may involve back-tracking. This may appeal to price sensitive leisure travellers, but not to business passengers, who may also be keen to amass frequent flyer points on the national carrier.

On the other hand, many of the city-pair markets served by low-cost carriers (LCCs) at Stansted can also be accessed at similar prices using different airport and surface trip combinations (see Figure 2.1).

A transfer to and from boat is also occasionally possible, for example at Venice Marco Polo Airport. For cargo, air to truck transfer is required for terminating and originating shipments (some of these trucks operate as feeder 'flights'). Although airports are built for air connections, surface-to-surface transfers are also possible. One coach company built a network of feeder services into Heathrow that allowed convenient bus-to-bus transfers, not altogether to the liking of the airport operator, BAA. Air France has an air cargo truck hub at Frankfurt Hahn Airport, where it has no flights, and British Airways did the same at Maastricht Airport.

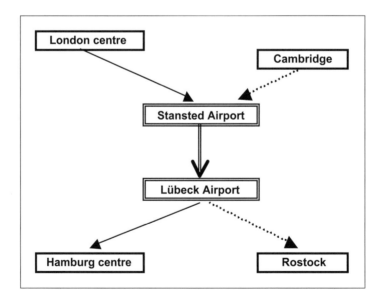

Figure 2.1   City-pairs served by LCCs at Stansted accessible at similar prices

Airports need to attract both airlines and surface operators to use them as one of the nodes in their networks. They provide the physical infrastructure to make that possible, and usually the management to operate the airport. Remote military airfields can and do handle commercial flights, but for these to be attractive to passengers, there needs to be some way for them to reach their final destination. Ryanair uses a number of these airfields in Europe, and often needs to make sure bus services are operated to the nearest city or towns. For example, their London Stansted flight to Lübeck competes in the London/Hamburg market by means of a bus service from Lübeck Airport to the centre of Hamburg, and a rail service from Stansted Airport to central London.

The European Commission's policy of encouraging airports to be intermodal hubs only emphasises something that airports and airlines have known for many years. Better access to the airport's hinterland expands the airport's catchment area and enables airlines to compete in more origin-destination markets. High-speed rail (HSR) is good at that for large city centres, but expensive and inflexible, requiring another interchange point before the passenger can travel on the first or final stage of the trip. The Commission's policy aims to move EU traffic from air to surface modes, with alleged benefits of time savings and reduced congestion and environmental damage.

The origin point of each trip involving air transport is usually the home or the place of work. Ideally, and other things being equal, private car or taxi would be preferred, with one non-stop link to the airport terminal (or close by). Some passengers will trade a longer non-stop car journey for changes at intermediate points. The originating passengers usually have much greater information of surface travel options than visitors.

Passengers arriving at a foreign airport might be collected by friends or colleagues, but more usually take public transport to city centres or local conurbations. They sometimes have little knowledge of either schedules or fares of local transport operators, and can make sub-optimal choices.

An airline can compete in more origin-destination markets if its services are part of a larger network, including good surface access links. It often has little control over these links, but does have the option to operate them itself. Integrated carriers such as FedEx offer door-to-door transport by operating its own fleet of delivery trucks (see also Gillen and Morrison, 2003). This differentiates it from traditional air cargo airlines. Few airlines have done the same for passengers, one exception being Virgin Atlantic, which included limousine door-to-door pick-up and set-down services for premium-class passengers.

The so-called low-cost carriers (LCCs) provide part of the network needed for price-sensitive passengers to travel from their trip origin to final destination. They let passengers make their own connections between flights, sometimes arranging for feeder services to be provided (by bus). But they do not take any responsibility for making sure their customers connect with other flights which they or other airlines operate, or with other modes of transport.[2] They do not schedule their flights to provide quick connections; nor do they provide through ticketing or baggage transfer services. Generally, the surface transport operators adapt their schedules to fit in with the LCC flights. In 2001, around 17 per cent of Ryanair's passengers transferred from one Ryanair flight to another, but did so at their own risk, and on two separate tickets.[3]

---

2   Only charter flights are sold together with surface transport connections included in the price of the tour, but usually only at the destination airport.

3   According to Tim Jeans, then Sales and Marketing Director Europe for Ryanair, in a lecture at the Air Transport Group, Cranfield University, on 14 May 2002. It should be added that those passengers connecting from one Ryanair flight to another at Stansted Airport would avoid paying the airport passenger departure tax twice if they could be through-ticketed (Ibarra, 2003).

In contrast, the so-called 'network carriers' encourage the transfer of passengers from one flight to another, especially between their own flights (online connections), and between theirs and alliance partner flights (interline connections). They adapt their schedules to make this as seamless as possible.[4] Airports that attract this type of airline, and for whom a large percentage of traffic is thus transfer traffic, will have more activity on the airside than the landside of their facilities (for example, Dallas/Fort Worth or Amsterdam Schiphol).

Airport Competition

Airports compete with other airports to attract airlines. Once the airlines decide to serve the airport, other service providers will also be attracted. These would include aircraft, passenger and cargo handlers, fuel suppliers, shops and car park operators, as well as taxis and bus companies. This is a key feature of a network: the need for critical mass to be reached before it becomes attractive to many users. This occurred with the internet, developed exponentially by offering free services (at least initially). It thus makes sense for airports to attract service providers through low charges, at least initially, and even paying airlines to base aircraft at their airport (for example, Ryanair at Gerona Airport in Spain).

This suggests that the main customer of the airport is the airline or surface transport operator, and not the passenger or shipper. These customers can be sensitive to changes in the relative price charged by one airport compared to that charged by a competing airport. Price here is effectively the average of a basket of services, including in some cases ground handling. An example of this price effect is the transfer by Ryanair of many of its Shannon flights to Kerry Airport in 2003 (European Commission, 2004, p.8).

Airports compete on service and price. Service will cover such elements as location, accessibility, and the quality and size of its aeronautical and related facilities. Airport management can influence all these variables except location, but even this can be improved by better surface access. Slot constraints will severely restrict an airport's ability to compete, as will environmental restrictions. Price will be reflected in various charges, the two major aeronautical ones being for aircraft landings and passenger departures. However, airport revenues can also be generated from the other service providers, and they may prefer to offer low charges to airlines and earn more from non-aeronautical revenues. These airport specific variables are important, especially for short-haul trips by air, but not the only ones that determine travel demand and the passenger's choice of airport (see Mandel, 1997 for a fuller discussion).

ACI Europe (1999) identified different forms of competition or 'perceptions of competition' between airports:

- competition to attract new services;
- competition between airports with overlapping hinterlands;
- competition for a role as a hub airport and for transfer traffic between hubs;
- competition between airports within urban areas;
- competition for the provision of services at airports;
- competition between airport terminals.

---

4   But, like the LCCs, they do not normally include surface connections in through-baggage arrangements. There are one or two exceptions to this, notably SWISS for connections by bus and rail within Switzerland, and Lufthansa with the rail services that it operates.

The last two are not competition between airports, but between service providers *within* one airport. The second and fourth bullet points above could effectively be combined as one, and only then with the proviso that the competition is actually amongst airlines and surface transport operators for the same origin-destination markets. The third point is essentially the same, with the reach of the network extending considerably by means of air connections. This leaves the first form of competition as the only one that is strictly between airports: to attract new services, to which one could add 'from both airlines and other transport operators'.

An airport that has a large local market might still compete with airports adjacent to other conurbations for new airline services. For example, an Asian airline might have some extra capacity that might be applied to new European services. It might have a ranking of attractiveness of various new points, and the airports might try to influence this ranking. The new point (airport) will offer the potential for local traffic and also other markets using the connecting flights of other airlines (often alliance partners). Thus the airline decision is about the best airport to serve as a new node in its existing network of flights. Competing airports will obviously offer different local traffic potential, but also access to many of the same additional markets.

Airports have limited influence on the establishment of major airline hubs, at least in Europe. The hub airports of British Airways, Air France and KLM are all in their respective capital cities. Lufthansa had some choice, with Frankfurt winning in terms of location and facilities,[5] and Munich the second choice. The airport does need to invest in sufficient runway and passenger terminal capacity to accommodate the successive waves of arrivals and departures.

Competition through overlapping hinterlands or catchment areas is usually expressed in terms of numbers of people living within a certain time's journey in a car from the airport. Various definitions are used by European airports (ATG et al., 2002, pp. 4–8):

- percentage of current traffic having origins or final destinations within a specified surface journey *distance* (for example, Copenhagen, Frankfurt, London Luton);
- percentage of current traffic having origins or final destinations within a specified surface journey *time* (for example, Cologne-Bonn, Lisbon, Malmö, Rome and Turin);
- populations within a certain journey driving time (for example, Basle, Brussels, Dublin and Milan).

The first two definitions describe the locations of the beginning and end of the journeys of probably the majority of their existing passengers who do not transfer to other flights. The third expresses the potential number of passengers within a certain driving time, which varied according to the type of traffic considered (for example, 30 minutes for domestic and two hours for charter services). This definition totally disregards other modes of transport (rail and air), and excludes key factors in airport choice: cost and frequency of connections (both surface and air). The cost of the surface trip may well be closely correlated with distance or time, but certain airports may exert a disproportionate pull through having airlines that offer a good choice of high-frequency and/or low-fare air services. This would give airports such as London Heathrow (as a major network carrier hub), Charleroi (with its growing number of very low-fare flights) and Hanover (with its choice of charter or leisure flights) much larger catchment areas than predicted from driving time contours.

---

5   This decision was made at a time when the German capital was Bonn, a relatively small city at which to base their hub.

The size of catchment area thus depends on the type of airline that the airport attracts: one offering only short-haul domestic flights might have a relatively small area, while one offering very low-fare or long-haul flights may have a much wider one. This is because consumers evaluate the time taken to access the airport in relation to total origin-destination trip time. They would also evaluate the access cost in relation to total cost. Thus a long-haul flight might be accessed from a wider range of airports, some further afield. The same might be true of charter flights, where the use of private cars allows a wider choice of holidays to be accessed from further afield because of a low value of time and low perceived incremental surface travel costs.

The potential in Europe for competition for local traffic using alternative airports is considerable: it was found that there were 32, 34 and 28 airports within one hour of each other by surface transport in France, the UK and Germany respectively (Fewings, 1999).

The Cranfield study (ATG et al., 2002) asked airports who they considered their main competitors to be. The responses suggested that they were reporting the airports that passengers thought of as close substitutes to their own.

Examples from the survey of competitors for low-cost airline passengers were:

- Brussels citing Charleroi as their main competitor;
- Charleroi citing Brussels;
- Amsterdam citing Brussels;
- Lisbon citing Madrid;
- Malmö citing Copenhagen;
- Zurich citing Geneva;
- Belfast City citing Belfast International;
- Stansted citing Luton.

While some were reciprocal, Luton Airport did not mention Stansted as its main competitor, indicating Heathrow for short-haul scheduled services.

Another question that was not put to airports was 'Which airports do you consider as competitors in attracting airlines to start new services?' This was identified above as true competition between airports, rather than between airports or other transport operators. The following are examples of such competition of a more substantial nature, since the intention is to attract a base operator and a large number of new services:

- Liège/Cologne (competition for express parcels carrier hub);
- Frankfurt Hahn/Stockholm Skavsta (competition for aircraft base for low-cost carrier);
- Liverpool/Manchester (competition for aircraft base for low-cost carrier);
- Munich/other larger German airports (competition for second hub for Lufthansa).

The above list identifies alternatives open to the airline, whether or not any formal competition by the airports in question took place. In fact, Manchester Airport did not have the margin of unused capacity that might have motivated it to seek low-cost airline services. The same would have applied to Düsseldorf Airport in relation to Lufthansa's second hub.

Airports have become more proactive in seeking new airlines and new services from their existing airlines. As will be seen in the subsequent section, they increasingly offer promotional packages that contribute start-up funds, discounts, advertising and market research support. This is not just the case in Europe, but also in North America. Airports such as Rickenbacker, Fort

Myers and Melbourne all offer free or reduced-cost landings and marketing support to new airlines (Schwartz, 2003, pp. 52–5).

Government as Owners and Regulators of Airports

An airport can bring considerable benefits to its surrounding area, and for this reason many airports in Europe are still owned by central or local government:

- owned by central government (for example, Spain, Sweden, Ireland and Greece);
- owned by central and local government (for example, many of the larger German airports);
- owned by local government (for example, most major Italian airports);
- mostly privately owned (for example, UK airports, with the notable exception of Manchester).

The Paris airports are owned by the central government, which reduced its stake from 100 per cent to 67.5 per cent in June 2006 by means of an initial public offer. In 2005, the 12 largest regional airports were converted to companies in which the central government controlled 60 per cent of the shares, the rest allocated to local Chambers of Commerce and government. At the same time the remainder of the regional French airports were transferred to local government ownership.

Those airports that are not government owned in some form are either fully privatised (for example, BAA) and quoted on a stock market (in BAA's case no longer since its assets were acquired by a consortium led by Ferrovial in 2007), or owned by a specialist airport operator. The latter more often has a long-term concession to manage the airport for a government owner (for example, London Luton). Even in cases where governments own the majority of an airport, they often corporatise the airport and bring in private-sector management.

Where the airport is majority owned by private interests, one would expect a more formal and transparent approach to economic regulation of airport services. That is certainly the case for the BAA in the UK, but less so for Copenhagen Airport. Even where the state owns the airport, economic regulation is carried out at either central or regional level by the appropriate arm of government. However, this may take the form of the approval (or rejection) of a revised scale of airport charges. The response may be given after reference to the inflation of prices in general, rather than supported by complex financial models and rate of return evaluation.

Government Support for Airports

EU Rules on Government Support for Airports

Article 87 of the EC Treaty provides that any aid granted by a member state that 'distorts or threatens to distort competition by favouring certain undertakings' is incompatible with the common market. A measure thus constitutes state aid if the following are shown to exist (ATG et al., 2002):

1. an advantage;
2. granted by a member state or through state resources;
3. favouring certain undertakings or the production of certain goods;
4. distorting competition;
5. affecting inter-state trade.

From the previous discussion on airport competition, such aids could potentially give one airport an unfair advantage over another in attracting airlines, whether they be low-cost, air cargo or hub carriers. Furthermore, such aids could be applied in such a way that a particular operator was favoured over others. For example, if a special rebate or landing fee holiday were granted to one airline for operating new routes, it should be available to all airlines starting such new services.

Governments can support airports economically or financially in a variety of ways. Support is described in EU legislation as 'state aids'. These can take the form of:

- tax exemptions or rebates;
- restructuring aid;
- privatisation through a trade sale;
- operating subsidies (for example, public service obligations or PSOs);
- regional aid;
- exclusive right concessions.

There are rules on the above that apply to both airports and airlines, with the market investor principle applied where there is doubt as to whether state aid applies (in the case of public capital injections and loan financing and guarantees). Restructuring aid has tended to apply to airlines, while the exclusive concessions have been more airport-related.

The EU makes a distinction between capital and operating subsidies. The European Commission has made considerable progress on removing operating subsidies from state-owned airlines, and these can only be applied on designated city pairs after an open public tender (PSO rules). However, little progress has been made on removing such operating subsidies received by state-owned railways. These are sometimes operated in direct competition to airlines, and they provide many of the surface links that increase the attraction of airports and airline networks (ERA, 2003). The European Commission's guidelines, which it published as a memorandum in February 2005, entered into force in December of that year, with their publication in the *Official Journal* (European Commission, 2005).

Operating Subsidies

Many airports are still state owned, and thus operating subsidies are provided where these airports lose money. Sometimes loss-making airports are combined with profitable ones in a state-owned airport holding company (for example, Aena Spanish airports). Cross-subsidies in such cases are usually applied so as not to favour any particular operator at the loss-making airport. However, they may distort competition between the latter and another in attracting airlines.

An example of this was the low airport charges that London Stansted Airport was able to offer to new services in direct competition with London Luton Airport. Stansted at that time was cross-subsidised from the two profitable London airports in the BAA group, Heathrow and Gatwick. In response to a complaint, the UK CAA ruled that the BAA was not abusing its dominant position, because it was reasonable to offer very low airport charges at an airport with a substantial margin of unused capacity. Furthermore, BAA's policy of low charges (for a limited period) was not intended to damage Luton Airport (ATG et al., 2002, pp. 2–7).[6]

Another example was the dispensation given to Amsterdam Schiphol Airport on the payment of corporation tax. The Dutch government agreed to change this after the investigation of a complaint by the European Commission and the subsequent ruling against them. The ruling was based both on the fact that Amsterdam competed with other airports, given its location and type and volume of traffic handled, and the unfair advantage that it had in competing for the acquisition of other airports.

Airports can also confer state aids on transport providers if they themselves are owned by the state. Thus a state-owned airport could offer one airline an unfair advantage if it offered special discounts or subsidies to an airline that was not open to other airlines that offered similar services. This is the argument being brought by the complainants in both the Charleroi and Strasbourg airport cases. Stockholm Skavsta Airport, on the other hand, is privately owned, but a similar issue applies to it insofar as the regional government provided financial support.

Capital Subsidies

The EU is less strict on capital support for airports. It has often made use of EU funds at concessionary rates of interest or repayment terms, and grants have also been available. Rail and high-speed rail have also benefited from such funds in the context of the Trans-European Networks. Access roads up to the airport perimeter are generally paid for by central government, and reduced congestion on these roads can considerably enhance an airport's competitiveness.

Manchester Airport has received loans from its local government owners on terms that might not have been available to a private company (for example, over a 35-year term). These were applied to expanding the airport infrastructure for the benefit of all airport users. A complaint made to the European Commission in 1998 alleged that this was illegal state aid, but this was rejected on the basis that state owners of airports are entitled to expand them, given that 'access to air transport services is of basic importance for the economic and social development of the region' (European Commission, 1999). The Commission is unable to intervene where national transport policies are being applied, and can only control such support where particular operators are placed at an advantage compared to others.

Capital subsidies can also distort competition between airports in attracting airlines. Paris Charles de Gaulle (CDG) Airport has a high-speed rail station, but less good connections to the city centre. Amsterdam, on the other hand, has excellent surface access, and an HSR station was expected to be opened in September 2009.

In 1999, the European Commission confirmed that a grant of Lire 1,000 million by the Tuscan regional authority for the modernisation of Marina di Campo Airport on the Island of Elba did not amount to state aid.[7] Reasons for this decision were the fact that the capital investment was part of the Italian government's regional economic development strategy and that, in any case, the airport when completed would be open to all operators on an equal basis. A year later, the

---

6 The UK CAA is now firmly opposed to any cross-subsidisation of Stansted by BAA's Heathrow and Gatwick airports.
7 European Commission, N 638/98 Aerelba, Italy.

Commission's competition authorities took the same view with regard to the Piedmont airports case. That case concerned the public investment in the Italian airports of Turin, Cuneo and Biella. The Commission's decision included a view that these airports did not compete with other airports, either for origin-destination traffic (no other airport having the same catchment area) or for hub interchange traffic (since the airports did not have a major hub carrier), and thus the subsidy did not result in a distortion of competition (ATG et al., 2002, Annex A28).

Government-Owned Airport Support for Airlines

Air Belgium and Sunair New Route Support: Ostend International Airport

The European Commission investigated in 1997 a complaint from the Belgium Tour Operators' Association concerning the support provided by the Flemish Region (owner of Ostend Airport) and Air Belgium and its associated tour operator, Sunair. This involved an advertising programme paid for by the Flemish Region that promoted both Ostend and Antwerp airports, and the flights in question. Second, a subsidy was also provided so that each passenger buying packages using these flights would obtain discount relative packages to the same destinations from Brussels Airport. Third, a subsidy was provided in 1994 for the additional cost of Air Belgium as a result of using Ostend Airport.[8]

The European Commission overruled the first complaint, in that the advertising was benefiting the airports in general, more than the specific flights. But the two other practices were found to constitute state aid, and the Flemish Region was ordered to withdraw them.

Ryanair New Route Support: Charleroi

According to the European Commission's *Official Journal*, Ryanair were offered the following by the Walloon Region (European Commission, 2004):

- cheap landing charge of €1 per passenger, rising to €1.13 in 2006 and €1.30 in 2010;

Brussels South Charleroi Airport (BSCA) offered the following, in addition to the above:

- €160,000 for each of the first 12 routes that Ryanair opens from Charleroi;
- €768,000 to subsidise recruitment and training of pilots and cabin crew;
- free offices and €4,000 for office equipment;
- up to €250,000 towards Ryanair's hotel and subsistence costs while it sets up its Charleroi operation;
- joint venture marketing company to which BSCA contributes €62,000 plus €4 per passenger.

The above total benefits have been estimated to be worth €11.57 per passenger in 2001 and €8.29 in 2002 (Lobbenberg, 2003). The same source contrasted the reduced landing fee payable by Ryanair for a B737-800 aircraft with an 80 per cent load factor of €151 with the full published fee of €1,448. Ryanair's average fare paid per passenger was €46.51 in the financial year ending March

---

8   Commission decision of 21 January 1998, in Air Transport Group (2002).

2003. This meant that the second year benefit per passenger at Charleroi was almost 18 per cent of the average system-wide fare paid. On the other hand, Ryanair consistently produces operating margins that are well over 20 per cent.

The European Commission ruled that parts of the deal were incompatible with the functioning of the internal market (European Commission, 2004) but this decision was overturned by the European Court of First Instance (2008) in December 2008. The initial ruling concerned the concessions awarded to Ryanair by the Walloon regional government to promote new services from Charleroi that were not transparent and were discriminatory. Other parts of the package such as help with marketing were largely approved. Ryanair was required to repay a proportion of the aid, but the carrier subsequently launched an appeal against the decision.

Ryanair estimated that the European Commission's decision on the Charleroi case would at worst cost them around €4.8 million annually since April 2001, and they would be required to repay the one-off launch aid of €2.5 million.[9]

Ryanair New Route Support: Strasbourg

Ryanair also received support from various public authorities to start new flights from Strasbourg Airport. An Air France subsidiary, Brit Air, filed a complaint in December 2002 to a French court (rather than the European Commission). The details of the deal that Ryanair obtained from the Chamber of Commerce have not been officially published, but Aviation Strategy revealed the major elements of the complex package:[10]

- one-off €150,000 for the launch of each new daily service from Strasbourg;
- annual subsidy of between €216,000 and €224,000 for each daily service operated;
- an annual sum of €492,000 from the Strasbourg Urban Community, the Alsace region and the *département* of the Lower Rhine for each daily service operated.

According to Aviation Strategy, a French government commissioner had earlier dropped the case against the Chamber of Commerce. However, the Strasbourg administrative court ruled that the contracts with Ryanair were illegal. Ryanair appealed to the French court against the decision, unsuccessfully, and did not move any flights back to Strasbourg that had been switched to nearby Karlsruhe-Baden Airport.

The total annual support amounted to around €1.4 million, or €6.34 per passenger based on 220,000 passengers per annum carried on two daily services at an 80 per cent load factor. The structure of the support was similar to the Charleroi case. Proponents of these types of support argue that Ryanair promotes the city of Strasbourg and the Alsace region on its website and brings new tourists and spending to the area. The European Commission may argue that this should have been done through a PSO, and in any case should have been through a public tender. However, the PSO is more targeted on a particular route, and does not support the general start-up costs that a carrier incurs when adding a new point to its network or operating a number of new services with aircraft based there. These costs help establish an airport's role in a network of air services, and are more akin to investment in intangible assets. The Scottish Executive has a route development fund that supports new business routes without the need for a PSO. This has been tapped by Ryanair, as well as by other airlines such as US carrier Continental Airlines (ATI, 2003).

---

9   Ryanair Holdings plc (2003), p.84.
10 Aviation Strategy, No.71, September 2003.

Ryanair New Route Support: Other Airports

Lobbenberg (2003) also identified a number of other airports and/or regional authorities that had supported new routes operated by Ryanair:

- Pau;
- Stockholm Skavsta;
- Klagenfurt;
- Girona;
- Birmingham;
- London Stansted;
- Aarhus.

The last three were discounts on airport charges available to any airline starting new services (Dublin also has such discounts), with the proviso in the case of Aarhus that they should offer air fares that were 50 per cent below the published fare for the route. None were likely to be as generous as the Charleroi and Strasbourg packages.

The Danish government has ordered Aarhus Airport to end the 50 per cent passenger charge subsidy paid to Ryanair.[11] This was being contested by the airport on the basis that the concession was available to any airline meeting their conditions.

A Swedish court has ruled against a 10-year SEK55 million marketing support package provided by the Nyköping regional government to Ryanair (Lobbenberg, 2003, p. 14). This suggests that financial support given to private airlines to establish services at a privately owned airport (Skavsta) can also be outlawed, as well as illustrating that the affected parties are seeking redress both in local courts and with the European Commission.

New Route Support by Airports: Other Airlines

It is becoming increasingly common for airports to try to attract new services by means of incentives. These usually take the form of published discounts on the aircraft landing fee element of airport charges and/or the passenger departure fee. They might also include money for promoting the new service together with the new destination city and region. The published discounts vary considerably in the way they can be earned, as the following examples of Manchester and Birmingham in the UK show.

Manchester Airport publishes discounts on its runway charges to encourage new destinations to be served, and additional frequency or capacity on existing routes. New destinations receive a 100 per cent reduction in the first year and 50 per cent in the second, with the full charge payable in year three.[12] The incentive for existing direct services is based on an increase in seats offered on the route in question, provided there is a net increase in the seats offered by that airline on all their routes from the airport (and Chapter 3 aircraft are used). The airline would then qualify for a 75 per cent rebate in the first year and a 30 per cent rebate in the second. Manchester also offers discounts for the operation of quieter aircraft.

---

11   *Travel Trade Gazette*, 22 September 2003, p. 80.
12   Hong Kong International Airport also offers new destination incentives, with a reduction of 50 per cent off the landing fee in the first year and 25 per cent in the second (Civil Aviation Department, 2001).

Birmingham Airport grants discounts to any airline achieving growth in passenger numbers and flights compared to the previous year. The number of years that these can be applied could be much longer than Manchester's case, but the method of calculation and duration are based on rather complex formulae.

Conclusions

Airports, particularly in Europe, have come a long way from being treated as government public services to become the commercial corporations of today. It is thus not surprising that airlines such as Ryanair treat them as just another supplier: they negotiate aggressively on airport charges, just as they do with their ground handling suppliers. However, the ground handler is usually one of a number of others providing very similar services at an airport; on the other hand, an airport may have a quasi-monopoly position in controlling access to a network as the alternatives would result in much higher trip costs (including a valuation of time).

Investigation into the degree to which airports compete is often confused by the examination of competition between airlines and other transport operators for end-to-end markets. The airport was rarely an end (or beginning) of any trip, and so cannot totally control any of the myriad origin-destination markets that operators are able to tap using its facilities. The airport does have certain spatial advantages, especially when it is close to a large local market.

Heathrow Airport has long enjoyed such an advantage, to the extent that airlines operating there were supposed to enjoy a significant yield advantage over Gatwick Airport. That advantage has eroded over time with lack of investment and congestion and delays both in the air and on the surrounding road network. It has also lost out to other London airports offering low-cost carrier service: Heathrow's share of the London–Dublin market has fallen from 100 per cent in 1985 to only 46 per cent in 1998 (Barrett, 2000).

Airports clearly compete with each other to attract airlines to add them to their networks. They do this either for the start-up of new routes or for basing aircraft at their airport and adding quite a number of new routes and services. They also offer incentives for those airlines that increase traffic on their existing services.

The European Commission and the courts in some EU countries have investigated such support made to airlines either by airports or the regional authorities in which they are situated. These have usually been in response to complaints from airlines that have not had access to such funding.

The way that the European Commission has tended to view this is as state aid to one airline that gives it an unfair advantage and is not available to all on a non-discriminatory basis. The provider of the support thus has to be a government authority, either local or national. It is of no consequence that the recipient of aid is a privately owned airline such as Ryanair, since it is the distortion of competition that is the point. The European Commission's decision on the Ryanair Charleroi case will result in more transparency, shorter-term support and a less discriminatory approach to support of this kind.

Some airports construct their published charges in such a way that only one particular airline or aircraft type operator can benefit. Frankfurt Hahn waives landing fees for aircraft of up to 90 tonnes, and Ryanair's largest aircraft weigh 75 tonnes (Lobbenberg, 2003, p. 11). Volume discounts can be offered which effectively restrict them to larger carriers, as in the case of Ryanair at Klagenfurt and the former Sabena at Brussels Airport.

It is inevitable that airports with a large amount of unused capacity will wish to make attractive offers to airlines, since their marginal costs are very low. Furthermore, the passengers they bring spend money in shops and car parks.[13] Where both parties to the offer are private companies (for example, BAA Stansted and Ryanair), it will always be difficult to discover the nature of the deal, but public authorities are obliged to inform their taxpayers and voters as to how their money is spent.

The European Commission has in the past tended to allow public support for airports, both on capital and operating account. It is often impossible for small regional airports to cover operating costs, especially immediately after the opening of new facilities. These could be covered by the state or other profitable airports until the airport reached a viable throughput. The airport would be open to any operator, and support regional economic policy. The same argument is applied to capital support, say, for rail or road access to the airport. However, it is difficult to argue that state support for a high-speed rail interchange at a major airport such as Amsterdam is part of regional economic policy. It may distort competition for attracting new airlines, but it is part of EU policy and the facilities will benefit both existing and new operators at the airport.

References

ACI Europe (1999). 'European airports: a competitive industry'. Policy paper submitted by the ACI Europe Policy Committee, 22 October.

ATG et al. (2002). 'Study on competition between airports and application of state aids rules'. Final Report for the European Commission DG TREN, volume 1, Air Transport Group, Cranfield University with Alan Stratford Associates, INECO, Gruppo CLAS and Denton Wilde Sapte, June.

ATI (2003). 'Continental adds Edinburgh route', *Air Transport Intelligence*, 8 September.

Barrett, S.D. (2000). 'Airport competition in the deregulated European aviation market', *Journal of Air Transport Management*, 6(1), January.

Court of First Instance of the European Communities (2008), Case T 196/04 Ryanair Ltd v Commission of the European Communities, State aid – Agreements entered into by the Walloon Region and the Brussels South Charleroi airport with the airline Ryanair – Existence of an economic advantage – Application of the private investor in a market economy test, Luxembourg 17 December 2008 http://eur-lex.europa.eu/LexUriServ/LexUriServ.do?uri=CELEX:62004A0196:EN:HTML.

ERA (2003). *Discrimination against Air Transport: Unjustified and Unjust*, European Regions Airlines Association.

European Commission (1999). SG(99)D/4235, 14 June.

European Commission (2004). 'Commission Decision of 12 February concerning advantages granted by the Walloon Region and Brussels South Charleroi Airport to the airline Ryanair in connection with its establishment at Charleroi', *Official Journal* L.137, 30 April.

---

13 Although adequate retail facilities need to be in place in order to capitalise on the increased traffic flowing through the terminal (Francis et al., 2003).

European Commission (2005). 'Community guidelines on financing of airports and start-up aid to airlines departing from regional airports. Communication from the Commission', *Official Journal* C.312, 9 December.

Fewings, R. (1999). 'Provision of European airport infrastructure', *Avmark Aviation Economist*, July.

Francis, G., Fidato, A. and Humphreys, I. (2003). 'Airport-airline interaction: the impact of low-cost carriers on two European airports', *Journal of Air Transport Management*, 9(4), July.

Gillen, D. and Morrison, W. (2003). 'Bundling, integration and the delivered price of air travel: are low cost carriers full service competitors?', *Journal of Air Transport Management*, 9(1), January.

Civil Aviation Department (2001). Airport Authority Ordinance (Chapter 483), Hong Kong, 19 January.

House of Commons Transport Committee (2006). Minutes of oral evidence taken before the Transport Committee relating to its inquiry into the work of the Civil Aviation Authority, 11 January, HC 809-ii.

Ibarra, G. (2003). 'European low cost carriers handling connecting passengers – a cost and benefit analysis', MSc thesis, Cranfield University.

Lobbenberg, A. (2003). 'The emperor awaits judgment'. ABN AMRO Research Paper, 18 September.

Mandel, B. (1997). 'The interdependence of airport choice and travel demand'. Paper to an ICAO symposium in Montreal, April.

Ryanair Holdings plc (2003). *Form 20F Report, filed with the US Securities and Exchange Commission*, 30 September.

Schwartz, A. (2003). 'Priming the pump', *Air Transport World*, October.

Chapter 3
Airport Entry and Exit: A European Analysis

Christiane Müller-Rostin, Hansjochen Ehmer, Ignaz Hannak, Plamena Ivanova,
Hans-Martin Niemeier and Jürgen Müller[1]

Introduction

There have been many changes in recent years in both the airline and the airport industries in Europe and elsewhere. These changes have been interpreted by some authors as an indication that the airport industry is tending to become a competitive industry. The so-called 'new view on airport regulation' (Gillen et al., 2001; Tretheway, 2001) argues that airports are no longer natural monopolies and that more competition would be preferable to traditional regulation. This chapter is written with more emphasis on the old, more sceptical tradition that airports are monopolistic bottlenecks: either regional natural monopolies or legal monopolies due to planning and other restrictions (Niemeier, 2004). The scepticism does not stem from the belief that airports are naturally monopolies. Changes in demand and supply might eventually lead to a competitive industry structure. Nor is the scepticism rooted in a distrust of competition. Of course, perfect and perhaps even less intense competition is superior to regulation. Rather, the scepticism is based on the belief that despite these recent changes, competition is still minimal and not sufficient to prevent airports from abusing their market power (Forsyth, 2006). We would like to stress that this is our opinion as there is little empirical evidence on the intensity of competition among airports.

As airport competition is a complex phenomenon, we prefer in this chapter to analyse two specific, but very prominent, aspects of competition: market entry and exit. While these have been well researched for other industries, this is not the case with the airport industry. We confine our analysis to Europe both for historical reasons and because, in light of the changes in governance structure and the density of economic activity, Europe is the first continent where we feel airport competition might work and where we could observe airports entering and exiting the market. During World War II, and the subsequent Cold War, an extraordinary number of military airfields were built which can be relatively easily converted into commercial airports. This eases entry for a potential entrant who compares the sunk costs of entry with the present value of post-entry profits. The commercialization and increase in privatization of European airports, together with continued growth – especially through low-cost carriers (LCCs) – has increased incentives for entry.

Our research aims to analyse the following research questions:

---

[1] We thank Doug Andrew, Peter Forsyth, Karsten Fröhlich, David Gillen, Vanessa Kamp, Peter Morrell and David Starkie for helpful comments. This chapter arose from the research project German Airport Performance (GAP) which is supported by the Federal Ministry of Education and Research. For further details see http://www.gap-projekt.de.

1. Have there been significant market entries and exits in the airport industry in Europe?
2. If yes, can we observe specific characteristics of these entries and exits? How does the corporate structure of the new entrants compare to that of the incumbents? What type of airport usually exits the market?
3. How do entries and exits in the airport industry compare with those in other competitive industries, in particular with the downstream airline industry?
4. Have entries forced incumbent airports to cut costs and lower prices? Have they become more efficient than airports in a market without entries?

In this chapter we focus on the first three questions. For the purpose of our analysis, we have selected a period from 1995 to 2005. This period was partially determined by lack of data, but it covers a long phase of boom and recession, which should contribute to entries and exits.

In the following section we outline the major changes in governance structure of European airports. This gives us not only the background but also an overview of certain factors such as licensing, price regulation and partial privatization, which might influence entry and exit. We then outline the theory of entry and exit in relation to the intensity of competition in the airport industry in order to gain an understanding of what could happen in the future. In the next section we summarize the literature on entries and exits in other industries. This gives us a potential benchmark for entry and exit in the airport industry. We then examine airport entry and exit on a country-by-country basis and close with a summary of our main findings and an outlook for further research.

Brief Overview of the European Airport Landscape

In this section we give a brief overview of major changes in the regulatory environment of airports and outline the possible implications for entry and exit. We start with an outline of privatization followed by recent trends in airport regulation.

Privatization Trends

In 1987 the British government privatized the three London BAA airports – Heathrow, Gatwick and Stansted – together with BAA's Scottish airports. Today the majority of UK airports are fully privatized (Graham, 2004). Though BAA's performance and its rising share prices were widely seen as a success, making it a kind of role model for the privatization of airports, most European governments were reluctant to give up control completely. In the 1990s a number of European airports were partially privatized. The crises in aviation from 2001 onwards stopped this trend almost completely and only recently the airports of Brussels, Budapest, Lübeck, Malta and Paris were partially privatized. Among the non-British airports, only Brussels, Copenhagen, Malta and Vienna are majority privately owned. No major airport in Continental Europe has been fully privatized without any ownership restrictions (Gillen and Niemeier, 2006).

According to Gillen and Niemeier (2006) privatization on the European Continent has not changed the nature of the industry as it has in the UK, but it has led airports to become more profit-oriented. The typical private airport in Europe is a partially privatized airport that tries to pursue a wide range of objectives, in addition to profits, such as regional development, job creation and tourism growth.

This pattern of ownership certainly has implications for entry and exit as it influences motivation and behaviour. The airport industry has unquestionably become more business-oriented so that airports are looking for new profitable business opportunities. This holds true for the minority of private airports, but also for public airports that have reorganized and became more businesslike. Municipalities see business opportunities in airports and in addition, see airports as an instrument for regional development. Both motives have been at work in Germany and France (Gillen and Niemeier, 2006).

Privatization has also created a market in airport business assets. Financial investors are now looking for business opportunities in the airport industry – something unheard of prior to privatization. The capital market in airport assets has two implications for entry and exit in the airport industry. Firstly, profit opportunities should be sought out so that in this respect the airport industry (slowly) becomes more like an ordinary industry. If profitable, new airports should be built. Secondly, the capital market is revaluing old stranded assets. Airfields and unprofitable airports that have been in public hands and not kept open might come on the market after being written off and revalued by the market.

Regulatory Trends and Their Impacts on Entry and Exit

In Europe airport charges have traditionally been regulated on a rate of return or cost-plus and a single till basis. The charges should generate just enough revenue to cover total costs, including the depreciation of capital and a normal rate of return on capital for the whole airport. The structure of charges should also be cost related.

However, UK price cap regulation has been copied by some European authorities. Today only a few airports (for example, Dublin, Hamburg, Malta, Vienna) are price capped (Gillen and Niemeier, 2006).

Cost-based regulation and price-cap regulation set different incentives for airport management. Unlike cost-based regulation, price caps do not regulate profits; but instead set incentives for cost reduction and revenue generation. The gains from cost reduction and additional revenues can be kept by the regulated airport within the regulation period, and might then be passed on to the users via lower charges in the next period. Cost-based regulation sets the incentives in the opposite direction. It leads to higher costs, expensive non-essential items and to price structures that do not increase passenger throughput.

Regulation might influence entry and exit in various ways. Firstly, price regulation in general prevents an airport from charging monopoly prices. This might lead to fewer entries than in an unregulated industry in which airports could set Cournot prices and in which the profitability becomes known to all potential market participants. Secondly, cost-plus regulation might be attractive for new investors who prefer a good and safe return; otherwise a cost-plus regulated industry offers only limited profitability. Thirdly, price-cap regulation should offer a relatively higher profitability and lead to higher entry rates than a cost-plus regime.

In short, the diversity of different European regulatory systems – with their implications for incentives – should lead to different patterns of entry and exit, given that other things are equal.

Theoretical Background: The Theory of Entry and Exit with a View to the Airport Industry

In this section we do not review the existing literature on entry and exit. The objective of this chapter is to use the theory as a heuristic tool for discovering and analysing potential market entry and exit behaviour in the airport industry.

One limitation of the heuristic power of the market entry and exit theory should be stated. The theory assumes rational behaviour on the part of the firms, namely that they maximize profits. However, airports are typically endowed with a much richer motivational structure. Some are run as public utilities trying to maximize welfare; others are run to maximize the regional impact of an airport. Creating jobs, securing rents for special groups, attracting tourists and providing political motives are further examples. Even for the fully privatized airports Starkie (2006) doubts the simple profit-maximizing assumption and argues that empire building and revenue-maximizing behaviour are relevant motives of airport management.

We start with a definition of entry and exit, briefly review the role of entry and exit for competition, look at barriers to entry and entry-deterring strategies in the airport industry and, finally, examine the effects of entry.

Definition and Forms of Airport Entry or Exit

Besanko et al. (2003, p. 298) state that a firm enters the market, when 'it starts production' and that it exits the market when 'it stops production'. We apply this definition to the airport industry. In our definition, an airport first enters the market when it is opened for commercial civil aviation activities (scheduled and charter flights). Hence, an airport exit occurs when commercial aviation activities at the airport cease. Here we define the production of an airport as its connection to commercial aviation and all the airport activities that are related to commercial flights.[2]

Entry and exit can take many different forms. An entrant may be a brand new firm, as in the case of Don Quijote International Airport near Madrid which entered the Spanish airport market in 2007. An entrant may also be an established firm that is diversifying into a new product market. Exits occur if a firm simply folds up operations or exits a particular market segment.

Role of Entry and Exit for Competition

The ease of entry and exit is perhaps the most important condition for effective competition. Curtis Eaton (1987, p. 156) summarizes this well by writing that:

> entry – and its opposite, exit – have long been seen to be the driving forces in the neoclassical theory of competitive markets. Long-run equilibrium in such a market requires that no potential entrant finds entry profitable, and that no established firm finds exit profitable. There is very little more to the theory of equilibrium in a competitive market than this simple yet powerful story of no-entry and no-exit.

While we do not think that the model of perfect competition can be applied to the airport industry in any way, the model nevertheless gives us an understanding that entry and exit matter for competition and welfare. In Cournot models of oligopoly, market entry decreases above normal profits by lowering market prices and reducing single firm output. In general (see subsequent section on the

---

2   We do not include freight or general aviation activities within our analysis. Of course competition might start earlier, i.e. when an airport having no scheduled services up to now competes with other airports to get low-cost services. But this kind of competition is extremely difficult to define; that is why for pragmatic reasons we concentrate on competition for airlines and passengers.

discussion on excessive entry) the beneficial effects of competition in terms of economic welfare and Pareto efficiency can only be expected with low entry and exit barriers.[3]

Entry Barriers and their Relevance to the Airport Industry

Barriers to entry enable incumbents to earn abnormal profits without attracting entry. A profit-maximizing potential entrant compares the sunk costs of entry with the present value of the post-entry profit stream. The first factor in this calculation, sunk costs of entry, may range from investment in specialized assets to government licenses. They occur when the new firm stops production and exits the market. The second factor, post-entry profits, will depend on demand and cost conditions as well as the nature of post-entry competition factors that can be influenced by the strategy of the incumbent.

Following Lipczynski et al. (2005) we distinguish between structural, strategic and administrative barriers to entry.

Structural Barriers

Structural barriers to entry are related to the technical aspects of production. Unlike strategic barriers, the incumbent has no direct control over these factors. In relation to airports we discuss three types[4] of structural entry barriers: sunk costs and economies of scale and scope, absolute cost advantage and network effect on the demand side.

Sunk costs and economies of scale and scope Economies of scale have been seen as an important factor, perhaps even the most prominent factor for airports. Airports may well be considered natural monopolies. Unfortunately there are discrepancies regarding the exact form of the long run average cost function. Estimates therefore differ substantially concerning the level at which economies of scale are exhausted. Average costs might decline up to a level of between 3 and 12.5 or even up to 90 million passengers, depending on the sample of airports and analysis performed. Hubs might even experience diseconomies of scale (Kamp et al., 2005).

To our knowledge there are no estimates on the strength of *economies of scope* in the airport industry. It is plausible that economies of scope might arise from the use of runways for scheduled, charter and cargo traffic as well as from jointly offering non-aeronautical services (Australian Productivity Commission, 2002, p. 102).

Since the demand for air transport services is still low in many small regional European markets, incumbent airports might enjoy a natural monopoly position, which, together with sunk costs, would make entry for newcomers unprofitable. The expected strong growth in air traffic volumes in Europe may change this in the coming decades. Usually market entry involves the building of a runway system and terminal facilities. These are largely specialized investments that cannot be recovered when exiting the market.

However, Starkie (2005) argues that Europe might be different because of a substantial number of former military airfields built during the Cold War era, which are no longer utilized by the

---

3 Entry is also important for dynamic concepts of competition along the lines of Schumpeter and the Austrian School of Economics. See Geroski (1991).

4 Natural product differentiation does not seem to be of any relevance for airports since they have not established a brand name. Advertising costs are relatively low compared to consumer products such as cereals or beer.

military and could easily be converted into civil aviation airports. The opening of such airports might involve less fixed and sunk costs than those involved in the construction of an airport built on a green field. However, military airfields are usually not located in densely populated areas. Therefore, their location is inferior to that of main airports, and in most regions public entities would have to invest in transport infrastructure to improve access to these airports. Since planning processes and infrastructure policies differ between the European member states, one would expect different outcomes depending on the region.

Absolute cost advantage A superior location plus subsidies might create an absolute cost advantage (Forsyth, 2006), but clearly many of the other factors, such as patents, which might work in other industries, are most likely of less relevance for airports. Airports with a history of being managed as private or public monopolies probably have cost disadvantages vis-à-vis privately managed airports due to managerial slack and investments in unnecessary facilities. This would create opportunities for a new entrant that is managed like a low-cost firm, much like LCCs in the downstream airline industries.

Positive network effects-hub operations at established airports There are multiple benefits for airlines that choose to concentrate their operations at one airport since this creates economies of density due to higher frequencies, larger aircraft and joint usage of common facilities such as lounges. Since part of these costs, such as building up hub operations, are sunk costs for the airlines, lock-in effects might occur and switching costs would probably be substantial. Although there are no estimates on positive network effects, they might 'create a more significant barrier to entry than do airport supply characteristics' (Australian Productivity Commission, 2002, p. 105).

Strategic Barriers

Strategic barriers to entry stem from the strategy an incumbent chooses to deter entry (Hüschelrath, 2005). The following strategies might be applied by incumbent airports, although there appears to be no empirical evidence so far for the first two of these strategies.

Excess capacity This might well be a rational strategy for airports since it provides a basis for a credible commitment to reduce post-entry prices of a new entrant. Short-run marginal costs are low, investment costs are sunk and in many European non-hub markets it may take several years before demand outstrips capacity. However, the problem of over-excessive investments exists. Over-excessive investments can be caused by cost-plus regulation because it sets incentives to under-price peaks and to produce too capital-intensively (Niemeier, 2004).

Limit pricing Limit pricing is a strategy applied before entry and it gives rise to the question of whether airports react to plans to build a new airport by lowering the airport charges or by credibly committing themselves to lower post-entry charges.

Predatory pricing While there have been a number of cases in the airline industry, to our knowledge no airport has ever been accused of applying predatory pricing (Forsyth et al., 2005). The notable exception to this rule is BAA's pricing of Stansted airport below marginal costs to compete against Luton airport (Starkie, 2004).

Raising rivals costs An incumbent can try to raise structural and legal entry barriers to deter entry. The high fixed-cost nature of airports could make it profitable to engage in such strategies. Furthermore, the incumbent airport is usually well connected with authorities and might therefore lobby for high-quality standards for new airport services or tough environmental and planning restrictions.

Legal Barriers

Legal barriers can be very effective entry barriers. The following three types of legal barriers are of relevance for the airport industry.

Monopoly rights The state can grant an existing airport a monopoly by not permitting other airports to be built and operated either in the close vicinity or within the country. In the case of a natural monopoly this might be efficient, as one airport operates with lower average costs than two airports do and there is still sufficient capacity even at peak times. Granting monopoly rights can be a legitimate policy, but it does not have to be. If demand is strong enough and no economies of scale and scope can be reaped, an industry structure with two or more airports can be sustained. In such a case granting monopoly rights to one airport might lead to inefficiencies and prevent effective competition. Starkie and Thompson (1985) were critical of the British government granting BAA a monopoly in the London region. The Office of Fair Trading (2007) has now taken up the case again. Malina (2007) also argues that some regional public authorities in Germany have misused their power through joint ownership of airports. Within privatization processes the objective of the Finance Ministry very often is to maximize the sale price by granting a monopoly to the new owner. In the case of Berlin, a small regional airport operator was not allowed to operate low-cost flights for technical reasons although it was technically feasible. This delayed market entry for a long period.

Planning and environmental restrictions The construction of a new airport, as well as the conversion of a military airport into a commercial airport, is subject to planning and environmental restrictions. As with the case of granting monopoly rights, there are cases in which planning and environmental restrictions are either welfare enhancing or welfare reducing. The decision to allow the entry of a new airport should be based on a cost-benefit analysis of the environmental and safety externalities of a new airport. Since demand is strong in metropolitan areas such as London, Berlin and Paris, negative externalities (noise, pollution, etc.) are also high relative to rural areas. Before investigation, it might be rational to expand an airport in a densely populated area instead of permitting the construction of a new airport. Unfortunately, these decisions are typically not based on a cost-benefit analysis (Niemeier, 2001). Furthermore, restrictions might reflect not so much on environmental concerns, but rather on the intention to shelter the incumbent airports, which sometimes may, as in the case of the regional airport Kassel, lobby against a new airport as purportedly useless in economic and environmental terms.

Bilateral air service agreements These agreements can limit the take-off and landing points for the air carriers in the two countries involved. Such a legal restriction is specific to the airline and aviation industry and may lead to first mover gains for large airports, as well as structural barriers to entry in the form of positive network effect (Gillen et al., 2000).

Effects of Entry

As outlined above, the possibility of entry into regional airport markets should lead to reactions of the incumbent to deter entry. However, if entry is successful we normally expect the following reactions:

- entry might drive down airport charges in markets with new entrants (example of cheaper secondary LCC airports);
- increased price differentiation by incumbent airports to compete with new entrant(s);
- reductions in access time for citizens;
- greater choice of airports and differentiated products;
- incumbent airports might be forced to reduce X-inefficiencies and to cut costs;
- incumbent airports might see their profit levels decrease if traffic is diverted to the new airport(s).

In general – and this underlies much of the demand for reform towards a more competitive airport industry – more competitors lead to more competition with lower prices and costs and hence to an increase in welfare. However, this reasoning assumes constant returns to scale which might not be appropriate for small markets with up to 3 million passengers and medium-sized markets with up to 12 million passengers in which the incumbent airport benefits from economies of scale and scope. With economies of scale, entry will increase the average costs of each airport and might lead to unnecessary duplication of fixed costs. This effect might counteract the effect of price competition and strike the balance towards a net welfare loss (Mankiw and Whinston, 1986). The welfare effects of entry must therefore be analysed on a case-by-case basis.

Entry and Exit in Other Industries

The aim of this section is to describe entry and exit patterns in other industries and to deduce similarities and differences with regard to entry and exit in the airport industry. The airline industry is highlighted separately due to its significance *for* and interdependence *with* the airport industry.

Entry and Exit Evidence in Other Industries

Entry and, to a certain extent, exit play a crucial role in most industries since they lead to changes in market structure, prices and competitive behaviour amongst firms. However, the majority of markets are not fully contestable and often entrants encounter substantial entry barriers, both structural and behavioural. In their meta-analysis of empirical studies on entry and exit, Siegfried and Evans (1994) found considerable empirical support for the existence of absolute cost barriers – most often sunk costs in machinery, buildings and other specific equipment (Kessides, 1990) – and some empirical evidence of an entry-deterring effect of multi-plant incumbent operations (due to incumbent cost advantages).

Exit barriers have not been studied in as great detail as entry barriers, but here as well Siegfried and Evans (1994) found some interesting empirical evidence. The largest exit barriers are sunk costs in both tangible and intangible durable specific assets. The higher the investment of a firm in specific machinery and buildings, the more likely it is to stay in the market, even if it is unprofitable. Other exit barriers include managerial hesitation to exit and long-term labour agreements with workers and suppliers. Exit is often induced by negative market growth and low or non-existent profits.

In general, a number of studies (Siegfried and Evans, 1994; Geroski, 1995; Dunne et al., 1988) have found that entry and exit rates are positively correlated with one another. Industries that show high rates of entry also exhibit high rates of exit and vice versa. In addition, entry and exit firms were usually smaller than incumbent firms. Evidence from Germany (Schwalbach, 1987) shows that firms will enter the market if:

- profits in the market are higher than in other comparable markets;
- the market growth rate is positive;
- the accumulated expertise (in the case of diversifying firms) can be transferred profitably to the new market.

Entry and Exit in the Airline Industry

The airport industry and the airline industry differ in one fundamental aspect: the degree of sunk costs. Therefore, it is likely that entry and exit patterns in the airline and airport industries will differ greatly.

Up until about 20 years ago, the airline industry was viewed as an example of a perfectly contestable market. Factors such as free entry and costless exit, access to the same production technologies, low or non-existent sunk costs and the threat of potential competition were all inherent to the airline industry. It was therefore concluded that the hit-and-run entry of a new carrier into a market was possible since incumbents would not have the time to respond by cutting prices or increasing capacity. This constant threat of entry in turn would force incumbent airlines to price their product at average costs to prevent entry (Baumol et al., 1982). However, evidence collected during the last 20 years has shown that there are many indications that the airline industry is not an example of a perfectly contestable market (Morrison and Winston, 1987; Borenstein, 1992). Although it is true that economies of scale are relatively small, this does not hold for economies of scope and density of hub-and-spoke networks. In addition, there are a number of entry and exit barriers obtaining gates and slots at congested airports, computer reservation systems, frequent flyer programmes and certain brand image and perceived safety advantages (Hüschelrath, 2005; Schnell, 2006). Predatory pricing behaviour was also observed, especially as a reaction to the rise of LCCs in the US and European airline market (Windle and Dresner, 1999; Forsyth et al., 2005).

Numerous studies have looked at price and output effects of entry into and exit from the airline market. Joskow et al. (1994) studied 375 city pairs regarding entry and exit. The results of their econometric analysis suggested that below-average fares lead to market exit, whereas above-average fares do not necessarily correlate with entry. They found that airlines choose to enter on certain city pairs because these routes fit well into their overall network; a route's profitability in this case is secondary. In addition, their results show that fares increased after exit and decreased after entry. Total output (measured in revenue passenger mile) increased after entry and decreased after exit. According to Vowles (2000) and Morrison and Winston (2001) the presence of LCCs has the highest effect on incumbent fares and increased welfare also on adjacent routes.

Evidence of Entry and Exit in the European Airport Industry

Of the 25 countries analysed, entry and exit only took place in 14 countries (see Figure 3.1). Our findings show that, in general, entry (and exit) was more predominant in already well-developed airport markets and that most countries exhibited one or a maximum of two entries (exits) between 1995 and 2005. The vast majority of new airports are planned in the developing markets of Eastern Europe.

Germany

Germany has a diverse airport structure. Airport ownership, for example, varies – some airports have been partially privatized, whereas others are still completely under public ownership. Although overall airport policy is decided on a national level, airport regulation and management are delegated to the federal states. Currently there are 18 airports that have the status of international airports. Two of these airports (Munich and Frankfurt) are Lufthansa's main hubs; Frankfurt-Hahn Airport and Cologne-Bonn Airport are mainly served by LCCs, whereas airports such as Düsseldorf, Hamburg and Stuttgart cater mainly to business travellers while also feeding traffic into the two Lufthansa hubs. In addition, there are 41 regional airports in Germany, most of which are located quite close to one another with overlapping catchment areas (ADV, 2005). According to Malina (2007) the airports of Hamburg, Frankfurt, Munich, Stuttgart and Berlin have strong market power as airlines have no well-located airports in these regions and moving hub operations would induce very high switching costs. Five airports in Germany (Hannover, Nürnberg, Leipzig, Dresden and Bremen) possess modest market power, while all other airports in Germany have little to no market power.

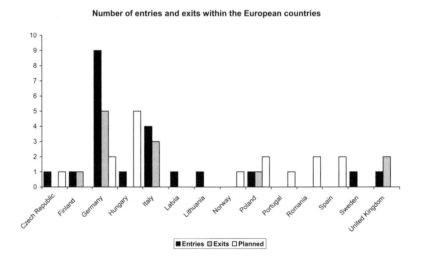

Figure 3.1   Airport entries and exits

There have been a number of entries into (but also some exits from) the German airport market since 2000 (Behnen, 2004). This expansion in – what some consider unnecessary – capacity has been criticized by both airlines and analysts, who feel that the construction of new civil aviation airports and/or the conversion of military airports into civil aviation airports are a waste of taxpayers' money (Heymann, 2005). In Germany, we have mostly observed the conversion of military aerodromes into civil aviation airports.[5] Only three airports have been built during the last 30 years (Paderborn-Lippstadt, Münster-Osnabrück and Munich). Nine airports entered the market between 1995 and 2005, whereas five airports exited (Figure 3.2). Entries occurred mainly in north-eastern Germany, where all new entrants have overlapping catchment areas with at least one other airport in the region. Additional entries occurred in the Ruhr area as well as south-western Germany – areas with an already high concentration of airports. The majority of exits also occurred in these two areas, strengthening the argument that airports with strong overlapping catchment areas relative to demand tend to cannibalize one another.

Figure 3.2   Entry and exit of German airports
Source: ADV (2005) and own research.

---

5   Examples include Altenburg-Nobitz, Rostock-Laage, Magdeburg, Nürnberg, Karlsruhe/Baden-Baden and Zweibrücken.

In 1994 Düsseldorf Airport took over 70 per cent of the shares of Mönchengladbach Airport to alleviate capacity problems at Düsseldorf Airport. Although traffic figures rose substantially between 1996 (146,586) and 1999 (226,408), they decreased from 2000 onwards, reaching a low of 41,244 passengers in 2004 – the year in which all scheduled and charter flights ceased to operate from Mönchengladbach Airport. The main problem at Mönchengladbach Airport was its short runway of 1,200 metres. Rostock-Laage Airport is an example of a new entrant that chose the cargo market as its niche. Although passenger numbers increased significantly between 1995 (year of entry) and 2006, it is the cargo and airmail segments that have seen the highest growth rates.

Another strategy often used by new entrants is the construction of an aeroplex (an airport business park and/or shopping facilities) near the airport to increase non-aeronautical revenues and to lessen the dependence on generating aeronautical revenues to turn a profit. Both Zweibrücken Airport (entry in 1998) and Karlsruhe/Baden-Baden Airport (entry in 1997) have airport business parks; however, only Karlsruhe/Baden-Baden has seen a significant increase in passengers at its airport since its entry in 1997. Passenger figures at Zweibrücken have more or less stayed constant.

The most favored strategy for new entrants, however, is the entry and positioning as an LCC destination and base. Many smaller regional airports in economically underdeveloped regions hope to achieve economic growth duplicating the success of Frankfurt-Hahn Airport. So far, only one new entrant, Airport Niederrhein-Weeze, has been able to establish itself successfully as an LCC hub. Since its entry in 2003, it has seen an increase in passenger numbers every year, and even the retreat of its home carrier had only a minor impact on passenger figures.

In general, new entrants in the German airport market face a number of hurdles. Firstly, and most importantly, a large number of airports already exist in Germany, many of them with overlapping catchment areas and similar business models. Secondly, a number of smaller regional airports are located in areas with low population densities and below-average income per capita, which in turn lessens the demand for air transport services. In addition, infrastructure links to these airports are usually quite limited. Thirdly, although most small airports hope to attract LCC services, the demands of these airlines are often not economically feasible for smaller airports. Despite these hurdles, many regional politicians in economically depressed regions actively lobby for expansion of old and construction of new airports, hoping to foster regional economic growth. These failed expectations have led to five market exits during the last ten years. The case of Mönchengladbach Airport was already mentioned above. In addition, Kiel, Siegerland, Augsburg and Kassel-Calden airports exited the market. Kiel exited the market in 2006 after Lufthansa's partner Cirrus Air gave up its last scheduled service from Kiel to Munich. Siegerland and Augsburg exited the market due to the cessation of regular scheduled services in 2003 and 2005 respectively.

Kassel-Calden Airport is an exceptional case. Here the airport was closed in 2005 so the runway could be extended. It plans to re-enter in 2011. However, this has caused criticism from both airlines and neighbouring airports. Memmingen Airport, located in the wider Munich region, started to offer scheduled flights by July 2007.

Italy

Italy has 48 commercial airports, of which 38 are under ENAV management (Ente Nazionale di Assistenza al Volo, the Italian Company for Air Navigation Services), including some of the most important airports in Italy – Rome Fiumicino, Milan Malpensa and Linate, Catania, Palermo, Cagliari, Bergamo, Turin, Bologna, Venice, Naples and Florence. The two major Italian airports (hubs) are Rome Fiumicino (over 30 million passengers in 2006) and Milan Malpensa (ADR, 2007).

Relative to other European countries, the Italian airport infrastructure is characterized by a lower average size of major international airports, a larger number of medium-sized airports and a wider dispersion of air traffic. Italian airports face strong traffic growth with limited opportunities to increase capacity. In the period 1995–2005, three new airports have entered the Italian airport market: Aosta (1996), Bolzano (1999) and Brescia (1999). At these airports, traffic has grown strongly due to LCCs. Finally, some airports started international routes (for example, Crotone in 2003 and Reggio Calabria in 2004). There are a few cases of very small airports which exited the market (at least temporarily). For instance, the airport of Taranto – Grottaglie – entered in 2003, dealing with about 35,000 passengers, but then stopped the flights one year later; the airport of Cuneo – Levaldigi – temporarily stopped operation during 2005. Also, the airport of Tortolì (in Sardinia) had an average traffic of about 50,000 passengers in 2003 and then stopped flights.

The United Kingdom

In contrast to other European countries, most airports in the United Kingdom are not owned and managed by central or local government entities. Indeed, UK government policy actively promotes and encourages private ownership of airports, and the majority of British airports are either partially or fully privatized.

There are over 50 airports with scheduled air services in the UK and over 20 airports with 1 million passengers or more per year. Many airports are located close to one another and compete with each other. Some airports handle primarily LCC traffic (for example, London Luton), others LCC as well as FSC/charter carrier traffic (for example, Birmingham). Airports with very little LCC traffic are London Heathrow, London Gatwick and other smaller regional airports. During the last 20 years, the United Kingdom has seen the conversion of many former military aerodromes into civil aviation airports, with many new airports entering the market and trying to attract LCC services. However, the question of profitability of these airports remains; with many airports offering just one flight per day, costs can hardly be covered. Starkie (2002), on the other hand argues that many of these smaller former military airports in the UK are quite profitable since they have low sunk costs, have been privatized and have diversified into different business areas such as General Aviation, cargo, or airport business parks. A number of these airports have also seen exponential traffic growth as they have been able to attract LCC traffic; whereas other airports, such as new entrant London City Airport, have been quite successful in attracting business flights or other niche markets.

Only one airport entry was observed in the timeframe under consideration. In 2004, Robin Hood Doncaster Airport entered the market and has so far proven to be quite successful with exponential traffic growth. Manchester initially opposed entry of Robin Hood Doncaster Airport since it feared intensified competition in the region (Manchester is the majority shareholder of Humberside Airport, which is in close proximity to Robin Hood Doncaster), but it was not successful in deterring entry. The entry of Robin Hood Doncaster Airport has also led to the expected exit of Sheffield City Airport, which cannot compete with Robin Hood Doncaster due to its shorter runway that does not allow large aircraft to take-off and land.

In addition only one airport exit was ascertained for the UK. In 2005, all commercial flights ceased from Swansea Airport and have not resumed to date. The airport is in need of a facility upgrade to be able to continue offering scheduled services.

Spain

AENA (Aeropuerto Españoles y Navegación Aerea) was created as a public entity under Article 82 of the General Budget Act 4/1990 on 29 June 1990 and manages 48 airports and five air navigation regional directorates corresponding to three Flight Information Regions. Airport development in Spain is centrally planned and controlled, with some cross-subsidization of infrastructure expenditure between individual airports. Thus, there is no competition amongst airports in Spain. No airport entries or exits have occurred. However, two private airports are planned and one of them has been completed.

Don Quijote International Airport is the first international airport in Spain that has been financed entirely through private investors. It is located near Ciudad Real in Castilla-La Mancha, about 200 kilometers (45 minutes) south of Madrid and 45 minutes north of Córdoba by high-speed train. Construction of the airport started in 2004 and was completed in 2007. Ciudad Real Airport is operated by a Spanish division of Vienna Airport, which owns an 18.7 per cent stake in the project. According to Davy (2004) it will be the first international airport in Europe that will have been financed entirely through private investors. Due to its proximity to Madrid, as well as the planned high-speed rail connection, it might prove to be a competitor for Madrid-Barajas Airport.

Another international airport, Aeropuerto de Castellón, is being built in the province of Castellón. Castellón is the only province on the Spanish coast that does not yet have an international airport. Fomento de Construcciones y Contratas, the second largest Spanish construction group, will build the airport together with the Aeropuerto de Castellón Company, which is 100 per cent government owned. The airport will be privately run, though, and will not be subject to AENA tariffs. It is not yet clear, however, when the construction of the airport will be completed. Due to its proximity to the airport of Valencia and its location in a popular vacation region, the Aeropuerto de Castellón might prove to be a competitor for Valencia Airport, especially if low-cost and charter carriers choose to operate from the new airport.

Poland

Poland currently has 12 airports that offer scheduled services. Three of these airports (including Poland's largest airport, Warsaw Frederic Chopin Airport) are managed by the Polish Airport State Enterprise, which also holds minority and majority shares in the remaining nine airports. Airport charges are set by airport operators after consultation with users and must be approved by the Civil Aviation Office. Besides Warsaw Frederic Chopin Airport, the other airports are marginal players serving their respective regional markets. Warsaw Airport plays an important role in connecting Poland and Eastern Europe with the European Union and North America.

Wizz Air, Central Wings and Ryanair are the dominant players at Poland's regional airports, offering point-to-point services mostly to Germany and Britain. In almost all cases, airports are regional monopolists. The only two exceptions are Warsaw and Lodz and Krakow and Katowice, having strong overlapping catchment areas.

Until 2004 there was a regular service on the Warsaw–Poznan–Zielona Góra route. This service was cancelled at the end of 2004 and the airport of Zielona Góra was without scheduled services for two years. This route was reinstituted at the beginning of 2006 as a public service obligation (PSO) service.

There are currently two airports planned. A second airport for Warsaw is currently being built on the base of a former military airfield in Modlin, 40 kilometers north of the Polish capital. Completion is expected by 2019. This airport is intended to serve mainly LCCs, but it should also be able to attract charter and cargo operators. Another new airport is planned for southern Poland in the city of Lublin. Lublin (340,000 inhabitants) is the biggest Polish city without an airport. At the end of 2006 the local government decided to build a regional airport with a 1,800-metre runway and a terminal capacity for 500,000 passengers.

Other Eastern European States

In general, very few entries and exits were observed in the Eastern European countries. Airport entries occurred only in Bosnia and Herzegovina, the Czech Republic, Poland and Hungary. In all of these countries former military aerodromes were converted into civil aviation airports during the 1990s. In terms of competition with established airports, currently only Pardubice could develop into a threat for nearby Prague-Rozny Airport, if it is able to attract sufficient LCC traffic.

There are a number of factors that might explain the low rates of entry and exit in the airport industry in Eastern Europe. The major factor is that the majority of Eastern European countries in this study are too small (in terms of both size and population) to be able to support more than one or two international airports. In addition, income levels in many of these countries are very low, which again adversely affects demand. Many airports in Eastern Europe are still state owned and operated and are often part of larger country-wide airport systems. Although a number of airports have been privatized in recent years, many governments are still reluctant to allow for more competition either by breaking up existing airport systems or by privatizing their individual airports.

Other Countries

For all other countries within this study, no evidence of entries or exits was recorded. However, there are a few findings we would like to highlight.

France did not experience any entries or exits between 1995 and 2005, which might be due to ample capacity at existing regional airports (Thompson, 2002). Currently, there are 113 airports in France that offer commercial services. Hub competition within France does not exist; however, Paris-Charles de Gaulle Airport faces some competition from other European hubs like, for example, Heathrow. Major regional airports do not compete with one another since they are situated too far apart; the only exceptions are Nice and Marseille airports, which have overlapping catchment areas. Some smaller airports, however, actively compete with other airports – for example, Carcassonne Airport with Toulouse Airport; Bergerac Airport with Bordeaux Airport; or, in the Paris area, Paris-Beauvais Airport with the other Parisian airports. These airports are mainly used by LCCs and have seen some exceptional growth in passenger figures during the last few years.

In Portugal, there was also no evidence of airport entries or exits; however, there have been talks about building a new airport for Lisbon near Ota, 35 kilometers north of Lisbon. A master plan was devised 15 years ago, and Ota was chosen as the site for the new airport. Construction was to start in 2007; however, the current government has shelved the plans to build a new airport due to the high construction costs, estimated at €2.3 billion.

The airport market within The Netherlands is already saturated, with Amsterdam Schiphol being the main hub and a number of smaller airports offering charter and LCC flights. However, there are also large numbers of passengers using airports located in northern and western Germany close to the Dutch border. Of special interest is the case of Niederrhein-Weeze Airport. The new entrant in the German airport market is predominantly used by people of Dutch nationality.

Summary and Agenda for Further Research

The theoretical review shows that entry and exit are crucial for intense competition. Economies of scale and scope, combined with the sunk cost nature of airport assets, might be very effective entry barriers if in the regional market the demand for air transport is low and does not come near a threshold of about 12 million passengers. Other structural barriers to entry might be effective as well. Absolute cost advantages in the form of superior location and subsidies, as well as economies of density due to airline networks, all favor large airports and might create significant barriers to entry – although the extent of these barriers has not yet been quantified. An indication of their existence and importance is the entrance of former military airfields in Europe. Such entrants might be successful especially if they are managed as low-cost firms without the typical extravagance of public facilities. Strategic barriers such as excess capacity, limit pricing and predatory pricing could all be effective in the airport industry, but they do not to seem to be very common. Incumbents can raise rivals' costs and create legal barriers to entry by making planning and permission processes more costly and by prohibiting new airports in the near vicinity. Furthermore, the availability of attractive sites in densely populated European regions might be limited by environmental factors.

In summary there could be a significant number of barriers to entry and exit that could effectively limit such activity in the airport sector. In general the effects of entry in the airport industry are welfare enhancing when they lower airport charges, increase price and product differentiation and reduce X-inefficiencies. In regional markets with decreasing average costs, however, new entrants might lead to unnecessary duplication of fixed costs and welfare losses.

Our empirical research supports the theoretical analysis. The airport industry is characterized by very low rates of entry and exit compared to other industries of similar size and structure. In the period of study the airline industry is characterized by the entrance of LCCs, which leads to intense competition with sophisticated market entry and exit strategies of incumbents and new entrants. However, it would be misleading to expect such an active entry and exit behaviour from airports as specific investments lead to high sunk costs, relative to the airline industry, if a new airport has to be built. Nevertheless the extreme low entry and exit rates indicate that the airport industry is characterized by substantial market entry and exit barriers. In the 25 countries analysed, 21 entries and 12 exits occurred in 14 countries. In addition, our findings show that in another two countries there might be entries in the form of new airports or converted military aerodromes within the next couple of years. Entries (and exits) have occurred mainly in well-developed air transport markets, with Germany (nine entries, five exits), Italy (four entries, three exits) and the UK (one entry, two exits) leading the group with the majority of entries and exits. With regards to type and form of entry, most new entrants were military fields converted into civil aviation airports. In some cases in Germany, existing General Aviation airports were opened for scheduled and charter traffic. In Germany, all new entrants are run and managed by public entities; in the UK we were able to observe the entry of one privately managed airport (Robin Hood Doncaster Airport). With this exception no other private airport entries occurred during our study period 1995–2005.

In our analysis, we also found that very few airports were being built from green field. The most notable exceptions to this, however, are the two Spanish airports that are being financed completely from private funds. In all other countries in which entries occurred, these were converted military aerodromes.

These findings indicate that, unlike in other industries, entry and exit in the airport industry is not so much driven by the profit nexus, but rather by the desire of public airports to increase economic activity for their region. Interestingly most of the new entries are airports serving only one European LCC, Ryanair.

In terms of profitability, very little information could be obtained. The majority of entrants have only been in operation for less than five years – and very few publish financial data. A number of airports in Germany are subsidized by regional and state governments; however, increasing traffic rates at Niederrhein-Weeze, Karlsruhe/Baden-Baden and Zweibrücken airports are positive growth indicators and might lead to an increase in profitability in the future. One new entrant, however, has been extremely successful: Robin Hood Doncaster Airport.

Most new entrants do not yet actively compete with primary airports; however, this may be apt to change in the future as some of these airports manage to attract more airlines and passengers. Again, there are some exceptions: Robin Hood Doncaster actively competes with Manchester for passengers in the holiday travel segment and the two Spanish airports, Don Quijote and Castellón, which are currently under construction, might very well prove to be competitors for Madrid-Barajas and Barcelona-Reus and Valencia respectively.

All new entrants have tried to position themselves within a certain segment of the market – either by establishing themselves as holiday and LCC airports, or as cargo airports. It remains to be seen if they will be successful with their respective strategies. Entry seems to be especially difficult in regions with excess demand such as London, Frankfurt and Paris. In such regions, entry at locations that would substantially increase competition with the major incumbent airports has not occurred.

Exiting airports usually left the market due to decreases in passenger numbers resulting mainly from the cessation of scheduled services.

Our findings also show that there were very few entries and exits in Eastern European countries. However, it will be interesting to follow the developments in Eastern Europe over the next couple of years as these countries integrate with the common European market.

We interpret our results as a further indication that up until now the European airport industry has not been behaving as a normal industry. The forces of entry and exit are too weak to discipline the monopoly position of incumbents. However this conclusion needs further research:

- What is the relative size of entry in terms of total and potential demand relative to existing capacity in each country? Has demand led to internal or external growth in the industry?
- How has the structure of the regional market been changed by entry and exit?
- What happens after entry? Does intensified competition between airports after entry have an influence on incumbent airport efficiency? Do costs and prices fall?
- Will the strong growth of air cargo lead to the emergence of specialized cargo airports, and will this intensify competition with other nearby airports which offer both cargo and passenger services?

References

ADR (2007), Traffic data of Aeroporti di Roma, available at: http://www.adr.it/datitraffico.asp?L=3&idMen=193&scalo=FCO&mese=12&anno=2006.
ADV (2005), 'Regionale Verkehrsflughäfen und Verkehrslandeplätze in der ADV', study commissioned by the Arbeitsgemeinschaft Deutscher Flughäfen, Berlin.
Australian Productivity Commission (2002), 'Price Regulation of Airport Services', Report No. 19, Canberra, AusInfo.
Baumol, W. J., Panzar, J. and Willig, R. D. (1982), *Contestable Markets and the Theory of Industry Structure*, New York, Harcourt.

Behnen, T. (2004), 'Germany's changing airport infrastructure, the prospect of "newcomer" airports attempting market entry', *Journal of Transport Geography*, vol. 12, pp. 227–88.

Besanko, D., Dranove, D., Shanley M. and Schaeffer, S. (2003), *Economics of Strategy*, 3rd edition, Hoboken, NJ, John Wiley & Sons.

Borenstein, S. (1992), 'The evolution of U.S. airline competition', *Journal of Economic Perspectives*, vol. 6, pp. 45–73.

Church, J. and Ware, R. (2000), *Industrial Organization: A Strategic Approach*, New York, McGraw-Hill.

Davy European Transport & Leisure (2004), *Late Arrival: A Competition Policy for Europe's Airports*, Dublin, Davy Research Department.

Dunne, T., Roberts, M. J., and Samuelson, L. (1988), 'Firm Entry and Post-Entry Performance in the U.S. Chemical Industries', Papers 0-88-4, Pennsylvania State Department of Economics.

Eaton, C. (1987), 'Entry and market structure', in Eatwell, J., Milgate, M. and Newman, P. (eds), *The New Palgrave: A Dictionary of Economics*, London, Macmillan, vol. 1, pp. 156–8.

Forsyth, P. (2006), 'Airport competition: regulatory issues and policy implications', in Lee, D. (ed.), *Advances in Airline Economics: Competition Policy and Antitrust*, Amsterdam, Elsevier, pp. 347–68.

Forsyth, P., Gillen, D., Mayer, O. G. and Niemeier, H.-M. (2005), *Competition versus Predation in Aviation Markets: A Survey of Experience on North America, Europe and Australia*, Aldershot, Ashgate.

Geroski, P. A. (1991), *Market Dynamics and Entry*, Oxford, Blackwell.

Geroski, P. A. (1995), 'What do we know about entry?', *International Journal of Industrial Organization*, vol. 13, no. 4, pp. 421–40.

Gillen, D. and Niemeier, H.-M. (2006), 'Airport Economics, Policy and Management: The European Union', Rafael del Pino Foundation, Comparative Political Economy and Infrastructure Performance: The Case of Airports, Madrid, 18–19 September 2006.

Gillen, D., Hinsch, H., Mandel, B. and Wolf, H. (2000), *The Impact of Liberalizing International Aviation Bilaterals on the Northern German Region*, Aldershot, Ashgate.

Gillen, D., Henriksson, L. and Morrison, W. (2001), 'Airport Financing, Costing, Pricing and Performance in Canada', Final Report to the Canadian Transportation Act Review Committee, Waterloo, Ontario.

Graham, A. (2004), *Managing Airports: An International Perspective*, 2nd edition, Amsterdam, Elsevier.

Heymann, E. (2005), 'Expansion of Regional Airports: Misallocation of Resources', *Deutsche Bank Research* (November).

Hüschelrath, K. (2005), 'Strategic behaviour of incumbents', in Forsyth, P., Gillen, D., Mayer, O. G. and Niemeier, H.-M. (eds), *Competition versus Predation in Aviation Markets: A Survey of Experience on North America, Europe and Australia*, Aldershot, Ashgate, pp. 3–36.

Joskow, A. S., Werden G. J. and Johnson, R. L. (1994), 'Entry, exist and performance in airline markets', *International Journal of Industrial Organization*, vol. 12, no. 4, pp. 457–71.

Kamp, V., Müller, J. and Niemeier, H.-M. (2005), 'Can we learn from benchmarking studies of airports and where do we want to go from here?', GARS Workshop Airport Competition and the Role of Airport Benchmarking, Vienna Airport 24–25 November 2005.

Kessides, I. N. (1990), 'Towards a Testable Model of Entry: A Study of the US Manufacturing Industries', *Economica*, vol. 57, no. 226, pp. 219–38.

Lipczynski, J., Wilson, J. and Goddard, J., (2005), *Industrial Organization: Competition, Strategy, Policy*, 2nd edition, London, Pearson Education.

Malina, R. (2007), 'Competition and regulation in the German airport market: an Empirical Investigation', forthcoming in Forsyth, P. (eds), *Airport Competition*, Farnham, Ashgate.

Mankiw, N. G and Whinston, M. D. (1986), 'Free entry and social inefficiency', *The RAND Journal of Economics*, vol. 17, no. 1, pp. 48–58.

Morrison, S. and Winston, C. (1987), 'Empirical implications and tests of the contestability hypothesis', *Journal of Law and Economics*, vol. 30, pp. 53–66.

Morrison, S. and Winston, C. (2001), 'Actual, adjunct and potential competition, estimating the full effects of south west', *Journal of Transport Economics and Policy*, vol. 35 (May), 239–56.

Niemeier, H.-M. (2001), 'On the use and abuse of Impact Analysis for airports: A critical view from the perspective of regional policy', in Pfähler, W. (ed.), *Regional Input-Output Analysis*, Baden-Baden, Nomos Verlag.

Niemeier, H.-M. (2004), 'Capacity utilization, investment and regulatory reform of German airports', in Forsyth, P., Gillen, D., Knorr, A., Mayer, O. G., Niemeier, H.-M. and Starkie, D. (eds), *The Economic Regulation of Airports*, Aldershot, Ashgate.

Office of Fair Trading (2007), 'BAA. The OFT's reference to the Competition Authority', OFT 912 London, www.oft.gov.uk/shared_oft/reports/transport/oft912.pdf.

Schnell, M. C. A. (2006), 'Existence and effectiveness perceptions of exit-barrier factors in liberalised airline markets', *Transportation Research Part E: Logistics and Transportation Review*, vol. 42, no. 3, pp. 225–42.

Schwalbach, J. (1987), 'Entry by diversified firms into German industries', *International Journal of Industrial Organization*, vol. 5, no. 1, pp. 43–9.

Siegfried, J. J. and Evans, L. B. (1994), 'Empirical studies of entry and exit: A survey of the evidence', *Review of Industrial Organization*, vol. 9, no. 2, pp. 121–55.

Starkie, D. (2002), 'Airport regulation and competition', *Journal of Air Transport Management*, vol. 8, pp. 63–72.

Starkie, D. (2004), 'Testing the regulatory model: the expansion of Stansted Airport', *Fiscal Studies*, vol. 25, pp. 389–413.

Starkie, D. (2005), 'Making airport regulation less imperfect', *Journal of Air Transport Management*, vol. 11, pp. 3–8.

Starkie, D. (2006), 'Investment incentives and airport regulation', *Utilities Policy*, vol. 14, pp. 262–5.

Starkie, D. and Thompson, D. (1985), 'Privatising London's Airports', Institute for Fiscal Studies, Report 16, London.

Thompson, B. (2002), 'Air transport liberalisation and the development of third level airports in France', *Journal of Transport Geography*, vol. 10, no. 4, pp. 273–85.

Tretheway, M. (2001), *Airport Ownership, Management and Price Regulation*, Vancouver, InterVISTAS Consulting Inc.

Vowles, T. (2000), 'The effect of low fare air carriers on airfares in the US', *Journal of Transport Geography*, vol. 8, no. 2, pp. 121–8.

Windle, R. and Dresner, M. (1999), 'Competitive responses to low cost carrier entry', *Transportation Research Part E: Logistics and Transportation Review*, vol. 35, no. 1, pp. 59–75.

Chapter 4
Airport Pricing

Eric Pels and Erik T. Verhoef

Introduction

With the ongoing privatization of airports, it is essential to know how airport pricing and airport competition influence the outcome of airline competition. It is generally accepted that private railway companies and bus companies, and also infrastructure operators like airports, have some degree of market power, although the source of market power may not be straightforward. In the case of airports, Starkie (2001) argues that (some) airports may be 'locational monopolies' rather than natural monopolies. An often used argument is that the introduction of competition prevents the abuse of market power. However, following Kreps and Scheinkman (1983), we can assume that even when airports engage in price (Bertrand) competition, the commitment to a certain level of capacity eventually yields Cournot outcomes. Thus market power is 'abused' in the sense that prices are set above marginal costs. Moreover, airports that face a derived demand function will demand airport capacity because passengers demand trips. Even when airports are prevented from abusing market power by some form of regulation, welfare will not be maximized when airlines are still able to set prices above marginal costs.

In this chapter we analyse airport pricing in a theoretical model. The key question is how privatized airports set profit-maximizing prices for services provided to airlines and other commercial activities. Starkie (2001) argues that airports may not have an incentive to charge airlines monopoly prices, because this might reduce the profitability of the airport's complementary activities. Starkie (2001) asks whether inefficiently high monopoly prices are worse than a regulatory outcome in which prices may be too low. In this chapter we determine the conditions under which the airport will price above marginal cost.

Apart from the market power effect, there is another externality that affects the economic efficiency of the transportation sector. If airport (or en-route) congestion is not accounted for in airport or ticket prices, efficiency is not maximized. Congestion pricing is suggested in the literature as a solution; a regulator sets a congestion toll in addition to the regular charges. Profit-maximizing airports and airlines, however, already take congestion into account, at least to some extent, when setting the prices. As a result, prices will typically be higher than the marginal (production) costs. The fact that the airport operator can charge prices higher than marginal production costs is in itself an indication that the airport operator has market power. But high prices do not necessarily imply that the airport actually abuses market power. Only when the airport overcharges the airlines (for congestion), can the airport operator be said to abuse market power.[1] Note that the regulator could simply set a number of slots, and allocate this number to the airlines. This would maximize welfare only if the slots end up with the airlines that attach the most value to them, and the frequency (i.e.

---

1   Another question is whether the revenue from the congestion charges should go to the airport operator. That is not addressed in this chapter.

the number of slots) is set at the welfare maximizing level. For any other number of slots, welfare is not maximized. Therefore, we consider only congestion pricing in this chapter.

A third market failure is that of market power of the airline (i.e. the company purchasing services/capacity from the airport operator and offering services to passengers). Airlines typically internalize the congestion they impose on themselves and their own passengers (Brueckner, 2002; Pels and Verhoef, 2004). But this does not mean that the output is set at the welfare-maximizing level.

In this chapter we analyse airport pricing in a Cournot model for airlines. We argue that airports face a derived demand function: airlines set their optimal output and, based on this optimal output, have a requirement for airport capacity. Direct competition for passengers between airports is not analysed in the model; the emphasis is on the implications of the derived demand function for airport pricing. Note, however, that in a model in which airports compete, the results will be very similar to the results in this chapter: also in a Cournot oligopoly, airports have market power, although less so than the monopolist. Key outcomes of the model will be oligopolistic airlines, commercial activities at airports, and congestion at airports.

The next section discusses the passenger optimization problem. In the section following the airline maximization problem is formulated. The penultimate section introduces airport pricing strategies, and the final section concludes.

The Passenger Optimization Problem

We consider a simple network with two airports and two airlines. Passengers travel between the origin and destination, and origin-destination flows are symmetric, so that routing is not an issue in this chapter. Passenger demand for return trips is given by the following inverse aggregate demand function:

$$D\left(\sum_{i=1}^{2} q_i\right) = \alpha - \beta \cdot \sum_{i=1}^{2} q_i \tag{1}$$

where q_i is the number of passengers transported by airline i, α (>0) represents the price at which demand is 0, and β (>0) is the demand sensitivity parameter.

The generalized price for an airline's service as incurred by passengers is:

$$g_i = p_i + vot_p \cdot \phi \tag{2}$$

where p_i is the fare charged by airline i, vot_p is the passenger value of time, and φ is the travel time, including congestion experienced by the passenger. The travel time φ depends directly on the frequency of service offered by the airlines in the market. The frequency of service offered by airline i (in both directions) is:

$$f_i = \frac{1}{\lambda_i} \cdot q_i \tag{3}$$

where the occupancy λ_i is the (exogenous) product of the load factor and the seat capacity. The frequency of service therefore depends on the number of passengers moved by the airline, and with fixed aircraft occupancies the relationship is proportional. The average time loss per passenger due to congestion at airport h is assumed to increase in proportion with flight frequency. Ignoring the flight time in absence of congestion, we obtain:[2]

$$\phi_h = \cdot_h \cdot \sum_{i=1}^{2} f_i = \cdot_h \cdot \sum_{i=1}^{2} \frac{f_i}{K_h} \qquad (4)$$

where η_h is the slope of the congestion function at airport h, and K_h is the airport's capacity (note that η_h includes the congestion incurred during one take-off plus one arrival at airport h). Total congestion experienced in a market is the sum of the congestion experienced at both of the airports used:

$$\phi = \sum_{h=1}^{2} \phi_h \qquad (5)$$

Passengers are in equilibrium when the marginal willingness to pay (determined by equation 1) is equal to the generalized price (determined by equation 2). Passengers have no incentive to leave the market, because the willingness to pay of the marginal passenger equals the generalized price. New passengers have no incentive to enter the market, because their willingness to pay will fall below the generalized price. The equilibrium condition for both airlines in the simple network implies the following relationship between fares and demand (q):

$$p_i = \alpha - {}^2 \cdot (q_1 + q_2) - \sum_h \eta_h \cdot vot_p \cdot \frac{q_1 + q_2}{\lambda \cdot K_h}, \quad i = 1,2 \qquad (6)$$

The equilibrium fares paid by the passengers and given on the left-hand side of (6) are used to determine airline revenue. Passenger payments for non-transportation activities at the airports are dealt with later in the chapter. In this model, there are no direct payments from the passengers to the airports or regulator.

The Airline Maximization Problem

Airline i incurs a cost c_i^q per (return) passenger and a cost c_i^f per (return) trip. In addition, airline i may experience congestion which airlines value at a unit shadow cost vot_k. Airline i also pays an airport charge r_h to the airport (valid for one arrival and one departure), and a toll t_h^c to the regulator for the use of airport h (valid for one arrival and one departure). Total operating costs for airline i are then:

$$C_i = f_i \cdot \left(c_i^f + \sum_{h=1}^{2} (t_h^c + r_h) + vot_k \cdot \phi \right) + c_i^q \cdot q_i - F_i \qquad (7)$$

where F_i is airline i's fixed cost under given passenger loads. Equation (7) may be rewritten as:

$$C_i = q_i \cdot \left[\frac{1}{»_i} \cdot \left(c_i^f + \sum_{h=1}^{2} (t_h^c + r_h) + vot_k \cdot \phi \right) + c_i^q \right] - F_i \qquad (7')$$

---

2 We assume that congestion occurs at airports only.

Note that the passenger's value of time would typically only reflect lost time. The airline's value of time (vot_k) may also include additional resource costs: vot_k would include expenditures on fuel, crew time, and so on.

The airlines act as Cournot duopolists; that is they choose an optimal output, taking the other airline's output as fixed. Airlines do not believe that by their actions they can affect the airport operator's charges or the regulator's tolls. Airport operators and airlines are playing a Stackelberg-type game, and so are regulators and airlines. In both cases, the airlines act as followers. As will become clear, regulators and airport infrastructure operators also play a Stackelberg game, with the regulator leading.

Airline i maximizes profits with respect to q_i, taking the competitor's quantities and all tax levels as fixed:

$$\max_{q_i} \pi_i = p_i \cdot q_i - C_i \tag{8}$$

The first-order necessary conditions for $i = (1,2)$ are:

$$\alpha - \beta \cdot (q_1 + q_2) - \sum_h vot_p \cdot \eta_h \cdot \frac{q_1 + q_2}{\lambda_i \cdot K_h} - q_i \cdot \left(\beta + \sum_h \frac{vot_p \cdot \eta_h}{\lambda_i \cdot K_h}\right) - \left[\frac{c_i^f + \sum_{h=1}^{2}(t_h^c + r_h) + \sum_h vot_k \cdot \eta_h \cdot \frac{q_1 + q_2}{\lambda_i \cdot K_h}}{\lambda_i} + c_i^q\right] - q_i \cdot \sum_h \frac{vot_k \cdot \eta_h}{\lambda_i^2 \cdot K_h} = 0 \tag{9}$$

From the first-order condition for profit maximization, it can be seen that airline i only internalizes the congestion incurred by itself and its passengers: only its own clientele q_i is considered (the last LHS-term and the fourth LHS-term respectively). Because the airlines have the same outputs in the symmetric equilibrium, it follows that the airlines internalize half of the congestion they cause (Brueckner, 2002). Furthermore, the airline exploits its market power, as indicated by the demand-related markup $\beta \cdot (q_1 + q_2)$. Imposing symmetry on airports and airlines and solving the first-order conditions yields the following tractable equilibrium outputs:

$$q_1^* = q_2^* = \frac{1}{3} \cdot \frac{\lambda \cdot [\alpha \cdot \lambda - 2 \cdot (t^c + r) - c^f - \lambda \cdot c^q]}{\beta \cdot \lambda^2 \cdot K + 2 \cdot \eta \cdot (\lambda \cdot vot_p + vot_i)} \tag{10}$$

which are positive when

$$\alpha \cdot \lambda_i > 2 \cdot (t^c + r) + c^f + \lambda \cdot c^q \tag{11}$$

The latter condition simply states that outputs are positive when the highest reservation price (α) exceeds the average cost of the service divided by the load (i.e. the average cost per passenger). This simply states that revenue should at least be equal to the costs of the service when output is positive. We now turn to the third group: the airports.

The Airport Maximization Problem

We assume that airports derive revenue from two activities: transportation activities and retail activities. For the former, an airport incurs a cost k_h for each transport movement (an arrival or departure), and charges a price r_h for such a movement. We assume that the potential scale of the retail activities depends linearly on the total number of passengers Q_h passing through the airport; with $Q_h = \sum_i q_i^*(r)$, where $q_i^*(r)$ is determined in (10).

For simplicity, we assume that the airport operates the retail activities, and incurs both the (total) costs and the revenues from offering these services[3]. To determine the retail revenues, we assume that each individual traveler has an identical inverse demand function for the non-transportation good:

$$d_y = \delta - \gamma \cdot y \tag{12}$$

where δ is the reservation price for the commercial activity, γ is the demand sensitivity, and y is the amount of commercial goods purchased by a representative passenger. We note that the demand for trips using an airport is assumed to be independent of the consumer surplus that can be derived from individually optimized shopping at the airport. Revenues derived from an individual traveler for the non-transportation good are:

$$R_y = (\delta - \gamma \cdot y) \cdot y \tag{13}$$

Since all passengers have identical inverse demand functions, and assuming that the only cost of offering the retail services is a charge t^{ret} imposed by the regulator on each product sold, we may determine the commercial profits as:[4]

$$\Pi_h^{ret} = \sum_i q_i^* \cdot (\delta - \gamma \cdot y - t^{ret}) \cdot y \tag{14}$$

Note that these profits are proportional to the number of passengers, which simplifies further calculations. To determine the revenues from landing charges, we rewrite (9) as:[5]

$$r_{a,i} + r_{b,i} = \alpha \cdot \lambda - (2 \cdot q_i + q_{-1}) \cdot \left(\beta + \sum_h \frac{vot_p \cdot \eta_h}{\lambda_i \cdot K_h} + \sum_h \frac{vot_k \cdot \eta_h}{\lambda_i^2 \cdot K_h} \right) - c^f - \sum_{h=1}^{2} t_h^c - \lambda \cdot c^q \tag{15}$$

where $r_{a,i} + r_{b,i}$ denotes the total expenditure on airport charges per flight by transportation company i. Equation (15) is therefore airline i's inverse demand function for slots at both airports (the subscripts a and b denote airports). The willingness to pay for the use of an airport decreases as

---

3 Alternatively, the airport could give a license to third parties to operate commercial activities. This would introduce another party into our model, which we would like to avoid for reasons of transparency. But, also with third parties, it is of course likely that airport profits from commercial activities would increase with the increase in number of passengers.

4 When the marginal costs of the non-transportation activities are positive and constant, they would appear in (14). Setting them equal to zero is therefore a normalization that is allowed as long as these marginal costs are constant.

5 The subscript $-i$ denotes the competitor to airline i. Note that, due to symmetry, $r_{a,i} = r_{a,-i}$.

congestion levels increase, as indicated by the second RHS-term. By rewriting (15), we can write the equilibrium condition for the charge for airport a paid by airline i as:[6]

$$r_{a,i} = \alpha \cdot \lambda - (2 \cdot q_i + q_{-i}) \cdot \left[\beta + \sum_h \frac{vot_p \cdot \eta_h}{\lambda_i \cdot K_h} + \sum_h \frac{vot_k \cdot \eta_h}{\lambda_i^2 \cdot K_h} \right] - c^f - \sum_{h=1}^{2} t_h^c - \lambda \cdot c^q - r_{b,i} \qquad (16)$$

Combining profits from retail and transportation activities, the airport's total profit function is (we assume identical airports, so we only discuss the objective function for 1 airport):

$$\Pi_a = \sum_i q_i \cdot \left(\delta - \gamma \cdot y - t^{ret} \right) \cdot y + \sum_{i=1}^{2} \frac{q_i}{\gg} \cdot r_{a,i} - \sum_{i=1}^{2} \frac{q_i}{\gg} \cdot k_h - w \cdot K_h \qquad (17)$$

where w is the cost of capacity K_h. The first RHS-term is retail profits. The second RHS-term is total transportation revenues. The last two RHS-terms are variable and fixed costs, respectively. The first-order conditions for profit maximization are obtained by substituting equation (16) in (17) and differentiating with respect to q_i, y, and K_a:

$$\frac{\partial \Pi_a}{\partial q_i} = -2 \cdot (2 \cdot q_i + q_{-i}) \cdot \left[\beta + \sum_h \frac{vot_p \cdot \eta_h}{\lambda_i \cdot K_h} + \sum_h \frac{vot_k \cdot \eta_h}{\lambda_i^2 \cdot K_h} \right] +$$
$$\frac{\alpha \cdot \lambda - c^f - \sum_h t_h^c - \lambda \cdot c^q - r_{b,i} - k_h}{\lambda} + (\delta - \gamma \cdot y) \cdot y = 0 \qquad (18)$$

$$\frac{\partial \Pi_a}{\partial y} = \sum_i q_i^* \cdot (\delta - 2 \cdot \gamma \cdot y) = 0 \qquad (19)$$

$$\frac{\partial \Pi_a}{\partial K_a} = \left[\frac{q_i \cdot (2 \cdot q_i + q_{-i})}{\lambda} + \frac{q_{-i} \cdot (q_i + 2 \cdot q_{-i})}{\lambda} \right] \cdot \left(\frac{vot_p \cdot \eta_h}{K_a^2} + \frac{vot_k \cdot \eta_h}{\lambda \cdot K_a^2} \right) - w = 0 \qquad (20)$$

The airport's optimal charge for airline i can be found by determining the difference between the airport's and airline's first-order condition for profit maximization – (18) and (9) respectively) – which is for airport a:

$$\frac{\partial \Pi_a}{\partial q_i} - \frac{\partial \pi_i}{\partial q_i} = 0 \Leftrightarrow$$

$$r_{a,i} = (2 \cdot q_i + q_{-i}) \cdot \left[\beta + \sum_h \frac{vot_p \cdot \eta_h}{\lambda_i \cdot K_h} + \sum_h \frac{vot_k \cdot \eta_h}{\lambda_i^2 \cdot K_h} \right] + k_h - \lambda \cdot (\delta - \gamma \cdot y) = 0 \qquad (21)$$

---

6 The expression for r_b is symmetric. The corresponding expression for airline $-i$ is similar. Due to symmetry, all expressions for r are the same. Although equation (17) may therefore be simplified further, the current notation is useful in the discussion that will follow.

When $r_{a,i}$ (i.e. the charge paid by airline i to airport a) satisfies (21), airline i gains an incentive to set its output at the level which maximizes the airport's profits. Note that the expression for $r_{a,i}$ does not depend directly on $r_{b,i}$, which is treated as a constant by airport a. However, the airlines respond to both airport charges – (equation (9)) – so that airport a is still affected by airport b's pricing decisions, namely via the airline's optimal outputs. From (21) it is clear that r is not set at the level of direct marginal costs k_h. The first RHS-term consists of three parts. The discussion that follows is based on a simplified model presented in Appendix 1. The three parts of the first RHS-term are:

(i) $\beta \cdot [2 \cdot q_i + q_{-i}]$ reflects airport market power, which increases when demand is less elastic (i.e. when β is relatively high). Airline i overcharges the passengers by $q_i \cdot \beta$ because it has market power (Appendix 2). Without congestion and airline $-i$ (i.e. when airline i is a monopolist), the airport would charge airline i the marginal costs per flight *plus* $2 \cdot q_i \cdot \beta$. The markup (per passenger) over the marginal costs that the airport charges the airline therefore is twice the markup that the airline charges its passengers. This is an instance of so-called 'double marginalization': the monopoly supplier of an intermediate good faces an inverse demand function which is derived from the profit-maximizing condition for the supplier of the final product. In our case, the monopoly airport delivers capacity to airlines which compete in Cournot fashion. An airline sets its profit-maximizing output at the point where marginal revenue is equal to the marginal cost. *Given* the marginal cost, the airline's marginal revenue curve may be interpreted as the demand function for airport capacity. The monopoly airport's marginal revenue is twice as steep as the demand curve it faces. As a result, the airline's markup over marginal costs is doubled by the airport.

In the present case, airline i has a competitor (airline $-i$). Any decision taken by airline i affects the airport directly due to a change in revenue from airline i. But since airline $-i$ will also react to airline i's decisions, the airport is also indirectly affected, via the revenue it generates from airline $-i$. The airport accounts for this in its charge for airline i; this is the second term in square brackets. If airline i raises its outputs, the output and/or revenue for $-i$ decreases. As a result, the airport's revenue from airline $-i$ also decreases. To compensate, the airport increases its charge for airline i.

(ii) $(2 \cdot q_i + q_{-i}) \cdot \sum_h (vot_p \cdot \eta_h)/(\lambda_i \cdot K_h)$ reflects the airline's firm-internal and external 'indirect' congestion costs (i.e. cost experienced via passenger's time losses). Airline i internalizes the congestion costs incurred by its own passengers; this results in a markup of $q_i \cdot \sum_h (vot_p \cdot \eta_h)/(\lambda_i \cdot K_h)$ over the constant marginal production cost $c^f/\lambda + c^q$ (Appendix 1). Again, because of double marginalization, the airport charge for airline i includes *twice* the airline's markup, multiplied by λ (because it is a charge per transport movement); the argument is analogous to (i) described above: the congestion term appears in the inverse demand function for airport capacity, and is thus part of the airport's profit-maximization problem. The airport abuses its market power by doubling the airline's congestion charge to maximize its own profits. Appendix 1 provides the explanation in a simplified example. Since airline i does not internalize the congestion costs it causes for airline $-i$'s passengers, the airport internalizes this part of the congestion experienced by the passengers. A profit-maximizing airport thus not only internalizes the congestion not internalized by the airlines (which increases welfare due to decreased congestion costs), but also charges the airlines for the congestion already internalized by them. The profit-maximizing airlines reduce congestion by charging the passengers for congestion. The ticket price is relatively high, so that demand is relatively low and the impact of congestion is relatively low. This *would* improve welfare, but the profit-maximizing airport doubles the airline's congestion charge to the passengers, so that passengers are overcharged for congestion (the airlines pass the airport congestion charge on to the passengers), so that welfare is decreased compared to optimal pricing.

(iii) $(2 \cdot q_i + q_{-i}) \cdot \sum_h (vot_k \cdot \eta_h)/(\lambda^2 \cdot K_h)$ reflects the airline's firm internal and external direct congestion costs. Because direct congestion costs (i.e. the congestion costs experienced by the airlines) do not appear in the passenger's generalized cost function, there is no double marginalization in this case. The airlines charge the passengers for the direct congestion costs (Appendix 2), and this charge is directly transferred to the airports via the airport charge. The airport thus receives the revenue from this specific charge, but this final allocation does not change the welfare-improving effect of the charge.

In effect, the airport charges airline i for all the markups over the constant marginal cost, $c^f/\lambda + c^q$, and the congestion and market-power externalities for airline $-i$.

The second RHS-term of (21) is the airport's marginal cost per flight, and the last RHS-term is the marginal revenue of commercial activities. Note that when these marginal revenues are large enough, the airport charge may be negative. For airlines it is therefore beneficial when (unregulated) airports have relatively strong commercial activities (with relative large marginal revenues), since this helps to bring down the airport charge. An additional benefit, not modeled explicitly, would be when strong commercial activities increase the demand for trips. Note, however, that revenue from commercial activities may be relatively high due to the exploitation of market power for these activities. A negative infrastructure charge (i.e. a subsidy) may in this case appear to be an instrument to stimulate demand, but it is in fact the result of a market failure.

Conclusion

This chapter investigated airport pricing when various externalities were present. First of all, congestion is an important externality. Even though many European airports are slot-constrained, passengers may still experience congestion (or need to include additional travel time) because, for instance, the terminal cannot handle the passenger flow. Given the current trend towards privatization, market power needs to be considered as a second market failure in the process of policy-making. In this chapter we analysed the problem of dealing with congestion on a privatized monopolistic airport. Regulation is not considered in this chapter, since we first need to develop insight into the behaviour of privatized airports and airlines when congestion occurs. A crucial aspect of this chapter is the practice of double marginalization. Airlines maximize profits, and set prices above marginal costs. Airline demand for airport capacity is derived from their own first-order conditions for profit maximization. Specifically, the airport's marginal revenue curve is twice as steep as the airline's marginal revenue curve, so that the airline's monopolistic markup is doubled by the airport. This means that with private airlines and private airports, overcharging of congestion is to be expected. Without congestion, the airport operator doubles the airline's markup over marginal cost. Passengers are overcharged because both the airport *and* airline set prices above marginal costs.

Profitable retail activities at an airport may help to bring down the airport charge paid by the airlines to the airport (Starkie, 2001). Although the airport charge may in fact be negative when retail activities are important (and revenues are high), airport charges may still reflect airport market power because i) the airport may also have monopolistic power in the commercial activities (i.e. car parking), and ii) the airport doubles the airline's markup over marginal costs.

The consequences are that a regulator who faces both private airlines and airports, and aims to maximize welfare using tolls, may find that the optimal toll is negative when the market power effect exceeds the congestion effect. Note that this chapter does not discuss economies of scale and density in airline networks, which are important in practice. As a result, the airlines cannot set their ticket prices at marginal costs unless they are willing to incur a loss. A profit-maximizing, unregulated airport will then set its airport charge above marginal cost.

An important question is what happens when airport competition is introduced, i.e. in a multi-airport region, when we assume that airports compete for passengers[7]. A first effect may be that the airport charges (excluding the retail component) are lowered because of competition. However, the airports still face the derived demand curve, and engage in Cournot competition, so that the airport charge will still most likely be set above marginal cost, albeit lower than in the monopoly case. The main contribution of this chapter lies in discussion of the derived demand function and the concept of double marginalization for airport pricing. The same principles will hold for extensive networks in which airports 'compete'. Because the airports face derived demand curves, airport pricing and airport competition are directly related to airline pricing and airline competition. The demand curve faced by the airports is derived from the airline's first-order conditions for profit maximization. Airports that act as Cournot competitors will set prices above marginal costs.

References

Brueckner, J.K. (2002), 'Airport congestion pricing when carriers have market power', *American Economic Review* 92, 1357–75.

Kreps, D.M. and J.A. Scheinkman (2003), 'Quantity precommitment and Bertrand competition yield Cournot outcomes', *Bell Journal of Economics*, 326–37.

Pels, E. and E.T. Verhoef (2003), 'The economics of airport congestion pricing', *Journal of Urban Economics* 55, 257–77.

Starkie, D. (2001), 'Reforming UK airport regulation', *Journal of Transport Economics and Policy* 35, 119–35.

Appendix 1

Consider the following situation. Consumers buy a homogeneous product from a monopoly supplier: firm A. In order to produce this product, the monopolist needs the services of another monopoly supplier: firm B. For each product sold, A pays an amount θ to B for services rendered. Furthermore, A pays a regulator the amount τ_A for each product sold, while B pays τ_B for each product sold to A.

The monopoly firm selling its product to the final consumer faces an inverse demand curve $\alpha - \beta z - \gamma z$, where γz is the congestion component.

The maximization problem for the monopolist A, selling final products to the consumer, is:

---

7   In many European countries, the passenger charge for transfer passengers is lower than the airport charge for origin-destination (o-d) passengers, because transfer passengers usually can choose between many different hubs. Also, airports invest a lot of money in 'quality' to attract passengers (o-d or transfer). When airports compete for airline serves (groups of flights), competition becomes a lumpy process. In this case, we would need to extend the model with the airline's choice of optimal network and which airports will be served.

$$\max \pi_A = z(\alpha - \beta z - \gamma z) - \tau_A z - \theta z \qquad (B1)$$

The first-order condition for profit maximization is:

$$\alpha - 2\beta z - 2\gamma z - \tau_A - \theta = 0 \qquad (B2)$$

from which we can derive the optimal price: $p = \theta + \tau_A + \beta z + \gamma z$. Note that the price exceeds firm A's marginal cost, $\theta + \tau_A$, by two terms: a congestion component that perfectly internalizes the congestion externality (γz) and a demand-related monopolistic markup (βz).

From equation (B2) we can derive the inverse demand curve for B:

$$\theta = \alpha - 2\beta z - 2\gamma z - \tau_A \qquad (B3)$$

which is twice as steep as the inverse demand curve faced by the producer of the final product. Profits for the producer of the intermediate good are:

$$\pi_B = z(\alpha - 2\beta z - 2\gamma z - \tau_A) - \tau_B z \qquad (B4)$$

(we assume that B's marginal costs are zero). The first-order condition for profit maximization is:

$$\alpha - 4\beta z - 4\gamma z - \tau_A - \tau_B = 0 \qquad (B5)$$

Subtracting (B2) from (B5) and solving for θ:

$$\theta = \tau_B + 2\beta z + 2\gamma z \qquad (B6)$$

which implies that the consumer price will be: $p = \tau_A + \tau_B + 3\beta z + 3\gamma z$. The markup over the marginal cost of B applied to A is twice as large as A's markup (which is the result of double marginalization).

The regulator aims to maximize welfare:

$$\varpi = \int_0^z \alpha - 2\beta x \, dx - \gamma z^2 \qquad (B7)$$

The first-order condition for welfare maximization is:

$$\frac{\partial \varpi}{\partial z} = \alpha - \beta z - 2\gamma z = 0 \qquad (B8)$$

which implies an optimal consumer charge (effectively a congestion toll) of $p = \gamma z$. Subtracting (B5) from (B8) and solving for τ_B yields:

$$\tau_B = -3\beta z - 2\gamma z - \tau_A \qquad (B9)$$

A has two markups over the marginal costs. The market power effect results in a markup βz, which is doubled by B. The final consumer thus pays a total markup of $3\beta z$. A internalizes congestion, so that, without B, the regulator would not need to set a congestion toll. However, B charges A for twice the congestion charge that the final passenger pays (double marginalization). The regulator compensates this by subsidizing B with the amount of $-2\gamma z$ (the amount that B overcharges). A still has a markup over the marginal cost for congestion, but this is exactly the congestion charge that the regulator would impose if A would not internalize congestion.

Substituting (B9) in (B6) yields:

$$\theta = -\beta z - \tau_A \tag{B10}$$

Since A pays τ_A to the regulator, and receives τ_A from B, τ_A is superfluous. Therefore, the regulator sets $\tau_A = 0$. Solving (B2), (B6), and (B9) yields the final solution:

$$\begin{aligned} z &= \frac{\alpha}{\beta + 2\gamma} \\ \theta &= \frac{-\alpha\beta}{\beta + 2\gamma} \\ \tau_B &= -\frac{\alpha(3\beta + 2\gamma)}{\beta + 2\gamma} \end{aligned} \tag{B11}$$

One may wonder why congestion appears in the optimal prices in a way that is symmetric to the slope of the demand function. The reason is that the inverse demand function faced by the downstream producer, when the original inverse demand function is $D(Q)$ and the average congestion cost is $c(Q)$, can be written as: $P(Q)=D(Q)-c(Q)$. Marginal revenues can be written as: $MR=P(Q)+Q\cdot(dP/dQ)=P(Q)+Q\cdot[dD/dQ-dc/dQ]$. In other words, the slopes of the inverse demand function and the average congestion cost function enter the downstream monopolist's profit-maximizing condition symmetrically and additively. This is consistent with the standard result, that a (simple) monopolist would perfectly internalize a congestion externality imposed by its consumers upon one another, in addition to charging a demand-related markup, which vanishes when demand becomes perfectly elastic. This can be verified by rewriting $P(Q)+Q\cdot[dD/dQ-dc/dQ]$ as $P^*(Q)=Q\cdot[dD/dQ-dc/dQ]$.

Appendix 2

Disregarding congestion tolls and airport charges, the following airline fare is derived from the first-order condition for profit maximization (9) and the generalized cost function (6):

$$p_i = q_i \cdot \left(\beta + \sum_h \frac{vot_p \cdot \eta_h}{\lambda_i \cdot K_h} + \sum_h \frac{vot_k \cdot \eta_h}{\lambda_i^2 \cdot K_h}\right) + \frac{1}{\lambda_i} \cdot \left(c_i^f + \sum_h vot_k \cdot \eta_h \cdot \frac{q_1 + q_2}{\lambda_i \cdot K_h}\right) + c_i^q \tag{A1}$$

The first RHS-term is the airline's marginal cost (including congestion). The second RHS-term is the markup over the marginal costs, consisting of the market power effect and firm-specific direct and indirect congestion costs. (A1) can be rewritten as:

$$p_i = \frac{c_i^f}{\lambda_i} + c_i^q + q_i \cdot \left(\beta + \sum_h \frac{vot_p \cdot \eta_h}{\lambda_i \cdot K_h}\right) + (2 \cdot q_1 + q_2) \cdot \left(\sum_h \frac{vot_k \cdot \eta_h}{\lambda_i^2 \cdot K_h}\right) \tag{A2}$$

where the third and fourth RHS-terms are the markup over the constant marginal production cost $c^f/\lambda + c^q$.

Chapter 5
Countervailing Power to Airport Monopolies

Kenneth Button[1]

Introduction

Delivering air transport services involves a complex set of individual components: airlines, airports, airport access, air navigation services, etc. Individual companies or agencies—some of which are motivated and disciplined by market forces and others through institutions and regulations—provide these services, and there is no template that applies to all countries. The ways these suppliers interact in the supply chain affects the nature of the air services supplied and the efficiency with which they are supplied. The focus here is on the interaction between airlines and airports, and in particular on the ability of airlines to limit the potential market power that airports may enjoy.

Airports are physically immobile assets that overall have high fixed costs and relatively low operating costs. This tends to allow them to enjoy some of the advantages of monopoly power even in circumstances where this is not engendered through institutional fiat or state ownership.[2] There have been extensive debates about the degree of monopoly power they enjoy, and how they may exercise it, in terms of having to confront competition from other airports and, increasingly regarding short-haul traffic, from other modes of transport such as high-speed railways. Technically, the public policy concern is that when there is a lack of adequate "original competition"[3] for airport services this leads to allocative inefficiency, whereby an airport with monopoly power can exploit customers by levying high prices, and also possibly results in them suffering from static and dynamic X-inefficiency with consequential higher unit operating costs and "gold-plating" of investments and services.[4]

The degree to which an airport does have this sort of power, and also importantly is allowed to exploit it, is traditionally looked at in terms of the physical geographical proximity of alternative airports that cater for similar types of airline services and have potentially overlapping catchment areas. But there is also some limited point-to-point competition between airports involving, for example, Amtrak rail services in the northeast corridor of the US, and from high-speed rail in some parts of Europe and in Japan and Taiwan, where the dimension of competition with these alternatives often has more to do with airport pairings than with access to the initial airport.

---

1 I would like to thank Henry Vega for his research assistance in preparing this chapter.
2 In some cases airport authorities own more than one facility. BAA, for example, owned and ran in 2009 airports that handled about 63 percent of passenger traffic in the UK, with this figure rising to 86 percent in Scotland and more than 90 percent in London. Control over a system of airports may have various economies of size associated with it, but it also reduces the power of original competition within the airport market.
3 The term "original competition" is the one used in discussions of countervailing power and signifies situations whereby neither party to a transaction enjoys market power. It is the type of competition discussed in the classical economics of Adam Smith, David Ricardo, Hume, and others, and embedded in the neo-classic economics of Alfred Marshall.
4 The economic implications and causes of X-inefficiency are discussed by Harvey Leibenstein (1979).

The extent to which monopoly power may in fact be desirable is less often considered in efficiency debates, but there are situations where it is important for the effective workings of markets. If there is excessive competition then it may be impossible for airports to recover their large fixed costs: the so-called problem of an "empty core" (Telser, 1978). Basically, the competitive pressure from other airports forces the prices that can be charged to airlines down to short-run marginal cost with no contribution to the recovery of investment outlays. The outcome tends to be unstable markets as airports struggle to be financially viable. Monopoly power provides the airports with the ability to mark up their fees so that full cost recovery is possible.

But even if an airport has extensive monopoly power this does not automatically mean that there will be inefficiency, although there may be a public perception of exploitation of users. If the airport can engage in significant discriminate pricing to extract different levels of consumer surplus from each customer according to willingness-to-pay, then efficiency is maintained. There is, of course, a transfer of welfare from the customers—essentially the airlines, and *ipso facto* their customers—to the airports. The issue then becomes much more one of normative considerations of equity than one of efficient supply and use; after all the government can always tax the airports' profits for redistribution to the airlines.[5] Technically, there is in traditional welfare economic terms a Pareto optimal situation, but the airport initially takes the social gains.

But what exactly constitutes traditional monopoly power for an airport? Geographical distance is an obvious parameter to explore; or, to be more precise, the accessibility of alternative airports since surface transportation is neither ubiquitous nor of a uniform standard. Work in the US on air–automobile mode choice indicates that about two hours' driving to an airport is the boundary before air transportation gives way to making the entire trip by automobile. For longer trips it is also the trade-off range between access time and airfares within which travelers will select the airport to use. The cut-off distances are shorter in Europe. But this type of golden-line approach is not always helpful when airports differ in the types of facility that they offer or, in the case of many international airports, the routes that they can serve.[6]

Government regulation of airports, including ownership—often by local authorities—has been the norm, with the idea of containing monopoly power as one of their main motivations. This approach, however, takes it as axiomatic that there are no other "non-original competitive" forces at work to circumscribe the actions of airports; original in this sense being the types of market pressures discussed in classical economics. In some instances, however, it is clear that the standard neo-classical economic model of a monopoly supplier confronting an atomistic market of small buyers is not what pertains in the airport context. There are, indeed, very large and powerful airlines that make use of many airports, and even smaller ones, and in some circumstances these may be able to exercise a degree of countervailing power to curtail excesses in pricing of such things as gates or slots.

Countervailing Power

The development of the idea of countervailing power is usually credited to Kenneth Galbraith (1952) and his book on *American Capitalism*. This very un-Galbraithian volume essentially argues

---

5   This sort of redistribution is practiced in New Zealand with respect to profits earned by its air navigation system, although its fees are only a partial reflection of willingness-to-pay.

6   It also seems unlikely that contestability forces, concern over potential new competitive capacity being built, affect the airport market given the extensive time taken to add capacity to the system.

that large-scale enterprises in oligopolistic or monopoly markets often have much less flexibility than conventional theory postulates to exploit their power because of the existence of monopoly power at other points in what we would now call the "value chain."[7] In particular, the interest was initially stimulated by the observation that large buyers often enjoyed significant discounts when purchasing from suppliers. While there may be cost reasons for this, such as economies of density in shipping and handling, the outcome may also be the result of the strong bargaining power of the purchaser.[8]

Essentially, what is seen is a situation where, rather than having original competition in a market, the competition is between the various buyers and sellers up and down the value chain—there are, according to this approach, bilateral monopolies or oligopolies at various points in the chain; a bilateral monopoly being a situation where there is a single buyer and seller of a given product in a market. The level of concentration in the sale of the product results in a mutual interdependence between the seller and buyer in price determination. The notions of price-taking inherent in more standard neo-classical economics are not upheld.

Countervailing power, Galbraith (1952: 12) argues, prevents a large business from fully exploiting customers. "In a typical modern market of a few sellers, the active restraint is provided by competition but from the other side of the market by strong buyers ... At the end of virtually every channel by which consumers goods reach the public there is, in practice a layer of powerful buyers." Basically, market power on one side of a market "create(s) an incentive to the organization of another position of power that neutralizes it."

Not all have agreed with this outcome, and the idea of countervailing power from the time of Galbraith initiating his ideas has been the subject of some skepticism. Academics such as Hunter (1958), George Stigler (1954), and Whitney (1953), for example, questioned in a variety of ways whether the ability of large purchases of factors of production from monopolists in the value chain really had the incentive to pass on any savings to final customers; why not keep the rent themselves? Stigler (1954: 9) was particularly strong in his criticism of the concept:

> it simply is romantic to believe that a competitive solution will emerge, not merely in a few particular cases, but in the general run of industries where two small groups of firms deal with one another suddenly all the long-run advantages of monopolistic behavior have been lost sight of in a welter of irrational competitive moves.

The theory is messy in the sense that it does not, except in very stylized situations, produce a neat equilibrium, optimal outcome, but rather results in prices and outputs determined by games played by the various parties involved and influenced by their power in the market. These may be short-term equilibriums but are not optimal. There is no canonical model. In particular, the degree to which the outcome approaches either monopoly or competitive outcomes depends on the strength and strategic aptitudes of the various parties involved. A variety of things can influence both of these factors.

The generic types of issues that can ensure in these conditions were first explored by Francis Edgeworth in the nineteenth century and have subsequently attracted considerable attention within the industrial organization fraternity with an interest in game theory.[9] Figure 5.1 provides a simple,

---

7   The concept of the value chain was brought into the management literature by Michael Porter (1985).

8   There is also the possibility that it results from second-degree price discrimination on the part of the seller if market segmentation according to purchase volumes is possible.

9   Edgeworth (1891) is the classic reference on the subject, but an important early contribution also came from Arthur Bowley (1928).

and extreme, example with a monopsony buyer (say, a large airline) negotiating landing slot charges with a monopoly airport.

Initially at one extreme, we treat the airport as a monopolist with numerous potential airlines wanting to use its slots. In this case we get the standard textbook situation with the demand for airport will set its output where marginal cost equals marginal revenue (Y_1) and set price to clear the market (that is, price will be P_1). In contrast, if an airline (or a strong alliance of airlines) has monopsony power, and the airport is confronted with competition from alternative facilities, then the price of a slot will be P_2. This is because the market supply curve for slots will be positively sloped with a market slot curve above it. The marginal revenue product curve of slots—the additional revenue the airline obtains by buying an additional slot - is downward sloping, reflecting diminishing marginal productivity. To maximize its revenue, the airline will buy slots up to the point where the marginal cost of a slot is equated to the marginal revenue it will generate (that is, Y_2 slots), for which the market clearing price is P_2.

The situation where there is bilateral monopoly generally leaves the outcome indeterminate, with a market price somewhere between P_1 and P_2.[10] The resultant price and output combination will depend on the bargaining skills of the two parties and that, of course, will be a function of their game-playing skills.[11] While one can relate to this in the case of a large hub-and-spoke airline negotiating slot rates at its main hub airport, there are other points in the value chain where these types of bilateral monopoly or oligopoly may emerge. Of particular relevance here is the monopoly power of labor unions in their dealings with airports, and the airframe manufacturers with the airlines.

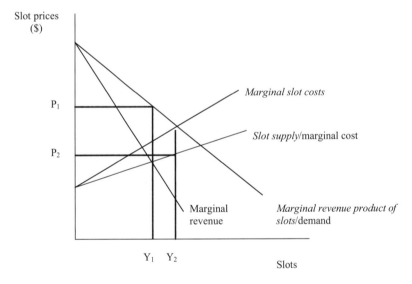

Figure 5.1   Price and output under a pure bilateral monopoly

---

10   In some situations where the bargaining position of one party is very weak, such a party may accept a situation off its contract curve to avoid the transactions costs of further bargaining.

11   There is a significant body of academic literature that provides stable solutions (Nash equilibrium) to this type of problem that can come about through negotiation. (It has, for example, been applied to the Cold War situation where two nuclear powers coexisted for over 40 years—Schelling, 2006.) The general point is that, whatever the outcome, the final price will be higher than that which would emerge with perfect competition on both sides of the market.

What is clear, irrespective of the power of airports and airlines in the example, is that the final consumer—passenger or shipper—does not benefit greatly from a bilateral monopoly situation further up the value chain. The sharing out of economic rents seen in Figure 5.1 between the two providers of intermediate services is an effective extraction of consumer surplus from the final users of air transportation services. The situation may also pertain in other cases where there are imperfections up the value chain that result in non-competitive prices being pushed down the chain and being imposed on passengers and freight consignors.

Testing for the existence of countervailing power in any market is challenging. It is difficult to discern, for example, whether the fact that a large-scale buyer of a product from a large producer gets a discount because of the scale economies of such transactions or because the buyer can drive a hard bargain by virtue of its purchasing power. Much of the general empirical work trying to ascertain if countervailing power has any significant impacts on markets uses a structure-conduct-performance framework, looking in particular at links between supplier profits, or markups, and on supplier concentrations, with the additional consideration of buyer concentrations. Limited work of this type, however, has been performed on the aviation sector. There has been some work trying to benchmark different airports' efficiency, generally using data envelopment methods (Oum et al., 2003), but normalizing for the particular characteristics of airports is difficult, especially when using programming techniques.

One of the problems in exploring countervailing effects is that the market power on either side of the market varies considerably by context. The situation depicted in Figure 5.1 of a monopoly confronting a monopsony is very seldom seen, and more common are oligopolistic bilateral structures where the market powers on both sides are defused by continual games being played between the buyers and suppliers themselves. Turning specifically to aviation, in some cases groups of clustered airports are under common ownership, as with the main London airports until 2009, giving them considerable regional monopoly power, whereas in other cases there may be alternative terminals in close proximity: in the US a two-hour drive between airports is generally seen as providing competition. Additionally, many airports have their rates regulated—for example the BAA is price-capped by the Civil Aviation Authority—making econometric analysis of market power complicated. Air services are also not homogeneous, and neither are the airlines providing them, making generalizations difficult.

Trends in Airport Economics[12]

One can look at the supply chain underlying the air transport market in several different ways, but here the concentration is on airports—in terms of their relationships with both suppliers and users. Airports clearly constitute part of the geographically fixed infrastructure of the air transportation industry and are also, in at least part of their operations, characterized by many of the economic characteristics of fixed assets. Economies of scale and scope are the most obvious but inevitably linked to these are issues of indivisibility, congestion, and the interface with other forms of infrastructure (for example, surface access and air traffic control in this case). They also supply what are almost purely private goods in the sense that potential users can be excluded, and that the use of the facility by one actor can affect the use of another—congestion can, and often does, arise. Despite this, they are often, and erroneously, seen as being the providers of public goods and thus in need of some form of regulatory protection or public ownership.

---

12 Rigas Doganis (2005) offers an overview of how airports function and of their basic economics.

Airports are also, however, multiproduct providers. They supply a series of services to air transport companies and to passengers, including:

- aeronautical services (rescue, security, fire-fighting, infrastructure supply, runway, and taxiway maintenance);
- aeronautical-related commercial services (catering, supply of fuel and lubricants, baggage handling, passenger and aircraft assistance);
- commercial services (banks, hotels, restaurants, car rental, car parking, retail shops, duty-free shops).

While many airports have control over activities located on their land, these are often provided by separate entities, and even publicly owned airports outsource many of them or develop concessionary arrangements. There is often strong competition between some of these elements provided on-site and those provided outside the airport perimeter; car parking, car rental, hotels, and restaurants are perhaps the most obvious examples.

While the broad economic features of airports are generic, they vary in their nature and the ways in which they are institutionally treated. What has impacted on them in recent years is the way in which they are viewed The deregulation of many airline markets—coupled with the commercial approach, increasingly involving privatization and concession, now being foisted on numerous airports—has led to an increase in competition and, in a few cases, contestability between the airports (Barrett, 2004). The environmental intrusion associated with airports, most notably noise, has led to difficulty in increasing capacity in many markets, especially in relation to some of the major hub airports. On the other hand the development of new types of air transport services, most notably those of the low-cost carriers, and their need for relatively uncongested airports to facilitate fast turnaround times, have made formerly little-used facilities commercially viable and led to many smaller airports attracting increasing amounts of traffic. This is particularly so in Europe, where route lengths are relatively short.

The larger airports in most economically developed countries earn a relatively high, if sometimes volatile, rate of return (Table 5.1), especially compared to the airline sector that, as a whole—and particularly in deregulated markets—hardly covers its operating costs (Button, 2003). In many cases airports also earn considerable returns from their concessions (the "second till") and this can bring their activities function into conflict with the airlines. To put it simply, airlines like to pass passengers through airports as rapidly as possible in a seamless fashion to maximize their traffic flow, whereas airports may have a motivation to encourage a slower throughput to stimulate expenditure at their concessionary facilities and thus the rent obtained from them. This, along with other factors, means that negotiations between airlines and airports are over a complex portfolio of attributes.

Airlines versus Airports

Neither airports nor airlines are homogeneous. Differences in markets, institutional structure, geography, and business models result in wide variations in each. The basis of any discussion of countervailing power cannot, therefore, be in anything but very broad terms, interspersed with a few case studies for illustration.

The ability of an airline to switch to an alternative airport is a crucial underpinning required for it to be able to exercise countervailing power. The ability to do this depends not simply on the

Table 5.1   Operating margins at European airports

Airport group	Operating margin (2001)	Operating margin (2002)
BAA plc (UK)	29.8%	30.6%
Fraport (Germany)	18.0%	15.8%
Aéroport de Paris (France)	6.0%	9.2%
Schiphol Group (Netherlands)	31.7%	32.0%
Luftartsverket (Sweden)	3.7%	9.1%
Flughafen München GmbH (Germany)	11.8%	3.7%
Avinor (Norway)	22.9%	17.1%
Aeroporti di Roma Spa (Italy)	16.8%	21.2%
SEA Aeroporti di Milano (Italy)	11.5%	10.4%
Manchester Airport Group (UK)	19.2%	19.3%

availability of alternative airports, but also on the business model of the airline concerned. There can be very significant costs of moving services even if there is adequate technical airport capacity available at an alternative location.

Superficially, one factor favoring countervailing power is that airlines' capital is relatively mobile, and hence has the potential to be deployed elsewhere. While airlines were described by Alfred Kahn, the former head of the US Civil Aeronautics Board, in the debates over US deregulation as "marginal costs with wings," the intra-marginal costs of moving entire airline operations can be costly. For example, overseas-based international airlines have the power to deploy their limited fleets to destinations in other countries, and services are withdrawn, or a carrier resorts to code-sharing, when services prove to be unprofitable because of airport costs. But these are normally individual services, or at most a small number. Many airlines do, however, invest in costs that become sunk at particular airports (for example, maintenance facilities, passenger-handling facilities, or terminals in the US case), thereby reducing their ability and the credibility of any threat to move their business elsewhere, at least in the short term.

Traditional Carriers

The service networks of the traditional, or full-service, airlines have largely evolved into hub-and-spoke systems, with carriers pushing on-line services through a limited number of large hub airports (Button and Stough, 2000). The primary gain comes from the economies of scale, and particularly those associated with scope and density, that allow the airlines to both employ larger equipment and benefit from high load factors. These were often well over 80 percent in the continental US in 2007. These economies have often been supplemented with a variety of outsourcing arrangements involving smaller, lower-cost carriers providing part of the feed to the main hubs, and through strategic alliances that have both reduced costs further and enhanced economies of market presence by allowing a larger network of services to be accessed seamlessly by potential passengers.

A significant feature of the larger traditional markets has been the emergence of "fortress hubs." There is no strict definition involved, but a fortress hub is normally seen as an airport dominated by a single airline that controls 70 percent or more of flights, making it difficult for new entrant carriers to penetrate the market. In the US many airports exceed this level of domination by a large margin; for example, Northwest Airlines carries over 75 percent of enplaned passengers out of both Detroit and Minneapolis airports, and Delta has over 75 percent from Atlanta. In the case of Chicago O'Hare there is a virtual duopoly situation, with both United and American Airlines using it as a major hub.

In many cases, and especially in Europe where few countries are large enough to have significant domestic networks, the hub-and-spoke networks have been developed around a major international airport, with the spokes being international services operated under restrictive bilateral air services agreements. While liberalization of many international air transport markets has tempered the influence of such institutional factors, the hard- and software associated with them has meant change has taken time. Table 5.2 offers an indication of the degree of hub dominance at major European airports.

These fortress hubs epitomize the classic bilateral monopoly situation, with the airports and airlines involved each having an input into pricing decisions and other decisions. The latter, for example, may entail the allocation of landing and take-off slots through local scheduling committees.[13] Neither in effect is the classic price taker of perfect competition or the price setter of classic monopoly theory.

Low-cost Carriers

Low-cost—or no-frills to use the European terminology—airlines have gradually been taking increasing shares of air transport markets since Southwest Airlines in the US developed its successful business model some 30 years ago, and since economic regulatory reforms have allowed freedom of market entry and fare setting. While there is no genuinely generic model of a low-cost carrier, there are certain characteristics displayed by many that reflect their relationships with airports. They normally operate short-haul routes, using standard equipment, provide limited on-board services unless a premium is paid, and often use electronic ticketing and avoid on-line services. They are certainly not all successful, and many enter and leave the market every year. Carriers such as Southwest, Ryanair, and easyJet have, however, proved themselves robust and influential in the way airlines and airports interact.[14]

While there are some definitional issues, the market share of low-cost carriers in Europe reached nearly 16.5 percent by mid-2006; some 2.4 percent higher than at the same point 12 months prior. The UK has the highest penetration of low-cost carriers that took over 28 percent of the market in the first half of 2006. Some 48 low-cost airlines operate in Europe out of 22 countries, but only 15 of these operate more than 50 flights a day—Ryanair and easyJet being the largest. Low-cost carriers are not always small undertakings, however, and many have a significant capitalization (Table 5.3) that provides them with a significant degree of market power.[15] This situation contrasts

---

13   In some cases, where there are controls of slot prices, either because of local regulations or because services have to conform to International Civil Aviation Organization (ICAO) rules, the bilateral debates are effectively over capacity rather than price.

14   There is also evidence that airports are not shying away from trying to attract low-cost carriers. In 2002, for example, 140 destinations in the US sought to have Southwest services.

15   One also observes in Table 5.3 the capitalization of some of the large cargo carriers—notably UPS and FedEx. These carriers operate hub-and-spoke style networks—for example Memphis is FedEx's main US hub—and, although freight is not discussed here, these hubs are effectively parts of bilateral oligopoly markets.

Table 5.2   Market share of passengers by airline at Europe's 10 largest airports (2002)

Airport	Carriers 1	Carrier 2	Carrier 3
London Heathrow	British Airways 41.6%	bmi 12.1%	Lufthansa 4.8%
Frankfurt	Lufthansa 59.4%	British Airways 3.6%	Austrian 2.9%
Paris Charles de Gaulle	Air France 56.6%	British Airways 5.15%	Lufthansa 4.9%
Amsterdam	KLM 52.2%	Transavia 5.5%	easyJet 4.3%
Madrid	Iberia 57.0%	Spanair 12.7%	Air Europa 7.1%
London Gatwick	British Airways 55.1%	easyJet 12.8%	flybe British European 5.6%
Rome	Alitalia 46.2%	Air One 10.0%	Meridiana 3.9%
Munich	Lufthansa 56.8%	Deutsche BA 6.6%	Air Dolomiti 6.5%
Paris Orly	Air France 64.2%	Iberia 8.2%	Air Littoral 3.6%
Barcelona	Iberia 48.5%	Spanair 9.4%	Air Europa 5.5%

Table 5.3   Market capitalizations of leading airlines ($ billions) in November 2005

Airline	Market cap ($ billion)	Airline type	Bankruptcy history
UPS	82	Integrated cargo	
FedEx	28	Integrated cargo	
Southwest	13	Low-cost	
Singapore	9		
Ryanair	7	Low-cost	
British	5.5		
Lufthansa	5.0		
Air France	4.3		
Gol	3.9	Low-cost	
American	2.3		
easyJet	2.1	Low-cost	
jetBlue	1.9	Low-cost	
Virgin Blue	1.3	Low-cost	
Air Tran	1.3	Low-cost	
Japan Airlines	1.0		
Alaska	0.9		
Continental	0.9		Yes, pre-2000
Westjet	0.4	Low-cost	
Delta	0		Yes
Northwest	0		Yes
Air Canada	0		Yes
United	0		Yes

Source: de Neuville (2006)

with many of the legacy carriers that have found their capital base eroded over successive business cycles and, in many cases in the US, reformations under bankruptcy laws.

Many low-cost carriers eschew the larger airports to avoid both high landing fees and the disruption that congestion can cause to their finely tuned schedules and turnaround times (Warnock-Smith and Potter, 2005). Their network patterns, although often exhibiting some of the simplest features of the hub-and-spoke configurations favored by the legacy carriers, are often linear or radial in nature; the latter involving radiating services from a focus airport but with no, or only limited, on-line services through the hub.[16] Many of the services are to second- or third-tier airports (Table 5.4) and, in the case of Europe carriers such as Ryanair, some are operate from former military airfields.[17]

Where there are alternatives, low-cost carriers thus commonly avoid congested airports and fly instead to secondary airports in metropolitan regions: notably Boston/Providence and Boston/Manchester (New Hampshire), Dallas/Love, Los Angeles/Long Beach, and Los Angeles/Ontario, Miami/Fort Lauderdale, and San Francisco/Oakland and San Francisco/San Jose in the US; and Brussels/Charleroi, Frankfurt/Hahn, London/Luton, and London/Stansted; Oslo/Torp, Rome/Ciampino, and Stockholm/Skavsta in Europe.

In effect, all else being equal, low-cost carriers prefer to avoid congested airports because this strategy permits them to achieve extraordinarily high productivity from their aircraft, compared to the legacy carriers. (Warnock-Smith and Potter, 2005) They aim to fly their aircraft as much as possible by minimizing unproductive time either on the ground or in the air. They achieve this by avoiding congestion that will keep the aircraft on the ground, either because they are waiting for air traffic control clearance, have to queue up for a open gate, or have roundabout taxi distances. This is a central aspect of their service, in the same vein as the more easily observable fact that they cut the turnaround time on the ground to a minimum (about 25 minutes in the case of Southwest, and between 30 and 45 minutes in the case of jetBlue, depending on its two current types of aircraft). In contrast, traditional carriers typically use one hour or more.

The limited services that the low-cost carriers provide to many airports, especially the larger ones, would seem to make their bargaining position relatively weak.[18] This is less so at the other extreme, where small airports located within a limited geographical area are often competing for business. As Barrett (2004: 36) puts it, "A parameter in business discussions between airports and low-cost airlines is that the penalty for non-agreement by the airport is likely to be the loss of the low-cost airline and its passengers to a competing airport." The situation in these circumstances is more akin to a monopsonist carrier, or at least a set of oligopolist low-cost carriers, buying into a market offering relatively competitive services in terms of slot capacity. The desire of the traditional carriers to build up their hub business left many smaller airports underserved, creating a vacuum for the low-cost carriers to exploit.

At existing airports, the main advantage for the low-cost airlines lies in their bargaining strength when they can attract significant new airside business and when they can pull in supplementary

---

16 This reduces repositioning costs of aircrew, overnight allowance payments, and allows for simpler scheduling of fleets and for quicker network "recoveries" after mechanical failures or adverse weather conditions. Some traditional carriers also operate some of their services in this way; for example Northwest Airlines in the US uses Indianapolis as a focus city, with numerous non-connecting services offered from it.

17 Fewings (1998) provides a broad inventory of the former military air fields in Europe that have become available, and estimated that in the late 1990s within the 10 European Union states that he examined, at least 131 of these were within one hour of surface travel from existing airports.

18 For example, Southwest withdrew services for a period from San Francisco in 2001, an airport in its network since 1982, because of higher fees that it could not negotiate down.

revenues for the airports from such things as concessions and car parking. The aim for the airport then becomes one of contracting with the low-cost airlines in such a way that volume makes up for the lower landing and take-off charges negotiated. As a strategic game, some airports hope that in subsequent negotiations they may be able to leverage up charges as the carrier finds itself committed to the facility because of sunk costs.[19] Evidence on repeat contracting is, however, limited, in part because many contracts are relatively new.

Table 5.4   Low-cost carriers and airports in metropolitan regions (2006)

Metropolitan region	Secondary airport	Low-cost airline
Boston	Providence	Southwest
Boston	Manchester, NH	Southwest
Brussels	Charleroi	Ryanair
Copenhagen	Malmo, Sweden	Ryanair
Dallas/Fort Worth	Love	Southwest
Frankfurt	Hahn	Ryanair
Glasgow	Prestwick	Ryanair
Hamburg	Lübeck	Ryanair
Houston/Galveston	Hobby	Southwest
London	Stansted	Ryanair
London	Luton	easyJet
Los Angeles	Long Beach	jetBlue
Manchester (UK)	Liverpool	easyJet
Melbourne (Australia)	Avalon	Jetstar
Miami	Fort Lauderdale	Southwest
Milan	Orio al Serio	Ryanair
New York	Islip	Southwest
Oslo	Torp	Ryanair
Paris	Beauvais	Ryanair
Rome	Ciampino	easyJet, Ryanair
San Francisco	Oakland	Southwest
Stockholm	Skvasta	Ryanair
Vancouver	Abbotsford	Westjet

---

19   Airports themselves may also incur sunk costs if they seek to attract more low-cost carrier business by investing in specific infrastructure. The need to cover short-term costs can then force them to offer these facilities at below long-run marginal cost.

The low-cost model, at least when there are many alternative airports that an airline can serve, is often akin to the wholesaler–retailer relationship originally examined by Galbraith. A low-cost airline has highly flexible and mobile assets, and is not encumbered by the need to optimize a hub-and-spoke network; it can therefore simply adjust its services if the costs of using an existing airport are too high. While not directed explicitly at this industry, the literature of countervailing power suggests that the buyer (the low-cost carrier) in these circumstances benefits from advantages of asymmetrical information. Basically, a low-cost carrier can bargain separately and simultaneously with a set of alternative airports, each of which, in the absence of full information about what the others are offering, views itself as marginal and thus unable to price above its costs. The airports could, of course, collude to move towards a bilateral monopoly situation, but this is likely to be difficult to sustain (Snyder, 1999).

In some cases, however, where there are repeated games, airlines may be at a disadvantage. In these circumstances, as airline/airport contracts are renewed, an incumbent carrier may have incurred costs in setting up its operations that could become stranded should it leave. This may put it at a disadvantage if the airport is cognizant of these costs when negotiations occur and uses the potential of the airline's loss if it leaves to push up rates.

Francis et al. (2003) look at a repeat game situation involving low-cost carriers at two small European airports. These examples offer insights but, as with all case studies, there are issues about the generality of the results that are found. In one case, the airport was near a major hub and had lost previous low-cost and charter business. To attract the low-cost operator the airport was forced to offer concessionary terms that were theoretically below low-run average costs. The shortfall is made up in terms of the non-aeronautical revenues that are generated. Subsequent attempts by the carrier to obtain better terms were rejected by the airport. In the second case, the low-cost carrier initially asked the airport for "marketing support"—free handling and some advertising costs—if it were to locate there. The situation was short-lived, however, because the airport wanted to renegotiate the terms of the ground-handling arrangement. The main difficulty with small airports is that they rely heavily on aeronautical fees for their revenues, and in the case of Europe their large number gives them limited bargaining power with the airlines. The countervailing effects of this are that such facilities have very limited opportunities to exercise monopoly powers.

Collective Airline Action

In some markets airlines are permitted to combine and to negotiate rates collectively with airports. In effect, a cartel can be seen as developing countervailing power to monopoly airports; and the analytics of the situation are the same, with the complication that the cartel must formulate its objectives across diverse actors. Whilst this is not always explicitly allowed, an airport may be reluctant to lose one carrier because of its links with others. The growth of the three major strategic alliances, oneworld, Star Alliance, and SkyTeam, may be seen as offering this sort of collective countervailing power.

This process can be more formal and, for example, is a feature of the New Zealand system through the Board of Airline Representatives (BARNZ), where the countervailing power may be exercised through "fiscal duress" (the withholding of payments), "reputation duress" (effectively the exercise of "voice" through such things as media campaigns), and legal, political, and regulatory sanctions (for example, through judicial review). Official reviews of the situation suggest that the force of such countervailing power can restrain the actions of airports (New Zealand Commerce Commission, 2002).

The Ability of Airports to Price Discriminate

One indicator of the extent of countervailing power is the ability of an airport to price discriminate between airlines. If there is price discrimination, at least one of the prices is not equal to marginal cost. Therefore, the existence of price discrimination indicates that there must be at least some degree of market power (Varian, 1989). Airports are often legally committed not to discriminate between users other than by cost, but in cases where price differentiation based upon demand is allowed, and practiced, then its presence would indicate that neither original competition nor the countervailing powers of airlines are fully limiting the airports' market power.

The ability to first-degree discriminate, involving charging different prices for each take-off or landing depending on willingness-to-pay, is rare and unlikely to be a practical proposition for an airport even if it does enjoy considerable monopoly power; the information and administrative requirements are too large. A more viable proposition to consider is third-degree price discrimination whereby the market for airport services can be segmented, and where the segments have different elasticities of demand. At the most basic level we can think in terms of segmentation between low-cost and traditional carriers.

We can take the simple case of a profit-maximizing airport that is serving both a low-cost carrier and a legacy carrier. In Figure 5.2 the airport decides its output by equating its overall marginal costs (MC_M) with its overall marginal revenue (MR_M). The price levied on the airlines will then be at the point on the average revenue curve directly above the result flights attracted (AC_M)—not illustrated in the figure. However, landing fees are normally a larger part of a low-cost airline's costs than a traditional, legacy carrier's. This makes the former more sensitive to the level of such fees. To increase its revenues, therefore, the airport can try to charge separate fees to its two potential customer groups. It would effectively do this by drawing a horizontal line through the $MC_M = MR_M$ point and carrying this over to the separate markets for low-cost and legacy carriers. This intersection gives the combined profit-maximizing output for the airport. Taking the resultant marginal cost across to the separate marginal revenue curves of the different categories of carrier optimally shares out slots between them. Prices for the low-cost carriers (P_{LCC}) and the legacy carriers (P_{LC}) are then determined by the average revenue curves corresponding to these slot numbers.

The degree to which airports can segment in this way depends on the extent to which the airlines are incapable of diluting the revenue being collected by the airport—essentially their

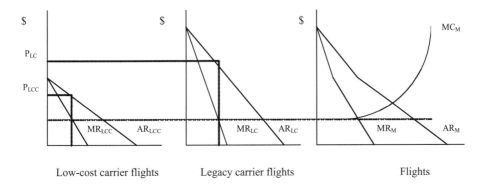

Figure 5.2 **Third-degree price discrimination between low-cost and legacy carriers**

ability to prevent differential fees. The extent to which airports third-degree price discriminate in practice has not been rigorous examined. A study by Jacobs Consultancy (2007) found differential passenger charges for low-cost terminals at Budapest, Kuala Lumpur, Marseille, and Singapore airports.[20] While airports may exercise some forms of price differentiation by carrier type, it becomes more difficult to separate out the effects of cost variations from charging policies based upon willingness-to-pay. Even in the cases cited, low-cost carriers are often based in separate terminals that offer passengers more limited, less grandiose, and generally less well-located facilities. For example, Richard de Neufville (2006) cites Geneva, Berlin, Kuala Lumpur, Paris Charles de Gaulle, Singapore, and Warsaw airports as situations where specific facilities for low-cost carriers has been provided

The fact that only a few cases of clear third-degree price discrimination seem to exist could be seen as suggestive of airlines being able to exercise a considerable degree of countervailing power. This would seem to be illusionary because of capacity problems at major airports and regulatory regimes that force airports to set charges based upon narrow accountancy costing principles. The former exists when, given the capacity of an airport, the nature of the demand curves for potential low-cost and legacy carrier users are such that profit maximization leads to a high price that only the traditional, legacy airlines are willing to pay.

Policy Approaches

Returning to the origins of the idea of countervailing power, Galbraith (1952: 136) observed that "Because of the nature of countervailing power has not been firmly grasped, the government's role in relation to it has not only been imperfectly understood bit also imperfectly played." In the aviation context, this begs the question of what the public policy attitude towards countervailing should be, and how it should be exercised.

While Galbraith (1952: 136) had the idea in the that "the support of countervailing power has become in modern times perhaps the major domestic peacetime function of the US government," whatever the validity of his general view, any countervailing power in the airline sector has been more of a by-product of other policy initiatives than a policy in its own right. The regulatory regimes that typified most markets until the 1980s certainly gave airlines and airports a large degree of monopoly rights, but regulations limited the extent to which this power could be exercised over each other.

The various forms of liberalization that have taken place since 1978 have profoundly, changed many element of this situation, but it is unclear as to how much government has deliberately tried to foster countervailing forces to airports, and even less clear as to its likely success. The extent to which countervailing power, if it exists, is a natural phenomenon in the market for airport services, determined by technical parameters or a function of institutional structures, is unclear. While some countries such as New Zealand see it as a useful and relatively powerful force that reduces the burden of regulation, the continuing plethora of regulations and controls put on airport charging policies elsewhere indicates that most others see it as, at best, a very weak mechanism that is no substitute for direct government intervention.[21]

---

20   There are also cases, as with Frankfurt Main and Frankfurt Hahn airports that are both operated by Fraport, where price discrimination may be seen to operate across a system of commonly provided facilities even if there are none within a single facility.

21   Although some of the regulatory agencies, such as those in the UK (Bush, 2005), seem to be moving to embrace the market power of airlines in the mechanisms by which they regulate air transport infrastructure.

But even if airlines did manage to curtail the market power of airports, it is not always clear that enhanced overall social welfare is obtained. Kenneth Galbraith's view on countervailing power is that "In all but conditions of inflationary demand, countervailing power performs a valuable—indeed indispensable—regulatory function in the modern economy." While there are such arguments that countervailing power does result in equilibrium outcomes in most cases, there are, however, equally strong arguments that outcomes are seldom optimal in the traditional neo-classical economic sense. In particular, while a large airline negotiating with a major airport may result in it keeping slot fees or passenger fees down, both parties will enjoy some economic rent. It is a matter of distribution of this rent, but it is at the expense of final consumers of air services. This loss may be exacerbated in situations where X-inefficiency accompanies the allocative costs of rent seeking. Basically, to exert countervailing power, one must have some market power oneself.

Some of the policy measures that have been introduced in various contexts, for example price capping of airport charges, can be seen as efforts to limit these social welfare losses; but equally some have been motivated by other factors, or have been counter-productive. But whatever the motivation for such measures, the underlying policy stimulus offers little support for the idea that countervailing power itself provides for social welfare maximization.

Conclusions

The systematic deregulation of airline markets beginning in the US in the late 1970s, and the resultant increased commercial pressures on carriers to reduce costs, has led to increasing interest in airport charging and investment policies.

Although much of the limited knowledge we have on the economics, countervailing power in aviation has tended to focused on the bargaining strengths of particular groups of buyers and sellers of the services—airports and airlines in our case. As Galbraith indicated, it may also come from other sources in the value chain; for example the countervailing power of labor may be seen as important in the market for air navigation services. These sources are often dependent on the institutional structure under which the services are provided. Institutions, both formal as set down in law and informal as inherent by governance practices, are important in determining relative market power and the types of situation where countervailing power may prove important in determining bargaining outcomes.

Markets are complex institutions.[22] Economists have long recognized that institutional structures can be as important in determining the way a market functions as are the technical nature of the product or service being produced—the form of the production function or elasticity of demand—and the availability of factors of production. These institutions transcend the narrow notions of legal frameworks to embrace such things as the business models and ways of doing business that suppliers and purchases adopted, and the motivations of the various actors involved.

In oligopolistic markets, countervailing power is often seen in terms of the relative market strengths of suppliers and buyers—the monopoly/monopsony struggle—within this institutional context. But whether this buyer power is in our case substantial enough to significantly counteract that of airports, or is simply a "mirage," as Peter Forsyth (2004) describes the situation in the Australian environment, has yet to be systematically tested. We move, unfortunately, very much

---

One may also argue that the legal reforms and international agreements that have allowed the formation of airline strategic alliances have implicitly increased the power of airlines when negotiating with airports.

22   Williamson (2000) provides an overview of the broader developments in institutional economics.

in the world of anecdotes, spiced with the views of vested interests and specific case studies when we debate the topic.

References

Barrett, S.D. (2004) How do the demands for airport services differ between full-service carriers and low cost carriers? *Journal of Air Transport Management*, 10: 33–9.
Bowley, A.L. (1928) Bilateral monopoly, *Economic Journal*, 38: 651–9.
Bush, H. (2005) Regulation of airports and air traffic control: an evolving story, paper to the Oxera Conference on the Future of Infrastructure Regulation, 1 March.
Button. K.J. (2003) Does the theory of the "core" explain why airlines fail to cover their long-run costs of capital? *Journal of Air Transport Management*, 9: 5–14.
Button. K.J. and Stough, G. (2000) *Air Transport Networks: Theory and Policy Implications*, Cheltenham: Edward Elgar,
de Neufville, R. (2006) Accommodating low cost airlines at main airports, Transportation Research Board presentation, summarized in *International Airport Review*, 1: 62–5, http://ardent.mit.edu/airports/ASP_papers.
Doganis, R. (2005) *The Airport Business* 2nd edn, London: Routledge.
Edgeworth, F.Y. (1881) *Mathematical Physics*, London: Kegan Paul.
Fewings, R. (1998) Provision of European airport infrastructure, *Avmark Aviation Economist*, 15 (July): 18–20.
Forsyth, P. (2004) Replacing regulation: airport price monitoring in Australia, in P. Forsyth, D. Gillen, A. Knorr, O. Mayer, H.-M. Niemeier and D. Starkie (eds) *The Economic Regulation of Airports: Recent Developments in Australasia, North America and Europe*, Aldershot: Ashgate.
Francis, G., Fidato, A., and Humphreys, I. (2003) Airport-airline interaction, *Journal of Air Transport Management*, 9: 267–73.
Galbraith, J.K. (1952) *American Capitalism: the Concept of Countervailing Power*, Boston: Houghton Mifflin.
Gillen, D., and Lall, A. (2004) Competitive advantage of low-cost carriers: some implications for airports, *Journal of Air Transport Management*, 10: 41–50.
Graf, L. (2003) Airport-airline interaction: the impact of low-cost carriers on two European airports, *Journal of Air Transport Management*, 9: 267–73.
Humphreys, I., Ison, S., and Francis, G. (2006) A review of the airport-low cost airline relationship, *Review of Network Economics*, 5: 413–20.
Hunter, A. (1958) Notes on countervailing power, *Economic Journal*, 68: 89–103.
Jacobs Consultancy (2007) *Review of Dedicated Low-cost Airport Passenger Facilities*, Dublin: Commission for Aviation Regulation.
Leibenstein, H. (1979) A branch of economics is missing: micro-micro theory, *Journal of Economic Literature*, 17: 477–502.
New Zealand Commerce Commission (2002) *Final Report of the Commerce Commission on the Study of Airfield Activities at Auckland, Wellington and Christchurch International Airports*, Wellington: Ministry of Economic Development.
Oum, T.H., Yu, C., and Fu, X. (2003) A comparative analysis of productivity performance of the world's major airports: summary report of the ATRS global airport benchmarking research report—2002, *Journal of Air Transport Management*, 9: 285–97.

Porter, M.E. (1985) *Competitive Advantage: Creating and Sustaining Superior Performance*, New York: Free Press.

Schelling, T.C. (2006) An astonishing sixty years: the legacy of Hiroshima, *American Economic Review*, 96: 929–37.

Snyder, C.M. (1999) Why do large buyers pay lower prices? Intense supplier competition, *Economic Letters*, 58: 205–9.

Stigler, G. J. (1954) The economist plays with blocs, *American Economic Review*, 44: 7–14.

Telser, L.G. (1978), *Economic Theory and the Core*, Chicago: University of Chicago Press.

Varian, H.R. (1989) Price discrimination, in Schmalensee, R. and Willig, R,D. (eds) *Handbook of Industrial Organization*, vol. 1. Amsterdam: Elsevier.

Warnock-Smith, D. and A. Potter, A. (2005) An exploratory study into airport choice factors for European low cost airlines, *Journal of Air Transport Management*, 11: 388–92.

Whitney, S.N. (1953) Errors in the concept of countervailing power, *Journal of Business of the University of Chicago*, 26: 238–53.

Williamson, O.E. (2000) The new institutional economics: taking stock, *Journal of Economic Literature*, 38: 595–613.

Chapter 6
Competition Between Major and Secondary Airports: Implications for Pricing, Regulation and Welfare

Peter Forsyth[1]

Introduction: The Problem

In recent years, both in Europe and the US, there has been increasing use made of secondary airports, which are often some distance from the main origin/destination city. This development has been especially associated with the growth of low-cost carriers (LCCs). These have sought to cut costs in whatever ways possible, and they have been prepared to bypass main airports if they have been able to negotiate good deals from secondary airports. At least, for LCCs, the secondary airports are providing some competition for the main airports.

Normally, we would expect that an increase in competition would be welfare enhancing. However, we need to be aware that this is not always the case. One situation where additional competition can be welfare reducing is outlined by Suzumura and Kiyono (1987). When extra firms come into an oligopolistic market, they may survive, but overall welfare can fall because of the loss of economies of scale. Another case, also directly related to the airport one, is outlined by Braeutigam (1979). A natural monopoly (for example a rail system) may be covering its costs using Ramsey prices. However, if competition develops for some of its product range (for example, because of the entry of road-based carriers), it will have to restructure its prices, to meet the competition. The result will be a move away from the initial Ramsey second best solution, to a new solution which will be inferior (though perhaps not by very much).

This situation can happen with airports, though this is not always the case. Sometimes, when secondary airports enter, they take traffic away from major airports which are facing excess demand. In such a situation, their entry is likely to lead to a more efficient allocation of flights to airports. This possibility is recognised, though it is not considered in this chapter.

However, when the major airport has ample capacity, the marginal cost of handling extra flights may be minimal (and well below price). The secondary airport will attract flights away from the major airport, and overall costs may increase, especially if the travelling costs of passengers to the secondary airport are included. The major airport might or might not be able to adjust prices to capture the traffic. If it can, an efficient allocation of flights to airports will come about, but the price structure of the major airport will be less efficient than before (though the welfare loss from this may not be substantial). Often, though, airports do not respond, perhaps because it is difficult to do so. One is left with a situation whereby flights which could be handled at minimal cost at the major airport are handled at inconvenient secondary airports at no less, and possibly greater, cost. As an example, Ryanair flights use Lübeck Airport (about 60 kilometres from Hamburg) for

---

[1] I am grateful to Thomas Immelmann of Hamburg Airport and Mathias Seidenstucker of Lübeck Airport for helpful discussions. Any errors are my own.

Hamburg-destined traffic, even though Hamburg Airport has ample spare capacity for most of the times Ryanair would like to fly.

There can be other effects, which are not explored here. The major airport may be a hub airport, which gains from linking flights to many different destinations. If the entry of the secondary airport leads to some traffic shifting away from the hub, the latter is able to offer fewer flights and destinations, which will make it less attractive as a hub.

In the next section, the options open to the major airport to adapt its pricing to meet the new competition are considered. Then, in the following section, the incentives for an airport to respond are considered – these depend on ownership and regulation. After this, the issue of why secondary airport charges are below those of the major airports is considered; do they have lower costs, or are there other factors present? Next, comments are made about the institutional arrangements under which airports compete – competition may not work well if these are poorly structured. Finally, conclusions are drawn, and some unanswered questions are identified.

Competitive Responses

Competition between a major city airport and a secondary airport is becoming a common feature in Europe, and to an extent in North America and elsewhere. Major city airports tend to be located within the city boundaries, and they are convenient for passengers. Until recently, they handled all or nearly all of the traffic for the city. In recent years, secondary airports, located at some distance from the city, have entered the market for airport services. These airports may have been small regional airports, or they may have been military airports. They are less convenient than the major airports, but they have been able to attract cost-conscious traffic by offering lower charges. These have particularly appealed to low-cost carriers (LCCs), whose passengers are price sensitive but willing to put up with inconvenience to save money. They also offer the LCC an opportunity to fly passengers to or from a city without entering head-to-head competition with incumbent full-service carriers (FSCs).

Competition between major and secondary airports for LCC traffic is, along with competition between secondary airports, perhaps the main form of active competition between airports (for a discussion of airport competition in general, see Forsyth, 2006b), except in densely populated countries such as the UK and Germany (see Mandel, 1999a, 1999b). With the growing market share of LCCs, this aspect of competition is becoming more significant. Major airports in different cities do not compete very much, except sometimes to attract hub traffic.

One key efficiency issue concerns allocation of traffic to airports. It is efficient that traffic goes to the airport which can serve it at lowest overall cost. For present purposes, the overall cost of using an airport can be thought of as including the costs of the airport (runway and terminal) use, the cost of access to the airport and the cost of related services such as car parking (see Hess, Chapter 11 in this volume). The users – airlines and their passengers – choose between airports on the basis of airport charges, access costs and prices of related services. These prices need not reflect costs – airport charges may be higher or lower than the marginal costs of airport use – because of a cost recovery requirement or subsidies, and the same may be true of prices of related services. The likely starting point for a major airport with ample capacity and a cost-recovery requirement is that the prices for airport use exceed the marginal cost. Introduction of competition to a situation where prices do not reflect marginal costs can lead to inefficient choices of airport.

Suppose a situation whereby an established major airport has excess capacity. It is possible that the marginal cost of a flight is quite low (though see Hogan and Starkie, 2004); however the airport will face large sunk costs associated with its construction, for example, in building the runways. Suppose further that the airport is operating under a cost-recovery constraint, and that this takes the form of earning a target revenue each year – this target is set by the costs allocated each year. Note that this is an artificial, accounting-based version of cost recovery – economic cost recovery only requires that the total cost be recovered over the airport's life span – and a more efficient pattern of cost recovery would involve loading revenue targets onto years when the airport is busy. In practice, however, many airports operate with this type of constraint. The airport implements a weight- or passenger-based schedule of prices which can be considered to be an (imperfect) approximation to Ramsey pricing (see Morrison, 1982).

When a secondary airport opens and starts attracting LCCs, the traffic these bring will to an extent be generated traffic, but some will be diverted from the major airport. If the major airport wishes to retain and capture some of this traffic, it will need to reduce its charges while at the same time continuing to cover its costs. It will have to restructure its price schedule. For example it might be able to lower prices for the category of user being attracted away. This will amount to an implicit form of price discrimination. Thus if the LCC is using Boeing 737-sized aircraft, it may be feasible to lower prices for aircraft with a weight or passenger load of this aircraft type. Charges for both larger and smaller aircraft would have to be adjusted upwards to meet the cost-recovery constraint. The result would be a rather artificial price structure, though it might work. It will be a less optimal one than in place before, since an additional constraint has been added to the Ramsey pricing problem. However, the welfare loss is not likely to be large.

Another form of implicit price discrimination is where lower prices are offered for categories of flights which happen to encompass the LCC's flights. An example would be discounts for new services – the LCC will be able to take advantage of these, though the FSC would only be able to make limited use of the discounts, since most of its flights would not be new. Some airports offer discounts for airlines newly operating from them.

At the extreme, an airport may simply practise explicit price discrimination by offering lower prices to the LCC. It is very difficult for one user to on-sell to another and, since the airport is a near monopoly, the FSC will just have to pay up whether it likes it or not. This said, many airports may not like to discriminate against its old customers, and they may not see doing this as profitable in the long run (and FSCs may set up their own LCC subsidiaries which can take advantage of concessions for 'new' services at airports, as Lufthansa has through Germanwings).

A number of airports are building 'low-frills' terminals to attract LCCs. This may or may not be a form of price discrimination. If the terminal is designed so as to not be attractive to FSCs, and if the prices charged for the low-frills terminal use are less than cost, then this constitutes market segmentation and price discrimination. It is a way of offering a lower price to self-selected customers. In some contexts, it may be akin to the Saturday night stay restriction on low-fare airline round trips. On the other hand, if the new low-frills terminal is simply a cost-effective means of providing services to airlines which are not prepared to pay for all the frills, it will not be price discrimination.

Price discrimination may be effective in enabling the major airport to keep traffic, and it may contribute towards an efficient allocation of traffic between airports. However, it may not be costless. Where otherwise unnecessary facilities are constructed to enable price discrimination, as can happen with low-frills terminals which are built so as to be inappropriate for FSCs, even when there is available capacity in existing terminals there is a deadweight loss. In addition price discrimination can lead to inefficiencies at the airline level, when competition between FSCs and LCCs in the same market is distorted through the latter obtaining cheaper airport facilities than the latter.

Another way in which an airport may attempt to address the problem is to change the price structure in such a way as to make it more attractive to the LCC. It could introduce peak/off-peak pricing, and give the LCC (and any other users) the option of reducing costs by using capacity in the off-peak. This option need not be particularly attractive to LCCs, since the LCC may need to keep its aircraft fully utilised, and it may not have the scope to fly mainly in the off-peak. This response could help a little if the LCC does have some scope to organise its flights into the airport such that they do occur mainly in the off-peak.

There can be other ways of targeting LCCs and their passengers. Many passengers on LCCs may be taking trips of longer duration than typical passengers on FSCs. For these passengers, car parking charges will be a significant proportion of their overall trip costs. One option for the major airport to offer a competitive package would be to reduce car parking rates for long-stay users. Car parking charges may well be above costs, and it may be easy to reduce charges to target cost-conscious travellers. Again this will be a form of price discrimination if the resultant prices for short- and long-stay parking do not reflect the cost of provision. These reductions in parking charges may help the major airport to win LCC traffic, though they can also pose a cost-recovery problem – to cover costs, it may be necessary to raise other charges, such as short-stay charges or aeronautical charges.

If competition from a secondary airport develops, the major airport can respond or not. If it does not respond, it will lose traffic. This poses a problem for cost recovery, though the airport will most likely be able to raise the overall level of its charges. If it responds, and it is not able to lower its costs, it will have to alter its price structure so as to attract LCC traffic. This is likely to involve explicit or implicit price discrimination – while it may work in that it enables the airport to attract the traffic, it will not be costless. In addition, price discrimination is not always easy to impose – the airport must devise ways of getting different users to pay different prices for essentially the same service.

Some airports may have the scope to reduce the overall level of charges, and for these, there may not be a need to alter their price structure. It is possible that the major airport has not been minimising its costs in the past, and that there is scope for cost reductions. The secondary airport may have been able to offer lower charges because it is more efficient. The competition which the major airport now faces forces it to review its costs, and it may be able to reduce them, and lower all charges, and thereby attract the LCC. While this may not happen overnight, it could be that competition from the secondary airport is the wake-up call to the airport, which has allowed costs to rise over the years during which it has faced no competition. In this situation, the additional competition from the secondary airport can have a positive effect on welfare through its impact on productive efficiency. This possibility has particular relevance for Europe, where airport unit costs and charges are high relative to those in other parts of the world (see Air Transport Research Society, 2008). The cost advantages of secondary airports could be based on superior efficiency, something which can be eroded over time.

When competition for a major airport develops from a secondary airport, it may lead to a loss of price-sensitive traffic. If the major airport price structure involves prices well above marginal cost, it is efficient that it respond by reducing charges to the price-sensitive traffic so that this traffic is retained. This is not always easy or feasible, and in many situations secondary airports have been able to gain a large share of the LCC traffic because the major airport with ample capacity has not altered its price structure. They have been able to do this even though it would be cheaper and more convenient to handle this traffic at the major airport.

Incentive Issues

As always, the responses to competition depend on the incentives the firm faces. Airports could be publicly owned, and could be set a cost-recovery target. Such airports are not likely to be interested in maximising profits. Airports could be fully or partially privately owned and subject to regulation. This regulation could take the form of rate of return regulation (for example this was the case for Düsseldorf; see Niemeier, 2004). Alternatively an airport could be subject to incentive regulation such as price caps (for instance Hamburg airport; see Niemeier, 2004) or profit-sharing/sliding-scale regulation.

The response of the publicly owned or the rate of return regulated airports is not likely to be one of attempting to capture the LCC's business. The publicly owned airport is not interested in profit opportunities, and the regulated airport cannot gain extra profit from it. When costs are fixed, revenues are fixed. When costs are variable, revenues will be allowed to increase only to the extent that costs increase, allowing no extra profit. Such an airport will be faced with low demand elasticity for its output and will have no difficulty in achieving cost recovery, and will not have an incentive to chase the LCC's business. If, however, the airport is a size maximiser, it will have an incentive to increase the cost base by attracting the LCC.

The airport operating under incentive regulation does have an incentive to gain the LCC's business. Suppose there is a price cap on average revenue per passenger (a common form). If it can adjust its price structure in some way to attract the LCC, it will be allowed to earn more revenue; and, assuming that the price cap is set at above marginal cost, it will earn more profits.

If cost reduction, to enable lower overall prices, is a feasible option, the airport may have an incentive to try doing it. An incentive-regulated airport may not have been minimising costs (incentives for cost reduction are rarely perfect, and most incentive regulation of airports is relatively new), but now it has a stronger incentive to keep costs down. The rate of return regulated airport will not see any point in reducing its costs because it will lose revenue if it achieves this. If it is a size-maximising airport, it will have even less incentive to respond, because reducing its costs means reducing its size in terms of inputs (labour, capital).

The upshot of this is that an efficient allocation of traffic to airports need not come about. Even when it is feasible for the major airport to alter its price structures to attract LCC traffic, it may not have any clear incentive to do so. Relatively few airports are subject to an ownership and regulatory environment which give them incentives to alter price structures, or improve efficiency, so that they can win traffic which they are potentially the most efficient supplier of.

Why Are Secondary Airports Cheaper?

In the previous discussion, it has been assumed that prices at the secondary airport are lower than those at the major airport. This is the typical case – LCCs have been induced to use secondary airports because they are cheaper. There are other factors which have been significant, such as the absence of on-ground congestion, which makes LCC operations cost less, and there is less head-to-head competition with FSCs from secondary airports. However, in practice, lower charges are a strong selling point for secondary airports.

However, given the normal patterns of airport costs, it might be expected that secondary airports would have higher, not lower, costs. If there are economies of scale present, smaller airports would have higher costs. If airports involve substantial sunk costs, for example in the provision of runways, the average cost of the larger airport would be lower than that of the smaller airport because the sunk costs would be averaged over more users. In spite of this, many secondary airports are able to offer lower charges, which is one of the reasons why they have been so successful in attracting LCCs. There must be other factors at work.

Greater Efficiency

Clearly, if airport costs are dominated by sunk costs, it would be difficult for a smaller airport to have lower costs than a major airport (since the marginal cost of sunk costs is zero). However, variable costs may be present, and they could be quite significant (Hogan and Starkie, 2004). The lower operating costs of an efficient small airport could outweigh the economies of scale gained by the less efficient major airport. If so, overall average costs could be lower.

If this were so, then it could be efficient if the smaller airport gained the LCC market, since costs overall, including costs of access and related services such as parking, could be lower. Competition from the secondary airport would then lead to a welfare improvement.

Subsidies

When subsidies are present, competition need not be welfare improving. Subsidies induce users to choose high-cost over low-cost producers.

Many secondary airports have been enabled to offer lower prices through being offered subsidies by local governments (Air Transport Group, Cranfield University, 2002; Morrell, Chapter 2 in this volume). (Some major airports have in the past been given subsidies, and some may still be in receipt of subsidies. If both secondary and major airports are currently receiving comparable subsidies, there is no misallocation of traffic.) Could these subsidies be warranted? From the perspective of the local region, this could be so. Suppose that subsidies are effective in bringing more tourists to the region. Suppose further that tourism brings economic benefits. For example, a tourist may spend $100 in the region, but the cost of supplying the goods and services consumed might only be $900 – there will be a net benefit of $100 per tourist to the region (for discussion of the economic benefits of tourism, see Forsyth, 2006a). If subsidies to the airport succeed in attracting LCCs and more tourists, they could be in the interest of the region.

While this may be correct, there will be negative impacts on other regions. The tourists who have been attracted to the region offering the subsidies spend less in other regions, such as that surrounding the major airport. Unless this region is congested, there will be a loss of tourism benefits in this region. Using subsidies to shuffle tourists from one region to another is not a welfare-enhancing exercise. What is rational from the perspective of the individual region offering the subsidy is not rational taking all regions together.

Lower Input Costs

Suppose that the secondary airport is located some distance away from the destination, but the main airport is located close by. Land prices in the remote location are likely to be lower, and to the extent that these are factored into the airport costs, the secondary airport will have lower costs. It can then offer a lower price schedule to the airlines. Some related services such as car parking may be quite land intensive, and secondary airports may be able to offer much lower-priced car parking. Labour costs could also be lower outside the city.

Passenger Ignorance

Some secondary airports are a long way from the main city. If passengers realise this, they will factor it into their assessment of the attractiveness of the LCC's product. Alternatively it is possible that passengers are not aware of the remoteness of the secondary airport, and that they would not choose to travel through it if they were aware. If ignorance is present, inefficient allocations can come about. Over time, this should become less of a problem.

Inefficient Bypass

When a sunk cost is recovered by setting a charge equal to average cost, inefficient bypass can come about. Infrastructure can be over-provided, and when attempts are made to cover costs by charging high prices, inefficient bypass is encouraged. Suppose that there is a large sunk cost associated with the construction of a major airport, but revenues must be raised to cover this cost. Suppose also that airport services can also be provided at a secondary airport at a variable cost less than the average cost, though not marginal cost, of the major airport. Under these cost conditions, it would not have been optimal to build the major airport to the scale that has been chosen, but this has already been done. If the major airport is required to cover costs by setting a price at average cost, the secondary airport will undercut it and win traffic from it. This results in an inefficient allocation of traffic, since traffic which can be handled at zero marginal cost will be handled at a positive variable cost. The bypass problem is a real one in telecommunications, and it stems from inefficient pricing of sunk assets. The problem can arise with airports too.

Different Asset Valuations

It is quite possible that the assets of major and secondary airports have been valued using different and inconsistent principles, and that this is leading to problems when they compete.

The major airport will often be required to cover costs, including the sunk capital costs of providing fixed assets such as the runways and terminals. When the airport is corporatised or privatised, the sunk assets will be valued, perhaps using replacement cost or some variant of it. These could be quite high valuations, and prices will have to be high to recover them.

By contrast, the asset valuations which form the basis of the secondary airport's prices may have been arrived at on a very different basis. The secondary airport could well be a military airport which is no longer in use – it might have been sold to a local authority at a nominal price. The capital costs could have been more or less fully written off. Suppose the local government wishes to recover the costs of its investment when it operates the airport as a commercial entity. It will be able to do this at a low price which does not factor in the sunk cost. Essentially, the secondary airport is able to offer services at a lower price than the major airport not because it has lower costs, but because of different accounting valuation conventions and different requirements for cost recovery. Competition between two airports, when one is required to cover historic costs and the other is not, is bound to give rise to misallocation of traffic. Getting around this problem is not easy to do in practice. It would not be feasible to require secondary airports to recover past costs incurred by an owner different from the current owner, or for major airports to write off all sunk costs and to set prices close to marginal cost.

Differential Service Quality

The secondary airport may be able to offer lower prices to the LCC because it is supplying a lower quality of service. In particular, it may have a low-cost terminal, while the major airport may have a costly, high-quality terminal. If the variable costs of terminal operation are lower at the secondary airport, then it is efficient for the LCC to be attracted to the secondary airport, granted that it is not prepared to pay for the higher service quality.

In the longer term, the major airport should be able to offer terminals of a quality which LCCs are willing to pay for. However, it is possible that the major airport has constructed a high-quality terminal with ample capacity which can be used at low marginal cost. If so, it would be efficient for the LCCs to use it; however, if the sunk costs of the terminal must be recovered by high use charges, this will not happen. It will be profitable for the secondary airport to build a new low-cost terminal for the LCCs – another example of inefficient bypass.

In summary, there are several reasons why a small secondary airport may be able to offer lower airport charges to the LCCs than the major airport does. It is necessary to determine exactly which of these reasons apply in a particular case. If the reason for lower charges is greater operational efficiency, it can desirable for the secondary airport to capture the LCCs' traffic. In addition, competition from a lower-cost competitor can put pressure on the major airport to lower its own costs, thereby increasing efficiency. If, on the other hand, the lower charges come about because of different asset valuations or subsidies, competition between the airports will result in an inefficient allocation of traffic.

Institutional Arrangements and Airport Competition

Competition works well when all the competitors operate under similar conditions – the level playing field assumption. This situation often does not occur with airports. Secondary airports can be, but often will not be, equally or more efficient suppliers of airport services as the major airports. We would not normally expect competition to always work well at airports, with economies of scale and significant sunk costs. When additional complicating factors, such as subsidies which are available to some though not all airports are allowed for, it is even less likely that competition will work well. Competition between secondary and major airports is now being observed in many instances, but it cannot be concluded that it is always welfare enhancing.

The pricing structure of airports is not conducive to competition. There are large sunk costs of constructing airports, and most airports operate under a cost-recovery constraint (and privately owned airports are regulated and allowed to set prices to cover measured costs). Furthermore, the cost-recovery constraint is invariably set in terms of arbitrary annual terms, which means that prices and marginal costs are driven further apart. Sunk costs are amortised, that is arbitrarily allocated to specific years, resulting in costs being recovered even when excess capacity is present and marginal costs are minimal. The upshot is prices well above marginal cost when excess capacity is present and prices well below efficient rationing levels when there is excess demand. When there is excess capacity, competition does not encourage an efficient allocation of traffic between airports, and there will be too strong an incentive to use the secondary airport. Ideally, prices could be set by regulators or government owners low when there is ample capacity – but this rarely, if ever, happens. In short, the ways in which pricing is actually handled poses problems for competition.

On top of this, the rules under which different airports operate are different. The secondary airport may be able to charge lower prices simply because it is being subsidised and the major airport is not. In other cases, the secondary airport's advantage may lie in different, and arbitrary, accounting treatments of sunk assets. The sunk costs associated with the secondary airport's assets may have been written off, while those of the major airport may be required to be recovered. Under these circumstances, competition between airports will not lead to an efficient allocation of traffic.

This has implications for the way we view the role of competition between airports. When there is a natural or locational monopoly, it is often desirable to expand the role of competition where possible, to reduce the dependence on regulation, given the problems associated with regulation. Competition can sometimes be used to discipline the price behaviour of the firm instead of regulation. In the airport case, the increasing presence of secondary airports need not be disciplining the price behaviour of major airports in an efficient manner.

While competition between airports can be counterproductive, it does not follow that it would be desirable to prohibit it. A preferable strategy would be to address the underlying environment as far as is possible, with a view to making competition work better. The elimination of subsidies is an obvious starting point – in this respect, the recent developments in Europe, requiring removal of some existing subsidies, are positive. It is also important to address the less apparent sources of distortion, such as the accounting treatment of assets. Ideally, secondary airports should only be in a position to offer lower prices than the major airports when their costs are genuinely lower. While much can be done to level the playing field, it will still not be level while sunk costs are recovered on an arbitrary annual basis, leading to inefficient price structures and the risk of costly bypass.

Conclusions and Unresolved Questions

Competition between major city and secondary airports is probably the main form of competition between airports, and certainly it is the most problematic. One critical issue is that of the allocation of traffic between airports. This will depend on the overall charges faced by users, including airport charges, access costs and costs of related services such as car parking. These charges need not reflect the underlying costs and, as a result, an inefficient allocation of traffic to airports can come about.

Secondary airports have been winning new traffic and also winning traffic from major airports. This might or might not represent an efficient outcome. It could be that the major airports are setting charges well above the marginal costs of handling traffic, to cover total costs, including sunk construction costs. If these airports do not respond to competition from secondary airports, an inefficient allocation of traffic will come about. While it is not always easy for the major airport to alter its price structure to compete, the ownership and regulatory environment within which airports operate often gives them little incentive to respond efficiently. Secondary airports may also be able to offer lower charges because they have been subsidised. If this is the source of their advantage, while it may be efficient from the perspective of individual regions to subsidise their airports, the allocation of traffic to airports which comes about will be inefficient from the overall national perspective.

On the other hand, if the source of the secondary airport's cost advantage is greater efficiency, competition will probably have a positive effect. Unless the less efficient major airports keep the traffic by cross-subsidising it from less price-sensitive traffic, the switch in traffic to the secondary airport will be efficient. In addition, competition from the secondary airport may induce the major airport to improve efficiency to enable it to compete. There is considerable evidence that European major airports are not as efficient as those of other continents, and the secondary airports could be more efficient.

Thus, overall, competition between major and secondary airports can have both positive and negative features. Competition can lead to better allocation of traffic to airports, and to pressure on inefficient airports to perform better. On the other hand, inefficiencies in allocation of traffic can come about when prices do not reflect costs, when major airports set prices above marginal costs to recover costs or when prices at secondary airports are kept low by subsidies. In the light of this it is desirable that competition between major and secondary airports be considered on a case by case basis, and where feasible, institutional arrangements be reformed so as to align competition with the achievement of efficiency.

This said, there are a number of unanswered questions which need to be explored. One of these concerns how it is that secondary airports are able to offer lower charges – is it because they are more efficient and face lower input costs, or is it because they are heavily subsidised? Another question concerns whether it makes sense, from a regional and from a national perspective, to subsidise airports. Many regions believe that it makes sense, but is there any evidence for this? Other questions concern the responses by major airports – will they be able to continue to offer LCCs lower charges by price discriminating, or will this break down when LCCs gain greater market shares? Furthermore, is this price discrimination a relatively efficient response, or are there hidden costs associated with it? These questions need to be answered before we can conclude how efficiently competition between airports is working.

References

Air Transport Group, Cranfield University (2002), *Study on Competition between Airports and the Application of State Aid Rules, Final Report*, vol. 1, September European Commission, Directorate-General Energy and Transport, Directorate F-Air Transport.

Air Transport Research Society (2008), *Global Airports Performance Benchmark Report On World's Leading Airports*, Vancouver, University of British Columbia.

Braeutigam, R (1979), 'Optimal Pricing with Intermodal Competition', *American Economic Review*, March, 69, pp. 38–49.

Forsyth, P (2006a), 'Tourism Benefits and Aviation Policy', *Journal of Air Transport Management*, 12, 1–13.

Forsyth P (2006b), 'Airport Competition: Regulatory Issues and Policy Implications', in D Lee (ed.), *Advances in Airline Economics, Vol. 1 Competition Policy and Antitrust*, Amsterdam, Elsevier, pp. 343–65.

Hogan, O. and D. Starkie (2004) 'Calculating the Short-Run Marginal Infrastructure Costs of Runway Use: An Application to Dublin Airport', in P. Forsyth, D. Gillen, A. Knorr, O. Mayer, H.-M. Niemeier and D. Starkie, *The Economic Regulation of Airports: Recent Developments in Australasia, North America and Europe*, Aldershot, Ashgate, pp. 75–82.

Mandel, B. (1999a), 'The Interdependence of Airport Choice and Air Travel Demand', in M. Gaudry and R. Mayes (eds), *Taking Stock of Air Liberalization*, Dordrecht, Kluwer, pp. 189–222.

Mandel, B.N. (1999b), 'Measuring Competition in Air Transport', in W. Pfähler, H.-M. Niemeier and O.G. Mayer (eds), *Airports and Air Traffic Regulation, Privatisation and Competition*, Frankfurt am Main, Lang, pp. 71–92.

Morrison, S.A. (1982), 'The Structure of Landing Fees at Uncongested Airports', *Journal of Transport Economics and Policy*, 16, 151–9.

Niemeier, H.-M. (2004), 'Capacity Utilization, Investment and Regulatory Reform of German Airports', in P. Forsyth, D. Gillen, A. Knorr, O. Mayer, H-M. Niemeier and D. Starkie, *The Economic Regulation of Airports: Recent Developments in Australasia, North America and Europe*, Aldershot, Ashgate, pp. 163–92.

Suzumura, K. and Kiyono, K. (1987) 'Entry Barriers and Economic Welfare', *Review of Economic Studies*, 54, 157–67.

Chapter 7
Airport Strategies to Gain Competitive Advantage

Dr Anne Graham

Introduction

The commercialisation and privatisation of airports in combination with the deregulation of many aviation markets has introduced new possibilities for competition amongst airports. This in turn has meant that potentially there may be increased opportunities for airports to develop new strategies to gain competitive advantage. It is thus the aim here to investigate the strategies which airports have actually developed. Firstly, this is done by undertaking a competitive analysis of the industry and assessing the extent of influence that an airport operator has over the factors which affect its competitiveness. The relevance of these factors is then examined within the context of the different strategies that airports have seleted to adopt, and the directions and methods that the airports have chosen to pursue these. Finally the case of low-cost carriers (LCCs), and the choices of strategic direction which are available to address the needs of these types of airlines, are considered.

An attempt has been made to relate the situation at airports to general competitive strategy theory. This is in order to assess whether the characteristics of the airport industry are so unique that this necessitates airports being still treated as a special case or whether, at this stage in the evolution of the airport industry, it is now relevant to look to successful practice in other sectors outside the airport industry when formulating competitive strategies.

Competitive Analysis of the Airport Industry

The traditional view of many airports acting as natural monopolies is increasingly being questioned, primarily because of the improved forces of competition which have occurred due to greater airline liberalisation and airport commercialisation. Therefore a competitive analysis of the airport industry has been carried out before attention is turned to the actual strategies which airports have adopted. This has been undertaken by applying Porter's widely used five forces framework of competitive analysis, which considers the threats of new entrants and substitutes, the power of buyers and suppliers, and the rivalry amongst existing organisations (Porter, 1980).

It is the airport's key role to sell aeronautical services and facilities (that is, the provision of runways, taxiways, aprons and terminals) to airlines. The airlines then sell their own product on to the passengers. The airport also sells services such as shops, catering and car parks, usually through concessionaires, direct to the passengers, local residents and other customers. It is useful to make this distinction when considering the five competitive forces. The analysis will start off by looking at the services provided to the airlines.

Threat of New Entrants

The threat of new competing airports is generally low because of the large investment which is needed for the new infrastructure and because of the long and complex planning and regulatory processes which frequently have to be followed in order for approval of any new development to be given. Moreover in many areas of the world it is increasingly difficult to find suitable locations for new competing sites, although specifically within Europe the existence of a number of obsolete military airfields, such as Finningley and Manston in the UK, have provided some opportunities for new airport development. In many other industries, barriers to entry also tend to be high because of the existence of increasing returns to scale. However, within the airport industry, some evidence suggests that economies of scale tend to disappear once the annual output of an airport reaches around 3 million passengers and as airports grow much larger there may in fact be diseconomies of scale (Pels et al., 2003). This is probably due to airport operations becoming that much more complex. Overall this suggests that, although there are certainly substantial barriers to entry, they are not related to economies of scale (Starkie, 2002).

Threat of Substitutes

There are two threats of substitutes. Firstly passengers may decide not to travel at all and opt for alternatives such as video-conferencing. Secondly passengers may choose to use an alternative mode of transport, with probably the greatest threat being high-speed rail. For regional airports, the introduction of high-speed rail services can have a significant impact on air services to major airports. However at major capacity constrained airports increased usage of high-speed rail for short-haul trips may free up capacity for other long-haul services – although this may have a detrimental impact on the airport's ability to act as a hub and attract transfer passengers. Improvements to the road and rail infrastructure to major airports may also reduce the necessity for feeder services from regional airports. However, the recent growth of the LCC sector has changed the economic balance between rail and air travel, and in many cases has made air the cheaper and much more attractive option. This is working in the opposite direction to encouraging more travel by train, which is the objective of many European governments, but may well be reducing the threat of substitutes for airports.

Power of Suppliers

Many of the services which make up the composite airport product, such as air traffic control, security, ground handling and commercial facilities, can be provided either by the airport operator or by a third party. The way in which they are offered, and whether there are competing services, can have an impact on an airport's competitive situation. An extreme case is the management of entire terminals by different operators or suppliers.

For some of the operational services, such as air traffic control and security, the airport operator may have no choice over suppliers as this will be determined by government policy. However, the airport operator may not have to pay for all suppliers, for example with security if it is provided by some state agency. In other cases, airport operators will be less restricted and may have a greater choice of suppliers. Looking specifically at ground handling there may be specific regulations, as with the EU Ground Handling Directive, which stipulates the number and nature of suppliers which must be used. Overall this situation means that the power of suppliers varies quite significantly depending on which aspect of airport operations is being considered, and also on whether the airport operator itself supplies any of the product. The situation is made even more complicated since in some cases the suppliers, for example, air traffic controllers or ground handlers, will charge the airline direct.

Power of Buyers

The relative strength of the airlines can vary significantly. This power may influence the charging practices of the airport either directly through governmental pricing control or more indirectly through some type of economic regulation process. In lesser developed areas a national carrier is often politically well connected and very influential – particularly if its broader role in encouraging trade or tourism is taken into account. Generally since there are only a limited number of airlines at any airport, they could be expected to possess considerable market power. Many small airports, particularly those serving LCCs, may depend on just one or two key airlines, which make the airlines very influential. Moreover as airline alliances become larger and more established, it could be argued that the balance of power is shifting much more in favour of the airline. However, the real relative power of the airlines and airports is not really related to the relative size of the organisations, but to whether the airline has the ability to switch to an alternative airport. For many network carriers this may not be possible, but it may well be so for the more footloose LCCs, charter and freight operators. For example, LCCs can do this because they do not rely on transfer passengers and they have price-sensitive passengers who are willing to bypass the nearest airport and travel further to the airport if the fares are lower enough.

Rivalry amongst Existing Airports

The amount of rivalry amongst existing airports varies considerably (ACI-Europe, 1999; Cranfield University, 2002; Forsyth, 2006). If airports are located on small islands or in remote regions, the scope for competitive rivalry is very limited or non-existent. It also tends to be weak at airports which have a high concentration of both short-haul and long-haul services. These airports appeal most to the traditional scheduled carriers who have networked services. In these cases it is difficult for other airports to provide effective competition. This is unless the airport is competing for the role as a hub by aiming to provide good flight connectivity and efficient passenger transfers; then the amount of rivalry will be more intense. However, if the airports are physically close, their catchment areas may overlap and the competitive rivalry will be greater, particularly for point-to-point services. This may be in an urban situation – for example London, which is served by a number of different airports – or in the regions, where catchment areas can be continually expanded or contracted depending on the nature of air services and surface links on offer at neighbouring airports. Tourist destination airports can also compete directly as often passengers on holiday may not have just one specific tourist destination in mind.

In summary the five forces framework for airline customers has shown that the threats of both new entrants and substitutes are comparatively low at airports. However, as regards the other three forces, it is impossible really to generalise, which is often a limitation of Porter's model. The supplier and customer power will vary – the latter being that much weaker if no alternative airport exists. As regards competitive rivalry, major airports serving a distinct catchment area with a wide network of services are also not likely to be subject to much competition unless they are competing as a hub. However, if the airports are physically close, the competitive rivalry will be more intense.

A different set of competitive forces exist as regards the commercial services which airports sell to passengers and other consumers. For example, if these are considered to be part of the general retail business, there are many retailers available to offer such products and many passengers to buy them. Moreover, airport operators are likely to be in a relatively strong position with their suppliers because of the attractiveness of their captive and often fairly affluent passenger market. However there are threats of substitutes, for example, from high street and internet shopping.

Thus a different picture emerges as regards the competitive forces affecting the provision of aeronautical and commercial services. However, these cannot be considered entirely independently since the services together contribute to the airport business and will both depend on the airlines offering flights at the airport as the passengers come to the airport to use the services of the airlines. It is the competitive strategies of the airlines which will primarily draw the passengers to the airport. At this final stage the competition between airports for the airline business will be hidden from the passengers as airport choice decisions will already have been made by the airlines.

Controllability of the Sources of Competitive Advantage

This competitive analysis can now be related to the amount of control or influence that an airport operator has over the two fundamental sources of competitive advantage, namely price and product characteristics. There are some areas where the airport operator has significant control and some where there is very little control. Meincke (2002) defined these as self-determinable and externally determined competition parameters.

Most Control

The area where an airport operator has the most control is in the pricing and provision of non-aeronautical services and facilities in the airport terminal and on the surrounding land. Hence most airports have become very experienced in taking advantage of the relative weak position of the suppliers and buyers and have exploited many of the non-aeronautical or commercial opportunities which exist. This has been achieved by increasing the number and mix of retail operators and caterers; introducing more branding and competition; expanding the space allocated to commercial facilities; and generally becoming more experienced in dealing with the growing level of expectations of the passengers and other airport customers.

Although these enhanced commercial facilities may well make the airport more attractive, they will play a very minor role in influencing the passenger's choice of airport. This is except in special cases, for example when duty-free products or prices are used to attract transfer passengers such as at some Middle Eastern airports. However non-aeronautical revenues will usually contribute significantly to an airport's overall financial well-being. If the airport operates under a single till regime a growth in commercial revenues may be compensated for by a reduction in, or pegging of, aeronautical charges – which may help the airport remain competitively priced. However, whilst single till regulation may be attractive to the airlines, it may not be in the best interest for the airport operator in the long term if it inhibits the pressures for diversification, innovation and business development which could potentially bring additional competitive advantage. Moreover, the inability of the single till to effectively allow pricing to allocate scarce resources may encourage congestion through low airport charges, which will ultimately make the airport less attractive for certain airline services.

Partial Control

The exact amount of control that airports have over the aeronautical aspects of operations is much more difficult to define. Not only do the forces of competition vary between different types of airports, but there are also numerous rules and regulations, which exist primarily for safety and security reasons, which airport operators need to follow.

As regards aeronautical charging, the amount of freedom that airports have in this area will depend on the extent of direct government control or the nature of economic regulation if it exists. Clearly this will be influenced by the real power of the airline. Moreover, the impact that pricing will have on demand, and hence ultimately on an airport's ability to compete, will also differ depending on what type of airline is being considered and its relative degree of price sensitivity. Airport charges can be substantially more important for short-haul operations as they are levied more frequency. For LCCs these can be even more significant because these airlines will have minimised many of the other airline costs. For these airlines, the availability of discounts on airport charges and financial help with marketing – particularly at the initial, most volatile stages of operation – can be very significant in influencing the airline's choice of airport.

An airline will consider all the costs at an airport – not just the airport charges, but also other charges, for example, for fuel and handling. The airport operator can also indirectly influence these costs through its choice of firm or organisation to provide such services, and also the amount of the competition it allows. It may even choose to undertaken some of the activities itself.

Least Control

Finally there are some aspects of the airport product which the airport operator has very little or no control over. This includes slot availability and also location, which is undoubtedly one of the most important factors which affects an airport's competitive position – for example, see Park (2003) for evidence in Asia. The factors which will determine the attractiveness of the location include the size of the population and its propensity to fly, the economic strength of the area and its importance as a tourist or business destination (Favotto, 1998). This means that the competitive position of specific hub airports which are developed primarily for transfer traffic, without a natural local market due to economic factors or tourism, may be fairly weak. The same may be the case of certain remote secondary airports which have been chosen for their operations by LCCs.

Although an airport cannot alter its location, its catchment area will not be fixed and will vary, for example, depending on whether short-haul or long-haul services are being considered. For short-haul travel to popular destinations there may be considerable competition from other airports and so the catchment area will probably be comparatively small. For less popular or longer-distance destinations there is likely to be less competition and so the catchment area will be extended over a greater area. The catchment area will also change depending on the air services being offered at the airport and its neighbours, and also on the relative nature and quality of the surface access links. LCCs have been particularly successful in attracting passengers from much wider catchment areas than other airlines (Barrett, 2000). For example, Dennis (2004) describes how Stansted, which effectively operated as a regional airport for East Anglia, has been transformed into a major airport serving the London area. Another example cited is Charleroi, where only 16 per cent of Ryanair passengers resident at the Charleroi end of services come from the natural catchment area of the airport. However Dennis also explains how catchment areas can contract as the provision of low-cost services increases – giving the example of the low-cost services to Barcelona from East

Midlands which began in 2002 but two years later had competing services from the nearby airports of Leeds/Bradford, Birmingham and Manchester.

Airport Strategic Options

Having assessed the influence that airport operators can have over the price and product features, it is now possible to consider how this relates to the actual competitive strategies that airports have developed. A starting point is to consider the relevance of Porter's generic competitive strategies. These acknowledge that price (or at least cost) and product characteristics or differentiation are the only two sources of competitive advantage, but also that the competitive scope of target customers can vary. This means that it is possible to provide products with a broad target which will appeal to most of the market or to pick a narrow target and focus on a niche within the market. This led Porter to define his key generic competitive strategies as being cost leadership, differentiation and focus. These ideas have come under increased criticism because of their simplistic nature and logic, particular in service industries, but nevertheless provide a useful starting point for considering strategy options.

Cost Leadership

The cost leadership strategy aims to place the organisation amongst the lowest cost producers. This is realised by reducing costs, such as low cost inputs, low distribution and location costs, by offering a standardised product, and by achieving high volume of sales and economies of scale. This will then enable the organisation to offer lower prices than its competitors. There is little evidence of such strategies within the airport industry. On the cost side, the ability of the airport to achieve many cost savings is fairly limited because of its fixed location, a large proportion of unavoidable costs being related to compliance with safety and security legislation, and because of the apparent lack of economies of scale beyond a certain size. More generally, the relevance of such a strategy to airports has to be questioned given the relative price insensitivity of many of the markets and thus the lack of competitive pressures to produce a reduction in costs. Moreover, the issue is complicated by the weak relationship between airport costs and prices at some airports: for example when public sector owners subsidise airport operations to achieve some broader objective such as economic development, or when an airport is operated as part of a group with uniform prices across the group which do not link very closely to the costs of the individual airports.

Differentiation

The next generic strategic option is differentiation – when an organisation will develop a product or products which are perceived as being different or unique from its competitors. This is more appropriate for a more price-insensitive market. This may be achieved through enhanced service features, brand image, promotion, technology, distribution or other dimensions. In other words it can be achieved if there are real (by product design) or perceived (by advertising) differences between its products and those if its competitors. Within the airport industry, there appears to be some scope to pursue differentiation strategies. For example, there could be the development of a new airport in a city centre position, with its 'uniqueness' being its proximity to the urban population. Alternatively the uniqueness could be the airport's design, which enables it to handle transfer passengers in a very short period of time; or its lack of environment restrictions compared to neighbouring airports, which enables it to operate for 24 hours. Interestingly Zurich airport actually rebranded itself as 'Unique Zurich Airport' in 2002 to reflect a new management structure, partial privatisation and expanded facilities, but does not really seem to follow a true differentiation strategy (and has now dropped this unique brand).

There may be a number of differentiated products to suit the needs of different customer groups. The 'fast-track' concept – when first-class and business passengers are given preferential treatment through various airport processes – is a good example which is used at BAA London airports and others. However, many of the business-class products such as fast check-in, airline lounges and other facilities are actually provided by the airline rather than the airport operator. As regards terminals, typically larger airports with more than one terminal have separate services for different types of passengers or airlines, for example, short-haul and long-haul, domestic and international, or by alliance members. However the quality standards do not usually vary significantly between terminals. There is also the option of having competing terminal products offered by different operators. In the USA, for example at JFK, the terminals are operated by different airlines; but practice elsewhere has been very limited, with the notable exceptions of Birmingham and Toronto. However in recent years there has been growing interest in increasing the amount of market segmentation which takes place. At one extreme this has led to the development of low-cost terminals which are discussed later in the chapter. At the other extreme there has also been the development of dedicated terminals for premium passengers, for example, at Frankfurt and Doha airports (Sobie, 2007).

Focus or Niche

The third generic strategy, in addition to cost leadership and differentiation, is the focus or niche strategy which is built around satisfying a particular small target market. With airports, this could be with a particular type of airline (for example, charter, low-cost, freight) or services to a particular geographic area. This is suitable for organisations which are not large enough to target the whole market. These strategies can either have a cost or differentiation focus. Within the airport industry, secondary airports which offer price deals to LCCs are examples of organisations pursuing a cost focus strategy which seeks a cost advantage in its target segment. There are also examples of airports that are following a differentiation strategy, for instance by providing specialist cargo facilities such as at Liège Airport in Belgium, or facilities for short-haul business such as at London City Airport.

Generally it is worth noting that since there are a number of airport operators that manage more than one airport, strategic options need to be considered at an overall operator level as well as at an individual airport level. For example at an operator level the Schiphol Group aims to differentiate itself somewhat by developing the concept of 'AirportCities', which it defines as a 24-hour dynamic hub for people and businesses, logistics and shops, information and entertainment. By contrast the PlaneStation Group followed a niche strategy in attempting to develop a global network of regional airports which previously had disused or underutilised facilities, and which would be particularly attractive to low-cost and dedicated all-freighter airlines. However, this airport operator went out of business in 2005 having invested in one of its key customers, EUjet, which suffered financial failure.

In summary whilst there are a number of airports or airport groups which can fall into differentiation or focus strategy, there are many airports which have too high a cost base to be considered a cost leader; too standardised a product to compete against 'differentiation airports'; and too broad an appeal to be considered as a 'niche airport'. Porter (1980) defined such organisations that do not conform to one of the strategies as 'Lost-in-the-Middle'. In reality this is very much related to the less competitive environment within which these types of airports operate.

Airport Strategic Directions and Methods

Strategic direction relates to which products and services should be developed and for which markets. The main directions are market penetration, market development, product development and diversification. Market penetration involves increasing market share of existing products in existing markets whilst market development entails introducing existing products into new market areas. Product development involves developing existing or new products for existing markets and diversification is concerned with developing new products for new markets. These directions can be summarised in the well known Ansoff positioning matrix which distinguishes between the degree of market and/or product development involved (Ansoff, 1987). At the one extreme, market penetration can be criticised for ignoring new opportunities whilst at the other extreme there may be many opportunities for diversification but at least some of these may come with a high degree of risk.

Internal Growth

For all these directions there are different methods by which a particular strategy may be developed. Firstly there is internal or organic growth. This is the approach that airports traditionally adopted since they really had no option because of public sector ownership and the strict regulatory air transport environment. Market penetration can be achieved internally with a typical example being a regional airport offering discounts on new regional services to encourage the use of the airport. For the passenger market there is the example of the development of loyalty cards (Jarach, 2005). Market penetration directions may be particularly relevant for airports following a niche strategy.

A product development strategy could be encouragement of long-haul services to an airport that has previously only offered short-haul services. Vienna airport is a useful example of an airport which is using financial incentives to encourage services to Eastern Europe and support its role as a West–East hub (Vienna Airport, 2004). Likewise Dublin airport in 2005 began offering financial help and marketing support for services specifically to non-EU destinations. Product development can also occur internally with the non-aeronautical side of the business. For instance many airports, including Changi Singapore and London Heathrow, have developed internet booking of various commercial services on offer, such as car parking or foreign currency. New product development strategies also had to be devised in Europe after the abolition of duty and tax free sales in 1999 with the introduction, for example, of the Travel Value concept (Freathy and O'Connell, 2000). For airports following differentiation strategies product development is necessary to maintain differences as competitors imitate previous innovations.

Market development can be achieved internally by encouraging better surface access to extent the catchment area. This could include supporting local road and rail improvements through the planning process or perhaps providing financial support to bus or rail surfaces. It could also be argued that the practice of certain airport companies such as BAA, Schiphol, Fraport, Aéroport de Paris and the Dublin Airport Authority in providing consultancy and management contracts in specialist areas such as engineering, construction, handling or commercial facilities is also an internal way of achieving market development.

The most common way that airports have internally followed related or concentric diversification strategies is through developing shopping, leisure or entertainment facilities for local residents and/or business facilities or services (such as conference facilities, offices and warehousing) for local businesses. Airports can then not only be considered as modal interfaces but attractions in their own right – Jarach (2001) calls such airports 'multipoint service-provider firms'. Schiphol is a good example of this with its AirportCity concept. BAA's operation of the Heathrow Express rail link could also be considered to be related diversification. Unrelated or conglomerate diversification is possible to achieve internally but is fairly rare because of the narrow role which public sector owners usually define for their airport operators. Dublin Airport Authority's ownership of the Great Southern Group of Hotels in Ireland (this was actually sold off in 2006) was an example of an unrelated or conglomerate diversification strategy which was developed internally.

Integration

Airport privatisation has enabled other strategic methods to be used. This is partly because it has made it possible for airports to buy other airports and also because it has given private airports more freedom to pursue other methods of strategy development. Horizontal integration, which occurs when organisations combine with other organisations in the same industry, is an important strategic development method which is used in other industries. Within the airport industry there is now the situation where established airport operators such as Fraport or Schiphol or new airport operators such as Hochtief, Ferrovial or TBI (acquired by Abertis in 2005) are now operating a number of airports (Koch and Budde, 2005; Graham, 2008). Such developments can bring about market development in that they involve introducing existing products (that is, the established commercial product) into new market areas which could include new geographical areas but may also allow for product development as new products will be acquired.

For many industries, including travel-related businesses such as hotels, travel agents and tour operators, clearly a key motive for horizontal integration is to create market power and brand strength and hence to reduce competition. This is not really the case with the airport sector. Indeed the benefits of horizontal expansion are less well defined but are likely to be more associated with knowledge transfer, risk spreading and perhaps some limited cost synergies through, for example, joint purchasing, training and insurance. Strategies which might be attractive to airlines such as standard contracts for the whole airport network, quantity discounts on charging or common agreements on the use of gates and other facilities have not been introduced. Neither are the airports usually marketed under a common brand.

There are a few situations where horizontal integration has enabled airports to achieve market penetration by taking control of neighbouring airports. For instance this is the case with Manchester, which has acquired both East Midlands and Humberside airports in the north of England. Fraport buying Hahn Airport (although now sold) and Schiphol likewise owning some of the small regional Dutch airports caused a similar effect. In this case the competitive advantages of horizontal integration will be much clearer in that such a strategy is reducing the number of competitors which exist.

This leads on to the whole issue of whether different airports in relative close proximity or in the same geographical region should be operated as individual entities or as a group or system. Arguments favouring group ownership and operations include economies of scale and enhanced career opportunities for employees, a stronger financial structure which can support the investment peaks and troughs at different airports and a more consistent strategic planning and investment policy. On the other hand, supporters of individual operations claim that separation would produce more competition and choice and superior local management. These issues were debated at length in Ireland in 2004 when it was decided that Aer Rianta (now Dublin Airport Authority) should be split into three more autonomous organisations. Similar discussions occur whenever an airport group is being considered for privatisation and the outcomes have differed. For example the BAA airports were privatised as a group whilst the major Australian airports were privatised individually. However, with the BAA case there have been a number of subsequent reviews post-privatisation as to whether the group should be split up and in 2009 the Competition Commission in the UK ordered that BAA should sell some of its airports (Competition Commission, 2009).

Then there is vertical integration, either forward integration when organisations seek increased control of distributors, or backward integration when organisations seek control of suppliers. This is not a common strategic method which the airport industry has decided to follow. However, in the UK there are examples of two regional airports, namely Cardiff and Norwich, which developed local travel agency business in an attempt to increase awareness and thus flight bookings through the airport. This forward integration can cause market development from a passenger perspective since new passengers may be persuaded to fly from the airport.

It could be argued that backward integration has always really existed within the airport industry by nature of the fact that some airport operators themselves will choose to supply part of the composite airport product, for example with handling or air traffic control services. However, there are very few examples of airport operators who have consciously decided to grow their company by buying into a vertical integrated organisation rather than just historically maintaining an interest in an activity which has always been provided by the airport company. BAA's development of World Duty Free (recently sold in 2008) was one such example.

Within the context of vertical integration, some airports have decided that it is favourable to adopt a strategy of developing much stronger links and partnerships with the airline – which is in effect 'the supplier' for the passenger product. This is common practice with LCC operations and also in the USA where long-term airport–airline contracts have been the norm for many years. Elsewhere airlines have usually not entered into long term pricing agreements with airports and instead have just tended to pay charges based on a published tariff with associated conditions of use. However there are now examples of airline–airport agreements in Copenhagen, some of the major Australian airports and Frankfurt, where there are now examples of other airline-airport agreements, for instance at Copenhagen airport (Copenhagen airport, 2009). Such an approach is seen by some to be a more attractive and appealing option than the often confrontational and defensive approach which tends to be adopted when there is a more formal regulation process.

A few airports have even gone one stage further than this with their relationship with their airline customers. For example in the UK, Norwich airport chartered its own aircraft to prove to tour operators that there was a demand for charter flights. In 2004 the airport group PlaneStation acquired the new LCC EUJet primarily to secure its existence at PlaneStation's Manston Airport but, as discussed in 2005, the airline, and subsequently the airport group, collapsed. Interestingly, in a somewhat reverse situation, the travel company TUI in 2004 acquired (for a short while) the operating lease for Coventry or West Midlands Airport, which was a base for Thomsonfly. There are not many other examples outside the US of airline involvement in airport investment and operations except British Airways and the Birmingham Airport Eurohub and, more recently, Lufthansa's involvement with the financing of the second terminal at Munich and its acquisition of a 10 per cent share of Fraport.

Alliances and Franchises

There are other methods of strategic development which provide some of the benefits of collaboration without the transfer of ownership or operational control. This includes alliances and franchises. Airports quite often informally get together, for example for information exchange or marketing support, but the Pantares alliance between Schiphol and Fraport was a rare example of a more official alliance. Unlike airline or other alliances, airport alliances cannot be driven by network effects, or with the aim of increasing market accessibility, but instead may reap benefits from shared knowledge, expertise and financial resources and joint bidding for international projects. However the far-reaching expected benefits for Pantares, which was established in 1999, were largely not realised. It was also felt that the alliance would reduce some of the competitive pressures which existed when the two airports were involved in the bidding process for privatised airports by getting these two former rivals to co-operate in this area; but again these effects were very limited. Schiphol now has an alliance with Aeroport de Paris but it is too early to access the impacts.

Franchising, which is used in other industries to gain the competitive advantage through benefiting from rapid market development of a well-known brand without the need for investment, is not a practice which is used in the airport industry. This development method is only relevant if it could be proved that being branded as part of a successful airport group can substantially improve the market potential and image of a smaller airport. This does not currently seem to be the case.

Retrenchment and Divesture

Since the airport business is still generally a growth industry which has yet to reach maturity, all of the strategic directions and methods which have been discussed so far are associated with growth. However clearly there are some instances when competitive advantage may well be maintained only if retrenchment strategies are adopted. Zurich and Brussels are two airports which had to follow retrenchment strategies following the collapse of their two major airlines, namely Swissair and Sabena.

Then there are divesture strategies, which again are not very common. A relevant example is BAA, which, having been privatised in 1987, had considerable freedom to diversify in many unrelated areas. This included developing hotels, investing in property and running designer outlet centres. However the airport company subsequently decided that it had taken its diversification strategies too far, sold off most of its interests in these new areas and sought to focus back on what BAA views as its core business. More recently it has also sold off most of its external interests in other airports and other companies – primarily this time to bring in much needed cash to cover the large debts of the company since it was taken over by Ferrovial in 2006. Other examples include TBI/Abertis, which sold its interests in its airport services division that provided services such as refuelling and baggage handling to concentrate on its core airport management business; and Dublin Airport Authority, which sold off its hotel group.

Competitive Strategies for LCC Customers

The rapid growth of LCCs has created new challenges for airports in the area of competitive advantage. In terms of the forces of competition, for the smaller regional and secondary airports which are highly dependent on this traffic, these airlines can have very strong bargaining power and a number of existing rival airports for the products very often exist. Such airports will be following a niche strategy – both in terms of offering low prices through deals negotiated with the LCCs and maybe also in terms of the simplified and uncongested product on offer. For these airports it makes much sense to attract this type of traffic as it will often use the spare capacity at under-utilised existing infrastructure and will probably be the only type of traffic which will have a considerable growth opportunity – at least in the initial years until perhaps some kind of maturity sets in. It may well act as a catalyst for additional air services and encourage economic development and tourism within the surrounding area, which will be of particular interest to publicly owned airports. However such a strategy will run into problems if demand grows to such a level that new facilities are needed but the LCCs are reluctant to pay for them.

At medium-sized airports which serve other conventional airlines as well, the LCCs' power is less as these airlines will make up a smaller share of the total traffic. Many of these airports have encouraged the development of low-cost traffic to supplement their other more traditional traffic base. Dublin and Cologne are both medium-sized airports which have benefited from the rapid expansion of low-cost services. However, there is a danger that the LCCs will instead act as substitutes for the conventional airlines. This will probably not be financially beneficial to the airport operator and may cause a loss in choice of hub links. Dennis (2004) cites the example of Belfast International Airport, where local accessibility has improved in terms of price but global accessibility very much deteriorated because of the shift from conventional to low-cost services.

There are other issues to consider for medium-size airports which are serving both conventional and LCCs. As discussed earlier the traditional approach of airports used to be to offer a standardised product for all their airline customers. However, the LCCs have very different needs from the other carriers, including quick turnaround time, quick check-in, fast handling services and no airline lounges (Barrett, 2004). This has caused a small number of airports to develop dedicated low-cost terminals, either from old passenger or cargo terminals as at Marseilles and Budapest, or from new, such as at Singapore and Kuala Lumpur (Falconer, 2006; Jacobs Consultancy, 2007). Whilst experience of these terminals is still very limited, they do raise some important issues. In particularly there is the question of how they can be funded and operated in a way that the main network carriers do not think is discriminatory; and also consideration has to be given as to whether they have a negative impact on commercial revenues, since such terminals can only provide a more limited range of retail outlets and perhaps not offer the right atmosphere and experience to encourage travellers to shop. For this latter reason some airports such as Glasgow and Amsterdam have provided some facilities for low-cost carriers, but passengers still have access to all the usual commercial facilities.

Another option is to have a competing terminal. Within this context, Ryanair has lobbied for a separate terminal for its operations in Dublin for many years. In 2005, the Irish government announced its approval for a second terminal, but it is now being built by the Dublin Airport Authority (McLay and Reynolds-Feighan, 2006). Finally for airport groups there is another option for coping with low-cost traffic – namely by directing the traffic to one specific airport, as was the case with Hahn in Frankfurt and Ciampino in Rome.

Conclusion

It can be concluded that the competing forces at an airport vary depending on whether an airline or passenger viewpoint is adopted, and these cannot be considered entirely separately because of the interdependence of airline and passenger demand. The non-aeronautical product for the passengers is the easiest to influence but even apparently fixed features of the airport product, such as location, can be affected by efforts to extend the catchment area.

A number of strategies which an airport can adopt to gain competitive advantage have been considered. However, it does appear that many airports have a relatively strong position in their core market and the ways in which they compete is fairly marginal unless niche markets are being considered. Certainly there is very little experience of the head-to-head competition which has been seen with many other industries, leading to substantial extra business being generated by the lowering of core prices or by significantly improving quality. Airport privatisation has allowed for options such as horizontal and vertical integration to be adopted, although the practical benefits of such developments specifically for the airport industry do not seem very clear.

It has been demonstrated that there are some characteristics of the airport product, for example, the existence of more than one key customer and the nature of composite product, which make the task of formulating competitive strategies that much more complex and difficult. However, there are many airports which still exist in a fairly uncompetitive environment and for whom the relevance of the whole concept of competitive advantage at this stage in time must be questioned. Hence, in general, the need to have extensive and effective competitive strategies within the airport industry may very well be less than in other sectors. Nevertheless there are now a significant number of airports that are now operating in a much more competitive manner and for whom knowledge of competitive strategy in other industries could provide additional insight for achieving that vital competitive edge.

References

ACI-Europe (1999), *Policy Paper on Airport Competition*, Brussels, ACI-Europe.
Ansoff, I. (1987), *Corporate Strategy*, revised edition, Harmondsworth, Penguin.
Barrett, S. (2000), Airport competition in the deregulated European aviation market, *Journal of Air Transport Management*, vol. 6, 13–27.
Barrett, S. (2004), How do the demands for airport services differ between full-service and low-cost carriers? *Journal of Air Transport Management*, vol. 10, 33–9.
Competition Commission (2009) BAA airports market investigation final report, London, Competition Commission.
Copenhagen Airport (2009) New charges agreement for Copenhagen Airport, press release, 14 September.
Cranfield University (2002), *Study on Competition Between Airports and the Application of State Aid Rules*, Cranfield, Cranfield University.
Dennis, N. (2004), Can the European low cost airline boom continue? Implications for regional airports, 44th European Congress of the Regional Science Association, Porto, August.
Favotto, I. (1998), Not all airports are created equal, *Airport World*, December, 2–3.
Forsyth, P. (2006), Airport competition: Regulatory issues and policy implications, in Lee, D. (ed.), *Competition Policy and Antitrust*, Amsterdam, Elsevier.
Freathy, P. and O'Connell, F. (2000), Strategic Reactions to the Abolition of Duty Free, *European Management Journal*, vol. 18, no. 6, 638–45.
Graham, A. (2008), *Managing Airports: an International Perspective*, third edition, Oxford, Elsevier.
Jarach, D. (2001), The evolution of airport management practices: Towards a multi-point, multi-service, marketing-driven firm, *Journal of Air Transport Management*, vol. 7, 119–25.

Jarach, D. (2005), *Airport Marketing*, Aldershot, Ashgate.

Falconer, R. (2006), The low-cost challenge for airports, *Communiqué Airport Business*, June–July 21–2.

Jacobs Consultancy (2007), Review of dedicated low-cost airport passenger facilities, Dublin, Commission for Aviation Regulation.

Koch, B. and Budde, S. (2005), Internationalization strategies for airport companies, in Delfmann, W., Baum, H., Auerbach, S. and Albers, S. (eds), *Strategic Management in the Aviation Industry*, Aldershot, Ashgate.

McLay, P. and Reynolds-Feighan, A. (2006), Competition between airport terminals: The issues facing Dublin airport, *Transportation Research Part A*, vol. 40, 181–203.

Meincke, P. (2002), Competition of airports in Europe – Parameters and types of competitive situations among airports, *Air Transport Research Society Conference*, Seattle, July.

Park, Y. (2003), An analysis for the competitive strength of Asian major airports, *Journal of Air Transport Management*, vol. 9, 353–60.

Pels, E., Nijkamp, P. and Rietveld, P. (2003), Inefficiencies and scale economies of European airport operations, *Transportation Research Part E*, vol. 39, 341–61.

Porter, M. (1980), *Competitive Strategy*, New York, Free Press.

Sobie, B. (2007), Stress free, *Airline Business*, December, 47.

Starkie, D. (2002), Airport regulation and competition, *Journal of Air Transport Management*, vol. 8, 37–48.

Vienna Airport (2004), Tariff reform brings major cost advantages for airlines, press release, 30 September.

Chapter 8
An Empirical Analysis of Airport Operational Costs

Eric Pels, Daniel van Vuuren, Charles Ng and Piet Rietveld[1]

1. Introduction

Following the deregulation of the US aviation markets in 1978, a number of studies on airline cost functions emerged (for example, Caves et al., 1984; Gillen et al., 1990; Brueckner and Spiller, 1994; and Baltagi et al., 1995). These studies all concluded that airlines operate under economies of density, and thus supported the theoretical result that density economies are an important driver for the success of hub-and-spoke networks.

After the airline markets were deregulated, focus turned to airports. While there are a relatively large number of empirical cost function studies, for example in the airline sector, and especially the rail sector, the number of airport cost function studies is quite limited. Notable exceptions are Tolofari et al. (1990) and Morrison (1983). Tolofari et al. (1990) estimated a translog cost function for London airports and concluded that London Heathrow operates under decreasing returns to scale (short and long run), and that smaller London airports operate under increasing returns to scale. Workload units were used as the output variable. Morrison (1983) concluded that airports in the US operate under constant returns to scale; the number of air carrier operations was used as the output variable.[2]

In a number of countries airports are corporatized or privatized. The literature that followed (or in some cases preceded) this trend mainly focused on airport efficiency. Gillen and Lall (1997) was one of the first studies to use data envelopment analysis to measure airport inefficiency. Performance benchmarking has, in recent years, become an increasingly important part of the toolkit for regulators (especially in utility industries) in setting price caps. A number of studies in this field were published (for example, Oum et al., 2003). Many of these inefficiency studies use deterministic non-parametric techniques, such as data envelopment analysis. Data envelopment analysis was originally devised to measure the efficiency of firms (or 'decision-making units') with a non-profit objective. It also allows the researcher to determine whether the 'decision-making unit' operates under constant, increasing or decreasing returns to scale. A drawback of this non-parametric linear programming approach is that statistical tests are not readily available. For instance, earlier studies have concluded that the 'average' airport operates under increasing returns to scale when 'producing' passengers (Pels et al., 2003). Although scale efficiency for individual airports can be imputed from this method, it is difficult to conduct statistical tests of hypotheses concerning the inefficiency and the structure of the underlying production or cost function. On the other hand, in a parametric cost function approach, like that of Tolofari et al. (1990), statistical

---

[1] The authors thank participants of the 7th Nectar Euroconference in Umea, Sweden, June 13–15, 2003 and the 9th Air Transport Research Society World Conference in Rio de Janeiro for helpful comments. The views expressed in this chapter are those of the authors, and do not represent the views of the Civil Aviation Authority (CAA) or the Netherlands Bureau for Economic Policy Analysis (CPB).

[2] This conclusion is based on 38 observations (a survey was sent to the 100 busiest airports).

evaluation is much easier. This chapter therefore uses a statistical approach to test whether the scale elasticity of the operational costs of airports differs significantly from unity.

When airports are under price cap or other types of economic regulation (empirical) knowledge of returns to scale is a useful piece of information for the regulator. For example, under an RPI-X price cap incentive regulation, the X-factor is based on the regulator's assessment of the potential productivity growth of the regulated firm. Thus, the regulator (as well as the regulated firm) will be interested not only in the technical efficiency (using minimum inputs to produce given outputs), but also in scale efficiency (potential productivity gain from achieving optimal firm size). It is sometimes mentioned that airports are natural monopolies because they operate under increasing returns to scale, although in theory this is not a sufficient condition.[3] Moreover, airport charges above marginal costs do not imply market power abuse when an airport operates under increasing returns to scale, but may be seen as 'second-best' prices. Likewise, high (variable) costs per unit of output do not necessarily imply that an airport is relatively technically inefficient, although it may be scale inefficient.

The chapter is organized as follows. Section 2 discusses theoretical aspects of cost functions. The longitudinal data set we use in our empirical analysis is described in Section 3, and Section 4 presents estimation results for the relationship between operational cost and various output and control variables. Section 5 concludes.

2. Background

This section discusses some theoretical and practical aspects of the estimation of cost functions (cost frontiers).

The Cost Function

In economic theory, the cost function is derived by minimizing costs conditional on an output target; the cost function tells us the lowest possible cost level of producing a given output bundle. Cost function estimation thus has an implicit underlying behavioural assumption: the airport operator minimizes costs.

As was mentioned in the introduction, it is often assumed that airports operate under an L-shaped cost curve which envelops the short run cost curves, as shown in Figure 8.1. The long run average cost curve envelops short run average cost curves; each short run curve is associated with a different level of fixed assets (capacity). An airport operating on short run average cost curve 1 will minimize *short run* costs at output level q_1. If output is expanded beyond this level, average costs will increase. Beyond the point of intersection between short run average cost curves 1 and 2 the average costs are lower when capacity is expanded, so that average cost curve 2 becomes relevant. There are, however, two difficulties in estimating the long run average cost curve. Firstly, airport capacity is quite 'lumpy' in nature (Niemeier, 2004). The construction of a new runway does not lead to a 'smooth' increase in capacity but may, in some cases, double capacity. This may lead to

---

3   A firm that faces strictly decreasing marginal costs also faces a sub-additive cost function (Tirole, 1992). Sub-additivity thus implies that the aggregate profits of multiple firms are always lower than the profits of a single firm (given the price). Note that this a stronger condition than increasing returns to scale, because a U-shaped cost curve may exhibit increasing returns while not being sub-additive. A firm is a natural monopoly when the profits obtained by a single firm in the market exceed zero, while the profits for multiple firms operating in the same market are negative.

discontinuities in the cost curve. Secondly, the long run average cost curve is very hard to observe in reality. To calibrate the long run cost curve, it is necessary that all inputs (including runways) have reached equilibrium values. This is unlikely to be the case. Both problems can be avoided by estimating a short run (variable) cost function in which certain inputs are considered to be quasi-fixed.

The theory behind Figure 8.1 assumes that firms (airports) have a cost-minimizing objective. The use of (variable) production factors (for example, labour) is determined by the relative input prices: given the output target, the factor prices determine the actual usage of the factors, and thus the total operational cost. Unfortunately, comparable input prices are difficult to obtain. Average wages may be derived by dividing the total expenditures on labour by the number of employees. However outsourcing, which allows airports to turn fixed into variable costs, may give a distorted view. Other input prices can be determined in a similar fashion; for instance, Tolofari et al. (1990) determine the price of 'equipment' by dividing the expenditures on equipment by the net value of airport property. The difficulty here is that such data are often difficult to obtain for an international panel of airports. Even if they were available, comparability would be a difficult issue as different countries may have different accounting principles and depreciation methods.

When information on input prices is not included in the estimations, one simply estimates a statistical relationship between cost and output levels (without assuming a cost-minimizing strategy by the airports).[4] Using panel data, the estimated fixed effects provide an indication of the relative performance of individual airports. It is, however, noted that these fixed effects may also contain other unobserved time-invariant characteristics. A second note is that the relative performance of airports may change over time, which cannot be captured by the fixed effects.

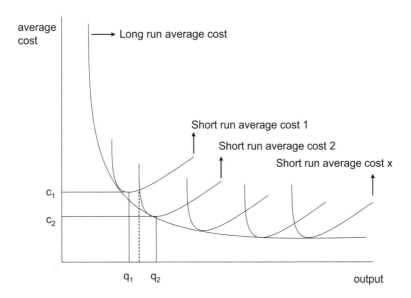

Figure 8.1   Long and short run cost curves

---

4   Such a model specification, in our view, is not unreasonable given the monopolistic elements of most airports. This has caused regulators to implement regulatory regimes that aim to provide incentives to airport operators who invest and operate efficiently and avoid unnecessary expenses.

3. Data

The data used in this study were collected by National Economic Research Associates (NERA, 2001), Transport Research Laboratory (TRL, various years) and by our own airport surveys. Similar questionnaires were used in order to make sure our data were compatible with the data collected by NERA (2001), both across airports and over time. We now turn to a discussion of the variables in our data set. Sample averages are reported in Table 8.1.

Operational Costs

Data on the operational costs (including staff costs) of airports were corrected by TRL for non-core activities, in particular car parking, utility sales, ground handling, consultancy and surface transport. This correction was done on a case-by-case basis, and is reported in more detail in NERA (2001) and NERA/TRL (2001). The operational costs are measured in 1993 $US. Table 8.1 shows a large spread in the operational costs faced by the airports in our sample. For instance, Heathrow – the largest airport in our data set in terms of total number of passengers – has an operational cost level about ten times higher than that of Berlin. When we consider this data in relative terms, the ratio between the operational costs and the total number of passengers appears to be more favourable for Berlin than for Heathrow. However, controlling for other variables that have an impact on operating costs could lead to very different insights on the relative efficiency of these airports. In this particular example, the fraction (proportion) of international passengers seems to play a crucial role. On the other hand, Munich and Miami airports appear to have relatively high cost levels, given their levels of passenger throughput and mix of passengers. This highlights the need for proper control of the various cost drivers and environmental factors when comparing the relative efficiency of airports. This is the focus of the next section.

Passengers

The number of passengers is available from Airports Council International (ACI), and represents the total number of passengers handled by the airports (international terminal passengers, domestic terminal passengers and transit and transfer passengers).

Fraction of Passengers to/from International Destinations

This variable has simply been computed as the number of passengers to/from international destinations divided by the total number of passengers. It is often mentioned that passengers travelling to/from international destinations are more expensive to handle because they spend more time at the airport and require more costly facilities (for example, security checks, passport controls and so on), although they may also generate higher revenue per passenger for the airports. In cases where we encountered missing values for this variable we have imputed values equalling the closest (in time) observation in our data set; if the lacking year was as close to two different observations (that is, in between two 'observed years') then the average of these observations was imputed. It is unlikely that this will affect our estimation results, as the fraction of passengers travelling to/from international destinations is a relatively stable variable over time (for a given airport). However, there is considerable variation between airports for this variable, as shown in Table 8.1. Airports serving the domestic market are typically found in larger countries such as Australia and

Table 8.1 Average values of four key variables, 1993–2000†

	Operational costs (x 1 million $US 1993)	Total passengers (x 1 million)	Fraction of passengers to international destinations (% total)	Fixed assets (x 1 million $US 1993)
Adelaide	10	4	6	n.o.
Brisbane	26	10	23	n.o.
Melbourne	32	13	18	n.o.
Perth	15	5	31	n.o.
Sydney Kingsford	68	21	35	n.o.
Vienna Int'l	92	9	n.o.	419
Calgary Int'l	28	7	24	70
Vancouver Int'l	70	13	44	315
Copenhagen	70	15	84	407
Frankfurt/Main	375	39	81	1543
Hamburg	60	9		162
Munich	248	16	63	2021
Hong Kong*	108	26		3348
Amsterdam Schiphol	230	29	100	1431
Auckland	24	8	59	411
Christchurch	11	4	25	115
Singapore Changi	126	24	100	1911
Cape Town	8	4	n.o.	35
Johannesburg	22	10	n.o.	152
Honolulu	70	23	n.o.	1314
Los Angeles Int'l	207	61		1270
Miami	313	33	44	1634
Ontario	35	6		380
San Francisco Int'l	135	38	17	1864
Washington Dulles	62	15	21	1076
Washington Ronald Reagan	50	15	n.o.	606
Nice Cote d'Azur	37	7	40	216
Berlin Tegel	58	9	40	104
Oslo Gardermoen	69	14		900
Lisbon	39	7	74	375
Glasgow	43	6	51	286

Table 8. 1   *Concluded*

Stockholm	66	15	60	263
London Gatwick	251	25	91	1371
London Heathrow	568	56	87	3708
Manchester	146	15	85	822
London Stansted	71	5	78	874

† Note that not all variables have been recorded in every year, so that some average values are computed over a sub-period of 1993–2000. The indication 'n.o.' applies to the case where a variable is not observed at all during this time period.

* Note that a new airport was opened in Hong Kong in 1999. The average values reported for this airport correspond to the period 1994–96.

the United States, while airports located in relatively small countries almost by definition serve the international market. Obvious examples of the latter are Singapore and Amsterdam Schiphol.

Fixed Assets

Estimation of a short run variable cost function necessitates the inclusion of quasi-fixed inputs; that is the inputs that are considered as fixed in the short run. If these quasi-fixed inputs were not included, then the estimates would be biased (Caves et al., 1981); the estimates are interpreted as if all inputs are at their long run equilibrium values, while they are not interpreted in this way in the short run. Tolofari et al. (1990) use the value of airport assets as an indicator of the capital stock. The problem is that most airports include financial assets in their fixed assets reports. We have tried to obtain data on intangible assets, operational fixed assets and investment properties, but unfortunately only a few airports have such data available. We can therefore only use the fixed assets as compiled by TRL. However, no reliable data could be obtained for the Australian and South African airports, so these airports are excluded from our subsequent analysis. Similar to the operational costs, fixed assets are measured in 1993 $US.

Service Quality

Service quality may be an important cost driver for airports. Schiphol Amsterdam (AMS), for example, constantly 'measures' its quality by means of passenger surveys, and invests accordingly to increase its attractiveness to passengers. Airlines also have an interest in high-quality service provided by an airport. According to readers of *Business Traveller*, the most important aspects of an airport are speed of check-in, baggage handling and lack of congestion. Airports that have low scores for these items are likely to incur additional costs on the airlines due to longer wait times (delays) and discourage passengers from using these airports. There is, however, no exact definition of airport quality. There are various ratings of airports based on passenger surveys. Such surveys include questions on passport control, luggage retrieval, customs clearance, safety, (duty-free) shopping, ambience etc. In addition to passenger satisfaction, airlines are also interested in the availability of gates and slots, ground handling facilities etc. Since the availability of slots and gates is a capacity issue, and capacity has already been (indirectly) included in the cost function estimation via the fixed assets variable, we therefore did not include other indicators of capacity. In order to assess the importance of airport quality, we use the 'IATA Global Airport Monitor', which contains the most complete listing of passenger satisfaction of airports. In the Airport Monitor, surveyed passengers are asked to rate different aspects of an airport on a scale of 1 to 10 (1 to 5 in 1999). Since the set of surveyed airports varies over time, it is difficult to give a complete overview of rankings. However, it is interesting to note that airport rankings, especially in Europe, are fairly constant over time. 'Passenger satisfaction' for 1994, 1995, 1996 and 1999 is available from various Monitors. Our initial analysis of the data for individual airports suggests that by using linear interpolation the missing data for 1997 and 1998 could be approximated.

Figure 8.2 contains a scatter plot for the logarithmic transformations of staff and operational costs and the total number of passengers served. Each dot corresponds to the time average value for a given airport. The straight line corresponds to an OLS regression performed on the set of time average values. Its estimated slope equals 1.19 with a standard error of 0.12, so that the average operational costs do not change significantly with output. In the following section we will present more substantiated evidence on scale elasticity, by making use of different observations over time for the various airports and controlling for observed and unobserved characteristics of the operational environment that are deemed to have an influence on airport operational costs. For example, Figure 8.3 shows the impact of international flights on cost levels. In the next section we will exploit the panel character of our data to better identify the scale elasticity of cost after controlling for the airport-specific effects.

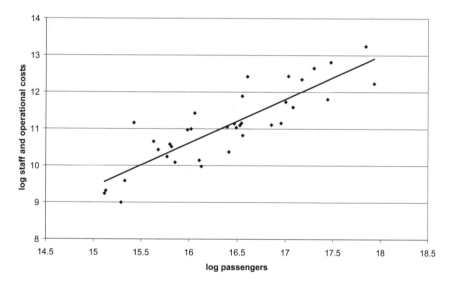

Figure 8.2 Cost as a function of the number of passengers (average values for each airport)

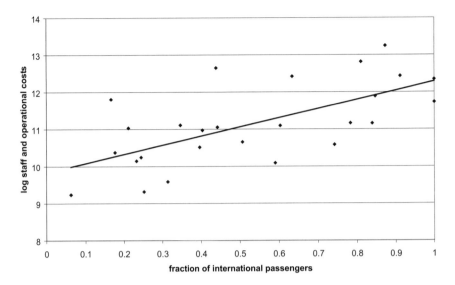

Figure 8.3 Cost as a function of the fraction of international passengers (average values for each airport)

4. Cost Function Estimation

Methodology

Indexing airport and time by subscripts i and t, respectively, and the explanatory variables by superscript j, we specify a statistical relationship of the translog form between cost and a set of explanatory variables:

$$\ln C_{it} = \alpha_0 + \sum_{j=1}^{k} \beta_j x_{it}^j + \sum_{j_1=1}^{k}\sum_{j_2=1}^{k} \gamma_{j_1 j_2} x_{it}^{j_1} x_{it}^{j_2} + u_i + \varepsilon_{it}, \quad (4.1)$$

with (the logarithmic transformations of) the explanatory variables standardized around their means:

$$x_{it}^j = \ln X_{it}^j - \mu^j, \quad (4.2)$$

with $\mu^j = \ln\left(\sum_{i,t} X_{it}^j / n_i\right) - \ln n_i$, i.e. the (logarithm of the) average of the time averaged values for X^j. Note that this is not equal to the 'overall' average for X^j (both across time and airports) because the number of observations in time differs across airports. The current definition of μ^j weighs each airport equally, irrespective of the number of observations across time, so that the cost function is approximated around an average that is not biased towards airports with many observations across time. The explanatory variables in x_{it} contain both outputs and control variables. Furthermore, the airport-specific effects u_i are estimated as fixed effects and the ε_{it} follows an independent and identical distribution. For symmetry reasons, the coefficients γ are constrained such that:

$$\gamma_{j_1 j_2} = \gamma_{j_2 j_1}. \quad (4.3)$$

Now it is easy to derive that within this specification the scale elasticity of cost with respect to output X^r equals (omitting individual and time subscripts):

$$\eta_r := \frac{\partial \ln C}{\partial \ln X^r} = \beta_r + 2\sum_{j=1}^{k} \gamma_{jr} x^j. \tag{4.4}$$

That is, for an airport with given explanatory variables $X^1,\ldots,X^k$ we can estimate the scale elasticity of cost with respect to output X^j as $\eta_r = \eta_r(x^1,\ldots,x^k) = \eta_r(X^1,\ldots,X^k)$. Standard errors for η_r can be easily computed with the Delta method (see, for example, Greene, 1993, p. 297). A familiar property of the translog specification in (4.1) is that evaluation of η_r at the vector of mean values of the explanatory variables (μ^j) gives an elasticity equal to β_r. Moreover, the linear specification of the elasticity implies that the average elasticity over all airports is equal to the elasticity computed at the vector with mean values of the explanatory variables.

Our model deviates from the translog cost model in Christensen and Greene (1976) in that we do not include factor prices, and hence do not have to impose the homogeneity conditions on the parameters involved.[5] Likewise, we do not simultaneously estimate the cost-minimizing share equations for different inputs. Ng and Seabright (2001) discuss several arguments for estimating a static cost specification rather than a dynamic one, and we follow their approach by not including one or more lagged dependent variables. Consequently, we identify the parameters of our model from the *within* dimension of the data.

Estimation

Estimation results for a number of selected specifications are displayed in Table 8.2. Each specification includes the total number of passengers and the number of air transport movements, and the related quadratic and interaction terms. Adding control variables to this specification will lead to an important loss of the degrees of freedom and the precision of the estimates. There are two reasons for this. First, not all control variables are observed in every time period for each airport; and second, every additional variable involves four to seven new parameters to be estimated as a result of the translog specification with its quadratic and interaction terms. It is illustrative to observe that the number of observations is almost halved if two additional variables are included (compare the third and fourth columns of Table 8.2). As a consequence, the addition of controls automatically implies that statistical tests lose much of their power as confidence intervals are widened.[6] The most robust finding is that the coefficient of the linear term of the total number of passengers is mostly around the value of 0.5 and significantly different from zero and one. We further find a positive coefficient for fixed assests. This may appear counter-intuitive, since it indicates that the marginal product of capacity is negative. It is, however, noted that this coefficient should not be interpreted as a causal effect, since our data did not allow us the opportunity to control for the utilization of these fixed assets. Oum and Zhang (1991) argue that variable costs decrease as fixed assets increase; but at some point, when there is large overcapacity, variable costs become independent of fixed assets. When there are firms in the sample with overcapacity,

---

5 To the extent that input prices influence operating costs, their exclusion from the model may lead to biased estimation when there exists correlation with other included explanatory variables. However, if input prices remain relatively stable over time for given airports, then this bias is filtered away by the fixed effects estimation.

6 Adding further control variables to an already 'over-fitted' translog model could also exacerbate the problem due to multicollinearity.

this may render the parameter for the fixed asset insignificant, or even cause it to have the wrong sign. In order to solve this problem, Oum and Zhang (1991) propose to control for capital stock utilization, and find that this changes the sign of the fixed factor from positive to negative in an analysis of airline costs. Since airport capacity is lumpy (see Niemeier, 2004), some airports will have excess capacity in the short run, and the current impracticability to include some measure for fixed assets utilitzation may well explain the positive parameter for fixed assets.

Table 8.2   Estimation results for a number of specifications of cost functions

Total number of passengers (PAX)	0.47*	0.43*	0.59*	0.09
	(0.17)	(0.18)	(0.19)	(0.53)
Air transport movements (ATM)	-0.04	-0.10	-0.16	0.23
	(0.19)	(0.19)	(0.19)	(0.51)
GDP growth (GDP)			-0.01	-0.01
			(0.01)	(0.02)
Fixed assets (FASS)		0.16*	0.17*	0.20
		(0.04)	(0.04)	(0.10)
Fraction international passengers (FINT)				2.29
				(2.20)
Quality of service (QUAL)				-0.45
				(0.24)
PAX*PAX	-0.56*	-0.12	-0.01	0.46
	(0.18)	(0.26)	(0.27)	(0.53)
ATM*ATM	-0.68*	-0.04	0.18	0.99
	(0.30)	(0.40)	(0.40)	(1.05)
GDP*GDP			0.01*	0.00
			(0.00)	(0.01)
FASS*FASS		0.02	0.00	0.01
		(0.04)	(0.03)	(0.06)
FINT*FINT				-7.05
				(4.26)
QUAL*QUAL				-0.04
				(0.22)
PAX*ATM	1.37*	0.34	0.09	-1.90
	(0.41)	(0.61)	(0.61)	(1.22)
GDP*PAX			-0.04	0.07
			(0.03)	(0.09)
GDP*ATM			0.01	-0.11
			(0.04)	(0.09)

Table 8.2   Concluded

FASS*PAX		-0.19	-0.16	-0.28
		(0.12)	(0.12)	(0.26)
FASS*ATM		0.05	0.00	0.42
		(0.11)	(0.12)	(0.33)
FASS*GDP			0.02	-0.03
			(0.01)	(0.03)
FINT*PAX				2.75
				(1.90)
FINT*ATM				-1.55
				(1.39)
FINT*GDP				-0.13
				(0.17)
FINT*FASS				0.26
				(0.33)
QUAL*PAX				-1.62
				(0.67)
QUAL*ATM				1.06
				(0.68)
QUAL*GDP				0.01
				(0.05)
QUAL*FASS				0.21
				(0.20)
QUAL*FINT				-0.79
				(1.90)
Observations		194	166	165	89
Degrees of freedom		189	157	152	62

Note: standard errors are reported in parentheses. The parameters concerning the interaction between variables indicate the complete effect. That is, for variables indexed by k and l, we report $2\gamma_{kl}$ (compare Equation 4.1).

* Significantly different from zero at 5 per cent confidence level.

Table 8.3 presents results on scale elasticities in more detail. It shows that the scale elasticity, with respect to the number of passengers, centres mostly around the value of 0.5. On the other hand, it appears that the scale elasticity with respect to the second output, the number of air transport movements, is never significantly different from zero. This suggests that the latter output is only of limited importance in explaining airport operational costs, whereas the number of passengers is a very important cost driver. The inclusion of the fraction of international passengers as a control variable brings down the point estimate of the elasticity of cost with respect to the number of passengers. This could be expected, as this variable is not a pure control variable but to some extent a cost driver itself. Likewise for airport quality, we find that the inclusion of the control variable concerned brings down the point estimate of the elasticity of operational cost in two out of three cases. Hence, the overall scale elasticities for the estimations with the fraction of international passengers and/or quality included may be higher than suggested by the table. It is nevertheless striking that the scale elasticity, with respect to the total number of passengers, is significantly lower than unity in all but one specification.

The distribution of estimated scale elasticities with respect to the number of passengers (PAX) and the number of air transport movements (ATM) for all airports are displayed in Figures 8.4 and 8.5, respectively. Note that, according to our specification, scale elasticities are allowed to vary by all variables included in the cost function (Equation 4.4). Both distributions for scale elasticities appear quite symmetric around the overall point estimates reported in Table 8.3. The same holds for the fixed effects distribution shown in Figure 8.6. It is interesting to test whether the scale elasticities vary for different sized airports. Looking at (4.4) again, we can do this by testing whether some second order derivatives are significantly different from zero. It appears, from Table 8.2 results, that this is never the case: airport size (whether measured by PAX, ATM or a combination of both outputs) does not significantly affect the respective scale elasticities.

Table 8.3   Average scale elasticities in different model specifications

Model specification (variable included y/n)											
GDP growth	-	x	-	-	-	x	x	x	x	x	x
Fraction of international passengers	-	-	x	-	-	x	-	-	x	-	x
Quality of service	-	-	-	x	-	-	x	-	-	x	x
Fixed assets	-	-	-	-	x	-	-	x	x	x	x
Scale elasticities (standard errors)											
Total number of passengers	0.47*+	0.62*+	0.32*+	0.62*+	0.43*+	0.49*+	0.71*	0.59*+	0.48*+	0.24+	0.09+
	(0.10)	(0.11)	(0.14)	(0.16)	(0.12)	(0.14)	(0.17)	(0.13)	(0.20)	(0.20)	(0.35)
Air transport movements	-0.04+	-0.14+	0.17+	-0.03+	-0.10+	0.05+	-0.03+	-0.16+	-0.12+	0.35+	0.23+
	(0.12)	(0.12)	(0.16)	(0.20)	(0.14)	(0.16)	(0.22)	(0.14)	(0.22)	(0.22)	(0.35)
Number of observations	194	193	157	114	166	157	113	165	129	102	89

* Significantly different from 0 at 5 per cent confidence level.
+ Significantly different from 1 at 5 per cent confidence level.

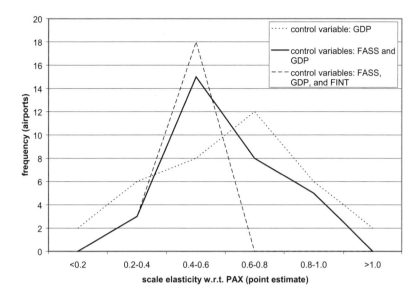

Figure 8.4   Distribution of scale elasticities (PAX) over different airports

Note: The histograms with controls for fixed assets (FASS) exclude Australian airports, while the histogram with controls for the fraction of international passengers (FINT) excludes Hamburg, Hong Kong, Oslo, a number of American and both South African airports.

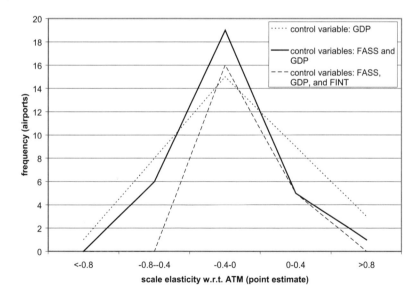

Figure 8.5   Distribution of scale elasticities (ATM) over different airports

Note: The histograms with controls for fixed assets (FASS) exclude Australian airports, while the histogram with controls for the fraction of international passengers (FINT) excludes Hamburg, Hong Kong, Oslo, a number of American and both South African airports.

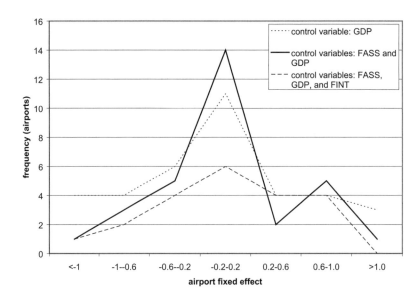

Figure 8.6   Distribution of fixed effects over different airports

Note: The histograms with controls for fixed assets (FASS) exclude Australian airports, while the histogram with controls for the fraction of international passengers (FINT) excludes Hamburg, Hong Kong, Oslo, a number of American and both South African airports.

5. Concluding Remarks

In this chapter we have estimated a statistical relationship of the translog type between the operational cost of an airport and various output and control variables. These estimations were performed on the basis of a longitudinal dataset of 36 airports from Asia, Australia, Europe, North America and South Africa, over the period 1993–2000. Input prices were not included in the analysis because comparable price data for individual airports was not available. In order to compute the necessary input prices from the airport financial data, all airports would have to provide a similar subdivision of operational costs (that is, labour costs, maintenance and equipment, residual costs) and accounting procedures. Unfortunately this was not possible and consequently we cannot assume or test the hypothesis that the airports included in our sample are operating under a cost-minimizing strategy.[7]

Our main conclusion from this analysis is that, for the 36 airports sampled, short run average operational costs decrease with output. Results suggest that the number of passengers handled at the airport is the most important cost driver, whereas the number of air transport movements is a relatively insignificant factor. When the fraction of international passengers is included as an explanatory variable, the elasticity of cost with respect to the number of passengers decreases. This is because the fraction of international passengers is an important cost driver: international passengers spend more time at the airport due to longer check-in times, customs and security checks, and so on. When the number of international passengers increases, the (variable) costs of these activities are spread out over a larger number of passengers.

---

7   See the discussion in Section 2 and footnote 5.

The elasticity of operational cost with respect to output appears to be robust enough to include different sets of control variables. An important control variable is the 'quality' of the airport, because passengers may value relatively short check-in times, better catering and shopping facilities etc., although this may lead to higher operational costs for the airports. In practice, however, it is difficult to define 'quality'. In this chapter we used IATA's Global Airport Monitor as the source of information on airport quality. In two out of three specifications that we estimate, the introduction of airport quality brings down the elasticity of operational cost with respect to output (passengers). This indicates that part of the average operational cost associated with quality may not decrease when output is increased. This may be relevant to regulators if there is reason to believe that certain regulatory regimes may lead to 'gold plating' by airport operators.[8]

Finally, we conclude that airport size has no significant influence on the scale elasticity of the short run variable cost function.

The policy implications of these findings are that i) marginal cost pricing may not be feasible for smaller airports (unless expenditures on fixed costs and additional operational losses are compensated); and ii) inefficient airports may be scale inefficient rather than technically inefficient. Scale inefficient airports may become efficient when they are allowed to grow, while technically inefficient airports have the wrong input mix. This is important information for investors: an airport that is not allowed to grow to an 'optimal' size may never become fully efficient and is therefore being deprived of the potential benefits from economies of scale.

In future research, the current estimates could be improved by including more time varying control variables, in particular input prices. In order to achieve this, more data is needed on a subset of airports for which comparable input prices can be obtained.

References

Baltagi, B., J. Griffin and D. Rich (1995), Airline deregulation: the cost pieces of the puzzle, *International Economic Review*, 36(1), 245–59.

Brueckner, J. and P. Spiller (1994), Economies of traffic density in the deregulated airline industry, *Journal of Law and Economics*, 37, 379–415.

Caves, D., L. Christensen and J. Swanson (1981), Productivity growth, scale economies and capacity utilization in U.S. railroads, *American Economic Review*, 71, 994–1002.

Caves, D., L. Christensen and M. Tretheway (1984), Economies of density versus economies of scale: why trunk and local service airline costs differ, *Rand Journal of Economics*, 15(4), 471–89.

Christensen, L. and W. Greene (1976), Economies of scale in U.S. Electric Power Generation, *Journal of Political Economy*, 84(4), 655–76.

Forsyth, P., D. Gillen, A. Knorr, O. Mayer, H.-M. Niemeier and D. Starkie (2004), *The Economic Regulation of Airports*, Aldershot: Ashgate.

Gillen, D. and A. Lall (1997), Developing measures of airport productivity and performance: an application of data envelopment analysis, *Transportation Research E*, 33(4), 261–74.

Gillen, D., T.H. Oum and M. Tretheway (1990), Airline cost structure and policy implications, *Journal of Transport Economics and Policy*, 24, 9–34.

Greene, W. (1993), *Econometric Analysis*, 2nd edition, New York: Macmillan.

---

8   Note that we are not implying that investments in quality are necessarily 'gold plating'.

Morrison, S. (1983), Estimation of long-run prices and investment levels for airport runways, *Research in Transportation Economics*, 1, 103–30.
NERA (2001), *CAA Benchmarking Study: Phase 2 Initial Work*.
NERA/TRL (2001), *Adjustments to the TRL Data Set. A Technical Supplement to 'The Application of Benchmarking to Airports Phase I'*.
Ng, C. and P. Seabright (2001), Competition, privatisation and productive efficiency: evidence from the airline industry, *Economic Journal*, 111, 591–619.
Niemeier, H.-M. (2004), Capacity utilization, investment and regulatory reform of German airports, in: P. Forsyth, D. Gillen, A. Knorr, O. Mayer, H. Niemeier and D. Starkie (eds), *The Economic Regulation of Airports*, Aldershot: Ashgate.
Oum, T.H. and Y. Zhang (1991), Utilisation of quasi-fixed inputs and estimation of cost functions – an application to airline costs, *Journal of Transport Economics and Policy*, 25, 121–38.
Oum, T., C. Yu and X. Fu (2003), A comparative analysis of productivity performance of the world's major airports: summary report of the ATRS global airport benchmarking research report, *Journal of Air Transport Management*, 9(5), 285–97.
Pels, E., P. Nijkamp and P. Rietveld (2003), Inefficiencies and scale economies of European airport operations, *Transportation Research Part E: Logistics and Transportation Review*, 39(5), 341–61.
Starkie, D. (2001), Reforming UK airport regulation, *Journal of Transport Economics and Policy*, 35, 119–35.
Tirole, J. (1992), *The Theory of Industrial Organization*, 5th print, Cambridge, MA: MIT Press.
Tolofari, S., N. Ashford and R. Caves (1990), *The Cost of Air Service Fragmentation*, Loughborough University of Technology, Department of Transport Technology, TT9010.
TRL (various years), *Airport Performance Indicators*, http://www.trl.co.uk.

Chapter 9
Competition Between Airports: Occurrence and Strategy

Dr Michael Tretheway and Ian Kincaid

Introduction

Until the 1980s the traditional view of airports, held by many governments, industry operators and academics, was that airports were monopolies (a view still held by many today). Airports were seen as, and generally operated as, monopoly providers of services to both airlines and passengers. Airports were not perceived as being subject to competitive forces. The commonly held view was that there was little an airport could do to increase demand for its services or divert demand from other airports. Airport marketing was viewed as an oxymoron.[1] Airports were largely passive service providers. The job of marketing and identifying new air service opportunities was left to the airlines.

The deregulation of the aviation industry in many parts of the world led to a change in the way airports were operated. Deregulation was largely focussed on the airlines, although many countries, notably the UK, Australia, New Zealand and Canada, have also divested or privatised their airports and air traffic control services.[2] As a result of deregulation, airlines have become much freer to operate out of any airport of their choosing. Deregulation was critical in the development of low-cost carriers (LCCs), which often operate out of secondary airports with lower costs and no congestion, challenging the notion that airports were absolute monopolies. The aggressive expansion of the LCCs, and their frequent choice of secondary airports, has resulted in airports, both primary and secondary, discovering that there may be great payoffs to more sophisticated and aggressive marketing strategies.

The increasingly competitive nature of the airport market is reflected in the increase in staff and resources employed in marketing functions for the airport. For example, as shown in Figure 9.1, a recent study of airport management illustrates the increase in marketing staff per passenger at selected regional airports in the UK between 1991 and 1997.[3]

---

1   See M.W. Tretheway 'Airport Marketing: An Oxymoron?' in G. Butler and Martin Keller (eds), *Handbook of Airline Marketing*, Washington, DC: McGraw Hill, 1998, pp. 649–56.

2   It can be argued that airport privatisation has resulted in additional regulation, as some governments have established regulatory bodies to specifically control potential abuses of market power by the airports. Nevertheless, this regulation has been introduced in the spirit of fostering (or simulating) competition, and is generally confined to larger airports, e.g., Heathrow, Gatwick, Manchester, Sydney (until recently) and so on.

3   Anne Graham, *Managing Airports: An International Perspective*, Oxford/Boston: Butterworth Heinemann, 2001. Although the marketing staff per passenger ratios fell at East Midlands and Newcastle airports between 1991 and 1997, the absolute level of staffing at both airports did increase over this period.

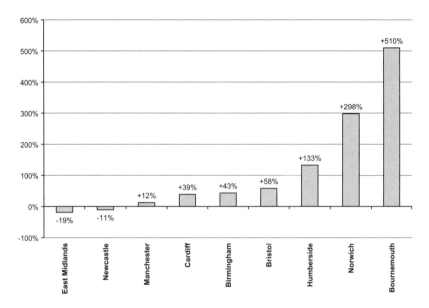

Figure 9.1 Percentage change in marketing staff per passenger at selected UK airports 1991–1997

Source: Graham, *Managing Airports: An International Perspective*, p. 179.

This chapter discusses the nature of competition among airports in many of the areas in which they operate. It then discusses strategies airports can adopt (and have already adopted) in the new millennium to compete with other airports.

Airports Operate in a Number of Competitive Markets

Many of the services provided by airports fall within competitive markets, and airports are (or should be) active players in competing for customers. Airports compete in the following ways:[4]

- competition for serving a shared local market;
- competition for connecting traffic;
- competition for cargo traffic;
- destination competition;
- competition for non-aeronautical services (retail, food and beverage etc);
- competition with other modes (e.g. Eurostar, TGV etc.).

The latter two forms of competition relate less to competition between airports. For example, airports compete for retail and food and beverage spending largely with non-airport providers of

---

4 In addition to competition between airports, there are also instances of competition between terminals at the same airport. For example, New York (JFK), LaGuardia (LGA), Los Angeles (LAX) and Toronto (YYZ) have terminals which are not operated by the airport authority. Instead they are operated by an airline, an airline consortium or by a non-airline group. These terminals can potentially compete on terminal fees, provision of office space or provision of other services.

these services. The first four types of competition between airports are discussed in more detail below.

Competition for Shared, Local Markets

Where airports are located in close proximity of each other, they compete for both passengers and air service. Many cities have two or more airports through which passengers can access air service. Examples include:

- Chicago: O'Hare and Midway[5]
- New York: JFK, Newark, LaGuardia, Islip (Long Island) and Westchester County
- Washington, DC: Dulles, Reagan National and Baltimore/Washington
- Brussels: Brussels Zaventem and Charleroi
- Hamburg: Hamburg and Lübeck
- London: Heathrow, Gatwick, Stansted, London City, Luton[6]
- Paris: Charles de Gaulle (CDG) and Orly.

In many cases, the airports have focussed on different market segments. For example, Heathrow dominates the long-haul market from London, while Luton and Stansted have become major LCC airports. However, many airports serving a single region offer competing services; all of the five London airports listed above have air service to Paris. There is increasing evidence that consumers will substitute between airports. A business traveller from London may choose to use London City Airport to travel to Paris due to its location and ease of check-in. That same traveller, when travelling for leisure purposes to Paris, may choose Luton instead to take advantage of lower fares offered there. Travellers will select the airport based on a number of factors, some of which the airport cannot control (e.g. distance),[7] and some over which the airport can exert some control or influence – price, service levels, check-in processes and so on.

The growth of the LCCs has provided smaller airports with increased opportunities to compete with more established airport competitors. The common view of these airports is that they are generally more remote than the established airport, and travellers choose to fly from the airport primarily due to the lower price offered.[8] However, it is our view that this is only one dimension of how consumers choose airports. Perhaps more important is that many of these secondary airports have 'primary' catchment areas of their own, for which the secondary airport is much more convenient than the region's primary airport. As an example, John C. Munro Hamilton International (YHM) Airport in Ontario, used by LCC WestJet, has developed as an alternative to Toronto's Lester B. Pearson International Airport (YYZ). The primary catchment area of YHM has a population of approximately 850,000, for whom YHM is closer than any other airport. When populations roughly equidistant from YHM and YYZ are considered, the population of the catchment area increases to just under 3 million. This is a population size greater than that of Vancouver, Canada's second busiest airport market. Prior to development of passenger air service at YHM,

---

5   In addition, Milwaukee County airport, roughly 90 minutes north of Chicago, claims it is the '3rd' Chicago airport, as do Gary, Indiana and Rockford, Illinois.
6   Heathrow, Gatwick and Stansted are all owned by BAA plc.
7   Even here, airports which invest in or advocate for highway or rail access projects may be able to reduce travel time to their airport, thereby influencing consumer choice.
8   For example, Charleroi is 46 km from the centre of Brussels, while Brussels Airport is about 15 km away.

travellers in this catchment area had to travel to YYZ. These travellers now have a more convenient alternative. Development of air service at Hamilton has attracted passengers from the airport's primary catchment area due to its locational advantage, and not necessarily due to its low fares. Another example is the rise of UK airports such as Nottingham East Midlands, Leeds Bradford and Liverpool John Lennon. These airports now provide air service to serve their own immediate local markets, where previously Manchester Airport had arguably been the dominant (and sometimes the only) choice of airports for travellers in the north of England and the Midlands.

Competition for Connecting Traffic

At many major airports, connecting traffic is a main component of the total traffic handled. At a number of major hubs, connecting traffic makes up more than half of the total passenger traffic handled by the airport. A number of secondary airports also participate in the connecting traffic market. Southwest Airlines' major connection points are almost all secondary airports such as Houston Hobby, Dallas Love Field, Baltimore/Washington, Chicago Midway and so on.

For almost all connecting traffic there are alternative connecting points. A traveller flying from Los Angeles in the US to Mauritius in the Indian Ocean can travel there via a point in Asia (Hong Kong, Singapore, Kuala Lumpur), Europe (London, Paris, Frankfurt, Zurich), the Middle East (Dubai) or Australia (Sydney). A passenger travelling from Aberdeen, Scotland, to Rome can get there via London, Paris, Amsterdam and so forth.

Connecting traffic can easily shift from one airport to another if cheaper, faster and/or more convenient connections become available. For example, Dubai has emerged as a major connecting airport as a result of Emirates' rise as a sixth freedom carrier and the airport's investment to facilitate this growth. While it may be tempting to view connecting traffic as fairly stable, shocks can occur which result in a quantum change in an airport's traffic. Consider the following:

- The collapse of TWA and its subsequent takeover by American Airlines in 2001 resulted in a large loss of American Airlines' service at St Louis International Airport.[9] American chose to consolidate its hubbing operations at Dallas Fort Worth and Chicago O'Hare and effectively demoted St Louis to a spoke airport.
- As a restructured US Airways reduced service at its traditional hub, Pittsburgh, that airport has lost noticeable levels of connecting traffic.
- KLM and Northwest have indicated that Paris may be developed as the major intercontinental connecting point for their combined networks, as a result of KLM's merger with Air France.[10]
- Hypothetically, if American Airlines were to enter Chapter 7 bankruptcy (i.e. liquidation), Houston Airport could emerge as the major connecting point for South American traffic from the US. through Continental's network, supplanting American's hub at Miami Airport.[11]

---

9 Following the takeover of TWA, America reduced its daily seat capacity by 62 per cent (Jan K. Brueckner, 'The Economic Impact of Flight Cutbacks at the St. Louis Airport: A Calculation of Job Losses', unpublished paper, University of Illinois, September 2003).

10 The carriers have indicated that Amsterdam is expected to remain a connecting point, but somewhat more focussed on intra-Europe traffic.

11 American Airlines acquired Eastern Airline's routes to South America in 1989 following the latter airline's collapse. American chose to continue using Miami as the hub for these services, as Eastern had been doing, rather than shift operations to one of its existing hubs.

While hub airports are clearly dependent on the operating strategies of the airlines, they influence these strategies through pricing, runway/terminal capacity and other factors.

Competition for Cargo Traffic

Cargo traffic can make up a major proportion of an airport's traffic base. Airports Council International (ACI) estimates that cargo accounts for approximately 17 per cent of annual airport revenue, on average.[12] Cargo traffic, both origin/destination (O/D) and transshipment/connecting traffic, is highly competitive. Today, much of the North American courier traffic to the Greater Toronto region flies to nearby Hamilton Airport and is then trucked to/from Toronto itself. Hamilton Airport handles approximately 100,000 tonnes of cargo each year that otherwise would likely have passed through Toronto Airport. As Hamilton has lengthened its runway, it is also now competing for intercontinental cargo flights.

Cargo traffic is highly price sensitive and can easily shift to alternative routings. If cargo rates for direct service from Amsterdam to Tokyo are too high, or if capacity at Amsterdam is limited, then the cargo can be trucked and flown via Brussels, Frankfurt, Paris or Liège. It can also be flown from Amsterdam to Hong Kong, Vancouver, San Francisco and so forth for onward carriage to Tokyo. Air cargo is notorious for being 'gateway competitive,' and airports must view much of their cargo traffic, even O/D cargo traffic, as subject to vigorous competition from other airports.

Destination Competition

Airports have a role in competing for destination traffic, as they are part of the overall tourism package offered by a destination. The quality, cost and scope of service offered at an airport impacts on the overall attractiveness of a destination. As the scope and frequency of air service to and from an airport increases, so does the overall attractiveness of the destination served by the airport.

One area where this is particularly important is the convention market. Convention planners consistently rank air service to a community as one of the top two criteria in choosing a site for a convention. Another example of destination competition is the cruise line market, one of the fastest growing tourism markets globally. Cruise line decisions on embarkation point will often depend on factors such as how well integrated the airport is with the cruise port, as well as the scope and competitiveness of air services. A number of cruise lines in the Mediterranean operate circle itineraries, and any port of call could be used by the cruise line as the itinerary embarkation point.

The Four P's of Marketing

The previous sections have argued that airports provide services in many markets that are competitive. This is not to say that there are no airport services that should be considered to be monopolies, but there are many airports and many airport services that are demonstrably subject to competition. Having discussed the nature of airport competition, we move on to examine strategies that airports can utilise to compete with other airports. To do this, the familiar paradigm used in the marketing field will be used. This involves strategies related to the classic 'four P's of marketing,' familiar to students of business administration and marketing:

---

12   http://www.aci-na.org.

1. product;
2. price;
3. promotion;
4. physical distribution.

Product

The first area of this marketing paradigm is defining the product to produce and sell. This includes issues such as the functionality, styling and quality of the product, as well as the support and accessories provided for the product. Ultimately, 'product' is about developing a good or service which will be useful and attractive to the market. For example, a few years ago Boeing developed the Sonic Cruiser as a product which provided shorter travel times than current aircraft. However, Boeing found that the market (that is, the airlines) did not so much want a faster aircraft, but rather a more cost-effective one, and switched development resources to the 787.

Price

This is the second dimension of marketing. Here the marketing issue is what price (or prices) the chosen products will be sold at; or, alternatively, what price segments of the market to pursue. Pricing the product too high could result in loss of sales to less expensive products that are reasonable substitutes. For example, if a burger chain prices its meals too high, it may lose sales to other, lower-priced burger chains, to alternative products (submarine sandwiches), or families may decide to save money by eating home-cooked meals. Equally, pricing too low can result in sub-optimal revenue levels. There may also be different consumers at different price points, and choosing which to serve is an important strategic decision.

Promotion

The third area involves creating awareness of the product and price with potential consumers. Some claim that promotion itself can create 'image' value, such as the image created by advertisements for certain sports cars.

Physical Distribution

The last P has to do with getting the product to the consumer. A manufacturer may offer a great product at a great price and may have succeeded in creating awareness in the consumer and getting the consumer to commit to a purchase; but if the product cannot be delivered where and when the consumer wants it, the sale will be lost. Traditionally physical distribution has been thought of as a 'place' value, but today the time value must also be considered. Physical distribution is relevant not only to manufacturing industries but also to service industries. For example, in the airline industry the product needs to be made available to consumers where and when they want it via computer reservations systems (CRS), websites and the like.

Applying the Four P's: Airport Strategies for Competing

The Airport Product

Each airport provides a product to air carriers and passengers with certain physical and operational characteristics. These characteristics can have a major impact on the type and quantity of traffic handled by the airport. Just like any consumer product, airports need to examine the package of features they want to provide in order to develop a product attractive to the market. Consider the package of features (both positive and negative) presented by a typical primary and secondary (low-cost) airport (Table 9.1).

Both types of airport may be attractive to different customers (both airlines and passengers), depending on how the air carriers weigh the trade-offs offered. The task for the airport management is ensuring that their product fits the market(s) they are targeting, and provides a competitive edge over rival airports.

Infrastructure The core element of the airport product is its infrastructure: runways, taxiways, terminals and so on. This infrastructure impacts on the airport's competitive position, and the level of provision of this infrastructure should reflect the marketing strategy of the airport. A number of regional airports in British Columbia, such as Cranbrook and Kamloops, are seeking government and private funding to extend their runways or construct other infrastructure that would allow the airports to accommodate longer range aircraft. This infrastructure upgrade is an important part of strategy to further develop these regions as international ski destinations (the current Dash 8 air service does not provide sufficient baggage capacity to meet the needs of heavily laden ski visitors, or a sufficiently low cost per seat).

Table 9.1   Feature presented by typical primary and secondary airports

Primary airport	Secondary airport
Closer to the city	More remote location to the central city (not always the case), but closer location to some parts of the metro region
High frequencies	Lower frequency
Wide range of non-stop destinations	Limited non-stop destinations
Enables connecting traffic	Focused
Higher fares	Lower fares
Wide range of retail and food and beverage	Limited retail and food and beverage offered
Capacity constrained (due to physical or political limitations)	Ample capacity, uncongested
May be subject to night curfew or noise quotas	Typically 24-hour operation
Wide range of handling equipment and facilities	Some handling equipment may be unavailable (e.g. wide-body main deck cargo loaders)
Higher airline operating costs due to long taxi times, congestion, higher labour rates	Lower airline operating costs due to short taxi times, lack of congestion, lower labour rates

Many airports are currently facing the decision as to whether to upgrade their facilities in order to accommodate the A380. Failure to do so could result in the loss or delay of connecting and O/D traffic. However, making this major upgrade when there is little likelihood of receiving A380 service also has a cost, as the investment would have to be recouped from existing traffic, reducing price competitiveness.

Passenger facilitation Passenger facilitation is an increasingly important area in which airports can achieve a competitive advantage. Airport design and processes affect the amount of time required for passenger connections. In addition, nowadays passengers at nearly all airports must pass through rigorous security screening, as must their baggage, and international passengers must clear customs and immigration. Meeting these requirements, while ensuring the efficient movement of passengers and their baggage through the airport, is receiving increased attention from airport managers.

Airports that manage to improve passenger facilitation processes to reduce processing and connection time can enhance their competitive position for attracting both O/D and connecting passengers. The airport will be more attractive to passengers due to the reduced time and hassle involved, particularly for connecting passengers. (This can also have an impact on passenger or travel agent booking choices, as discussed in the following section physical distribution) The airport will also be more attractive to airlines as it may achieve faster turnarounds and fewer missed connections for passengers, which have both customer service and cost benefits for the airline.

In Canada, a number of major airports provide 'preclearance' services, namely Vancouver, Calgary, Edmonton, Winnipeg, Toronto, Ottawa, Halifax and Montreal.[13] Preclearance involves US customs, immigration and other inspection processes taking place at the Canadian origin airport rather than at the US destination.[14] For originating passengers, there is a real convenience to US preclearance as, once they arrive in the US, they are processed in the same manner as a US domestic passenger, without any need for an international arrivals process. If the passenger is connecting at a US airport, their connection time will be considerably shorter, while destinating passengers can immediately leave the airport rather than queue up for US customs and immigration processes. The programme may also enable air service between a Canadian preclearance airport and a US airport that does not have customs and immigration facilities.[15]

As a further product development, Vancouver International Airport pioneered a 'transit preclearance' process, whereby international connecting travellers are able to go directly to Vancouver's US preclearance facility without needing to clear Canadian customs and immigration.[16] This eliminated one whole customs process, simplifying and reducing costs for baggage handling, and allowed the removal of almost half an hour of connection time. By removing the need to pass through two sets of customs and immigration, transit preclearance puts Canadian airports on a level competitive footing with US airports for competing connecting traffic.

---

13 Bermuda, Bahamas and Aruba have similar preclearance arrangements with the US for their major airports. Dublin and Shannon airports in Ireland currently have *immigration* preclearance arrangements with the US, but passengers from these two airports must still pass through customs and agriculture inspections upon arrival in the US.

14 US customs and immigration processes are handled by Customs and Border Protection, part of the US Department of Homeland Security.

15 Although it is also possible for US airports to provide a preclearance service for flights to Canada, to date, no US airport has asked for the establishment of a Canadian preclearance facility.

16 Canadian customs and immigration processes are handled by the Canada Border Services Agency.

Baggage processing is another area in which an airport can achieve a competitive edge. Airports are starting to explore the use of Radio Frequency Identification (RFID) tags to track baggage as it makes its way through the airport. This technology enables security officials to quickly identify and locate baggage that they wish to inspect further. If implemented effectively, RFID tags could allow aircraft-to-aircraft movement of baggage for international arriving passengers connecting to a domestic flight. Typically customs inspection authorities only wish to inspect a small fraction of the baggage of such connecting passengers. With RFID deployment it is possible to quickly find and deliver the small amount of baggage to the customs inspector without the need to deliver all connecting baggage to the customs facility. By eliminating the need for customs hall delivery of baggage, there are considerable savings in airport capital cost (baggage delivery systems and space requirements) and connection times.

Another example of improved passenger and baggage processing enhancing destination competition is the cruise line market. Maritime cruise ship operators make choices of embarkation city in part depending on factors such as how well integrated the airport is with the cruise port. Vancouver International Airport, working with the cruise lines, the Vancouver Port Authority, airlines, and Canadian and US customs and immigration, has developed a number of unique and innovative schemes to enhance cruise passenger processing. These include:

- *Straight to ship baggage transfer.* Passengers arriving by air to Vancouver for purposes of boarding a cruise ship can check in their bags at their home airport and the bags will be transferred all the way to their ship.
- *On-ship check-in.* Towards the end of their cruise, passengers who will fly out using Vancouver Airport can check in their bags while still on ship. Vancouver Airport has invested in technology that transfers check-in information from the ship to the airport which is then routed to the airlines. Once the ship docks, the bags are transferred directly to the aircraft at the airport. Passengers are then free to make their way (either directly, or they can sightsee in Vancouver) to the airport unencumbered by their baggage.
- *Direct transfer.* Vancouver put in place a scheme that allows passengers to be transferred from ship to the airport, bypassing both Canadian and US customs and immigration. In effect, the passengers are 'bonded' when transferred from the ship in the port to the airport. The cruise ship is arriving from a US port (in Alaska), and the flights are to a US destination. Without direct transfer, the passengers would have to be admitted into Canada at Port Vancouver, and then admitted back into the US at the US preclearance facility at Vancouver Airport. With the sterile transfer process, the passengers effectively stay within the US for customs and immigration purposes. As approximately 80 per cent of the cruise ship market in Vancouver is US citizens, this provides a much more streamlined process for many passengers.

Vancouver International Airport has led the way in developing these processes and has invested considerable amounts of money in technology and facilities to enable the processes. This has allowed Vancouver to enhance its competitiveness with US cruise embarkation/disembarkation ports, and has enabled the airport to handle large volumes of cruise passengers over the summer months, while minimising the pressure on terminal capacity. The airport estimates that approximately half a million passengers have been diverted to Vancouver Airport, who otherwise would have used a US airport as their air gateway to the Alaska cruise.

A final example of passenger facilitation is use of dedicated or streamlined facilities for shuttle or commuter air services. Typically these facilities have dedicated ticket counters with aircraft gates immediately behind to minimise walking time and baggage transfer time. These facilities result in shorter dwell times for passengers (particularly attractive to business travellers) and shorter turnaround times for airlines. By doing so, the airport positions itself in the market as the commuter or business airport. Airports can augment this with other conveniences for business travellers, such as curbside valet parking or remote check-in.

Flexible airport design   Airports can design their terminals and other facilities to allow them to be adapted to new and changing traffic demands. In this way the airport can respond to changing market conditions in a cost effective manner and ensure that its product stays relevant to the changing market. For example, newly expanded Ottawa Airport has developed a system that enables it to adjust the number of gates provided for domestic and international air service simply by opening and closing partitions, moving the wall that separates the two types of traffic.

Service provision and third party vendors   Airlines require a number of services when using an airport: a runway to land on, a terminal to process passengers, a warehouse to hold and process cargo, ground handling, fuelling, maintenance etc. An airport must decide how it is going to provide these services. One option would be for the airport to provide all of these services, as a one-stop shop for the airline. It may be perceived that there are economies of scale to the airport providing all the services required by the airline. However there may also be the temptation to extract monopoly rents as the sole provider of, say, ground handling services.

Increasingly though, airports are recognising the benefits of allowing third party providers of services such as ground handling, fuelling, warehousing etc.[17] In doing so, the airport can create a competitive market for the provision of these services, resulting in lower rates and higher service levels for the airlines. This can be achieved either through allowing two or more providers to actively compete or through a contract with a sole third party provider that is subject to regular review. In addition, this strategy allows the airport to focus on those services in which it specialises, and in which it has a comparative advantage, for example, landings and passenger processing.[18] As discussed in the price section below, this is another strategy in which the airport competes by attempting to minimise the airline's cost.

Curfews and noise quotas   Night curfews on airport operations clearly impact on the airport product. In particular, curfews can limit an airport's ability to handle certain types of long-haul traffic. For example, carriers like to operate late-night flights from North America to Asia in order to allow passengers to connect with the morning bank of flights out of Asian airports. Curfews also make the airport far less attractive to cargo operators and integrators, who require night operations. In some regions, airports at more remote locations which are not subject to a curfew have exploited this advantage by positioning themselves as 24-hour airports, and have successfully attracted cargo operations from noise-constrained major airports. For example, Hamilton Airport has attracted cargo operations away from Toronto Airport.

While there may be little an airport can do once a curfew has been imposed, there are strategies airports can adopt to avoid having one imposed. Since the pressure for a noise curfew generally comes from communities living close to the airport, cultivating good relations with the local

---

17   In the EU, many airports (broadly speaking, those handling more than 1 million annual passengers) are legally compelled to allow additional ground handling providers (airlines and third party providers) to operate at the airport.

18   The airport normally still obtains some revenues from these third party services, through rents, fees or royalties paid by the third party provider to the airport.

community can go some way towards mitigating this pressure. Open days, educational visits and sponsorships are used by some airports as part of this strategy, along with developing good links with the local and national press. Airports will often conduct economic impact studies as a means to demonstrate the economic and employment contribution of the airport to the community. These approaches are used, in part, to gain support for future development and reduce pressure for limits on the airport's operations. In addition, an airport can attempt to communicate and work with local government to avoid residential or commercial developments being located in areas that may conflict with current or future airport operations. Reconsidering flight paths and profiles for night operations can reduce the number of households impacted by night noise, both reducing the noise impact as well as improving community relations.

Airports subject to noise quotas face an interesting dilemma as how to best maximise value within the quota. Generally, these quotas give greater weight to night flights than daytime flights: one night-time cargo flight may be weighted the same as two daytime passenger flights (these weights can also vary depending on the noise certification of the aircraft). At first glance, this would appear to favour allocating as much of the quota as possible to daytime flights in order to maximise the total number of flights. However, the airport may find that a smaller number of night-time flights are more lucrative, or provide the airport with a competitive edge over rival airports, and choose to devote more of their quota to night operations, even if this results in a smaller number of total flights.

Cargo traffic Historically, air cargo has grown at a significantly faster rate than air passenger traffic: see Figure 9.2, which illustrates the growth of air passenger and air cargo traffic alongside the growth in the world economy. This trend is expected to continue, and even accelerate, as air freight rates continue to decline and the world economy further integrates. Air cargo has its own particular set of needs based on the market characteristics and economics, which an airport needs to address if it is to attract cargo operations. The airport's air cargo product can be enhanced in the following ways:

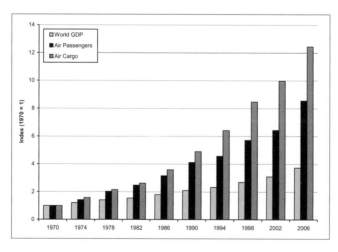

Figure 9.2 **Growth in world GDP, passenger traffic and cargo traffic**

Source: International Monetary Fund (IMF) and International Civil Aviation Organization (ICAO). World GDP is in constant prices; air passenger traffic measured in revenue passenger kilometres (RKP); and air cargo in freight tonne kilometres (FTK).

- *24-hour operations*. As discussed above, air cargo often requires 24-hour operations in order to achieve overnight delivery and to meet the needs of clients in different time zones, who themselves may have 24-operations.
- *Customs processing*. The development of facilities that can process and hold cargo requiring customs clearance can attract cargo operations to an airport.
- *Transshipment facilities*. As mentioned previously, transshipment cargo is highly time sensitive. A recent study by Ohashi, Kim, Oum and Yu examined the factors affecting the choice of transshipment airport for air traffic to/from north-east Asia.[19] They found that the connecting time was the most important factor in the choice of transshipment airport, more so than landing fees and other airport charges. A one-hour reduction in the connecting time at the transshipment airport was found have the same impact as a US$1,361 reduction in the total airport charges (based on a 747-400 cargo aircraft). Therefore, airports wanting to attract transshipment air cargo traffic need to focus on minimising connect times.
- *Value-added/distribution/free trade zones*. By developing value-added or distribution facilities on airport grounds, an airport can generate a critical mass of operations that enable the airport to become a major air cargo player. This strategy has been adopted by airports such as Amsterdam Schiphol and Hong Kong International, which have encouraged the development of a wide range of business activities at or near the airport (this strategy also provides the airport with a diversified revenue base).

Airports such as Hong Kong, Amsterdam and Frankfurt (Fraport) have developed facilities and related activities in order to establish themselves as major players in the cargo market for both O/D and transshipment cargo. Other airports, such as Anchorage, Alaska, have specialised in transshipment cargo, which is exploiting its unique geographic position between Asia and North America. Some airports have developed niche products to attract a specific segment of the market. For example, Calgary Airport has invested in facilities for handling livestock. These facilities are designed to minimise stress on the livestock and prevent the animals becoming 'spooked,' which greatly increases the reliability of the air service (for safety reasons, aircraft cannot take off if the animals are deemed to be unsettled). In addition it has a processing configuration which allows veterinary inspection and segregation of animals. This prevents shipments from being delayed because a diseased animal has been mixed into a herd. As a result, Calgary airport attracts livestock shipments from markets all across North America.

Price

Pricing is another tool airports use to compete with other airports. When considering airport pricing, not only should the airport's own fees and charges be considered, but other costs paid by the airline should also be considered. Any means of reducing costs for the airlines are a means to lower the 'price' they pay for using the airport.

Airport fees and charges: two-stage pricing? The general consensus is that airlines are price inelastic for airport services.[20] This would suggest that it may be irrational for airports to reduce

---

19 Ohashi, H., Kim, T., Oum, T. and Yu, C., *Choice of Air Cargo Transshipment Airport: An Application to Air Cargo Traffic to/from Northeast Asia*, CIRJE series, Faculty of Economics, University of Tokyo, July 2004.

20 Very few studies have been conducted which examine the price elasticity of airlines with respect to landing fees. One study – Gillen, D., Oum, T. and Tretheway, M., 'Airport Pricing Policies: An Application to

fees and charges as a means of competing with other airports. However, it may be useful to consider whether airline elasticities are different for different services or at different times. In particular, they may be price elastic when choosing a new destination to add to their network, but once that service is established and the airline has sunk marketing and facility costs at the chosen airport, the airline may become price inelastic with regard to airport fees and charges.

This suggests that the pricing of airport services (landing fees, terminal charges etc.) should be viewed as following a similar pattern to that of services associated with durable goods. For example, at the time an individual is considering purchasing an automobile, they may be fuel price elastic, but once the durable good (the auto) is purchased, the consumer becomes fuel price inelastic. Some have suggested that computer printer manufacturers understand this phenomenon all too well. Printers are sold for very low prices, possibly below cost, with the manufacturer expecting to exploit the post-purchase inelasticity with respect to the price of printer ink cartridges and toner.[21]

When starting a new air service, airlines can be highly price sensitive (price elastic, in economic terminology), as the economics of the route can initially be marginal. Even modest changes in fees and charges can have a major impact on route viability and thus strongly influence airline service decisions. As a result, it may be good economics for airports to discount their fees and charges, as well as provide other incentives (such as contributions to marketing costs, free office space, assistance with pilot training and staff accommodation etc.) to attract new air carriers or new routes. However, once the service has established itself and becomes profitable (the airline is 'locked in'), the airline appears to become less price sensitive (price inelastic). The airport is then able to increase its charges (i.e., return charges to normal) without losing the air service to a competitor airport. This behaviour can be seen as consistent with the two-stage elasticity phenomenon. Even the European Commission guidelines on state aid for regional airport services are consistent with this economic theory.[22]

The issue for the airport is at what point the airline starts to become price inelastic, the degree of inelasticity and whether elastic behaviour applies to all airlines all of the time – that is, are LCCs willing to abandon a successful route?[23] Also, the success of two-stage pricing is dependent on whether the airport is being undercut by another airport. Does the airport risk losing the air service to another airport once it attempts to increase the rate?[24]

Facilitating airline efficiency to reduce 'price' Airport design, layout and processes play a critical role in airline operating costs. For example, measures that enable airlines to achieve faster

---

Canadian Airports', *Journal of the Transport Research Forum*, vol. XXIX no. 1, 1988, pp. 28–34 – estimated the elasticity to be between –0.02 and –0.08, considerably below unity. The degree of sensitivity will likely depend on the type of air service operated. In general, landing fees represent a higher proportion of airline operating costs for short-haul services than for long-haul services, and so, operators of short-haul services, such as LCCs, are likely to be more price elastic in regard to landing fees.

21   Another example, all too familiar to parents, would be gaming systems such as PlayStation and Xbox, where the consoles are virtually given away, but the games themselves are priced at a premium.

22   *Communication from the Commission – Community Guidelines on Financing of Airports and Start-up Aid to Airlines Departing from Regional Airports*, 2005/C 312/01, 9 December 2005.

23   For example, Ryanair has reacted to airport fee increases (or failure to reduced fees) by switching service to other airports. One instance of this was Ryanair's decision in 2005 to operate increased service at Nottingham East Midlands rather than at Birmingham in protest to Birmingham Airport's higher fees.

24   Airports must also be wary of not upsetting existing airline customers by offering discounts to new entrants, from both a business and legal point of view.

turnaround times can have a major impact on an airline's bottom line and ultimately enhance the airport's competitive position. Faster turnarounds can have major benefits for the airlines:

- *Lower unit costs.* Increased utilisation of aircraft enabled by faster turnarounds mean that the capital and fixed costs of the aircraft are distributed over a greater number of passengers.
- *Higher revenue generation.* Faster turnarounds enable an airline to increase the amount of the time the aircraft is in the air, generating greater revenue.

There are a number of ways in which airports can facilitate airline efficiency:

- *Dual boarding bridges.* These bridges allow for much faster loading and unloading of passengers from aircraft. In April 2003, WestJet installed a duel boarding bridge (also known as an over-the-wing bridge) at Calgary Airport to be used for WestJet's 737 aircraft, reducing unloading/boarding times by roughly ten minutes.
- *Taxiway design.* Airports can design the runway and taxiway layouts in order to minimise time spent taxiing.
- *Swing gates.* Airports can design terminal facilities and gates (or at least some gates) that can be switched from domestic to international services, enabling airlines to switch an aircraft from one service to another without towing the aircraft between gates.
- *At-gate hydrant fuelling.* This enables the airline to reduce fuelling times relative to a fuel bowser system.
- *Ground power.* This can reduce the airlines' fuel costs while at the gate.
- *Dedicated facilities for shuttle services (or other types of services).* As already mentioned in the product section, dedicated facilities can reduce aircraft turnaround times.

Incentive pricing Airports can become more creative in how they charge for services, develop pricing schemes that meet the needs of the airlines and incentivise airlines to make better use of airport facilities. For example, rather than charging airlines a standard price per use of a gate, the airline could be offered the option to rent the gate for the whole day for a fixed daily amount. Doing so would provide airlines with an incentive to increase utilisation of the gate in order to reduce unit costs – spreading the fixed daily charge of the gate over a greater volume of traffic. If designed properly, the airport would still receive the same revenue per day from the gate, but will have enabled the airline to reduce its unit costs. Incentive pricing would allow better use of airport capacity, reducing the need to expand capacity (or at least delaying it). The latter outcome has cost benefits for both the airport and the airlines. In addition, lower unit cost may eventually be passed on in lower fares, stimulating the airport's traffic.

This pricing scheme may not be attractive to all airlines for all services. For example, an airline operating a long-haul service that uses the gate once in a day may prefer to pay per use, while an LCC operating a short-haul service may prefer a daily gate rental. The important point is for the airport to provide its customers with a greater choice of pricing options in order that the airline can match the pricing scheme with its own economics.[25]

---

25 A parallel, if imperfect, example would be mobile phone providers, who offer a wide range of pricing schemes in order to attract a greater range of customers.

Promotion

In the increasingly competitive environment that airports operate in, promotion has become an essential part of airport marketing.

Air service development Many airports now have very active Air Service Development (ASD) programmes to attract new air carriers to the airport and expand existing air services. ASD programmes are a highly targeted approach which connects an individual airline's needs and opportunities with the offerings of the airport. These programmes attempt to demonstrate to air carriers that there is sufficient demand, and suitable airport facilities, to profitably operate a route from the airport. In effect, airports are attempting to divert airline capacity from other possible routes or airports.

Passenger marketing While many airports now commit resources to promoting the airport to airlines, it is also necessary to ensure passenger, travel agency and shipper awareness of available airport services. This is especially important for secondary airports in a region and for secondary international gateways.

For secondary airports, residents in the airport's primary catchment area may have a general awareness that some services are available, but may continue to patronise the major regional airport due to lack of awareness of the full spectrum of flights. They may not be aware of the expanding range of destinations, increased frequency on existing routes and especially of one-stop/connection opportunities. Consider a major airport (AAA) in City A with hourly flights to City B and flights to many other destinations. A secondary airport (let us call it AAZ) may only have three daily flights to B. However, when ground access time, check-in, parking etc. are also considered, flights from AAZ may be of similar convenience to those from AAA. While flights from AAZ may be less frequent, savings in drive time, parking time, check-in time and security processes may make this option equally or more attractive than using AAA. In addition, the three flights from AAZ to B may have convenient connections to destinations C, D etc., making AAZ a competitive alternative to driving to AAA for a non-stop flight to C etc.

Secondary airports must communicate flight options to tour operators, convention planners, travel agents, freight forwarders and the general public. This can involve sending representatives to aviation industry and tourism conferences, putting on roadshows and advertising in the trade, local and national press. The aim of these efforts is to create an awareness of the airport's specific flight opportunities. Often this promotion will focus on the attractive characteristics of the airport. For example, a secondary airport may place an advertisement in the local paper highlighting the airport's short drive time (perhaps to combat the perception that it is remote) or the availability of lower-cost and more convenient parking.

Airports need to create awareness not just in the local market but also at the spoke ends. This is a more challenging task, but is particularly important for secondary airports and those attempting to establish themselves as secondary international gateways. Continuing the City A to City B example, the challenge for Airport AAZ is to create awareness among consumers and travel agents in City B that AAZ is a) an option they might not have been aware of for getting to City A; and b) more convenient for certain parts of the City A region.

For secondary international gateways, the challenge is to create awareness of connection opportunities at spoke ends on another continent. For example, Vancouver Airport routinely conducts visits with travel agents and major businesses in Asia to make them aware of the benefits of using Vancouver as an alternative to Los Angeles or San Francisco for onward travel to destinations in

North America. (Vancouver provides a shorter routing to many North American destinations than airports further south such as LAX and SFO.)

Integrated marketing approaches For nearly all air passengers, the aviation product is just one part of the total travel experience. Therefore, it makes sense for the airport to integrate its marketing efforts with those of other members of the travel supply chain, such as national and local tourism authorities, airlines, hoteliers, cruise lines and other relevant players.

One opportunity is for the airport to take a role in co-ordinating the marketing approach of the various market players. For example, the tourism authority may be focussing its resources on attracting more tourists from the North American market while the air carriers are increasing capacity in the Asian market. These two efforts do not mutually reinforce each other. Greater impact could be obtained if the tourism authority were to direct advertising and marketing dollars in those markets to which air carriers have deployed new capacity. The airport can work with the two groups in order to ensure that all parties are working in the same direction.

Naming the airport[26] Considerable attention is paid by some airport management to the name of the airport. The name gives an immediate indication of the type of service and major city served by the airport. The most common manifestation of this is the use of the word 'International' in the airport's name to demonstrate the scope of the airport's services, even if it only serves one international destination. For secondary airports, another approach is to include the name of the nearby major city, even if it is some distance away, for example:

- *London* Heathrow (the same for Gatwick, Luton and Stansted);
- *Brussels* South Charleroi Airport;
- Baltimore/*Washington* International Airport;
- *Hamburg* Lübeck Airport.

As discussed below, this naming convention, when combined with how the airport is linked to city names, can also have implications for reservation systems. The naming of airports is, of course, controversial. In 2003, Ryanair was ordered by a German court to refrain from using the word 'Düsseldorf' for Weeze/Niederrhein Airport, 70 km from the city of Düsseldorf.[27]

Branding Airport branding has also developed as a practice used by some airports over the last few years. Through naming, logos, styling and merchandising airports have attempted to develop a consistent and recognisable brand for the airport. However, there is little evidence that this branding gives the airport any competitive edge. As pointed out by Anne Graham, this branding may actually confuse the consumers due to the prevalence of airline and airline alliance brands that also exist in the industry.[28]

---

26   Some of this discussion draws from Chapter 7 (The Role of Airport Marketing), in Graham, *Managing Airports: An International Perspective*.
27   We note that Ryanair continues to use the word 'Düsseldorf' for this airport.
28   Graham, *Managing Airports: An International Perspective*, p. 170.

Physical Distribution

Even with an outstanding product that is competitively priced and effectively promoted, an airport still needs to ensure that its product gets to the final consumer.

Computer reservation systems   Whether through a travel agent, the airline or through the internet, virtually all airline tickets are booked through a computer reservations system (CRS). When a user makes a search for a flight, the CRS will rank the available flights based on a number of factors, including price, flight times and connections. Ranking can have a major impact on consumer decisions as to which flight to book as, typically, travellers select from only the top few flights listed. The CRS will generally increase the ranking of connecting flights with shorter connect times.

As has already been discussed above, airports can adopt a number of strategies to minimise passenger connect times and thereby move their airport up the rankings on the CRS. For any airport wanting to compete for connecting traffic, minimising connect times must be a core focus. Airport naming and registration can also have an impact in CRS systems. If a traveller searches for flights from Washington, DC to New York on travelocity.com, options appear for flights from Reagan National, Dulles and Baltimore/Washington, despite the latter airport being 53 km from Washington, DC.

Travel agents   Despite the increased use of the internet, travel agents still book a considerable share of airline tickets. As a result, secondary airports need to promote the airport to travel agents (both locally and at the spoke ends) and develop close relations with them. Unless a local travel agent is aware of the service provided by a secondary airport, they are highly unlikely to direct travellers to use that airport. Airports will routinely send out information on routes, fares, facilities and timetables to travel agents. Some airports have gone even further and taken matters entirely into their own hands by buying up travel agents; both Cardiff and Norwich airports in the UK have followed this strategy.

Airport websites   For many travellers, and even travel agents, the internet has become the primary source for obtaining information relating to their travel and for making bookings. Although airport websites play a part in airport promotion, they can also have a role in distribution. The website can be used to direct travellers to the airlines operating from the airport so that they can make bookings. Some more sophisticated sites (such as that for Hamburg Airport) actually allow the user to make bookings on the airport website for flights from the airport.

Airport websites can also act as a 24-hour information source on the airport. As well as flight information, they tell travellers how to reach the airport, and give details of parking arrangements and the range of services provided at the airport. In some cases, passengers can also pre-book car parking, car rental, foreign exchange and other services through the airport website.

Conclusions

This chapter has attempted to show that airports do compete with each other in a number of major business segments. The emergence of secondary airports has created intense competition for some major airports. In addition, major airports compete among themselves for connections, for gateway traffic and even for destination.

The chapter outlined a number of dimensions in which airports can compete. The product offered by an airport will affect both carrier and consumer choice. Of special importance is designing the airport product to reduce operating or capital costs for air carriers. Price competition was discussed, with emphasis on a) the two stages of air carrier price elasticity – leading to different pricing policies for the different stages; b) finding means to lower airline costs, such as through competitive choices for various services; and c) investigating use of incentive prices which allow air carriers to lower their unit costs by making more efficient use of the airport's infrastructure. Promotion of the airport is especially important for secondary airports in its catchment region and for secondary international gateways. Promotion does not imply airports should simply advertise. Rather, promotion should focus on creating awareness of specific services at an airport and advantages such as reduced driving times for some parts of the region. Finally, physical distribution was discussed, that is, the means by which the airport's services are delivered to consumers and carriers. Airports must pay attention to how they appear in CRS systems and on the internet so as to access the greatest number of passengers possible.

Chapter 10
Airport Competition for Freight

Dr Michael W. Tretheway and Robert J. Andriulaitis

Introduction

Air transport has involved the transport of both passengers and cargo from its earliest days. Yet most of the discussion and focus by media, policy makers, airports, and even airlines is on the transport of passengers. Cargo is often overlooked, despite the important role it plays for many carriers around the world. For example, air cargo represents a significant proportion of total revenues for carriers such as Lufthansa (12.2 per cent), Air France/KLM (12.6 per cent) and cargo leader Korean Airlines (28.7 per cent).[1] Moreover, air cargo plays a critical role in international trade. Although only about 2 per cent of international trade by weight moves by air, this represents more than 40 per cent of international trade by value.[2]

As a result of the focus on passenger transport, general understanding of air cargo tends to be low and perceptions of the nature of the industry, particularly the nature of competition, are coloured by passenger issues and considerations. Freight transport, however, differs from passenger transport in ways that affect the nature of competition between airports. In particular, the ability of airports to successfully compete as gateways for cargo traffic is not fully appreciated. This chapter offers a short, high-level review of those issues which affect the degree of airport competition for freight.

The chapter begins with a brief look at some of the main characteristics of air freight and then considers the elements which affect airport competition for freight.

Characteristics of Air Freight

Supply Side: Freighter versus Combination Air Cargo Services

Air cargo capacity is provided in two main types of services. 'All-cargo services' are provided by freighter aircraft whose entire capacity is dedicated to air cargo. 'Combination services' are passenger aircraft that carry passengers on the main deck and provide cargo capacity in their bellyholds. Boeing estimates that 47 per cent of the world's air cargo is carried by freighters and 51 per cent is carried in the bellyholds of passenger aircraft (with the remaining 2 per cent of the traffic carried on combi aircraft).[3]

---

1   *Source*: 2007 financial reports of the respective carriers.
2   The International Air Cargo Association (2003), 'It's Time to Set Air Cargo Free', *TIACA Times*, Winter.
3   Boeing, World Air Cargo Forecast, 2006/07.

Combination Service Air Cargo is a By-Product of the Transportation of Passengers

Air cargo capacity in combination service is a by-product of passenger service. Indeed, the design of passenger aircraft is dictated by passenger needs; space for cargo is simply what is left over in the otherwise unusable space below the main deck that is not required for passenger luggage and that exists simply because of the aerodynamic needs for a tubular shape for the fuselage. Air cargo is also a by-product in the sense that it is generally transported by air to where people want to travel, which is not necessarily where the cargo needs to go. Combination carrier route networks are predicated on where passengers want to travel. Even combination carriers that operate dedicated freighters tend to operate these aircraft to the same airports as their passenger services because of operational and cost synergies.

Marginal Cost Pricing is a Major Factor in Setting Cargo Rates for Combination Services

As cargo supplied in combination services is a by-product of the transportation of passengers, such carriers tend to focus on passenger revenue to cover the cost of the service, with cargo revenues viewed as incremental. The result is that much combination service cargo is often priced at rates that do not cover a (weight-based) proportional share of total flight costs. Rates may need to cover the incremental fuel and handling costs of carrying the cargo, but might not be expected to cover fully allocated costs. An examination of the Swedish air freight industry found that: 'airlines offer competitive pricing against the integrators and dedicated cargo carriers, since their cost is marginal on the capacity of the pax-belly operations'.[4]

The tendency toward marginal cost pricing by combination carriers has an impact on freighter rates as well. Freighter operations must recover sufficient revenue from cargo to cover all flight costs. But the fact that competing bellyspace is priced low serves to limit pricing for many freighter shipments. This puts very high competitive pressures on freighter operations. The International Air Cargo Association (TIACA) noted that '[f]reighter operators consistently complain that the viability of their operations is undercut by low cost belly operations.[5] In this environment, airport costs become a significant factor for air freighter operations, pushing this competitive factor faced by freighter operators down onto airports as well.

There are some limitations to the impact marginal cost pricing by combination carriers has on freighter operations, as bellyhold cargo may be viewed as lower in quality than main deck space. One reason is that bellyspace is dimension limited in what it can accommodate. Main deck space offers greater flexibility for palletized and unitized cargo, not to mention oversized cargo. Combination carrier capacity also moves at times and on routes dictated by passenger needs, whereas freighter capacity can be customized to the routes and times preferred by shippers. There is also a service reliability shortcoming of combination carrier cargo capacity. The fact is that the available bellyspace may be prioritized for passenger luggage, and cargo shipments may be

---

    4   Efstanthiou, Efstanthios and Niclas Anderson, *The Swedish Air Freight Industry*, Logistics and Transport Management Master Thesis No 2000:34, Graduate Business School, School of Economics and Commercial Law, Göteborg University, p. 38. This was a study initiated by the Swedish Institute for Communication Analysis and the Swedish Civil Aviation Authority. The UK Department of Transport, following the publication of the Integrated Transport White Paper, conducted a series of research projects and found a similar situation. The first report assembled data to describe the industry, and noted that '[t]he sale of freight capacity is essentially at a low marginal cost to passenger airlines,' *UK Air Freight Study Report*, http://www.dft.gov.uk/adobepdf/165217/ukairfreightPDF, p. 67.

    5   TIACA, 'It's Time to Set Air Cargo Free'.

'bumped' and required to await the next available flight. Nevertheless, a broad range of shipments may still find the service quality of combination carrier capacity to be acceptable (for example, the alternative may be much longer shipping time by ocean carrier) and this will constrain the prices which can be charged for freighter services.

Air Cargo Yields Have Steadily Declined Over Time

The International Air Transport Association (IATA) and others that track yields show significant drops in actual yields over time.[6] Improved aircraft performance has dropped unit costs, development of unit load devices has lowered handling costs, and a more open and competitive environment, with ever increasing numbers of freighters and combination carrier belly capacity being added to the world's fleet, have added supply pressure to lower rates. This has led to new markets developing (both in the type of goods now being moved by air and new service points), improved the viability of formerly marginal services, and increased competition between airports, routes, carriers, and even other modes of transport such as trucking.

Air Cargo Tends to Moves in One Direction

Unlike passengers, who tend to return to their origin point, cargo typically does not. Thus traffic on routes is not directionally balanced. While combination flights are confined to the needs of passengers, freighters can follow routings based on the needs of shippers. This makes issues such as fifth freedom rights important for freighters as an impediment or barrier to entry (where they do not exist) or as an opportunity (where they do). As a result, freighter carriers can operate more convoluted routes, and can serve non-traditional service points, which can add traffic to lower-volume legs. This increases the amount of competition faced by airports.

Product Differentiation is on the Rise

Air cargo is sometimes viewed as a commodity. However, there are not only differences between bellyspace and main deck space that are relatively well recognized; today, even among main deck operations, there are also a variety of services offered. These include temperature control, specialty products (pharmaceuticals is a key one), next day time-definite service, deferred time-definite service, non-time-definite service, and so on. This gives shippers much greater options to match service and price to their specific needs, and increases the overall demand for air cargo. While there is a large portion of the market for which cargo lift of different operators is readily substitutable, there are some shippers willing to pay a premium for differentiated services.

Night-time Operations

Unlike most passengers, who prefer to travel during the day, the preference for shippers is to move air cargo, particularly packages, at night.[7] This allows them to ship their entire day's production at the end of the working day, for delivery at the beginning of the next day. It also allows them

---

6   The Air International Air Cargo Association notes that cargo yields have fallen by 50 per cent in the last 20 years, and are forecast to continue to fall at 1 per cent per year. (Air Cargo in the 21st Century, http://www.tiaca.org/articles/2001/04/24/3072ACDCB6254A11AAB7FC6CE8FFE89D.asp).

7   Unfortunately for air cargo, night-time operations tend to be viewed negatively by those living in the vicinity of airports.

to receive their inputs first thing in the morning in time for production. This creates specialized demand for cargo services on routes where combination carriers likely do not offer them.

We now discuss in greater detail how these factors influence the amount of competition among airports for freight.

Key Elements of Airport Competition for Freight

A typical view of air cargo is that air carriers control the freight market, routing cargo according to their operational needs. The role of airports, in this viewpoint, is as passive recipients of air cargo flights. There are, however, several key factors and developments that suggest competition among airports for freight is much higher than perceived, and that this competition limits the ability of service providers to control prices.[8]

Shift in Market Power

A key element in understanding competition for air cargo is a relative shift in market power from air carriers to freight forwarders. Historically, the freight forwarding industry was characterized by a very large number of small local operators providing similar services in a highly competitive environment. There were global operators, but even the largest had a small market share. Individually, freight forwarders had limited influence over air carrier decisions regarding routes and service times, and, due to their large numbers and high degree of competition, collective action was simply not viable. Three developments have occurred that are changing the balance.

First, the freight forwarding industry has undergone considerable consolidation. Historically there were some large global companies, but none had the scale to really exercise market power. Today, there is a handful of large, international forwarding operations that control large amounts of traffic. Certainly there are still, and will continue to be, many small and mid-sized operators, as they can offer a degree of specialization or a combination of scope and flexibility that the large operators cannot provide. Nevertheless, the industry had never before seen such a degree of forwarder concentration as exists today. Given the amount of traffic these large freight forwarders control, they can have considerable influence over air carrier routing decisions, and are using that influence to develop cargo hubs at locations they want.

Second, some freight forwarders have gone beyond merely booking space on passenger and freighter flights of carriers. They are now offering their customers dedicated freighter services. Panalpina pioneered this trend in the 1990s by offering dedicated freighter services between Luxembourg and Huntsville, Alabama – the 'Dixie Jet.' (The actual flying is done on a charter basis from existing air carriers such as Cargolux and Atlas, though Panalpina controls the whole capacity of the aircraft.)

Third, key freight forwarders have merged with key air service providers to blur the traditional lines that separated these parties. Unlike the situation above, where forwarder and air carrier remain separate organizations, though closely aligned, we are also getting cross-organizational blending. For example, the European freight forwarder Schenker has acquired US BAX Global, and freight

---

8   For example, the International Civil Aviation Organization (ICAO) convened a conference in Montreal in September 2008 to discuss options for enhanced economic oversight of airports. While the ICAO does acknowledge that airport commercialization and privatization have increased competitive pressures, it evidently still considers airports to have significant market power.

forwarder Danzas has merged forces with global integrated carrier DHL to become major forces in global air cargo (see Table 10.1).

Historically, airports seeking to develop air cargo activity focussed their energies on proposals to attract *air carriers* to begin service. The shifts in market power that are taking place, however, suggest that airports need to also focus on attracting *freight forwarders* to their airport. Freight forwarders bring with them their market – the hundreds, if not thousands, of shippers who rely on them to arrange their transportation, distribution and, increasingly, logistics functions. By attracting a forwarder, the activity the forwarder handles in turn will be a key element in attracting (or even providing) air services. This opens the door to non-traditional international gateways and hubs based on changing forwarder market needs rather than relatively unchanging airline passenger needs. This is especially the case given the increasing use of road feeder services by forwarders, discussed in the following sections.

Table 10.1   Snapshot of the making of a logistics giant: Deutsche Post AG

1999: Deutsche Post AG acquires Danzas, one of the top European freight forwarders with 29,000 employees
1999: Air Express International, the leading US provider of international freight with branches in 135 countries, is acquired as a Danzas subsidiary
2001: DHL, a key global integrated carrier, joins the family
2003: Airborne Express is acquired, though the air operations had to be spun off to keep from violating US foreign ownership laws regarding air carriers
2005: Exel, a key UK freight forwarder, with 111,000 employees in 135 countries, is acquired
2006: Express and logistics services earn 40 billion Euros (US$59 billion), twice the size of FedEx

Gateway Proliferation

There traditionally have been a limited number of primary international air cargo gateways in Europe, Asia and North America. Many of the historical gateways (Frankfurt, Amsterdam, Paris, London Heathrow, Hong Kong, Narita, Seoul, Singapore, New York, Los Angeles, Chicago and Miami) primarily developed as major passenger gateways, and are still important factors in this role.[9]

However, as medium-capacity, long-range passenger aircraft came into service to supplement the original jumbo jet services, air carriers expanded the number of international gateways they served. As a result, air cargo now enters a continental market from a greater number of competing gateways. This adds to the competitive pressures faced by airports as forwarders and shippers have a greater choice of viable gateway options. Table 10.2 shows the top European cargo gateways in 2007.

---

9   According to Airports Council International (ACI), the global rank of these historical passenger gateways in total tonnage of cargo is Frankfurt (8th); Amsterdam (14th), Paris (6th), London (18th), Hong Kong (2nd), Narita (7th), Seoul (4th), Singapore (11th), New York (16th), Los Angeles (12th), Chicago (17th) and Miami (10th). *Air Cargo World*, July 2008.

Table 10.2 Top European cargo airports

Airport	Code	Country	World rank	Cargo in tonnes (2007)	Change
Paris de Gaulle	CDG	France	6	2,297,896	7.8%
Frankfurt	FRA	Germany	8	2,169,025	2.0%
Amsterdam	AMS	Netherlands	14	1,651,385	5.4%
London Heathrow	LHR	UK	18	1,395,909	3.9%
Luxembourg	LUX	Luxembourg	23	856,740	14.0%
Brussels	BRU	Belgium	26	728,689	2.1%
Cologne	CGN	Germany	29	710,244	2.8%
Liège	LGG	Belgium	38	489,746	20.6%
Milan Malpensa	MXP	Italy	40	486,169	16.0%
Copenhagen	CPH	Denmark	46	395,506	4.1%
Madrid Barajas	MAD	Spain	53	356,427	1.0%
Istanbul	IST	Turkey	55	341,514	14.7%
East Midlands	EMA	UK	60	304,853	1.2%
Zurich	ZRH	Switzerland	63	290,653	3.4%
Munich	MUC	Germany	69	265,607	0.7%
London Stansted	STN	UK	80	228,759	6.9%
Vienna	VIE	Austria	86	205,045	1.6%
London Gatwick	LGW	UK	95	176,807	19.7%
Manchester	MAN	UK	98	166,438	10.0%
Rome	FCO	Italy	104	154,439	6.1%
Helsinki	HEL	Finland	108	139,328	3.3%
Milan Linate	BGY	Italy	112	133,941	4.0%
Moscow Domodedovo	DME	Russia	113	133,662	5.8%
Moscow Sheremetyevo	SVO	Russia	114	128,152	5.4%
Stockholm Arlanda	ARN	Sweden	119	122,922	10.5%
Athens	ATH	Greece	120	118,959	1.0%
Hahn	HHN	Germany	124	111,687	0.5%
Ostend-Bruges	OST	Belgium	126	108,952	10.6%
Dublin	DUB	Ireland	127	107,921	7.5%
Barcelona	BCn	Spain	129	100,009	1.3%
Oslo	OSL	Norway	132	97,310	7.9%
Paris Orly	ORY	France	134	94,920	0.4%
Lisbon	LIS	Portugal	135	94,693	4.8%
Hamburg	HAM	Germany	141	86,997	5.3%
Leipzig	LEJ	Germany	144	85,361	NA
Lille	LIL	France	163	68,413	8.0%
Göteborg	GOT	Sweden	171	61,790	3.6%
Keflavik	KEF	Iceland	172	61,534	0.4%
Budapest	BUD	Hungary	176	58,885	9.6%
Billund	BLL	Denmark	178	58,612	4.7%

Expansion of Catchment Areas through Road Feeder Services (RFS)

Improved trucking reliability and service quality (e.g., the ability to track shipments, lower theft/damage/pilferage rates), coupled with rates lower than air carriers can offer, has put trucking in a position where it both competes with, and complements, air services. This means that airports in Brussels or Amsterdam can (and do) readily serve as gateways for more distant points. The converse is also true: more distant points such as Cologne, Frankfurt and even emerging Eastern European airports can serve as gateways for the Benelux market. A glance at any air cargo carrier's timetable will show how prevalent this is: many of the 'flights' operated between European points are actually road feeder services. As an example, the July 2008 Official Airline Guide (OAG) shows that Lufthansa flight 7387, operated each weekday from Frankfurt to Stockholm, is actually operated by truck. In fact, the OAG shows that in the month of July 2008, Lufthansa operated 10,069 one-way 'flights' by truck around the world.

The ability of trucks to move goods quickly and economically between competing gateways also means that minor price variations between air services at different airports will result in traffic shifts, often on a day-to-day basis. Goods will move to/from more distant gateways if they offer services at lower prices than the local gateway. Relatively small differences in air cargo rates can justify forwarders trucking consolidated shipments to an alternate airport. In effect, trucking acts as an arbitrage service, allowing shippers to keep rates competitive between points. It is thus incorrect to view, for example, the 'New York–Germany' market as simply what moves between JFK and Frankfurt airports. It also has to take into account services between points such as Newark, Toronto, Chicago, Philadelphia, Montreal, Boston and so on, and Cologne, Paris, Amsterdam, Luxembourg, Brussels and so on. Trucking is the great equalizer of air cargo rates among gateways.

Catchment Area Airport Competition

Even within a community's catchment area, there can be multiple airport choices. A number of military airfields have been reborn as cargo airports in Europe and the US, and China is considering the concept. Much as the operations of Ryanair at secondary airports tends to decrease passenger fares at the primary airports, the availability of air cargo capacity at low-cost secondary airports can act as a cap on cargo rates at the major airports.[10]

Thus while major passenger airports such as Paris, Hong Kong, Dallas and Toronto remain key cargo hubs, the congestion getting in and out of these major hubs means that cargo activity is also moving to nearby airports such as Vatry, Shenzhen, Fort Worth Alliance and Hamilton.

Importance of Night Operations

As noted previously, the preference for many shippers is to move air cargo, particularly packages, at night. Airports thus are competing on the basis of their operating windows. Given the significant growth rates for air cargo (they exceed passenger growth rates) cargo airports need to operate on a 24-hour a day basis to be successful in this industry.[11] The growing concern about environmental impacts of aviation, including noise, opens up significant possibilities for airports that are located

---

10   For a review of the literature on the effect of service at secondary airports on passenger air fares, see M. Tretheway and I. Kincaid (2005) 'The Effect of Market Structure on Airlines Prices: A Review of Empirical Results,' *Journal of Air Law and Commerce*, vol. 70 (3): 467–98.

11   Boeing's 2007 Current Market Outlook indicates annual passenger growth of 5 per cent from 2006 to 2026, but cargo growth of 6.1 per cent per year over the same timeframe.

in areas that allow them to guarantee 24-hour operations. Secondary points in the US and Canada, and points in Eastern Europe are becoming more viable options for cargo services. Airports that have curfews or noise quotas are less attractive for cargo operations. Some existing air cargo hubs such as Frankfurt and Brussels have had issues regarding air cargo operations because of this. Frankfurt had to limit night operations to secure approval for a new runway, and Brussels lost its DHL hub to Leipzig, Germany, because of a night-time operations ban.

Airport Cargo Fees

Air cargo carriers tend to have lower operating costs than combination carriers since they have lower crewing costs, reservations systems costs, terminal costs and so on, on a per flight basis. Airport landing fees and charges thus can represent a relatively higher proportion of total operating costs for cargo carriers than for combination carriers.[12] This in turn means airport charges have a relatively higher impact on cargo carrier routing decisions. Airports thus have an opportunity to attract services, and are in competition with each other to offer the best package possible to air carriers. For example, Ethiopian Airlines' recent decision to move from Brussels to Liège was a combination of costs, speed of customs clearance and inspections, and the specialized handling services available at Liège.[13] This in turn stimulates the competitiveness of the industry as a whole.

Air navigation fees also play a role. In Canada, NavCanada charges a modest overflight fee on flights from the US to overseas points in Europe and Asia. In contrast, cargo flights that land in Canada, even if they stop while en route between the US and Asia/Europe, are charged much higher terminal and en route fees. While airport operators have shown themselves willing to compete by offering carriers reduced fees or other incentives, this is not the case for air navigation service providers. Within the EU, the move to a Single European Sky should eliminate anomalies such as this.

Facilitation

In addition to the improved operational ability of trucking to replace air services, the ability for trucks to effortlessly cross borders (at least in most of Europe) has made the whole air cargo industry more competitive. Easier cross-border flows expand the number of competing gateways for any given shipment as foreign gateways become a more viable alternative. In Europe, the Schengen free travel area makes trucking between competing airports much easier. With the 8 November 2007 decision to expand the Schengen agreement to include Poland, the Czech Republic, Estonia, Lithuania, Latvia, Malta, Slovakia, Slovenia and Hungary, the competitive appeal of Baltic and Eastern European airports for cargo has been increased.

Post 9/11, this is a more difficult proposition in North America. However steps such as the Smart Border Accord, the Customs-Trade Partnership Against Terrorism (C-TPAT) and the Free and Secure Trade (FAST) programme between Canada and the US are designed to speed up commerce between the two countries, while still meeting security and safety concerns.

---

12 It should be noted that as fuel prices continue to escalate, this will impact some freighter operations relatively more heavily than combination carriers, as many freighters are relatively old and the heavier payloads translate into higher fuel burn. In this situation, airport fees become a smaller component of total operating costs. Airport location starts to play a larger role.

13 *Air Cargo Week*, 7 July 2008.

Air Cargo Facilities

The provision of air cargo facilities by airports is becoming an increasingly significant competitive factor. Particularly in the case of freight forwarders, the provision of infrastructure by the airport and of facilities by a third party provider (or even the airport itself) will be a key to securing significant cargo activity. In the past, the need to make capital investments at an airport may have hindered developments at all but the largest gateways, since investors would have to be assured of long-term commitments on airport charges and facilities in order to recoup the capital costs. Sinking capital into an airport ties the investor to that airport, making change more difficult.[14]

Other than the integrated carriers, which have dedicated highly automated facilities, most airport cargo facilities are basic, low-productivity facilities. While this has sufficed in the past, global supply chain pressures are suggesting a need for high-productivity centres. Few operators, however, can justify the investment in a high-productivity facility based on their individual traffic volumes. For them, a multi-tenant facility is likely the more viable choice, and an airport that offers this a relatively more attractive choice. This leaves airports with a competition vs consolidation dilemma. On the one hand, competitive facilities and operations would be desirable from the perspective of the cargo customer as it would help keep rates down. On the other hand, to get sufficient volumes to justify an investment in a high-productivity facility may require that the airport consider consolidation of cargo terminal operations. This is likely to require airports to take on a higher stake in facilities investment and absorb more of the risk. Undoubtedly, different airports will move along different paths; but if sufficient airports are interested in such facilities, the end result is likely to be an overall more competitive environment. The alternative is for airports to encourage a third party, such as a 3PL or freight forwarder, to make the investment, which would reduce the risk to the airports but leave the investor with some leverage over the airport.

Limits to Competition

The previous section reviewed some of the elements that are driving increased competition for freight among airports. Here we discuss some elements which might serve to limit the effectiveness of competition among airports.

Network Effects

Earlier it was noted that secondary airports serving the same community as a major hub have become competitive forces for air cargo. This, however, is limited by type of aviation activity.

Air carriers enjoy economies of density, in part from concentrating services at certain airports. It generally makes little economic sense for them to set up duplicate stations at separate airports in the same community, unless capacity at the primary airport is an issue or if ground transport congestion creates a marketing opportunity for a submarket in the wider urban region. Pure freighter operators, whether general heavylift or integrated carriers, are free to select any airport in the vicinity of the community they want to serve that makes economic sense for them, since they are not tied to passenger interest in the principal airport. They can, and do, use airports such as Paris Vatry instead of Paris Charles de Gaulle. Combination carriers, on the other hand, are tied to

---

14 It should be noted that sometimes some shifts do take place. DHL's move from Brussels to Leipzig is a recent example.

airports that passengers want to travel to. In order to capitalize on economies of density, they will co-locate passenger and freighter operations at the passenger hub. This limits the ability of airports such as Vatry, Shenzhen, Fort Worth Alliance and Hamilton to attract combination carriers, even for their freighter operations.

Specialized Investment

Heavylift air carriers tend to have relatively basic facility requirements. They generally use common-use facilities, unless their volumes are such that it warrants investment in a dedicated facility. As a result, they generally have little or no sunk costs in a particular airport, and can move relatively easily and quickly to a competing location.

Integrated carriers, on the other hand, require significant investment in highly specialized facilities, particularly for their major sorting stations. The high capital investment required may lock integrated carriers into their chosen hubs. It should be noted, however, that DHL has already moved its hub from Brussels to Leipzig and FedEx has announced it will relocate its German gateway from Frankfurt/Main to Cologne/Bonn in 2010, so this is not necessarily a binding constraint. In particular, the DHL example shows that airport competition can occur over significant distance. Nevertheless, this does serve to limit how much competition established integrated carrier hubs will experience from competing airports.

Government Policy

The traditional system of international air bilateral agreements has historically limited foreign carrier access to a specified airport(s). Thus air carriers had limited, and in some cases no, choice in which airports they could serve in a given country. Potentially they could access a market from airports in adjacent countries, but the impact of this policy would be to significantly limit the amount of competition faced by airports.

The opening up of the EU's internal aviation market and the trend towards open skies agreements globally has, however, increased access to competing airports, and enabled air carriers to select from a wide range of potential sites.

Conclusions

There are a number of factors related to airport competition for freight which suggest that the air cargo market is highly competitive.

- *Shift in market power.* The growing power and influence of freight forwarders has shifted the relative bargaining power from air carriers to freight forwarders. Unlike air carriers, forwarders are not tied to existing passenger gateways or hubs, and are open to new alternatives that better serve their customers. This has opened up new and more competitive options for shippers, such as the Panalpina Luxembourg–Huntsville Alabama service, and enables new airports to more effectively compete for traffic.
- *Gateway proliferation.* The addition of numerous new international passenger gateways provides shippers with new routing options, and increases the competitive pressure on established and developing cargo gateways.
- *Expansion of catchment areas through RFS.* The expansion of the geographic scope of an

airport's catchment area through expedited trucking services means numerous combinations of gateways/air service providers are now available to a wider range of shippers. With cost- and service-effective alternatives available, airports are not able to exploit 'captive' traffic, since traffic is not captive.

- *Catchment area airport competition.* The development of secondary airports in or near existing gateways gives air carriers/forwarders/shippers new options to consider that can be more cargo friendly, and less costly. This puts pressure on the existing gateways to control costs and offer a high level of service, particularly for pure freighter operators. This may, however, have limited effect on integrated carriers who have significant investment in hubs, and combination carriers who are tied to the major passenger hubs.
- *Importance of night operations.* The growing need for night-time capabilities for cargo operations opens the door to other airports that can offer 24-hour service. As cargo operations come under increasing pressure at existing hubs (for example, Brussels, Frankfurt), the opportunity is there for other airports to compete for and capture that traffic.
- *Air cargo fees.* Airport competition based on fees and charges can have an influence on air carrier routing decisions, particularly in air cargo where margins are low and airport costs make up a relatively significant portion of airline costs. Aggressive pricing by airports again enhances competitive choice.
- *Trucking facilitation.* Border facilitation initiatives such as Schengen in Europe and FAST/C-TPAT in North America serve to improve the competitiveness of long-haul trucking. By making borders disappear, competing gateways in other nations effectively become 'local' airports, and can offer competition for many other locations.
- *Air cargo facilities.* The provision of air cargo facilities by third party providers or airports themselves is an effective competitive tool for attracting service. Airports willing to take on that risk become effective competitors to existing and developing gateways.

PART B:
Traveller Choice and Airport Competition

Chapter 11
Modelling Air Travel Choice Behaviour

Stephane Hess[1]

The number of studies aiming to model air travel choice behaviour has increased over recent years, as discrete choice methods for analysing disaggregate travel behaviour have become more popular and widely applicable. However, despite this increased interest in this area of research, there is still a general lack of appreciation of the complexity of the choice processes undertaken by air travellers. This chapter aims to provide an in-depth look at the choice processes undertaken by air travellers, as well as discussing ways of modelling these processes. Additionally, the chapter discusses a number of issues that need to be addressed in modelling air travel behaviour, most notably the requirements in terms of data.

1 Introduction

Over the past few years, there has been an increase in the number of studies aiming to model air travel choice behaviour, mainly with the help of discrete choice models. For a comprehensive and up-to-date review of existing work in this area, see Pels et al. (2003) and Hess, Adler and Polak (2007).

Despite the increased interest in this area of research, a lot of work remains to be done. Indeed, the choice processes undertaken by air travellers are complex, involving decisions along a multitude of dimensions and influenced by a very high number of factors, making the modelling of such behaviour a non-trivial task. While authors are gradually acknowledging this in their work, and while the use of advanced model structures has allowed for more realism, the majority of studies still rely on stringent assumptions that unduly simplify the choice processes. This is partly because of the general poor quality of the data available (see Section 4.1), but can in many cases also be attributed to the use of inappropriate model structures. Additionally, it should be said that there is still a general lack of understanding of the actual choice processes undertaken by air travellers, a fact that is not helped by the dynamic nature of the problem at hand, as witnessed, for example, with the advance of low-cost carriers (LCCs).

Rather than presenting yet another empirical analysis, this chapter looks at the issue from a more theoretical perspective, and aims to give an overview of the choice processes undertaken by air travellers, along with discussing ways of modelling these processes on the basis of available data.

The remainder of this chapter is organised as follows. Section 2 provides an overview of the choice processes undertaken by air travellers, with a particular emphasis on the interdependencies

---

1 The author acknowledges the financial support of the Leverhulme Trust, in the form of a Leverhulme Early Career Fellowship. This chapter was partly written during a guest visit in the Institute of Transport and Logistics Studies at the University of Sydney. The author would also like to acknowledge the input of John Polak in earlier stages of this work, and would like to thank John Rose for useful feedback and comments.

between the various choice dimensions. This is followed in Section 3 by a discussion setting targets for practical research in this area while also acknowledging certain simplifications that are necessary in such analyses. Section 4 is concerned with data issues, while Section 5 looks at the question of model structure. Finally, Section 6 provides a brief summary of the chapter.

2 Choice Behaviour

2.1 General Framework

Broadly speaking, outside a mode-specific context, and without aiming to define the order of choices, travellers can, for a given trip, be seen to take decisions along four main dimensions of choice:

- destination;
- timing (time and date);
- mode of travel;
- route choice.

It can be seen that the latter two choices are strongly interrelated; in many cases, the choice of a given mode prescribes a specific route and vice versa, a principle that applies especially in the case of trips involving intermediary stops. Trip timing is clearly influenced by outside factors such as work commitments. Additionally, however, the choice set in terms of possible departure times (and in some cases even departure dates) depends on mode-specific attributes for all but self-operated modes (that is, car as driver, walking, cycling). Finally, choice set formation along the mode choice dimension is highly dependent on origin and destination, as well as on socio-demographic characteristics (for example, driving licence). The choice of destination can clearly be seen to have a very significant impact on the other three dimensions of choice. However, even within this general framework, it can occasionally be argued that the choice of destination is not in fact made a priori, but is itself a function of other choices. As an example, a traveller who takes a decision to rely on public transport (or is forced to do so by socio-demographic characteristics) will be limited in the number of potential destinations. An even stronger example is given by the case of people refusing to travel by air, or by sea. Clearly, such factors come into play mainly in the case of leisure travel, where, depending on the circumstances, they can play a major role.

2.2 Dimensions of Choice in Air Travel

In the case of air travel, the situation becomes significantly more complicated. Indeed, the various dimensions of choice listed above are not only strongly interrelated in the case where air travel is a possibility, but also have several subcategories.

Essentially, on top of the choice of destination and the choice of air as the main mode, the choices made by an air traveller can be divided into three main subcategories (origin side, destination side and air side), which we will now look at in turn in Sections 2.2.1 to 2.2.3 before discussing the interdependencies between the various choice dimensions in Section 2.3.

The choices in this section are described for the outbound leg of a return journey, where no attention is paid to trip duration. In general, for passengers on their return leg, the majority of journey factors are predetermined by the choices made on the outbound leg, although some factors, such as timing and possibly also routing, are determined separately. Finally, for passengers on one-way journeys, the choice process is very similar to the one described below for the outbound journey of return passengers, though outside factors and personal priorities may change significantly.

Here it should be noted that in the remainder of this chapter, we work on the assumption that travellers make a conscious choice along all dimensions of choice. This is clearly a major assumption, as some travellers may consistently ignore certain dimensions of choice. Furthermore, a different rationale might apply in the case of passengers who rely on travel agents for booking their tickets. Such factors are not taken into consideration in the remainder of this discussion, but appropriate simplifications of the framework are possible.

2.2.1 Origin side The choices on the origin side of an air journey can be divided into two main parts: the choice of a departure airport and the choices made for the ground level journey to this departure airport.

In many cases, the choice set for the departure airport is very limited, and dominated heavily by the airport closest to the passenger's ground level origin. However, for passengers living near major urban centres, there will often be a choice between a number of airports located at similar distances from a given passenger's ground level origin. In some rare cases, passengers may even be faced with a choice between airports located in separate nearby multi-airport regions, such that there is a choice of departure city.

Additionally, passengers take multiple decisions along the access journey dimension which are dominated by the choice of access mode, or combinations thereof. Depending on the mode chosen, and especially in the case of a combination of multiple-access modes, there is the additional choice of a route; while for journeys involving, and especially terminating with, a car journey there is often a choice to be made between self-drive and drop-off, and, in the former case, a choice between different parking options. Additionally, passengers do make a choice of departure time, which, although dependent on personal preferences, is highly influenced by the departure time of the actual air journey.

2.2.2 Destination side In many ways, the destination side choices are the mirror image of those made at the origin side. Aside from the actual choice of ground level destination, these include the choice of destination airport, and ground level transport between this airport and the final destination.

However, there are some subtle differences. Indeed, from the point of view of a passenger on the outbound leg, there is generally an issue of a lower level of knowledge than at the origin side. This relates partly to the geographical location of the different possible destination airports (and distance to the ground level destination), but also to choice set formation along the egress journey dimension, in terms of ground level transport modes as well as routes. Here, another point needs to be taken into account in that, for the majority of travellers, private car is not an option at the destination end, but is, for at least some of these travellers, replaced by rental car.

Aside from the above discussion about different choice set formation in the ground level dimension, the point about a lower level of information would suggest a less rational behaviour from an outside perspective,[2] except for the more regular traveller. Additionally however, it should be noted that the set of priorities at the destination end may be different from those at the origin end. As such, it is possible to imagine a higher reluctance to accept a longer egress journey, even in

---

2   This does not per se suggest irrational behaviour. It simply means that, had the traveller been in possession of all information, they might have been expected to behave differently.

exchange for fare reductions, than might be the case for the departure end, given the wish to arrive at the ground level destination more promptly.

2.2.3 Air side   Except for the questions of origin and destination, and ground level transport, the air side category contains all remaining choices describing the journey. Aside from spontaneous choices made at different stages of the journey (such as what to do while at the airport, or during the flight), these choices all relate to the actual travel from the origin airport to the destination airport. These in turn can be subdivided into three very much interrelated dimensions of choice.

The first choice is that of an airline operating a route to the chosen destination. In most cases, passengers will travel on a single airline for the duration of their journey. However, on some routings – for example those involving short-haul feeder flights to regional airports, or complex international routings – there is the possibility of a combination of airlines. This situation has in recent years increased in complexity, given that a large number of routes are now operated under code share agreements. The choice of an airline is one of the factors that make air travel different from other areas of transport analysis. Indeed, while in many areas passengers can be seen to make a choice between different modes of transport, the additional within-mode choice seen in the case of air travel is quite specific and, for example, does not apply to the same level in areas such as rail or coach travel.

The choice of an airline or combination of airlines is strongly related to the choice of a routing. The first level along this dimension of choice divides flights into direct and connecting flights, with the possibility of a third category for flights involving a stopover without a change of aircraft. The second level applies only to connecting flights, and involves the choice between a number of different possible routes, which includes a decision on the number of connections and the choice of connecting airports. It is also here that the possibility of multi-airline journeys arises.

The final dimension of choice for the actual air journey is that of timing, that is the choice of a departure time and a departure date. For some passengers, the most important factor will be the departure time, while for others it will be the arrival time. In practice, this equates to the choice of a specific flight.

There are ways to consider further subdivisions of choice dimensions along the air side category. One example includes the choice of aircraft type. Indeed, some passengers have an inherent dislike for turboprop planes, and use this as a determining factor in their choice process. However, such factors can in fact be seen as an attribute of a specific flight, which is thus accounted for by the other dimensions of choice listed above and, as such, can be included in models as a simple explanatory variable.

2.3 Choice Process in Air Travel

The above discussion has illustrated that the process of putting together a trip from a ground level origin to a ground level destination, with an intermediary air journey, is a complicated one, involving decisions along a multitude of dimensions. What makes the analysis of these choice processes more complicated is the fact that there exists a highly complex structure of interdependencies between the various dimensions of choice, and that the order of priorities amongst dimensions of choice is highly likely to vary across individuals as well as across situations. We will now look at these interdependencies, starting from a very basic situation and slowly moving up to more detailed interactions.

2.3.1 Base scenario In the most basic scenario, shown in Figure 11.1, only the most obvious of dependencies are shown for the actual air journey.[3] As such, the choice of access mode(s) and the choice of route clearly depend on the choice of departure airport, with a similar conclusion for the egress journey and the choice of arrival airport. This accounts for the fact that not all airports will be accessible by all ground level modes or combinations thereof. Interactions with individual-specific characteristics, such as car ownership, are not taken into account in this discussion; they are to be treated at an individual-specific level in the models. The other dependency shown for the actual air journey in Figure 11.1 relates the choice of connecting airports to the question of whether a direct or a connecting flight is chosen. Additionally, it should be noted that the impact of the choice of destination (ground level) and the decision to travel on flight-specific attributes (e.g. airline, time and date) is allowed for by the link from these nodes through the travel-by-air node.

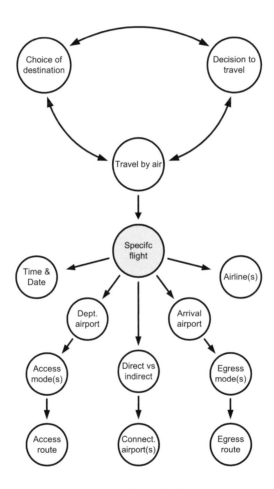

Figure 11.1 Main choice processes of an outbound air journey

---

3 A link from choice dimension A to choice dimension B indicates that the decision taken in choice dimension A can have an impact on choice dimension B in two ways: it can lead to changes in the attributes of the alternatives along this dimension, but it can also lead to a change in the composition of the choice set along this dimension.

Figure 11.1 shows an additional set of interactions outside the main air journey, relating the decision to travel to the choice of destination and the decision to travel by air. Each one of these links is shown as being reciprocal, a fact that needs some further explanation.

The decision to travel and the choice of destination are inherently linked, and the order of decisions depends on the situation at hand. The case where the decision to travel precedes the choice of destination applies mainly in the case of leisure travel, where it needs no further explanation (e.g. holidays). However, it can also be seen to apply in the case of business travel, as for example in the situation where a decision is taken to hold a meeting before deciding on an actual location. Conversely, the case where the choice of destination precedes the decision to travel applies more directly in the case of business travellers, where the a priori determined geographical location of a meeting or conference will have an impact on the decision to attend, and hence to travel. For leisure travellers, it is similarly possible to imagine cases where a respondent may make the decision to travel to a certain leisure event (e.g. joining friends on holiday) dependent on the location of the event.

The link between the choice of destination and the decision to travel by air also acts in both directions. Indeed, the fact whether air travel is a viable alternative obviously depends on the chosen destination, where there are clearly also cases where the choice of a destination effectively eliminates all other modes from consideration. Conversely, a respondent who takes an a priori decision to travel on a specific mode clearly imposes limitations on the set of possible destinations, where the same reasoning applies for travellers who decide to exclude a specific mode from consideration.

The link between the decision to travel and the decision to travel by air is less straightforward to explain. Nevertheless, it can be seen that a traveller deciding to travel may take the a priori decision to travel by air before choosing a destination, something that applies principally in the case of leisure travel. On the other hand, the link between the decision to fly and the decision to travel in the first place can be explained most readily in conjunction with the choice of destination. As such, if a destination is imposed by outside factors (e.g. meeting, conference), and if this implies a need to fly, some passengers may decide not to travel.

2.3.2 Detailed interactions between choice dimensions   We now turn our attention to the detailed interactions between the various choice dimensions. For this, the choice dimensions are again split into two parts, namely the upper and lower half shown in Figure 11.1, representing the upper level choices and the choices relating to the actual air journey respectively. We will first look at the interdependencies of the various choice dimensions describing the air journey, before turning our attention to the links between these choices and the upper level choices.

The interactions between the various choice dimensions relating to the actual journey between a ground level origin and destination are shown in Figure 11.2. Some minor simplifications are used here in that the nodes relating to the access journey (mode and route) in Figure 11.1 are now replaced by a single access journey node, with a corresponding approach for the egress journey. Additionally, the separate nodes for direct vs indirect and connecting airport(s) are now replaced by a common routing node. Finally, the specific flight node has been eliminated, to show the direct relationships between individual dimensions without passing through this node.

From Figure 11.2, it can be seen that, aside from the two ground level journey nodes, all decision processes are connected to each other, in both directions. This shows the high level of interdependency between the various choices. We will first turn our attention to the interdependencies between the five central nodes, before returning to the access/egress journey nodes to justify their special treatment.

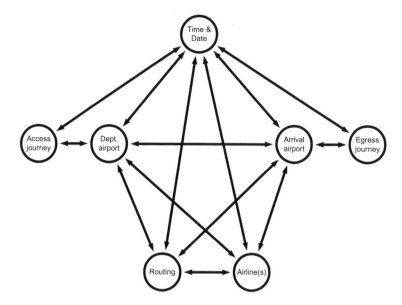

Figure 11.2 Interactions between air travel choice dimensions

It can easily be seen that the choice of a specific departure airport will reduce or affect the set of options open to the traveller along at least some of the other dimensions of choice, such that it possibly has an impact on the choice of arrival airport, departure time (and to a lesser extent date), airline and flight routing. The same reasoning applies in the case where a traveller makes their choice dependent on a specific arrival airport. Similarly, it can be seen that a respondent who chooses to travel at a given time and/or on a given date affects their options along the other dimensions of choice, with analogous reasonings in the case of a choice of a specific airline, or routing. As such, it is clear that the links between these five main nodes are justifiably set to act in both directions.

A special treatment applies in the case of the access journey and egress journey nodes. They are linked directly (and reciprocally) only to the respective airport node, and to the time and date node. It is clear that the decision to travel at a given time or on a given date will, in some cases, have an impact on the choices made for the access journey. As an example, a very early departure may in some cases remove the possibility to travel by public transport. This link clearly also acts in the opposite direction, in that travellers who are determined (or forced) to use public transport may only be able to travel at certain times or even on certain dates. It can also be seen that the choice of a specific departure or arrival airport has an impact on the choice made along the access journey and, respectively, egress journey dimensions. Here, the reciprocal link might be less obvious, but there are clearly cases in which the choices made along the ground level dimensions can take precedence over those choices made in terms of departure and, respectively, arrival airport (and hence over the three other central dimensions by extension). As such, for passengers who decide to travel on a given mode or combination of modes, the choice set in terms of departure airports may be reduced. While these links between the access and egress journeys and the respective airports, as well as trip timing, are thus well defined, it cannot easily be argued that there is a direct link (in either direction) between the choice of a specific airline or routing and the choices made for the access or egress journey.

Interactions with upper level choice dimensions We now turn our attention to the interactions between the upper level choices and the choices relating to the actual air journey. We will first concentrate on the downwards interactions before looking at the potential impact of lower level choices on upper level decisions.

The downwards interactions are shown in Figure 11.3, which additionally reproduces the level-specific interactions, as previously shown in Figure 11.1 and Figure 11.2. The only justifiable direct impact of the decision to travel is on the time and date of the journey. Any remaining impacts are indirect, either through the destination node or the travel by air node. For the impact of these two nodes, an intermediary node, relating to the choice of a specific flight, is introduced again. Through this, the choice of destination and the decision to travel by air have an impact on all the main characteristics of the air journey, and, by passing through the appropriate airport nodes, also the access and egress journey. Additionally, the two decisions, however, potentially also have some direct impacts on lower level choices, which are taken prior to selecting a specific flight, or at least carry special weight in the selection of a specific flight. Such interactions are shown as direct links in Figure 11.3. For the choice of ground level destination, two such direct links are shown. Indeed, it is clear that the precise geographical location of the chosen ground level destination has a strong direct impact on the choice of arrival airport in the case of a destination served by more than one airport. Similarly, it also potentially has a direct impact on the choice of egress mode and route at the destination end. No direct link is included between the choice of destination and the time and date of travel; this process can pass either through the decision to travel node or the choice of a specific flight. For the decision to travel by air, two main direct impacts are identified, namely to a specific departure airport and a specific airline, reflecting airport and airline allegiance. All

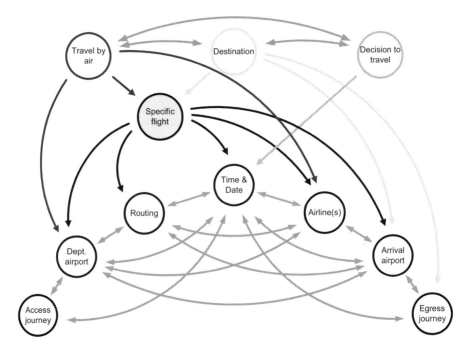

Figure 11.3 Downwards interactions between upper level choices and air travel choice dimensions

other links are represented indirectly, through the specific flight node. It could be argued that the decision to travel by air is in some cases coupled directly to a decision to choose only direct flights; however, this may not always be possible, and it is not clear whether there is a significant share of travellers who would in this case decide not to travel by air. In any case, the indirect link still allows for such effects.

Even though the choices shown in the upper level of Figure 11.1 are in general seen as taking precedence over those shown in the lower level, this is not always the case. The various possibilities of an upwards effect are shown in Figure 11.4, where it must be said that some of these links apply primarily to leisure travellers, as mentioned at appropriate stages in the following discussion.

The most obvious of the upwards effects are the three links associated with the time and date node; clearly the a priori choice of a date (and to a lesser extent time) for travelling has an effect on the choice of destination. Similarly, if a date or time for a specific event is imposed on the decision-maker through a third party, then this does potentially have an effect on the decision to travel, as well as on the decision to travel by air.[4] Another obvious effect that needs to be represented is the relationship between the available flight options in terms of airline, airports, route and fare on the decision to travel by air, as well as the decision to travel in the first place. This is represented in Figure 11.4 by appropriate links from the departure airport, routing and airline nodes to the two travel decision nodes, where the combination of these nodes reflects specific air travel options. The additional dimension of flight fares is not represented here; although there is clearly a choice in terms of class of travel, it can be argued that passengers do not generally make an additional conscious choice of fare type for a specific flight. The actual effects of differences in fares across different flights are implicit in the links from the various nodes listed above (especially the airline and routing ones).

In addition, some of the links already discussed above – as well as the remaining links shown in Figure 11.4 – can be explained in a different way, which can broadly be seen as a type of airport or airline allegiance, which, however, applies most directly in the case of leisure travellers. As such, it can be argued that some travellers make an a priori choice of departure airport before organising the remainder of their journey. This is most easily understood in the context of regional airports. In such cases, the choice of destination clearly depends on the destination on offer from this airport, and if no attractive destinations (or indeed flights) are available, the traveller may decide to travel on another mode, or not at all. A similar phenomenon can be observed in the case of strong allegiance to a given airline, where passengers make the choice of destination, as well as the decision to travel, conditional on what is on offer from this specific airline. While this may in some cases occur as an effect of airline allegiance in the classical meaning (that is, frequent flier account), it can be seen to apply even more clearly in the case of low-cost airlines; these operators can be seen to induce new demand, such that some of their passengers would not travel at all (or at least not by air) were it not for the presence of the specific airline.

The above discussion suggests that the product offered by the chosen airport, or the chosen airline, cannot only determine the potential destinations, but, if no destination appeals to the passenger, also the decision to travel in the first place. Although these factors are becoming increasingly important with the increased popularity of low-cost carriers, they can be seen to also apply in the case of flights offered in conjunction with package holidays, where passengers may be rather flexible in terms of their precise destination. However, as already alluded to earlier on, the

---

4   The latter applies principally for the case of short-haul trips, where the decision to travel by air as opposed to ground level modes is very much dependent on scheduling flexibility, but also on the advance notification.

principle also applies in the case of business travellers, where the possibilities in terms of departure airport and airline can play a deciding role in the choice for a meeting place.

The final link shown in Figure 11.4, namely that from the choice of arrival airport to the choice of ground level destination, again applies mainly to leisure travellers, and specifically the case where people book a flight before sorting out the detailed ground level destination within the wider surrounding area of the destination airport.

2.4 Discussion

The above description of the choice processes made by air travellers has shown that such journeys not only involve decisions along a multitude of choice dimensions, but that there exist complex interdependencies between these various choice dimensions. Given that a large number of the links have been shown to act in both dimensions, it is clearly inappropriate to attempt to model the decision-making as a sequential choice process, but rather, that simultaneous analysis is required in the absence of information on the relative level of priorities across travellers.[5]

The discussion has also shown that, while the top upper level decisions have major direct and indirect effects on all lower level decisions, these links potentially also act in the opposite direction. This leads us to an important observation. Indeed, the discussion so far, along with the point of view taken in the various diagrams, would suggest that the question of main mode

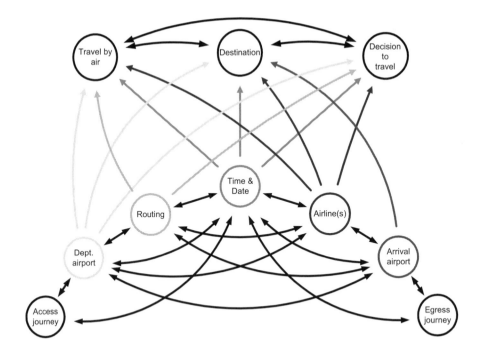

Figure 11.4 Upwards interactions between air travel choice dimensions and upper level choices

---

5   The use of such information is highly hypothetical in any case, as it is not clear whether there are situations in which it is possible to define a clear sequential choice process involving decisions along all of the above listed dimensions.

of travel is taken at a separate level. As such, travellers would gauge the overall product offered by the various modes, and then make a choice of main mode before moving on to mode-specific choices (e.g. airline, route and so on). Clearly, for some journeys, a mode choice decision is taken a priori, such that the above discussion holds. This applies most notably in the case of destinations or circumstances where only a single mode is a viable alternative. As such, in the case of aviation it can be seen that, beyond a certain distance, the mode choice question becomes obsolete. The problems arise in the case of short-haul to medium-haul distances, where the competition from car and ground level public transport needs to be taken into account, especially with the increase in high-speed rail routes. In such cases, it is not necessarily clear whether passengers make an a priori choice between air and rail before, if applicable, making within-mode decisions in the case of air travel. Rather, it can be imagined that, for at least part of the travelling population, the high-speed rail alternative appears on the same level as the various air travel alternatives. As such, the alternatives are evaluated in parallel, with all intra-mode considerations taken into account at the same time as the cross-mode comparison.[6]

Although such a parallel analysis does not pose any major problems from a methodological point of view in terms of model structure, it does come at the cost of increased data requirements. As such, it is now necessary to obtain detailed data for ground level modes in addition to the air travel data, the procurement of which already causes problems on its own, as described in Section 4.1. Additionally, it should be noted that, for some routes, the number of possible ground level options (in terms of combinations and routes) is so high that the data requirements can become insurmountable. As such, it should come as no surprise that the majority of studies make an assumption of an a priori decision to travel by air. While this does not per se invalidate the results of the analyses in question, it is important to acknowledge the potential shortcomings, at least in the presence of routes where there is potential high inter-mode competition which is not characterised by an a priori choice of mode before proceeding to intra-mode decisions.

3 Modelling Practice: Goals and Limitations

This section provides some guidance on several critical points that analysts should pay attention to in modelling air travel choice behaviour, while also looking at some of the simplifications that are required to enable large-scale analysis.

3.1 Goals

3.1.1 Recognise the multi-dimensional nature of the choice process It should be clear from the discussion in Section 2.2 that air travellers take decisions along a multitude of choice dimensions. Not recognising this in practical work can potentially lead to biased results. This is most easily illustrated on the basis of an example, say, with a study that does not account for the fact that passengers choose an airline in addition to a departure airport. It is clear that someone with a strong allegiance to airline X may not be interested in other airlines. As such, in the case where airline X operates only from an outlying airport with low overall frequency (across all airlines),

---

6   It should be noted that the interactions between the choice dimensions shown in the diagrams do allow for such comparisons, although not described explicitly. Indeed, the fact that the mode-specific characteristics are connected upwards to the decision to travel by air does allow for the rejection of an air alternative in favour of an alternative on another mode.

this respondent can be seen to choose an airport with lower frequency to the desired destination. From this point of view, it is important to work on the basis of frequencies specific to airport-airline pairings, and not airport-specific frequency, as has commonly been the case in existing work. A similar reasoning applies in the case of the other dimensions of choice.

A straightforward approach for the representation of such multidimensional choice processes can in general terms be described as follows. A combined alternative chosen by a given traveller is composed of D elementary alternatives, one along each dimension of choice. Some of the attributes of the combined alternative are specific to a single one of the choice dimensions, while others are specific to a given combination – say, for example, an airport–airline pairing. The availability of a combined alternative is a function of the availabilities of the elementary alternatives. Each combined alternative counts as a separate alternative, such that the choice process is indeed represented in a simultaneous rather than sequential fashion.

3.1.2 Account for correlation along different choice dimensions By understanding the multidimensional nature of the choice process, it becomes evident that some of the combined alternatives share the attributes of other alternatives along one or more of the choice dimensions. In the context of the combined choice of airport, airline and access mode, let K, L and M define the number of airports, airlines and access modes respectively, and let us assume that all combinations of airports, airlines and access modes are possible, leading to a total of KLM combined alternatives. It can then be seen that a given alternative shares the same airport (and hence the related attributes) with LM −1 alternatives, the same airline with KM −1 alternatives, and the same access mode with KL −1 alternatives. Furthermore, an alternative shares the same airport and airline with M −1 alternatives, the same airport and access mode with L −1 alternatives, and the same airline and access mode with K −1 alternatives. While some of these commonalities can be accounted for through the attributes included in the modelled part of utility, it is almost inevitable that there is also some correlation between unobserved utility terms along these dimensions. Here, it is important to account for this correlation, as described in Section 5.2.

3.1.3 Use highly disaggregate level-of-service data Another major problem with existing studies has been the use of an insufficient level of disaggregation in the level-of-service data. This is highly correlated with the decision in a lot of previous work to use simplifications along a number of choice dimensions. While some aggregation is almost inevitable, for example by working on the basis of average fares across flights on a given day, the use of an excessive amount of aggregation can lead to biased results. This can, for example, arise in the case of studies making use of weekly (or even monthly) data instead of daily data, hence ignoring the often significant variations in level-of-service attributes between different days of the week, especially in terms of flight frequencies.

3.1.4 Account for differences in behaviour across travellers One of the most important points to take into account in modelling air travel choice behaviour is the variation in behaviour across respondents. No two respondents are exactly the same, in terms of sensitivities as well as priorities. Not taking such variations in tastes (and hence behaviour) into account will inevitably lead to lower-quality results and may even lead to significant bias. This issue is addressed in more detail in Section 5.3.

3.2 Limitations of Practical Research

In practical research, it is not generally possible to jointly model all the choice dimensions described in Section 2.2, and some simplifications are required. There are some subtle differences here between revealed preference (RP) and stated preference (SP) data. As such, in the case of RP data, respondents did indeed perform a multidimensional choice process,[7] but the quality of the data often prevents an analysis along all the choice dimensions. On the other hand, in the case of SP data, information is available on all choice dimensions, but respondents are generally only faced with choices along a rather small subset of the dimensions listed in Section 2.2. As such, in both cases, only a subset of the choices will be modelled. We will now briefly look at the various dimensions of choice and discuss what simplifications are generally used.

The choice of destination and the actual decision to travel are not generally modelled, where a reservation applies for the latter in the case of SP surveys giving respondents the option not to travel. Research thus relies on the assumption that these decisions are taken at an upper level, prior to putting together the exact details of the journey.

The decision to travel by air is also generally treated as having been taken at an upper level. Although this can cause problems in the case of short-haul destinations, this simplification is almost unavoidable, given the heightened data requirements as well as added modelling complexity. As such, it is important to acknowledge that the estimates obtained from such models relate to the part of the population that has decided to travel by air, and are not representative of the overall population.

The majority of studies of air travel choice behaviour look solely at the choice of departure airport and ignore the choice of arrival airport, where the latter choice can, however, be modelled on the basis of non-resident travellers on the return leg of their journey (excluding the possibility of open-jaw tickets). Again, this simplification is primarily due to data issues, but also the relatively low number of routes connecting two multi-airport regions.

In studies that model ground level journey related choices in addition to air travel choices, it is generally only possible to model the choice of access mode to the departure airport on the current leg; an analysis of the egress journey related choices would lead to a requirement for detailed level-of-service data for the ground level transport network at the destination end. Furthermore, most studies are only able to look at the choice of main mode, ignoring the possibility of trip-chaining, as well as the choice of different routes. The effects of these restrictions are dependent on the geographical context.

In RP models, trip timing cannot generally be modelled due to a relative lack of information on preferred departure time and flight availabilities. These issues do not apply in the case of SP models.

Flight routing is generally left untreated in RP studies, which often look only at the choice between direct flights; this can have major effects in the case of destinations with a large share of connecting passengers. As with most other factors, SP studies, in theory, are able to model the choice of routing, where this is again dependent on the design of the SP surveys.

Given the great role that airline allegiance plays for some travellers, advanced studies increasingly model the choice of airline in addition to the choice of airport. However, combinations of airlines are generally not allowed in such models, and additional issues can arise in the case of code share flights. Again, the main issue is one of data quality.

---

[7] The possibility exists for some individuals not to make a conscious choice along some of the dimensions.

4 Data Issues

As the above discussion has shown, the analysis of air travel behaviour is a complicated undertaking, and a number of issues need to be dealt with it in the process. While the majority of these issues have been addressed in some form or other in the above discussion, two main remaining issues need to be addressed in more detail. They both relate to the data used in estimation, and are concerned with overall data quality and the selection of destinations to be used in the analysis. These two issues are now looked at in turn.

4.1 Data Quality

Almost certainly the single biggest issue that needs to be faced in the analysis of air travel behaviour is that of the quality of the available data. The issue can be divided into several sub-issues as follows.

4.1.1 Availability and attributes of unchosen alternatives The first major data problem that needs to be faced is the relative lack of information on the choices that travellers are faced with. Indeed, in an RP context, adequate information will generally be available for the majority of choice-affecting attributes for the alternative that was actually chosen. This does, however, not extend to the unchosen alternatives. While it is generally possible to obtain some information on such alternatives in terms of substantive attributes, the lack of data on availability of flights as well as of specific fare classes on a given flight significantly complicates the characterisation of unchosen alternatives. A similar issue applies along the access mode dimension, where information on the availability and attributes of unchosen modes is often not available.

4.1.2 Fare data Another very significant problem, which, although strongly related to the issues of unobserved attributes discussed above, deserves special attention, is that of the fare data. Air fares, and especially relative air fares, can be expected to play a major role in air travel choice behaviour. However, in the majority of studies of air travel choice behaviour, it has not been possible to recover a meaningful marginal utility of fare changes. This can almost certainly be explained on the basis of the poor quality of the fare data. Indeed, while for many attributes of the unchosen alternatives it may be possible to produce a fairly good approximation to the true value, this does not apply in the case of fare data. In most cases, only the average fare charged by a given airline on a specific route will be available. This clearly involves a great deal of aggregation, as no distinction is made between the fares paid across different travellers (that is, in terms of travel classes as well as booking classes).

In addition to the obvious loss of information, the aggregation of fares leads to another important problem, given that there is generally also no information on the availability of specific fare classes at the time of booking, as opposed to the mere availability of actual flights. Indeed, this essentially leads to the need for an assumption of equal ticket selling speeds across all flights (routes as well as departure times), which is clearly not necessarily the case. The increasingly dynamic nature of air fares makes the use of aggregate fares even less reliable.

Even though some progress can be made with the help of bookings data, issues of aggregation do remain. In fact it can be seen that, in RP studies, disaggregate choice data is used in conjunction with aggregate level-of-service data, for at least some of the attributes. While, for some characteristics, this may be acceptable, it does, as described above, create significant problems in the treatment of air fares, and flight availability by extension.

4.1.3 Frequent flier information It is well known that passengers are heavily influenced in their choice of airline by their membership in frequent flier programmes, either on a personal basis or as part of a company-wide scheme. Unfortunately, information on frequent flier memberships is generally not collected in passenger surveys. As such, this potentially crucial influence on choice behaviour cannot usually be taken into account in RP case studies (as opposed to SP studies). In the case of datasets including a large number of international flights operated by a variety of airlines, there is, however, an alternative way of modelling travellers' loyalty behaviour: by analysing their allegiance to their national carrier.

4.1.4 Influence of other attributes Aside from the most commonly used attributes, such as access time, flight time, frequency and fare, a host of other factors potentially play a role in travellers' choices. The problem is that, much like in the case of frequent flier programmes, the procurement of data on these attributes is often difficult. A possible exception is that of on-time performance, which can be expected to play a role, especially for experienced travellers. However, again, the data is often not available in a sufficiently disaggregate form. Finally, a number of other attributes that potentially have an influence on choice behaviour are of a highly qualitative nature, making their inclusion in models difficult from a data as well as methodological point of view. Examples include factors such as comfort, quality of the food and quality of the in-flight entertainment.

4.1.5 Survey design issues and inter-dataset compatibility An additional problem relates more directly to the nature of the available data. The main input into an analysis of revealed air travel choice behaviour comes in the form of passenger survey data collected at the departure airport. The fact that, in many cases, different people (or even institutions) are responsible for the design of the survey and the subsequent modelling analysis creates obvious problems. Furthermore, the survey data need to be complemented by level-of-service data, which can lead to major problems of inter-dataset compatibility.

4.1.6 Stated preference data The above discussion has highlighted the issues with using RP survey data in analysing air travel choice behaviour. This in turn could suggest that the way forward would be the use of SP data collected on the basis of tailor-made surveys, designed specifically for use in advanced modelling analyses. The main advantage of SP surveys in the present context is the fact that the modeller knows precisely what information the respondent was faced with when making his or her choice, especially with regards to the availability and attributes of unchosen alternatives – something that is not generally the case in RP data.

Given these theoretical advantages of SP data in the present context, it should come as no surprise that studies making use of SP data have been much more successful in retrieving significant effects for factors such as air fares and airline allegiance (cf. Adler et al. 2005, Hess, Adler and Polak 2007). Despite the apparent advantages of SP data over RP data in this context, it should not be forgotten that the use of SP data does pose some philosophical problems in terms of how the behaviour differs from that observed in RP data. Additionally, important issues need to be faced at the survey design stage, notably in terms of questionnaire complexity.

In closing, it can be said that both approaches have advantages and disadvantages, where the relative merit of each approach needs to be measured on a case by case basis, and depends heavily on the quality of the level-of-service data in the RP context. An interesting approach in this context is to combine RP and SP data, as done by Algers and Beser (2001), hence correcting for the bias inherent to models estimated on SP data. The problem in this case, however, is one of obtaining compatible RP and SP datasets.

4.2 Choice of Destinations

A major issue falling within the wider field of data quality is the question of which destinations to include in a study. This is clearly heavily influenced by the choice data used and the destinations represented therein.[8] As such, only destinations with a sufficient number of observations in the choice data can be included in the analysis.

The next point that needs to be taken into account is that only destinations that can be reached from at least two airports in the study area should be included, for obvious reasons. Furthermore, it is of interest to avoid including destinations that are served from more than one airport, but where the frequency at all but one of the airports is negligible. Additionally, if a decision is taken to include only destinations reachable by direct flight, then this clearly reduces the number of eligible destinations further.

The various requirements listed above can in some cases lead to an insufficient number of destinations/observations, depending on the choice data at hand, such that the requirements may have to be relaxed to guarantee an acceptable sample size. In the case of the analysis of a multidimensional choice process, this may, for example, be achieved by the inclusion of some destinations that are only reachable from one airport, but where a choice still exists along the other dimensions.

One main complication, however, still needs to be addressed, namely the treatment of destinations that are themselves located in multi-airport regions. Indeed, for such destinations, passengers not only make a choice of departure airport, but must also be expected to make a choice of destination airport; and it is not clear which choice in the study area takes precedence. This applies especially in the case of passengers who are on the return leg of their flight.

To conduct a proper analysis, the choice of arrival airport would in this case normally have to be included explicitly in the modelling framework; this, however, not only leads to a very significant increase in the complexity of the analysis, but also greatly increases data requirements, notably in terms of ground level transport data for the different destination airports. Given these complications, destinations located in multi-airport regions should thus, ideally, be excluded from the analysis if no treatment of the choice of destination airport is to be included in the modelling framework. However, this is not always possible, given the often high representation of such destinations in air passenger surveys. In this case, it is, however, important to maximise the probability of there being a conscious choice of airport in the study area, for residents as well as visitors. Furthermore, modellers should at least acknowledge the fact that their results may be influenced by the presence of multi-airport destination regions.

5 Model Structure

One of the most important questions arising in the analysis of air travel choice behaviour is that of what modelling approach to use. Given the important differences across travellers, both in terms of behaviour as well as choice context, the use of a disaggregate modelling approach is clearly preferable to an aggregate one. In the area of transport analysis, discrete choice structures belonging to the class of random utility models (RUM) have established themselves as the preferred approach

---

8   We ignore the case where the modeller themself is responsible for the collection of the choice data, and hence in a position to define quotas, so as to obtain an adequately sized sample for an a priori defined set of destinations.

for such studies over the past 30 years. An in-depth discussion of these modelling approaches is beyond the scope of the present chapter, and readers are referred to the excellent overview provided by Train (2003). In this section, we merely discuss the use of discrete models to represent correlation (Section 5.2) and taste heterogeneity (Section 5.3) in the context of air travel behaviour research. This is preceded by a brief introduction of some common notation.

5.1 Basic Concepts

In a discrete choice experiment, a decision-maker n chooses a single alternative from a choice set Cn, made up of a finite number of mutually exclusive alternatives, where the choice set is exhaustive and the ordering of alternatives has no effect on the choice process undertaken by the decision-maker. Each alternative $i = 1, \ldots, I$ in the choice set is characterised by a utility U_{in}, which is specific to decision-maker n, due to variations in attributes of the individuals, as well as in the attributes of the alternative, as faced by different decision-makers. The use of the concept of utility, along with the need for a decision-rule, leads to the single most important assumption in the field of discrete choice modelling, namely that of utility maximising behaviour by respondents. As such, respondent n will choose alternative i if and only if $U_{in} > U_{jn}$ for all $j \neq i$ with i, j.

In an actual modelling analysis, the aim is to express the utility of an alternative as a function of the attributes of the alternative and the tastes and socio-demographic attributes of the decision-maker. Here, the limitations in terms of data and the randomness involved in choice-behaviour mean that, in practice, modellers will only be able to parameterise part of the utility. As such, we have:

$$U_{in} = V_{in} + \varepsilon_{in} \tag{1}$$

with V_{in} and ε_{in} giving the observed and unobserved parts of utility respectively. Here, V_{in} is defined as $f(\beta_n, x_{in})$, where x_{in} represents a vector of measurable (to the researcher) attributes of alternative i as faced by decision-maker n,[9] and β_n is a vector of parameters representing the tastes of decision-maker n, which is to be estimated from the data. The function $f(\beta, x_{in})$ is free from any a priori assumptions, allowing for linear as well as non-linear parameterisations of utility. The inclusion of the random utility term ε_{in} means that the deterministic choice process now becomes probabilistic, leading to a random utility model (RUM), with the probability of an alternative being chosen increasing with the modelled utility of that alternative.

It can be seen that the probability of decision-maker n choosing alternative i is now given by:

$$P_n(i) = P(\varepsilon_{jn} - \varepsilon_{in} = V_{in} - V_{jn}, \forall j \neq i) \tag{2}$$

With the unobserved part of utility varying randomly across respondents, the mean of this term can be added to the modelled part of utility, in the form of an alternative-specific constant (ASC). The vector $\varepsilon_n = (\varepsilon_{1n}, \ldots, \varepsilon_{In})$ is now defined to be a random vector with joint density $f(\varepsilon_n)$, zero mean and covariance matrix Σ, and, by noting that the probability of alternative i in equation (2) is the cumulative distribution of the random term $\varepsilon_{jn} - \varepsilon_{in}$, we can write:

$$P_n(i) = \int I(\varepsilon_{jn} - \varepsilon_{in} < V_{in} - V_{jn}, \forall j \neq) f(\varepsilon_n) d\varepsilon_n$$

---

9 The vector $x_{i,n}$ potentially also includes interactions with socio-demographic attributes of respondent n.

where $I(\cdot)$ is the indicator function which equals 1 if the term inside the brackets is true and 0 otherwise. The probability is now given by a multidimensional integral which only takes a closed form for certain choices of distribution for ε_n, where the choice of $f(\varepsilon_n)$ has a crucial impact on the behaviour of the choice model. The most basic model structure, the Multinomial Logit (MNL) model, assumes that the error terms are distributed identically and independently, while more advanced model structures allow for correlation between error terms across alternatives and variation in the error terms across individuals, leading to a representation of variable inter-alternative substitution patterns and random inter-agent taste heterogeneity respectively. These two concepts are discussed in Section 5.2 and Section 5.3 respectively.

5.2 Correlation between Alternatives

As mentioned in Section 3.1.2, it is clearly a major and probably unwarranted assumption to rule out the presence of heightened correlation in the unobserved utility terms along any of the choice dimensions. In order for such an assumption to be valid, any commonalities between two alternatives sharing a common component along one or more of the choice dimensions would need to be explained in the modelled part of utility. This is clearly not possible in general, mainly for data reasons. The likely resulting correlation in the unobserved part of utility makes the use of the MNL model almost surely inappropriate, especially in forecasting.

Three quite different treatments of the correlation in the unobserved part of utility are possible in this context: the use of Generalised Extreme Value (GEV) structures; the use of an error components Logit approach; and the use of Probit structure. While potentially having advantages in terms of flexibility, the latter two are not applicable in the present context, simply on the basis of the added cost of estimation due to simulation requirements. As such, the discussion in this section looks solely at the case of GEV structures. Here it should be noted that this discussion looks only at the correlation along dimensions and not at the correlation across dimensions, such as the correlation between different airlines. This remains an important avenue for further research. Finally, the discussion in this section centres on the case of the combined choice of an airport, airline and access mode. Extensions to higher-dimensional choice processes are straightforward.

The Generalised Extreme Value family of models, introduced by McFadden (1978), is a set of closed form discrete choice models that are all based on the use of the extreme-value distribution, and which allow for various levels of correlation among the unobserved part of utility across alternatives. This is done through dividing the choice set into nests of alternatives with increased correlation, and thus higher cross-elasticities, between alternatives sharing a nest. As such, alternatives sharing a nest are more likely substitutes for each other. The use of such a nesting structure means that GEV models are most easily understood in the form of trees, with the root at the top, elementary alternatives at the bottom and composite alternatives, or nests, in between.

The MNL model is the most basic member of this GEV family, using a single nest of alternatives, resulting in equal cross-elasticities across all alternatives. While in the MNL model the error terms are distributed iid extreme value, in the general GEV formulation, the error terms follow a joint generalised extreme value distribution; the individual error terms follow a univariate extreme value distribution, but the error terms associated with alternatives sharing a nest are correlated with each other. This structure leads us away from the diagonal variance-covariance matrix of the MNL model.

The most basic GEV approach that can be used in the analysis of air travel choice behaviour is a simple two-level Nested Logit (NL) model. In this model, alternatives are grouped into mutually exclusive nests, where, for each nest, a structural (nesting) parameter is estimated that relates to the level of correlation in the error terms of alternatives sharing that nest. In the context of the present discussion, three main possibilities arise in this case: nesting together alternatives either by airport, by airline or by access mode. The resulting structures allow for correlation along a single dimension of choice, accounting for the presence of unobserved attributes that are specific to a given airport, or, in the two other structures, a given airline respectively a given access mode. In each of the three structures, separate nesting parameters are used with individual nests to allow for differential levels of correlation across nests.

As an example, the appropriate structure for the NL model using nesting by airport is shown in Figure 11.5, with K mutually exclusive nests, one for each airport, and where each nest has its own nesting parameter, λ_k. Only a subset of the composite nests and of the triplets of alternatives is shown in the graph. The corresponding structures for the models using nesting by airline and nesting by access mode are not reproduced here, being simple analogues of the structure shown in Figure 11.5.

The NL structures described thus far have only allowed for the treatment of inter-alternative correlation along a single dimension of choice. This is clearly a major restriction, and inappropriate in the case where correlation exists along more than one dimension. As such, the two-level NL structures should at best be seen as a tool for testing for the presence of correlation along individual dimensions; but, in the case where different structures indicate correlation along different dimensions, the two-level structure becomes inadequate for use in forecasting, and potentially even for the calculation of willingness-to-pay indicators.

The NL model can be adapted to allow for correlation along more than one dimension, by using a multi-level structure. A common example in the case of air travel is to nest the choice of airline within the choice of airport. It is important to stress that this should not be seen as representing

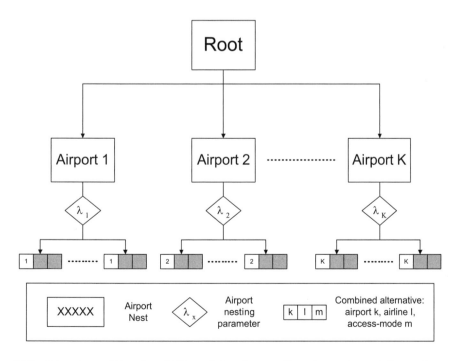

Figure 11.5 Structure of two-level NL model, using nesting along airport dimension

a sequential choice process. Rather, it means that there is correlation between two alternatives that share the same airport, but that the correlation is larger if they additionally share the same airline. The structure of such a model is illustrated in Figure 11.6, where λ_k is the nesting parameter associated with airport nest k, and π_l is the nesting parameter associated with airline nest l. Again, only a subset of the composite nests and of the triplets of alternatives is shown.

By noting that a model nesting airport choice above airline choice is not the same as a model nesting airline choice above airport choice, it can be seen that six possible two-level structures arise in the present context.

While NL structures can, in this form, thus be used for analysing correlations along two dimensions of choice, it should be noted that multi-level NL models have two important shortcomings which limit their potential for the analysis of choice processes of the type described in this chapter.

The main shortcoming in the present context is that the structures can be used for the analysis of correlation along at most $D-1$ dimensions of choice, with D giving the total number of dimensions. Indeed, using the example shown in Figure 11.6, it can be seen that, by adding an additional level of nesting by access mode below the airline level, each access mode nest would contain a single alternative, as the airline nest preceding the access mode nest would contain exactly one alternative for each access mode. This means that the model could not be used to explore correlation along the access mode dimension. The same principle applies in the case of the other possible four-level structures, where, in each case, the lower level of nesting becomes obsolete.

While a three-level NL model can in this case be used to analyse the correlation along two out of the three dimensions of choice, the second shortcoming of the structure means that problems arise even with this task. In fact, it can be seen that the full extent of correlation can only be taken into account along one dimension, with a limited amount along the second dimension. Indeed, by nesting the alternatives first by airport and then by airline, the nest for airline l inside the nest for airport k will only group together the options on airline l for that airport k. The same reasoning applies for other nests. As such, the model is not able to capture correlation between alternatives using airline l at airport k1 and alternatives using airline l at airport k2, which is clearly a restriction. This problem also applies in the other multi-level nesting approaches. Aside from being a major shortcoming, this is also another reflection of the above comment – that the order of nesting matters.

These deficiencies of multi-level nesting structures can be addressed through the use of a Cross-Nested Logit (CNL) model, as discussed by Hess and Polak (2006) in an analysis of the combined choice of airport, airline and access mode in Greater London. In the present context, a CNL model is specified by defining three groups of nests, namely K airport nests, L airline nests and M access mode nests, and by allowing each alternative to belong to exactly one nest in each of these groups. As such, the structure addresses both of the shortcomings described above for the three-level NL model. The structure is not only able to accommodate correlation along all three dimensions of choice, but does so in a simultaneous rather than sequential fashion. This means, for example, that the model is able to capture the correlation between all alternatives sharing airline l, independently of which airport they are associated with. At the same time, the correlation will be higher between alternatives that additionally share the same airport.

An example of such a model is shown in Figure 11.7, where, in addition to the previously defined λ_k and π_l, Ψ_m is used as the structural parameter for access mode nest m. Again, only a subset of the composite nests and of the triplets of alternatives is shown. Additionally, the allocation parameters, governing the proportion by which an alternative belongs to each of the three nests, are not shown in Figure 11.7.

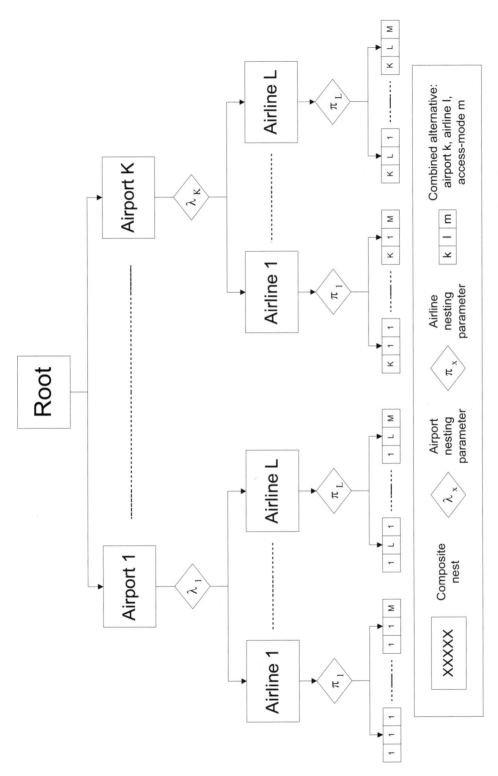

Figure 11.6 Structure of three-level NL model, using nesting along airport dimension and airline dimension

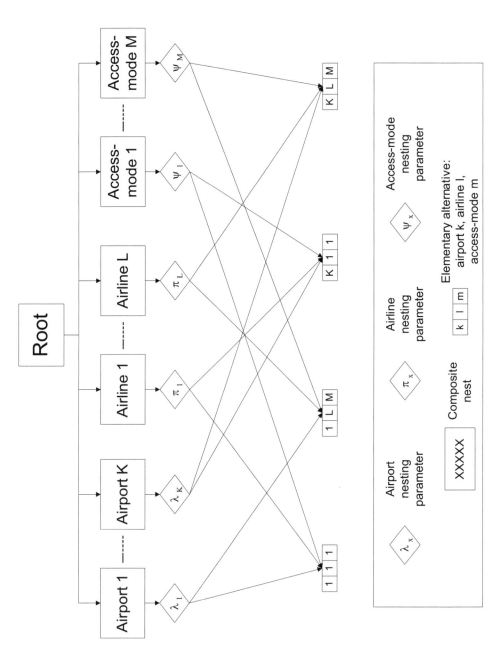

Figure 11.7 Structure of CNL model for the joint analysis of correlation along the airport, airline and access mode dimensions

5.3 Variation in Behaviour across Respondents

It should be clear from the outset that no two travellers are exactly the same in terms of their sensitivities to changes in attributes defining the alternatives. This applies most obviously in the case of factors such as air fare and access time, but also extends to the willingness to pay for flying on a certain airline or on a certain type of aircraft. On this basis, the assumption of a purely homogeneous population cannot be justified, and efforts need to be made to account for taste heterogeneity in the analysis of air travel choice behaviour if major bias in the results is to be avoided

Existing treatments of taste heterogeneity fall into three main categories, namely (in order of increasing complexity):

1. discrete segmentations;
2. continuous interactions;
3. random variations.

Discrete segmentations form the most basic approach for representing variations in sensitivities across respondents, either in the form of separate models for different population segments, or segment-specific coefficients within the same model. In the context of air travel behaviour research, possible segmentations would, for example, split the travellers by trip purpose or by residency status (residents vs visitors). While straightforward to apply, they lack flexibility as they still assume homogeneity within groups. Additionally, there is the issue of defining the groups, which can be very arbitrary in the case where continuous measures such as income are split into discrete segments.

Continuous interactions can be used to allow for a variation in a given sensitivity as a function of a continuous socio-demographic measure such as income. In the context of air travel behaviour research, this would, for example, allow for a continuous interaction between a given traveller's income and their sensitivity to increases in air fares. While offering much more flexibility than discrete segmentations, continuous interactions impose greater estimation cost, and are only used sparsely in practice.

The gains in popularity of mixture models such as Mixed Multinomial Logit (MMNL) have meant that modellers increasingly rely on a purely random (as opposed to deterministic) representation of taste heterogeneity. While offering very high flexibility, continuous mixture models are not only costly to estimate, due to the requirement to use simulation (see Train 2003), but also lead to a number of specification issues, such as the choice of random distribution (Hess et al. 2005b). An alternative approach to the representation of random taste heterogeneity replaces the continuous mixture distributions with a discrete mixture approach (cf. Dong and Koppelman 2003, Hess, Bierlaire and Polak 2007). While the issue of a choice of distribution does not arise in the case of discrete mixture models, there is now the problem of deciding on a number of support points to be used. Finally, in both cases, there are big issues in interpretation, as, without a posterior analysis, it is not possible to link the taste heterogeneity to socio-demographic information.

As shown in this section, the various methods for representing taste heterogeneity are characterised by differences in terms of flexibility, ease of implementation/estimation and ease of interpretation. While a decision of what approach (or combination of approaches) to use needs to be made on a case by case basis, modellers should always be conscious of the fact that the flexibility of the advanced methods comes not only with a hefty price tag in terms of the cost of estimation, but crucially also leads to issues in interpretation.

5.4 Discussion

The modelling approaches described in Section 5.2 and Section 5.3 have quite separate aims: the analysis of inter-alternative correlation; along multiple dimensions; and the representation of deterministic and random variations in choice behaviour. It is quite reasonable to expect that both phenomena play a role in specific datasets. To minimise the risk of confounding between the two phenomena (see Hess et al. 2005a), it is in this case important to attempt to model the two phenomena jointly, with the help of an advanced GEV mixture model; or, although this is less appropriate in the present context (for numerical reasons), an error components Logit formulation or a Probit model (cf. Train 2003).

6 Summary and Conclusions

The discussion in this chapter has highlighted the complexity of the choice processes undertaken by air travellers, and has shown how they differ from behavioural processes in other areas of transportation. While highlighting the fact that travellers take decisions along a multitude of dimensions, the discussion has also noted that, in practical research, it is almost inevitable to use some simplifications of the choice process, partly because of modelling complexity, but mainly because of data issues. In this context, the advantages of SP data make the use of such datasets an important avenue for further research, potentially in conjunction with compatible RP data.

In closing, it is worth offering some guidance to modellers. On the basis of the discussions in this chapter, researchers are advised to:

- explicitly model the multi-dimensional nature of the choice process;
- allow for deterministic, random and continuous variations in choice behaviour;
- use advanced structures for correlation along all dimensions of choice;
- avoid over-aggregation in level-of-service data.

References

Adler, T., Falzarano, C.S. and Spitz, G. (2005), 'Modeling Service Trade-offs in Air Itinerary Choices', paper presented at the 84th Annual Meeting of the Transportation Research Board, Washington, DC.

Algers, S. and Beser, M. (2001), 'Modelling choice of flight and booking class – a study using stated preference and revealed preference data', *International Journal of Services Technology and Management* 2(1–2): 28–45.

Dong, X. and Koppelman, F.S. (2003), 'Mass Point Mixed Logit Model: Development and Application', paper presented at the 10th International Conference on Travel Behaviour Research, Lucerne.

Hess, S., Adler, T. and Polak, J.W. (2007), 'Modelling airport and airline choice behavior with stated-preference survey data', *Transportation Research Part E: Logistics and Transportation Review* 43: 221–33.

Hess, S., Bierlaire, M. and Polak, J.W. (2005a), 'Confounding between substitution patterns and random taste heterogeneity', CTS Working Paper, Centre for Transport Studies, Imperial College, London.

Hess, S., Bierlaire, M. and Polak, J. W. (2005b), 'Estimation of value of travel-time savings using mixed logit models', *Transportation Research Part A: Policy and Practice* 39(2–3): 221–36.

Hess, S., Bierlaire, M. and Polak, J.W. (2007), 'A systematic comparison of continuous and discrete mixture models', *European Transport* 36: 35–61.

Hess, S. and Polak, J.W. (2006), 'Exploring the potential for cross-nesting structures in airport-choice analysis: a case-study of the Greater London area', *Transportation Research Part E: Logistics and Transportation Review* 42(2): 63–81.

McFadden, D. (1978), Modelling the choice of residential location, in A. Karlqvist, L. Lundqvist, F. Snickars and J.W. Weibull (eds), *Spatial Interaction Theory and Planning Models* (Amsterdam: North-Holland), Chapter 25, pp. 75–96.

Pels, E., Nijkamp, P. and Rietveld, P. (2003), 'Access to and competition between airports: a case study for the San Francisco Bay area', *Transportation Research Part A: Policy and Practice* 37(1): 71–83.

Train, K. (2003), *Discrete Choice Methods with Simulation* (Cambridge, MA: Cambridge University Press).

Chapter 12
Airport Choice Behaviour: Findings from Three Separate Studies

Stephane Hess and John W. Polak[1]

This chapter summarises the findings of three parallel studies on air travel choice behaviour conducted by the authors in the past few years. The three studies make use of different datasets, with two studies using Revealed Preference (RP) data and one study using Stated Choice (SC) data. This chapter has three main findings. Firstly, the results give a clear indication of the advantages of Stated Choice data in retrieving significant effects for a number of conceptually important attributes, such as air fares and frequent flier benefits. Secondly, all three studies show that advanced model structures do lead to better performance, but that these gains come at the cost of a very significant rise in estimation cost. Finally, all studies show the importance of airport access time in that passengers have a strong preference for departing from the airport closest to their ground level origin.

1 Introduction

As highlighted in the discussions in Chapter 11, the number of studies of air travel choice behaviour has increased significantly over recent years, with a particular focus on studies of airport choice behaviour.

Over the last three years, the authors of this chapter have been involved in three separate such studies, making use of RP data collected in the San Francisco Bay (SF-bay) area and Greater London, and SC data collected via an internet-based survey in the US. This chapter summarises the findings of these three case studies, with more details on the individual studies being given in the relevant publications; see Hess and Polak (2005a, b; 2006a, b) and Hess et al. (2007).

While the overall approach used in the three studies is very similar, there are some differences between the three datasets that change the scope between case studies. Detailed description of the three datasets are presented in the respective sections and related papers; the aim of the discussion that follows is simply to describe the approach taken along the various dimensions of choice with the separate datasets.

Although the discussions in Chapter 11 have highlighted the fact that air travellers take decisions along a multitude of dimensions, with potentially important interactions between them, in practice, it is generally only possible to look at a subset of these choice dimensions. This is

---

1   The material described in this chapter is based on work conducted by the main author during his time in the Centre for Transport Studies at Imperial College, London. This chapter was partly written during a guest visit in the Institute of Transport and Logistics Studies at the University of Sydney. The support of both institutions is gratefully acknowledged. The authors would also like to thank Thomas Adler for his input to the SC study, and would similarly like to thank the Metropolitan Transport Commission and the Civil Aviation Authority for data support.

mainly a reflection of the formidable requirements in terms of data that would arise in the case of a study looking jointly at all the choice dimensions. However, it should also be noted that, for an adequate analysis of the interactions between choice dimensions, dynamic model structures would almost certainly be required, in conjunction with repeated choice data.

The three case studies described in this chapter look at the joint choice of airport and airline, where the two RP studies additionally look at the choice of access mode. While the detailed study and modelling of the interactions between choice dimensions is an important avenue for future research, it should be clear that this is a learning process, and that, before attempting such an analysis, there is a requirement to first look at the joint modelling of even a subset of these choices.

More generic choices, such as the decision to travel, the choice of destination, trip timing and the decision to travel by air, are not modelled in any of the three case studies, primarily for data reasons. A similar reasoning applies for any unmodelled choices along the air travel and ground level choice dimensions, where the issues discussed in Chapter 11 apply. Finally, in the case of flight routing, an important difference arises between the two RP case studies and the SC case study. In the former two, the topic is left untreated, and the studies look only at the choice between direct flights, a decision based on the low share of connecting passengers in the data and the lack of detailed data on connecting flights. In the SC dataset, connecting flights are included in the choice set, allowing for an analysis of the relative valuations of direct and indirect flights, where the choice of air routing is, however, not modelled.

The remainder of this chapter is organised as follows. Section 2 looks at the findings of the SF-bay study, while Sections 3 and 4 look at the Greater London and SC studies respectively. Finally, Section 5 briefly summarises and compares the findings of the three studies.

2 San Francisco Bay Area Study

This section describes the case study conducted in the San Francisco Bay area, which is served by three major airports: San Francisco International (SFO), Metropolitan Oakland International (OAK) and Mineta San José International (SJC). In this area, there is strong geographical captivity, with each of the three airports being in relatively close proximity to one of the main urban centres in the region – something that applies especially in the case of SJC.

The remainder of this discussion is organised as follows. Section 2.1 describes the data used in the analysis, while Section 2.2 discusses model specification. This is followed by a discussion of the results in Section 2.3 and a model validation process in Section 2.4.

2.1 Description of Data

Data on passengers' choice behaviour were obtained from the Airline Passenger Survey conducted by the Metropolitan Transport Commission (MTC) in August and October 1995. This contained information on over 21,000 departing air travellers. Passenger interviews were conducted at the three main SF-bay area airports, as well as at the minor Sonoma County airport (STS), which was not included in the present study. For a detailed description of the survey, see Franz (1996).

A total of 14 destinations served by direct flights from all three airports on every day of the week during the study period were included in the analysis. This includes some destination airports that are themselves located in multi-airport regions; this was unavoidable, given the available data, and issues resulting from this are discussed in more detail in Hess and Polak (2006a). The final sample contained 5,091 observations, as summarised in Table 12.1. Here, SJC is over-sampled, while OAK is under-sampled, an issue taken into account at the estimation stage using weighted maximum likelihood. The sample was split into two parts: a dataset used in the actual analysis (4,582 observations); and a 10 per cent sample retained for later validation of the models (509 observations).

The passenger survey data was complemented by appropriate level-of-service data for the chosen and unchosen alternatives need, using air travel data obtained from BACK Aviation Solutions,[2] and ground-access data obtained from the MTC in the form of origin-destination (O-D) travel time and cost matrices for the 1,099 travel area zones (TAZ) used for the SF-bay area.[3]

In the final analysis, a total of eight airlines were included; along the access mode dimension, six modes were used, namely car, public transport, scheduled airport bus services, door-to-door services, taxi and limousine.

2.2 Model Specification

The final sample contains data on three departure airports, eight airlines and six access modes, leading to 144 distinct triplets of alternatives. The study uses six population segments, grouping travellers into residents and visitors, as well as dividing them across three purpose groups, namely business, holiday and visiting friends or relatives (VFR).

For each model, attempts were made to include coefficients showing travellers' sensitivity to various attributes of the airports, airlines and access modes. The set of potential explanators used in the specification search included factors such as flight frequency, flight time (block time, which indirectly takes into account airport congestion), fare and aircraft type (jet vs turboprop), as well as access time (in-vehicle), walk time to access mode (for example, to public transport station), wait time for access mode and access cost. To account for airport allegiance, inertia coefficients were included in the utility functions, associated with variables giving the number of flights that a given traveller had taken out of a certain airport over the last 12 months. In addition to direct effects, cross-effects were allowed for by including inertia coefficients associated with SJC and OAK in the utility of SFO, as well as including a coefficient associated with SFO in the utility of

Table 12.1 Summary of choice data for SF-bay area case study

Dept. Apt.	Burbank, CA	Chicago, OHare, IL	Dallas, Ft. Worth, TX	Denver, CO	LasVegas, NV	LosAngeles, CA	Ontario, CA	Orange County, CA	Phoenix, AZ	Portland, OR	Reno, NV	Salt Lake City, UT	San Diego, CA	Seattle, WA	Total
SFO	55	89	36	65	57	199	35	37	128	140	1	42	258	213	1,355
SJC	167	58	91	71	163	367	111	247	133	106	156	61	248	169	2,148
OAK	211	1	25	9	68	381	135	177	51	101	39	43	139	208	1,588
Total	433	148	152	145	288	947	281	461	312	347	196	146	645	590	5,091

2 http://www.backaviation.com.
3 http://www.mtc.ca.gov/maps_and_data.

SJC. Some potentially important influences, such as carrier loyalty, could not be explored due to lack of data (for example no information on frequent flier programmes), while, in the presence of national flights only, the notion of allegiance to the national carrier does not apply. Similarly, it was not possible to identify a significant direct effect of the on-time performance of airlines or airports on the respective choice probabilities.

Both linear and various non-linear specifications of the various explanatory variables were tested. The best results were obtained with the use of a logarithmic transform; this, however, only led to an improvement in model fit when applied to flight frequency and the inertia variables. Finally, attempts were made to segment the population by income, for example, in order to show different values of time in different income classes. Three income groups were defined, segmenting the population into low income (< $21,000 per annum), medium income (between $21,000 and $44,000 per annum) and high income (above $44,000 per annum).

2.3 Model Results

In this section, we summarise the findings of the model fitting analysis. Here, the emphasis is mainly on the substantive findings, where a detailed description of the results is given in Hess and Polak (2006a). We first look at the findings in terms of what attributes have an effect on choice behaviour in the various population segments. This is followed by a discussion of the differences in the results across model structures.

2.3.1 Factors affecting choice behaviour The estimation on the 1,098 observations for business trips by residents revealed significant effects of walk access time, access cost, in-vehicle access time, flight time and frequency. Additionally, a significant negative effect could be associated with turboprop flights. No meaningful and significant effect of fare could be identified, even after taking into account income. This can mainly be explained by the poor quality of the fare data, but could also signal indifference to fare increases on the part of business travellers. Finally, a low level of fare differentiation for the routes used in the study also plays a role. It was possible to segment the sensitivity to walk time and access cost by income, where the results show lower sensitivity to cost for people with higher income, along with higher sensitivity to increases in walk time. In terms of the airport inertia variables, the estimates show positive direct effects for all three airports.

The estimation dataset contains information on 1,057 business trips by visitors. Just as in the case of resident business travellers, no significant impact of fares could be identified. In-vehicle access time and access cost are again significant, and negative, with increasing sensitivity to in-vehicle access time with higher income, and lower sensitivity to cost with higher income. Whereas it was not possible to estimate a significant effect of wait time for resident business travellers, a significant negative effect could be identified for their non-resident counterparts. However, the estimate for flight time is no longer significant at the conventional 95 per cent level, and it was not possible to include an effect of aircraft type, as flights using turboprop planes were never chosen. Also, with this model, no effect could be associated with access walk time, while frequency again has a strong positive effect.

The model estimated on the 831 observations for residents' holiday trips suggests a lower utility for flights using turboprop aircraft, negative impacts by access cost and in-vehicle time and a positive effect of flight frequency. Finally, for this group of travellers, a negative effect could be identified for fare (although of lower statistical significance), while no effect could be associated with flight time and access walk time.

For the 534 visitors on holiday trips, no significant effect of fare could be identified, and the effect of access cost, although of the correct sign, is not significant at the 95 per cent level. In-vehicle time has a significant negative effect, as has flight time, while increases in frequency lead to increases in utility. Finally, the aircraft type coefficient had to be excluded from the model (turboprop flights never chosen), while no effect could be identified for wait time.

The estimates for the model fitted to the sample of 641 residents on VFR trips show significant negative effects of access cost, in-vehicle time and air fare, along with positive effects of flight frequency. Aircraft type could not be included and no effects could be identified for walk time, wait time and flight time, while segmentations by income led to a loss of information in the model.

The final subsample used in the estimation contains information on 421 VFR trips by non-residents. The results show negative impacts of fare in the medium- and low-income classes (with higher sensitivity in the low-income class), while the effect for high earners was not significant and was dropped from the model. In-vehicle time and flight time have a negative effect, with a positive effect for frequency increases. No effect could be associated with access walk time, wait time and access cost, and the turboprop coefficient had to be excluded.

2.3.2 Model structure We next turn our attention to the findings in terms of model structure. In addition to simple Multinomial Logit (MNL) models, the analysis also looked at the estimation of Nested Logit (NL) and Mixed Multinomial Logit (MMNL) structures. For the NL models, only the three possible two-level structures (nesting by airport, airline or access mode) led to satisfactory results, while, for the MMNL models, the random structure of the model was used solely for expressing random taste heterogeneity across respondents,[4] and not heteroscedasticity or inter-alternative correlation. In this discussion, we look mainly at the differences in terms of model performance, using the adjusted ρ^2 measure which takes into account differences in the number of parameters. More detailed results are again given in Hess and Polak (2006a), while the differences are also highlighted in the comparison of substantive results which follows. The final specification of the utility function was the same across model structures (MNL, NL and MMNL).

The results in terms of model performance are summarised in Table 12.2. The results show that, with the exception of the first two NL models for non-resident VFR travellers, in each of the six population segments, the advanced model structures outperform the basic MNL model in terms of the adjusted ρ^2 measure. In the three population segments for residents, the best performance is obtained by the NL model using nesting by access mode, while, for non-residents, a less consistent pattern is observed. Overall, however, the differences in model performance are very small, especially when taking into account the heightened estimation cost of the more advanced models. This is especially true for the MMNL model. Two very distinct reasons can be given for

Table 12.2 Model performance on SF-bay area data in terms of adjusted $\rho^2(0)$ measure

	Resident			Visitor		
	Business	Holiday	VFR	Business	Holiday	VFR
MNL	0.5861	0.5112	0.5046	0.4379	0.3725	0.5044
NL (by airport)	0.5873	0.5148	0.5088	0.4481	0.3826	0.5038
NL (by airline)	0.5890	0.5141	0.5105	0.4394	0.3770	0.5018
NL (by access mode)	0.5930	0.5207	0.5224	0.4390	0.3774	0.5149
MMNL	0.5871	0.5127	0.5073	0.4389	0.3766	0.5068

---

4 Using the Normal distribution for reasons of simplicity.

the relative lack of improvement in model performance obtained with the more advanced models. The advanced models differ from the MNL model in that they explain processes taking part in the unobserved part of utility, either in the form of inter-alternative correlation or in terms of random taste heterogeneity across respondents. The first explanation interprets the similarity in the performance of the various models as an endorsement of the MNL models, suggesting that the (observed) utility specification used in the models captures almost all of the behaviour, reducing the scope for the advanced model to capture any behaviour in the remaining unobserved part of utility. An alternative explanation is based on the reasoning that the advanced structures used are little better that the MNL model in capturing the true phenomena captured by the unobserved component of utility. It is not clear from the empirical results alone which of these potential explanations is most appropriate, especially given the inability to estimate multi-level NL structures or CNL structures (cf. Hess 2004). The limited success of the MMNL models is also shown by the small amount of random taste heterogeneity retrieved by these models (cf. Hess and Polak 2006a). As such, while the models for resident business travellers and non-resident VFR travellers manage to retrieve random variations in the sensitivity to four attributes, this reduces to a single attribute for the two remaining resident models and the model for non-resident holiday travellers.

Even though the differences in model fit between the five structures are relatively modest, it is conceivable that the actual substantive results vary more significantly across models. To illustrate this, a brief analysis was conducted with the aim of comparing a common trade-off across models, as well as across population subgroups. The only two coefficients that are included in every single model are frequency and in-vehicle time, allowing the computation of the willingness to accept increases in access time in return for increases in flight frequency. The results of this analysis are summarised in Table 12.3, where, given that frequency enters the utility under a log-transform, K is used to represent the inverse of the current frequency, and where a sign change has been used to represent the willingness to accept increases in access time in return for increases in frequency. For the closed form models, the trade-off is simply given by the ratio of the two point-estimates, multiplied by K. The same applies in the MMNL model for holiday trips by visitors, where both frequency and access time were treated as fixed coefficients. However, in the remaining five population segments, the variation in the coefficients needs to be taken into account, especially for the access time coefficient, which forms the denominator of the trade-off. Here, a basic simulation approach was used, where the aim was simply to produce an estimate of the mean value for the trade-off, such that a censoring approach could be used to deal with the tails of the distribution, hence avoiding extreme values for the trade-off. Given the use of this censoring approach, the estimated standard deviation for the trade-off is unreliable, such that only the mean values are presented in Table 12.3.

The first observation that can be made from Table 12.3 is that, overall, the results show a higher willingness to accept increases in access time for residents than for visitors. The differences are especially significant in the case of VFR trips, where the relative value of frequency increases is at its highest for residents, while it is at its lowest for visitors. In terms of purpose-related differences, the results suggest higher relative sensitivity to frequency for business travellers than for leisure travellers in the models for visitors, while, for residents, frequency is valued less highly for holiday travellers. To some extent, these conclusions are, however, potentially influenced by the quality of the access journey level-of-service data.

Some interesting differences also arise when comparing the results across model structures. Here, it is important to note that no major issues with parameter significance arose in any of the models for the coefficients used in the trade-offs, increasing the reliability of the comparisons. The results show that, although there is some overall consistency in the trade-offs, there are also some differences, for example when looking at the results for MMNL, which overall give greater weight to the frequency coefficient, highlighting the effects of accounting for random taste heterogeneity. However, there are also some differences between the MNL model and the NL models, and also across NL models, such as, for example, in the case of the model using nesting by access mode for resident VFR travellers. Overall, these findings show that, although the differences between models in terms of LL may be relatively modest, the actual substantive conclusions are quite different. This highlights a relative flatness of the LL function, but also shows the impact of model structure on substantive results, making it an important issue in policy-oriented research.

In closing, it is worth noting that, at the average observed flight frequency of ten flights, the resulting value of K (0.1) leads to very low willingness to accept access time increases in return for increases in flight frequency. This should however be put into context by noting that the average observed access time was only around 30 minutes. Finally, the high values for K in the case of routes with low frequency (e.g. at $f = 2$, $K = 0.5$) imply a willingness to accept significant increases in access time in return for increases in flight frequency on routes with big gaps between individual flight departures.

Table 12.3 Trade-offs between flight frequency and access time (min/flight) in models for combined choice of airport, airline and access mode ($K = 1/f$, with f giving current frequency)

MNL

	Resident			Visitor		
	Business	Holiday	VFR	Business	Holiday	VFR
common	25.28 K	22.3 K	29.47 K	-	14.01 K	10.38 K
Inc. < $21,000	-	-	-	26.32 K	-	-
Inc. > $21,000	-	-	-	15.93 K	-	-

Nesting by airport

	Resident			Visitor		
	Business	Holiday	VFR	Business	Holiday	VFR
common	24.27 K	19.81 K	26.51 K	-	12.74 K	10.25 K
Inc. < $21,000	-	-	-	21.41 K	-	-
Inc. > $21,000	-	-	-	13.32 K	-	-

Nesting by airline

	Resident			Visitor		
	Business	Holiday	VFR	Business	Holiday	VFR
common	26.35 K	23.59 K	32.98 K	-	16.81 K	11.38 K
Inc. < $21,000	-	-	-	27.58 K	-	-
Inc. > $21,000	-	-	-	16.74 K	-	-

Nesting by access mode

	Resident			Visitor		
	Business	Holiday	VFR	Business	Holiday	VFR
common	20.08 K	19.05 K	16.52 K	-	12.45 K	7.7 K
Inc. < $21,000	-	-	-	24.69 K	-	-
Inc. > $21,000	-	-	-	15.16 K	-	-

MMNL

	Resident			Visitor		
	Business	Holiday	VFR	Business	Holiday	VFR
common	34.13 K	25.03 K	38.78 K	-	12.5 K	13.6 K
Inc. < $21,000	-	-	-	28.63 K	-	-
Inc. > $21,000	-	-	-	17.63 K	-	-

2.4 Model Validation

To complete the analysis, the five sets of models were applied to the validation sample of 519 observations. For each of the models, the final coefficient values produced during the estimation process were used in the apply runs. On the basis of this, the validation approach produces, for every observation, a choice probability for each of the 144 triplets of alternatives; this can be used to calculate the average probability of correct prediction for the actual chosen alternative across respondents. Aside from this probability for the choice of the actual triplet of airport, airline and access mode, it is also of interest to look at the probability of correct choice for just the airport, just the airline and just the access mode. These probabilities can be obtained through summing the probabilities of the single elementary alternatives falling into the given group.

The results of the validation process are summarised in Table 12.4. The first observation that can be made from this table is the surprisingly high probability of correct prediction of the actual chosen alternative, where, with an average of 31 available alternatives, the correct prediction probabilities range from 27 per cent to 48 per cent.

In terms of the correct prediction of airport choice, the probabilities range from 68.51 per cent to as high as 85.39 per cent. This performance compares well with the results of other studies, where, as an example, in one of the more recent studies in the SF-bay area, Basar and Bhat (2004) obtain an average correct prediction rate of 72.9 per cent on their validation sample.

The performance in terms of the choice of access mode is also very good, although it is significantly lower than the performance along the airport dimension for residents on VFR trips; while it is also slightly lower for visiting holiday travellers. On the other hand, it is marginally better than the performance along the airport dimension for visiting VFR travellers. The variation in performance does suggest that the choice process is less deterministic in some segments than in others.

The performance of the models in terms of correctly predicting the choice of airline is poorer than that for the choice of airport and access mode; however the values still always exceed 50 per cent, despite the complete absence of a treatment of airline allegiance, and the low quality of the fare data.

In terms of differences between population segments, the best average performance across all choice dimensions is obtained for resident business travellers, where an argument can be made that such travellers behave in a more rational manner (from the modeller's point of view), due to better information. The comparatively poor performance of the models for holiday trips can partly be explained by the fact that at least some of the travellers on such trips have purchased a package holiday or special flight deal, where the choices of departure airport and airline are potentially influenced by factors not included in the models.

Given the relatively modest differences in performance between the five model structures in the actual estimation processes, it should not be surprising that there are no systematic differences in prediction performance on the validation sample. Even though there are some outliers, such as the performance of the NL model using nesting by access mode in the models for residents on VFR trips, the average differences in performance are too small to come to any conclusions in terms of advantages for one of the model structures. This is further reinforced by the fact that it is not directly clear what measure of error should be associated with these probabilities.

Table 12.4 Prediction performance on SF-bay area validation data

MNL

	Resident			Visitor		
	Business	Holiday	VFR	Business	Holiday	VFR
Choice	47.13%	30.56%	36.58%	34.33%	27.21%	36.83%
Airport	84.04%	69.58%	80.83%	70.69%	69.53%	73.20%
Access mode	84.04%	67.72%	66.47%	70.18%	63.22%	77.08%
Airline	60.68%	54.93%	60.26%	55.39%	53.31%	60.97%

Nesting by airport

	Resident			Visitor		
	Business	Holiday	VFR	Business	Holiday	VFR
Choice	48.02%	31.39%	36.74%	36.19%	28.97%	36.81%
Airport	83.69%	69.16%	80.07%	70.69%	68.51%	73.13%
Access mode	85.22%	68.91%	67.50%	72.39%	66.41%	77.25%
Airline	61.06%	55.03%	60.08%	55.90%	54.34%	60.73%

Nesting by airline

	Resident			Visitor		
	Business	Holiday	VFR	Business	Holiday	VFR
Choice	47.90%	31.82%	36.50%	35.00%	27.78%	36.93%
Airport	84.18%	70.24%	80.36%	71.21%	68.61%	73.26%
Access mode	84.92%	68.64%	67.26%	71.08%	64.24%	76.96%
Airline	60.30%	54.79%	59.41%	55.27%	51.60%	60.52%

Nesting by access mode

	Resident			Visitor		
	Business	Holiday	VFR	Business	Holiday	VFR
Choice	48.41%	31.38%	39.60%	34.65%	27.78%	37.83%
Airport	85.39%	70.98%	84.97%	71.11%	72.41%	74.46%
Access mode	83.76%	67.29%	66.16%	70.25%	62.11%	76.98%
Airline	61.33%	55.46%	61.36%	55.49%	53.49%	61.04%

MMNL

	Resident			Visitor		
	Business	Holiday	VFR	Business	Holiday	VFR
Choice	47.53%	30.89%	37.45%	34.52%	27.73%	37.59%
Airport	84.21%	69.60%	81.59%	70.71%	68.78%	73.81%
Access mode	84.09%	67.77%	66.80%	70.21%	63.18%	77.15%
Airline	60.85%	55.07%	60.77%	55.52%	53.08%	60.69%

3 Greater London Study

This section describes the case study conducted in Greater London, an area which has by far the highest levels of air traffic in Europe, with, in 2002, some 117.13 million passengers using the five main airports. The area is dominated by Heathrow (LHR), the world's busiest airport by international passengers, and the main hub in Europe. Additionally, a large number of routes are offered from Gatwick (LGW), the world's busiest single-runway airport; while Stansted (STN) and Luton (LTN) act mainly as bases for holiday and low-cost operators. Finally, the centrally located London City (LCY) airport caters primarily to business travellers and, due to its short runway, is restricted to short-haul flights operated by turboprop planes and small jet aircraft.

3.1 Description of Data

For the present analysis, data from the 1996 passenger survey were obtained from the Civil Aviation Authority (CAA 1996). The original sample obtained from the CAA contained responses from 47,831 passengers for 31 destinations (reachable by direct flights from at least two of the five London airports) and 37 airlines. After data cleaning (missing data, compatibility between datasets), a usable sample of 33,527 passengers was obtained. This compares favourably to the sample of 5,091 available for the SF-bay analysis. In the present discussion, we look only at resident business travellers, where 5 per cent of the 8,704 observations were removed from the original sample to use in later model validation. A total of 31 destinations were used in the analysis, all of which are served by a single main airport, avoiding the problem with multi-airport destination areas faced in the San Francisco study. Any competition with high-speed rail on short-haul routes is not taken into account in the present study, where we work on the basis of an a priori choice of main mode.

Air-side level-of-service data were again obtained from BACK Aviation. The main item of information missing from this dataset is that of the fares for the different routes and airlines. Such data were compiled from two sources: the International Passenger Survey (ONS 1996) and the fare supplement of the Official Airways Guide for 1996 (OAG 1996). Information on travel class as well as ticket type (single or return) was taken into account in assembling the data. As was the case with the fare data used in the SF-bay study, the resulting dataset is of highly aggregate nature, leading to similar problems in the estimation of the marginal utility of air fares. Again, no information is available on frequent flier programmes.

For the analysis of the ground-level choice dimension, data from the National Airport Access Model (NAAM) were obtained for the base year 1999 (Scott Wilson Kirkpatrick 1999), and corresponding cost information for 1996 was produced with the help of the retail price index, while assuming that relative travel times have on average stayed constant. Six different modes are considered in the analysis: private car, rental car, public transport (rail, bus, local transport), long-distance coach, taxi and minicab (MC). No combinations of modes were considered in the present analysis, and the final mode indicated in the survey was used as the chosen mode. This is a major simplification of the actual choice process, given the high incidence of access journeys using a combination of different modes. However, in the absence of detailed route choice information, this simplification was unavoidable.

3.2 Model Structure and Specification

With the use of 5 departure airports, 37 airlines and 6 access modes, a total of 1,110 combinations of airports, airlines and access modes arise. However, not all airlines operate from all airports, and the total number of airport–airline pairs is actually 54, which reduces the number of alternatives (airport, airline, access mode triplets) to 324, compared to 144 in the SF-bay area study. The number of available alternatives for specific individuals in the estimation sample ranges from 6 to 58, with a mean of 31. As was the case in the SF-bay area study, weights were again used in the specification of the log-likelihood function, to account for the quota used in data collection, which are not representative of the population level.

A comprehensive list of variables was used in the initial modelling analysis, including attributes relating to the air journey, such as frequency, fare, flight time, aircraft type and seat capacity, and attributes relating to the access journey, such as access cost, in-vehicle access time, out-of-vehicle access time, wait time, number of interchanges and parking cost. The analysis made use of three separate model structures, namely MNL, NL and CNL, where, for NL models, it was again only possible to estimate the three different two-level structures, with any multi-level structures either failing to converge or reducing to a two-level structure.

3.3 Model Results

The modelling analysis showed that only a small set of the attributes listed above have a statistically significant impact on choice behaviour, at least with the present sample and model specification. Indeed, no effect could be identified for parking cost, seat capacity, out-of-vehicle access time, wait time and the number of interchanges. Furthermore, aircraft size, in the form of a dummy for turboprop planes, showed no effect; here, however, the highly correlated flight time attribute had a significant negative effect. A significant effect of air fare could not be identified for the present population segment; this is again at least partly due to the poor quality of the data. Finally, the analysis showed that the use of the combined fuel and depreciation cost for car journeys is preferable to the use of fuel cost on its own. For resident business travellers, four attributes were thus found to have a significant effect: access cost, in-vehicle access time (IVT), flight frequency and flight time, where a log-transform was used for all four attributes, and where the list of significant attributes stays identical across model structures.

Before looking at the substantive results, we first look at the differences in performance across the five structures used in the analysis, as shown in Table 12.5. This shows that all three NL models outperform the MNL model, with the best performance offered by the model using nesting by access mode. The same is true for the CNL model, which also outperforms all three NL structures, and where the total improvement of the CNL model over the MNL model is bigger than the combined improvements in the adjusted ρ^2 measure for the three NL models. This highlights the fact that the CNL model indeed offers significant improvements over the NL models, and suggests that the combined analysis of the correlation structure along the three choice dimensions can offer great benefits.

Aside from the differences in model fit, it is again of interest to look at the differences across models in terms of substantive results. Given the use of the log-transform in nominators as well as denominators, the various trade-offs were calculated separately for each individual,[5] and summary statistics were then calculated across respondents.

The results of this process are summarised in Table 12.6. In each case, the tables present the mean value of the respective trade-off, along with the associated standard deviation, and show the

Table 12.5 Model performance on London data

Model	Final LL	Parameters	Adjusted ρ^2
MNL	-14,945.3	55	0.3445
NL by airport	-14,896.1	59	0.3465
NL by airline	-14,870.7	74	0.3469
NL by access mode	-14,816.7	60	0.3499
CNL	-14,603.9	91	0.3578

---

5   With a logarithmic transform, the ratio of partial derivatives of the utility function is no longer given simply by the ratio of marginal utility coefficients, but involves the attribute values themselves.

values across the five different models estimated. From these results, it can be seen that the first three models produce roughly similar results, while those produced by the CNL model and the NL model using nesting access mode are more extreme (when compared to the three first models), something that is especially true when looking at the value of travel time savings (VTTS) measures for the model using nesting by access mode.

Another observation that can be made for the trade-offs is that the VTTS measures are markedly lower than those reported, for example, by Pels et al. (2003), although they are still higher than in other contexts, which can be explained partly by concepts of risk-averseness, as discussed, for example, by Hess and Polak (2005b). Travellers are willing to pay for a reduction in the risk of missing their flight, where this risk clearly increases with access time. While the lower values (compared to the SF-bay studies) could be explained on geographical grounds, it seems more likely that the use of a non-linear specification is the main reason for the lower (and it should be said more realistic) values; indeed, much higher values, together with a lower model fit, were obtained when using a linear specification. Finally, the still high values should also be put into context by noting that the average access journey in this population segment was measured as 57 minutes.

Table 12.6   Model results for London data

	IVT vs. access cost (£/hour)	Freq. vs. access cost (£/flight)	Freq. vs. IVT (hours/flight)	Flight time vs. IVT
MNL				
Minimum	1.18	0.02	0.01	0.04
Mean	16.24	1.56	0.11	1.07
Maximum	143.38	231.05	4.06	7.43
Standard deviation	25.44	4.85	0.18	0.7
NL by airport				
Minimum	1.3	0.02	0.01	0.04
Mean	17.85	1.63	0.11	0.97
Maximum	157.65	242.38	3.87	6.72
Standard deviation	27.98	5.09	0.17	0.63
NL by airline				
Minimum	1.29	0.03	0.01	0.04
Mean	17.76	1.79	0.12	1.13
Maximum	156.8	265.11	4.26	7.85
Standard deviation	27.83	5.57	0.19	0.74
NL by access mode				
Minimum	0.99	0.02	0.01	0.04
Mean	13.52	1.11	0.1	1.05
Maximum	119.35	164.74	3.47	7.31
Standard deviation	21.18	3.46	0.15	0.69
CNL				
Minimum	1.16	0.01	0	0.04
Mean	15.96	0.9	0.07	0.95
Maximum	140.89	132.96	2.38	6.59
Standard deviation	25	2.79	0.1	0.62

3.4 Model Validation

The final part of the analysis is concerned with model validation. For this, the validation sample of 353 observations was used in application runs using the models presented in the preceding section. The results of this analysis are summarised in Table 12.7.

The results show relatively little variation between the three model structures, which was to be expected, when comparing the differences in model fit to the base LL. Furthermore, it is not clear a priori what measure of error should be associated with these measures, such that no inferences on differences between models should be drawn on the basis of these results. Without touching on the differences between models, it is of interest to compare the results to those obtained in the SF-bay study. The aggregate average probabilities of correct prediction are well below those obtained in the SF-bay study. This, however, needs to be put into context by noting that the choice set used in the SF-bay area was considerably smaller (3 airports, 8 airlines and 6 access modes). Furthermore, the exceedingly high market share for car made the analysis of access choice behaviour in the SF-bay area almost trivial. Finally, it seems that airport captivity plays a much bigger role in the SF-bay area than in London, where the levels of competition are much higher. This suggests that the models estimated on this data yield very satisfactory performance, even though they should still only be seen as a first step in the search of an optimal specification. Further gains can be expected by allowing for random taste heterogeneity inside a Mixed Generalised Extreme Value (GEV) framework; this is the topic of ongoing work.

Table 12.7 Prediction performance on London validation data

	Elementary alts. (324)	Airport (5)	Airline (37)	Access mode (6)
MNL	16.01%	61.47%	48.01%	39.27%
NL by airport	16.50%	62.88%	48.62%	38.58%
NL by airline	16.17%	61.34%	47.71%	39.54%
NL by access mode	16.03%	61.19%	47.84%	39.51%
CNL	16.48%	62.44%	47.79%	39.23%

4 Stated Choice Study

The two RP studies discussed thus far in this chapter have again highlighted the issues faced with this type of data, notably in terms of the problems with retrieving a meaningful effect of changes in air fares. The study presented in this section aims to illustrate how these problems can be overcome with the use of SC data.

4.1 Description of Data

The survey data used in this analysis were collected via the internet in May 2001 from a sample of around 600 individuals who had made a paid US domestic air trip within the 12 months prior to the interview taking place (Resource Systems Group Inc. 2003, Adler et al. 2005).

The survey uses a binary choice set, with ten choice situations per individual. In each choice situation, the respondent is faced with a choice between their recent observed trip and an alternative journey option, compiled on the basis of the information collected in the RP part of the survey. These two alternatives are hereafter referred to as the RP alternative and the SC alternative respectively.

Aside from the actual airline and airport names, from which access times can be inferred, the attributes used to describe the two alternatives in the SC survey include flight time, the number of connections, the air fare, the arrival time (used to calculate schedule delays), the aircraft type and the on-time performance of the various flights. Access cost was not included in the surveys (in the absence of an actual specification of the mode-choice dimension), and no choice is given between different travel classes.

The final sample contains data collected from 589 respondents; with 10 choice situations per respondent, a sample size of 5,890 observations is obtained, split into 1,190 observations by business travellers, 1,840 observations by holiday travellers and 2,860 observations by VFR travellers. Further segmentations, for example by employment status, did not provide additional insights. Given the small sample sizes, especially for the business segment, and the high number of explanatory variables, the decision was taken to include all observations in the estimation process rather than waste some of them on a validation sample.

4.2 Model Specification

A large number of variables were included in the initial specification search. Aside from the continuous variables such as flight fare, flight time, access time, on-time performance (in %) and early and late schedule delay (SDE and SDL), which need no further explanations, a number of discrete variables were also included.

As such, dummy variables were associated with different airports and airlines on the basis of the ranking provided by respondents in initial questioning, where appropriate normalisations were performed. Attempts were also made to estimate a constant associated with the airport closest to the passenger's ground-level origin. Three additional dummy variables were included in the base specification, to account for the effects of frequent flier (FF) membership, where these were associated with standard membership, elite membership and elite plus membership. Similarly, dummy variables were included for flights with a single or double connection, while dummy variables were also associated with the different types of aircraft included in the survey. Finally, attempts were made to account for respondent inertia or habit formation with the help of a number of variables. Aside from an alternative specific constant (ASC) for the RP alternative, airport and airline inertia constants were included in the utility of the SC alternative in the case where the RP airport or airline was reused in the SC alternative.

In addition to making use of non-linear transforms (log-transforms) where appropriate, this analysis also makes use of continuous interactions between socio-demographic attributes and taste coefficients, as described in Chapter 11. The socio-demographic attributes used in these interactions were income and travel distance (in the form of flight time for the RP alternative). Interactions with other factors, such as trip duration or party size, were not found to be significant.

4.3 Model Results

This section describes the findings of the estimation process. In the current work, only basic MNL structures were used. Nesting structures are not applicable given the nature of the choice set, while the use of mixture models, such as MMNL, was avoided with the aim of attempting to explain taste heterogeneity in a deterministic fashion. A separate analysis which made use of MMNL structures showed little additional gains in model fit, with the main advantage coming in a treatment of the repeated choice nature of the SC data. In this section, we only present a summary of the actual estimation results, with more details given by Hess et al. (2007).

4.3.1 Business travellers   The findings from the analysis using the 1,190 observations for business travellers revealed effects for all the main continuous variables, including access time, air fare, flight time and early and late arrival. Except for the early arrival penalty, the analysis showed that the use of a log-transform leads to significant gains in model performance, suggesting decreasing marginal returns for the associated attributes. The results further show positive effects of improvements in on-time performance. Initial results showed a reduced sensitivity to on-time performance on longer flights, but this resulted in problems with significance for the actual on-time performance coefficient. Efforts to use a power formulation for the on-time performance attribute (allowing for a much stronger dislike of very late flights) led to problems with parameter significance. In terms of interactions, the estimates additionally suggest a reduced sensitivity to early arrival on longer flights, as well as reduced fare sensitivity with higher income.

The final part of this discussion looks at the findings for dummy variables. Here, a significant positive ASC was found to be associated with the current alternative, capturing inertia as well as a host of other effects. The estimation further shows a strong effect of frequent flier membership on the utility of an alternative, where, due to insignificant differences, a common factor was used for elite and elite plus membership, where the estimates show this to be over twice as large as for standard frequent flier membership. The fact that none of the airline dummy variables (linked to ranking) was found to be significant suggests that, for business travellers, airline allegiance is primarily limited to membership in frequent flier programmes. In terms of airport allegiance, a significant effect could only be associated with the second and top-ranked airports, where the former one was significant only at the 81 per cent level. The estimated dummy variables for flights with one and two connections were indistinguishable, leading to the use of a common factor, where this can in part be seen as a result of the low incidence of flights with double connections in the data. The final set of dummy variables, associated with aircraft type, show that single-aisle jets are clearly preferred over turboprop planes and regional jets.

4.3.2 MNL model for holiday travellers   The findings from the analysis using the 1,840 observations for holiday travellers revealed significant effects of access time, air fare and flight time, where a log-transform was again found to be appropriate for all three attributes. The first difference with the business models arises in the treatment for schedule delay, where the use of linear effects was found to be preferable, and where, given the small differences between the effects for early and late arrival, a common coefficient was used. The results again show positive effects of improvements in on-time performance, where the associated interaction term suggests that holiday travellers' sensitivity to on-time performance increases with flight distance. This can be explained, for example, by the notion that holiday flights are often pushed to the edges of the off-peak periods, where sensitivity to on-time performance may indeed be greater, and especially so for very long flights.

Other interactions again show reduced fare sensitivity with higher income, although the confidence level for the associated term is very low. The interaction terms also show that, for holiday travellers, fare sensitivity increases with flight distance. It is important to put this into context by remembering that a log-transform is also used on the fare attribute. As such, the results simply suggest that, at a given fare level, increases are valued more negatively in the case of longer flights. A possible explanation for this could be the higher secondary costs associated with longer flights in the case of holiday travellers.

As expected, frequent flier benefits play a much smaller role in this segment of the population, where it was only possible to estimate a common dummy variable for all levels of membership, which in addition only attains a very low level of statistical significance. On the other hand, a significant positive effect is associated with the top-ranked airline.

In this sample, the effect associated with flights with two connections is also significantly larger than for flights with a single connection, and the scale of the difference (factor of 3) supports the decision not to use a linear effect, but to use two separate dummy variables. Finally, for the aircraft type dummies, the results suggest that holiday travellers do not distinguish between single-aisle jets, regional jets and turboprop planes, with the only aircraft dummy with a modestly significant value being that for wide-body aircraft, which are seemingly given a slight preference over single-aisle jets.

4.3.3 MNL model for VFR travellers   The findings from the analysis using the 2,860 observations collected from VFR travellers reveal an important difference when compared to those for business and holiday travellers. Indeed, while access time and flight fare again enter the utility function under a log-transform, the specification search indicated that it is preferable to treat flight time in a linear fashion. Early and late arrival penalties are treated separately in this model, and both enter the utility in a linear form, where the penalty associated with late arrival is lower and attains a very low level of statistical significance.

The non-linear interactions retrieved from this data show heightened fare sensitivity on longer flights, along with reduced fare sensitivity with higher income and lower sensitivity to access time on longer flights, which would support a decision to shift long-haul flights to outlying airports, where the issue of point-to-point passengers on the required feeder-flights would however need to be addressed separately.

In this segment, it was not possible to estimate a significant effect associated with frequent flier programmes, while the dummy variables associated with the two most preferred airlines are positive and significant at high levels of confidence. Airport allegiance also seems to play a role; but there is, however, essentially no difference between the estimates of the dummies associated with the two top-ranked airports. Finally, unlike in the other two population segments, it was also possible to identify a significant positive effect associated with the airport closest to the passenger's ground-level origin.

A common effect was again used for flights with single and double connections, while, in terms of aircraft type, the difference between single-aisle jets and regional jets is significant only at the 87 per cent level, where the results further indicate a significant dislike for turboprop flights and a significant preference for wide-body jets over single-aisle jets.

4.4 Comparison of Results across Population Segments

The description of the MNL model fitting exercises has highlighted a number of differences between the specifications used in the three population segments, notably in terms of non-linearities and interactions with socio-demographic variables. These differences in model specification need to be borne in mind when comparing the substantive results across the three population segments. The calculation of the trade-offs, and hence the comparison of results across groups, is further complicated by the high number of non-linear terms in the utility functions, where the simple ratio between coefficients is no longer applicable. The situation becomes more complicated again in the case of coefficients interacting continuously with income or flight distance, where any trade-off involving such coefficients will vary across individuals as a function of the associated attribute.

In the present analysis, the comparison was limited to two main sets of trade-offs, looking at the willingness to accept increases in fare and access time respectively, in return for improvements in other determinants of choice. In each case, the trade-offs are presented for the average flight distance and household income in that population segment, meaning that the interaction terms become equal to 1.

The results are summarised in Table 12.8 for the willingness-to-pay indicators and Table 12.9 for the willingness to accept increases in access time. In each case, several coefficients used in the trade-offs were not significant at the 95 per cent level, as pointed out in Sections 4.3.1, 4.3.2 and 4.3.3, and this is indicated appropriately in the presentation of the trade-offs.

Consistent with a priori expectations, the results show a much greater willingness to accept higher fares in return for shorter access times for business travellers than for holiday or VFR travellers, by a factor of just over 2. The models also indicate a higher willingness by business travellers to pay for reductions in schedule delay and for improved on-time performance. Interestingly, the models suggest that, except for holiday travellers, respondents are more sensitive to early than to late arrival – a finding that should, however, be put into context given the small differences and high associated standard-errors.

The models suggest that business travellers are willing to pay $125 to fly on an airline where they hold an elite frequent flier account. Even though this figure decreases to $49 in the case of standard membership, the figures are still much higher than for holiday travellers, while no such effects could be identified for VFR travellers. In these latter two groups, the results however show a certain willingness to pay a premium for flying on either of the top-ranked airlines.

Table 12.8   MNL trade-offs, part 1: willingness to pay ($)

	Business	Holiday	VFR
Reduction in access time (1hour)	75.40	35.80	35.48
Reduction in SDE (1hour)	13.27(*)	2.61(*)	3.68
Reduction in SDL (1hour)	11.08		2.25(*)
On-time (+10%)	10.39	7.02	5.57
FF elite or elite-plus *vs* none	125.24	11.44(*)	-
FF standard *vs* none	49.12		-
Top airline *vs* worst	-	25.07	21.06
2nd airline *vs* worst	-	18.16(*)	15.27
3rd airline *vs* worst	-	20.09(*)	4.77(*)
Top airport *vs* worst	83.22	53.97	55.73
2nd airport *vs* worst	30.56(*)	41.42	54.63
3rd airport *vs* worst	-	18.54(*)	25.89
Airport closest to home	-	-	28.02
No connection *vs* one connection	44.15	19.60	18.98
No connection *vs* two connections		62.21	
Jet *vs* wide-body	29.86(*)	-	-
Jet *vs* regional jet	79.51	-	10.59(*)
Jet *vs* turboprop	96.94	1.79(*)	17.77
Wide-body *vs* jet	-	13.45(*)	27.84
Regional jet *vs* jet	-	1.31(*)	-

(*) Coefficient used in numerator of trade-off not significant at 95 per cent level.

Table 12.9 MNL trade-offs, part 2: willingness to accept increases in access time (min)

	Business	Holiday	VFR
Reductions in fare ($1)	2.14	4.61	4.57
Reduction in SDE (1 hour)	17.38(*)	8.25(*)	12.24
Reduction in SDL (1 hour)	17.00		7.49(*)
On-time (+10%)	13.60	22.16	18.53
FF elite or elite-plus vs none	163.97	36.10(*)	-
FF standard vs none	64.31		-
Top airline vs worst	-	79.11	70.08
2nd airline vs worst	-	57.31(*)	50.81
3rd airline vs worst	-	63.40(*)	15.88(*)
Top airport vs worst	108.96	170.29	185.43
2nd airport vs worst	40.01(*)	130.69	181.78
3rd airport vs worst	-	58.49(*)	86.14
No connection vs one connection	57.81	61.86	63.15
No connection vs two connections		196.29	

(*) Coefficient used in numerator of trade-off not significant at 95 per cent level.

In terms of paying a premium for direct flights, the results again suggest a higher willingness for business travellers, although the different treatment in the case of holiday travellers results in a higher value for the trade-off in the case of flights with two connections in this group. A difference arises between the three population groups in the trade-offs looking at the willingness to pay for flying on a specific type of aircraft, where the differences in the most-valued type of aircraft led to a different base type.

The findings for the trade-offs looking at the willingness to accept increases in access time do, overall, show a lower willingness for business travellers than for holiday and VFR travellers, which is to be expected. The main exception again comes in the case of frequent flier benefits, where the results suggest that business travellers are willing to fly out of more distant airports in return for flying on an airline whose frequent flier programme they are a member of.

Some of the trade-offs presented in this section are very high; this could potentially be a reflection of the well-established notion that in SC studies, there is a tendency for respondents to exaggerate their responsiveness to changes in attributes (for example, Louviere et al. 2000, Ortúzar 2000).

5 Summary and Conclusions

The results of the three studies discussed in this chapter are not directly comparable, given the geographical differences, the differences in the age of the data and the differences in survey design as well as data type (RP vs SC). Nevertheless, some conclusions can be reached.

From a model structure point of view, all three case studies have shown that the use of more advanced model structures can lead to improvements in model fit. However, although the improvements are statistically significant, they are too small to lead to any major differences in model performance. Nevertheless, the advanced model structures provide further insights into choice behaviour, and there are also differences in the substantive results between the various models.

The main observation that can be made in the comparison of the results across the three studies is the greater ability of the SC models to retrieve significant effects for a range of variables that are generally not well estimated in RP studies, such as air fares, schedule delay and airline and airport allegiance. This is an illustration of the complications that arise with the use of RP survey data in the analysis of air travel choice behaviour, where there are issues of data quality in relation to air fares and availabilities; while information on a number of other attributes, notably the membership in frequent flier programmes, is not generally available in such datasets. On the other hand, it should be remembered that there is a risk of bias when relying solely on the use of SC data, making the joint estimation on RP and SC data an important avenue for further research, as discussed by Algers and Beser (2001).

The one common observation that can be made from the three case studies is that the results do suggest that access time plays a major role in the choice process, with passengers having a strong preference for their local airport. As such, the attractiveness of outlying airports depends heavily on good access connections, unless there are other incentives such as low air fares. This is reflected in the fact that only low-cost carriers find it relatively easy to attract passengers to outlying airports that are not served by convenient and fast ground-level services. It is conceivable that the sensitivity to access time decreases with flight time,[6] such that moving long-haul services to outlying airports would seem wise; this, however, causes problems as the associated (and necessary) short-haul feeder flights will also carry point-to-point passengers, who will again have a preference for more centrally located airports.

References

Adler, T., Falzarano, C.S. and Spitz, G. (2005), 'Modeling Service Trade-offs in Air Itinerary Choices', paper presented at the 84th Annual Meeting of the Transportation Research Board, Washington, DC.

Algers, S. and Beser, M. (2001), 'Modelling choice of flight and booking class – a study using stated preference and revealed preference data', *International Journal of Services Technology and Management* 2(1–2): 28–45.

Basar, G. and Bhat, C.R. (2004), 'A Parameterized Consideration Set model for airport choice: an application to the San Francisco Bay area', *Transportation Research Part B: Methodological* 38(10): 889–904.

CAA (1996), 'UK Airports Passenger Survey, CAP 677 – Passengers at Birmingham, Gatwick, Heathrow, London City, Luton, Manchester and Stansted in 1996', The Civil Aviation Authority, London.

Franz, J.D. (1996), 'Metropolitan Transportation Commission Airline Passenger Survey: Final Report', prepared by J.D. Franz Research for the Metropolitan Transport Commission, Sacramento, CA.

Hess, S. (2004), 'A Model for the Joint Analysis of Airport, Airline, and Access-mode Choice for Passengers Departing from the San Francisco Bay Area', paper presented at the European Transport Conference, Strasbourg.

---

6   As suggested by the results for VFR travellers in Section 4.3.3.

Hess, S., Adler, T. and Polak, J.W. (2007), 'Modelling airport and airline choice behavior with stated-preference survey data', *Transportation Research Part E: Logistics and Transportation Review* 43: 221–33.

Hess, S. and Polak, J.W. (2005a), 'Accounting for random taste heterogeneity in airport-choice modelling', *Transportation Research Record* 1915: 36–43.

Hess, S. and Polak, J.W. (2005b), 'Mixed logit modelling of airport choice in multi-airport regions', *Journal of Air Transport Management* 11(2): 59–68.

Hess, S. and Polak, J.W. (2006a), 'Airport, airline and access mode choice in the San Francisco Bay area', *Papers in Regional Science* 85(4): 543–67.

Hess, S. and Polak, J.W. (2006b), 'Exploring the potential for cross-nesting structures in airport-choice analysis: a case-study of the Greater London area', *Transportation Research Part E: Logistics and Transportation Review* 42(2): 63–81.

Louviere, J.J., Hensher, D.A. and Swait, J. (2000), *Stated Choice Models: Analysis and Application*, Cambridge University Press, Cambridge.

OAG (1996), *World Airways Guide Fares Supplement 1996*, Reed Elsevier, Dunstable.

ONS (1996), *International Passenger Survey*, The Office for National Statistics, London.

Ortúzar, J. de D. (2000), *Stated Preference Modelling Techniques: PTRC Perspectives 4*, PTRC Education and Research Services Ltd, London.

Pels, E., Nijkamp, P. and Rietveld, P. (2003), 'Access to and competition between airports: a case study for the San Francisco Bay area', *Transportation Research Part A: Policy and Practice* 37(1): 71–83.

Resource Systems Group Inc. (2003), *Air Travelers 2003: The New Realities?* Annual Air Survey project report.

Scott Wilson Kirkpatrick (1999), 'National Airport Accessibility Model: Model Status Report', prepared for the Department of Transport by Scott Wilson Kirkpatrick and Co Ltd, Basingstoke.

Chapter 13
Improved Modelling of Competition among Airports through Flexible Form and Non-Diagonal Demand Structures Explaining Flows Registered within a New Traffic Accounting Matrix

Marc Gaudry[1]

1 Introduction: A Demand Side Emphasis

The identification of the determinants of airport hub demand, unavoidable in the recent context of the Air France-KLM integration treaty, is clearly of a general interest because competition among airports is widespread and measurable, as the detailed studies by Mandel (e.g. 1999a, 1999b) have shown for Germany. As our focus cannot be that of a survey of the literature on airport competition, we simply ask to what extent key features of current air demand modelling practice and its trends can be useful in understanding the general question of air market development simultaneously with that of the competition among airports.

Why Neglect the Supply Side?

We do not deal with supply side modelling developments, a notoriously difficult task that is a good example of thought market failure. Indeed, when Alfred Kahn successfully proposed air deregulation in 1978 – thereby ensuring that he would be the last chairman of the United States Civil Aeronautics Board (CAB), abolished in 1984 – nobody, not even Professor Kahn himself, foresaw the future shape of the air route structure that would progressively ensue, either in the United States or internationally (Kahn, 2003).

This massive thought failure in forecasting performance by all academic,[2] policy and business experts occurred despite the existence of the successful Federal Express (FedEx) freight hub, initiated in 1973 by Fred Smith in Memphis, Tennessee, and clearly profitable since 1976. And some form of the same failure of vision occurred everywhere: even in perfectly hexagonal hub-

---

1 The author thanks Professor Jaap de Wit for comments on an early version of this chapter invited for presentation at the Airneth Workshop on Network strategies of multi-hub airlines and the implications for national aviation policies, The Netherlands Directorate-General for Civil Aviation and Freight Transport, The Hague, 28 October 2005. He also thanks Dr Benedikt Mandel of MKmetric GmbH, Karlsruhe, for his comments on this final version produced for the German Aviation Research Society (GARS) Workshop on Benchmarking of Airports and EU-Liberalization, Hamburg, 23 February 2006.

2 In 1993, interestingly, Mandel (1999a) estimated correct price curvatures of the demand curves but concentrated on issues raised by raising prices (the 50 DM Hamburg premium), not on the size of demand if one lowered prices with such curvatures which clearly could have correctly forecast LCC demand. No one thought prices could fall so much.

shaped France, the beginning of the structuring of the Air France hub at Charles de Gaulle airport awaited 1996, a process not completed until perhaps 2003.

Of course, this industry is full of similar supply side market forecasting failures: despite the exactly linear growth of Southwest Airlines over time since 1971, it has now been retiring pilots for about seven years. Who could have forecast the rise of the new stable low-cost carrier (LCC) business model firms challenging incumbent full-service airlines, at least on short, high-density markets, clearly benefiting in the short run from the obvious accumulated excess supply of subsidized airports? Not even the supply siders. We therefore concentrate on the more successful side of competition modelling: the demand side.

Structure of the Demand Side Approach

Our discussion of competition among airports emphasizes their hub character and neglects (except implicitly in the definitions of land-based access variables) the changes in competitive advantage that may arise from improvements in complementary land mode links.[3] It then proceeds in four steps: an accounting step and three sections of econometric modelling diagnostics.

Indeed, as our discussion needs a framework to pinpoint relevant flows and airport performance, we first provide a new accounting framework that pinpoints the relevant flows to be explained by air demand and airport choice models. In addition to definitional precision, this framework, as a side benefit, makes it possible to point out pitfalls of demand analyses and forecasts based on simple formulations relating intermediate network flows to final demand, without duly taking into account the necessary spatial features of inter-industry economics required for transport flow analysis.

We then proceed to an analysis of econometric demand models in three parts. First, we state that two questions should be asked about any demand model: whether it was estimated with fixed or variable mathematical forms and whether it allows for substitution and complementarity among alternatives. We then analyse Generation-Distribution models in current use from that perspective: it is not possible to discuss continental airport demand sensibly without this structure that determines total flows by all modes. Third, we analyse the Mode/Path choice component of models from the same double-barrel perspective.

Throughout our critical report, we focus on the crucial features of demand modelling practice that have particular bearing on airport competition and hub stability: the built-in IIA-axiom (Independence from Irrelevant Alternatives), consistency of typical "diagonal" structural demand and path choice procedures. We argue that the IIA property must be avoided because it is now particularly challenged by the fastest growing component of passenger air demand, non-business trips, and by the clear need for reference alternatives in mode and path or airline company choice procedures. Because flows are interdependent, the utility of alternatives cannot be defined only by reference to own (matrix diagonal) transport conditions.

Endogenous Form is Necessary but Insufficient

On this critical point, we summarize how Standard Box-Cox endogenous form specifications contribute to a much improved representation of the role of transport conditions within prevailing

---

3   For a formal discussion of microeconomic complementarity and substitution between air and land modes, notably high-speed rail, see Gaudry (1998). Postal TGV trains in France show that complements occur also for many air services.

IIA-consistent structures, but argue that freedom from "diagonal slavery" requires more: to wit spatial correlation processes in Generation-Distribution models and Generalized Box-Cox specifications in Mode and Path choice analysis. In both cases, IIA consistency is avoided in realistic ways by making parsimonious use of off-diagonal terms and permitting in principle the establishment of complementary alternatives, in contrast to the currently forced substitution straightjacket imposed on independently specified alternatives.

2 Flow Accounts, Textbook Demand Equations and Airport System Components

As mentioned above, it is necessary to pinpoint the flows of interest. We do this by extending the standard input-output framework and recalling how demand forecasts cannot simply relate network flows to changes in "final demand" without taking due account of the spatial structure of flows and of the multiple and necessary ways of specifying technical coefficient matrices in transport, in contrast with the standard unique specification used in unspatialized inter-industry economics. We complement this set-up by explaining where the equilibrium levels of service come from.

2.1 Pinpointing Flows Identified within Accounts and Explained by Typical Equations

Key modelling developments and hub stability If market reality speaks, can academic research say anything to increase the clarity of the message? I will select a few key new features of models and ask whether or not they are more likely to imply hub stability. In a formal sense, this question raises a problem of air company/air path choice but I will treat it without being very specific about issues such as detailed scheduling and flight coordination (Burghouwt and Wit, 2003) critical for hub performance, but within a representative schematic demand generating framework.

We define our interest as that of the total flow of air passengers using a hub location (airport), and composed of both direct and circuitous elementary flows. That total number, for a certain trip purpose g is the sum of the *origin-destination (O-D) passengers*, for whom the hub h is the origin or the destination city of the city pair considered; and of the *transfer passengers*, passing through h on their path between any O-D city pair:

[Air flow through h] = [Direct O-D traffic between h and any j]
+ [Circuitous traffic through h]   (1)

or $\quad T^g_{air, h} = T^g_{air, hj} + T^g_{air, ihj}, \quad i, j \neq h$ (2)

where for simplicity all flows are assumed to be bi-directional between city h and any city j, as well as between any city pair ij whose air passengers choose to go through hub city h. Also, for similar reasons and without loss of generality, we associate the airport cities with the origin or destination zones.

To clarify, Table 13.1 presents an accounting system to visualize the hub flows of interest in (2). Part A of Table 13.1 is a Traffic Accounting Matrix (TAM) reporting on four flows (D1, D2, C1 and C2) that *enter Airport J of City 4* and follow either a direct or a circuitous path to three other cities, as indicated in Part B. In this TAM, known O-D flows correspond to the bottom right-hand quadrant of Part A. Passengers are registered when they *enter airport J*, when they *use a flight* and when they *leave an airport for the outside*. One can apply such a TAM to a complete airport system such as that reported on in Table 13.2 for the same four cities listed in Table 13.1 but without the direct flight from Airport J to Airport B.

Table 13.1   Traffic accounting matrix (TAM) representation of the physical transport network flows

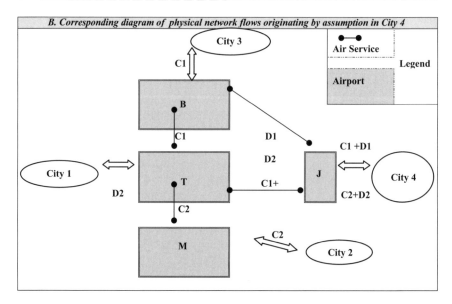

Note first that the system O-D matrix is asymmetric for reasons of realism. Also, adding up registrations yields **FE**, the vector[4] of flows entering an airport from any outside zone, and **NE**, the vector of flows entering an airport from any other airport. One can also compute **NL** and **FL**, the corresponding vectors of flows leaving an airport for another airport or for outside zones, respectively. Combining these, the total entering flow vector is given by **TE = FE + NE** and the total leaving flow vector by:

$$\mathbf{TL} = \mathbf{NL} + \mathbf{FL} \tag{3}$$

The accounting identity of a TAM As passengers do not accumulate in the airports, the accounting identity of the airport system states that the flow coming through an airport from the outside or from other airports must be equal to the flow leaving that airport for other airports or for the outside:

$$\mathbf{TE}' = \mathbf{TL} \tag{4}$$

where the (') denotes the transpose operator. We emphasize that the accounting identity of the system *is not* that the entering and leaving vectors **FE** and **FL** are the same. In the Table 13.2 example, the entering and leaving vectors differ due to the asymmetry of the O-D matrix. But, of course, if **e'** denotes the transpose of the unit vector, summing the flows entering all airports from the outside and the flows leaving all airports for the outside must yield the same total number of air trips for the purpose of interest g, as expressed in:

$$\mathbf{e}'\mathbf{FE}' = \mathbf{e}'\mathbf{FL} = 2175 = T^g_{air}. \tag{5}$$

A TAM is a doubled up input-output matrix In such an accounting system, it is possible to pursue the obvious analogies with economic input-output (I-O) tables made up of N and L sub-matrices (extended with the lower-level Entering and Origin-Destination matrices in Table 13.3) and show, for instance, the general lack of proportionality between forecasts of the leaving demand vector **FL*** and forecasts of network leg flows between airports **N***.[5] This is done by first defining the familiar matrices of direct (and, later, of indirect) input-output coefficients of inter-industry economics now available in two potential formats instead of one, as can be demonstrated with the help of Table 13.3 where a TAM is written in simplified format.

There are two possible matrices of technical coefficients because, after an explicit distinction is made between entering and leaving demand structures E and L, there are two ways to define the operation required to compute direct technological coefficients. Network flows N can be combined either with the leaving flows or with the entering flows; this possibility is absent from input-output tables (Leontief, 1941) that effectively contain only N and L matrices[6] and thereby impose a unique definition of the matrix of technological coefficients A in the familiar way:

---

4   In the remaining running text of Section 2.1, vectors are denoted in **bold**, but not matrices or scalars.
5   See Gaudry (1973 or 2007) for further elaboration.
6   Although, after some hard work, Quesnay's open economy *Tableau Oeconomique* (1758) "balanced" expenditures among four population sub-segments (farmers, farmhands, landlords and artisans), it is easier to rewrite it as a five-sector Leontief transactions table in order to make the foreign sector explicit (Brems, 1986). Crucially, Quesnay did not distinguish between intermediate and final flows as did Leontief by specifying N and L parts while keeping the N matrix square for inversion purposes. After Blankmeyer (1971), we simply interpret L as reporting on flows "towards the outside", which naturally calls for the "from the outside" counterpart matrix E and for their O-D combination in the remaining quadrant of Table 13.3. This new spatial

Table 13.2 Flow matrix for the four-city system without direct connections from Airport J to Airport B

Flow Matrix II		O-D	\multicolumn{4}{c}{Destination airport}			\multicolumn{4}{c}{Destination city}		Total	
			B	T	M	J	NL	1	2	3	4	FL	TL
Origin airport	B			150 500 75			725			25 200 25		250	975
	T		25 200 25		400 500 50	75 300 100	1675	250 150 100				500	2175
	M			300 200 100			600		400 500 50			950	1550
	J			25 400 250			675				75 300 100	475	1150
	NE		250	2000	950	475	T_I T_J	500	950	250	475	2175	
Origin city	1			100 25 50			175		50	25	100		
	2				100 200 300		600	100		200	300		
	3		150 500 75				725	150	500		75		
	4					25 40 250	675	250	400	25			
	FE		725	175	600	675	2175						
Total		TE	975	2175	1550	1150							

extension differs greatly from the standard spatialization of Leontief systems expounded by Moses (1955) and adopted in transport (e.g. Cascetta and Di Gangi (1996), where a diagonal spatial trade matrix T is used to multiply a redefined diagonalized matrix A and suitably redefined output vectors).

Table 13.3 Relationship between input-output (I-O) and traffic accounting matrices (TAM)

N	L
E	O-D

$$A = N \cdot B^{-1}, \text{ or } N = A \cdot B, \tag{6}$$

where B is a diagonal matrix with elements made up of components of the leaving demand vector **TL**:

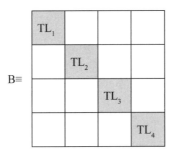

so that:

$$NL = A \cdot TL \tag{7}$$

and the usual operations can be effected, first by substituting (7) into (3), and so on.[7]

Matching TAM components to traffic modelling steps Independently from its extended input-output interpretation, the TAM found in Table 13.2 can be used to focus visually on the different objects to be explained by particular transport models. For instance, the trip generation step of the four-step transport planning format involves explaining the vectors T_i and T_j noted there in contoured symbols. Our interest is in Generation-Distribution models (matrix O-D) and in split mode/company/path choice models (matrix N).

In this analysis, we first match cities to airports, which imposes a particular structure of the network entering and leaving vectors **NE** and **NL** because (without loss of generality) flows are not allowed to enter or leave an airport from or towards more than a single zone. We then isolate the hub values of interest by underlining them or putting them in bold in Table 13.2. The underlined entries circumscribe the O-D passenger flows and isolate them from the *transfer passenger* flows: both are to be explained by econometric models.

---

7 From **TL** = A•**TL** + **FL**, one collects terms and writes **TL** = (I-A)⁻¹ **FL** thereby deriving C = (I-A)⁻¹, the famous matrix of indirect coefficients. Forecasting N* involves the reverse path from **FL*** to **TL*** and then to N* from N* = A•B* in (6).

A reference equation structure to explain flows of interest We explain them by associating representative analytical expressions to the two kinds of flows using hub h, and consequently explicate (2) as follows:

$$T^g_{air,h} = \{Pop_h^* \cdot \{\Sigma_j [Pop_j^*]^\alpha \cdot [U_{ij}]^\beta \cdot [U_{air,ij} / \Sigma_{m,ij} U_{m,ij}] \}\}, \qquad j=1,\ldots, Z, \qquad (8)$$

$$+ \Sigma_i \Sigma_j \{[Pop_i^* \cdot Pop_j^*]^\alpha \cdot [U_{ij}]^\beta \cdot [U_{air,ij}/\Sigma_{m,ij} U_{m,ij}]\} \cdot [U_{air,ihj}/\Sigma_{air,ikj} U_{air,ikj}], \qquad i, j \neq h, \qquad (9)$$

where (8) refers to O-D passengers and (9) to transfer passengers; α and β are parameters; there are modes $m = 1, M$, and zones $i, h, j = 1, Z$, and:

Pop_j^* : income and activity weighted measures of population at location j; (10)

$[U_{ij}]$: the aggregate utility of the modes from i to j built from the denominator of: (11)

$[U_{air,ij}/\Sigma_{m,ij} U_{m,ij}]$: the *air* modal market share from i to j; (12)

$[U_{air,ihj}/\Sigma_{air,ikj} U_{air,ikj}]$: the share of air trips from i to j going through hub h (13)

Note that a possible definition of a successful hub is a location, created for strategic reasons or because the market requires it, where the ratio of (9) to (8) is high. Note also that the quality of airport services is hidden in the definition of the utility terms consisting in "supply side" level of service or price variables found in modal utility functions defined in more detail below. Although we will not model the supply side, it is useful[8] to make clear how the levels of variables used in the demand models arise.

2.2 Airport Demand, Performance and Supply Components

This supply side notion requires development because it is insufficient without the further introduction of the notion of performance, a notion of great practical import in the determination of air service levels and, in consequence, in the explanation of competition among hubs. Why and how?

In our view, the simplest way of thinking of transportation is not to adopt a two-level Demand-Supply formulation. It is to add a third level: the determination of Performance that depends on both Demand and Supply, as researchers effectively do in structural transport analysis.

Gaudry (1976, 1979) introduced the *three-layer* structure to capture the fact that realized transportation service levels often differ from supplied service levels and are best modelled through a third and explicit level between the classical supply and demand levels. In the new structure, "costs" refer to *realized* money, time, crowding or *safety* levels. We also estimated with monthly time-series data a complete simultaneous three-layer, bi-modal urban model system on these lines (Gaudry, 1980), within the system definition schematized in Table 13.4, to which we now turn because it is essential to understand airport performance elements that enter the utility functions.

Using a D-C-S system instead of the classical D-S system gives rise to previously unheard of equilibria, such as the new "Demand-Generalized Cost" equilibrium, distinct from the "Demand-

---

8 As suggested by Professor Jaap de Witt. This section may be skipped by the reader interested only in demand.

Supply" equilibrium defined within the same three-layer system. To make the enriched formulation[9] more accessible within the wide transportation planning subculture, we subsequently relabelled the triplet D-C-S as D-P-S (Demand-Performance-Supply) and changed the notation (Florian and Gaudry, 1980, 1983) to that used in Table 13.4 (most recent version: Gaudry, 1999).

In the actual format of Table 13.4, the achieved **Performance [P, C]** contains realized queues, levels of congestion and risk, as well as other forms of flexible modal bearing ratios (effective capacity, occupancy or load factor and crowding etc.) conditional on both actual **Demand D** and given **Supply** actions **[S, T, F]**. For instance, in a network equilibrium there is a set of values of **P, C** and **D** that simultaneously satisfy the demand functions and (for given supply) the conditions required by the performance procedures. For our purposes here, money and time performance by origin-destination pair on the network have to be consistent with the demands generated under these network conditions, a non-trivial problem because the dimensions of the demand functions (from i to j) are not the same as those of the transportation performance conditions on individual network links a. And of course demand depends on the whole of the performance vector elements.

Airport performance and travel time In consequence, it is important for our understanding of the airport system problem to distinguish between the three layers because Demand, Performance and Supply jointly determine airport passenger and freight transfer and delay times that are themselves critical objects of analysis, along with "demand" and "supply" quantities. Airport performance elements partly define levels of service used in the representative demand functions that we presently concentrate on: e.g. the coordination of flight required for hub-and-spoke networks creates congestion peaks (Burghouwt and de Wit, 2003). Paradoxically, such time nuisances are immediately internalized because the peaking involves planes of the same company: higher peaking reduces the need for congestion charges, if not for terminal use charges.

Table 13.4   Market and network analysis: a three-level approach

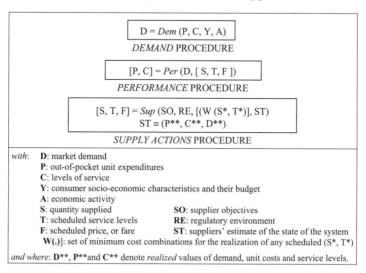

---

9   At the time, the supply side vector typically contained road infrastructure capacity and transit vehicle service characteristics pertaining to congested urban networks simultaneously used by fixed route public and flexible route private vehicles. But it is general.

3 Consistency with IIA and Structure of Core Equations Determining Hub Flows

Our analysis of econometric demand models generating hub demands proceeds in three parts. First, we state the two key questions to be asked about any demand model: whether it was estimated with fixed or variable mathematical forms; and whether it allows for substitution and complementarity among alternatives. Second, we analyse Generation-Distribution models in current use from that perspective: it is not possible to discuss continental airport demand sensibly without a structure that determines total flows by all modes. Third, we analyse the Mode/Path choice component of models from the same double-barrel perspective. This step is important not only to explain continental air traffic flows and itinerary choices, but also to model intercontinental air company and path choices, notably those that matter for competing hub demands.

3.1 Two Questions to be Asked of Each Model Component

Each of (8) or (9) can usefully be broken down between a Generation-Distribution part explaining total flow (by all modes or passenger paths, depending on the case); and a Split choice term explaining shares of alternatives (a mode, company or air itinerary, depending on the case). We note that the two parts are coupled through a total utility term U_{ij} which we assume here (keeping formulation (9) simple at the cost of incompleteness) to be built from the denominator of the split term (a mode, company/path or combined hierarchical nested Logit choice model). In consequence, each part contains transport conditions and other, typically socio-economic, variables; but we focus on the former.

For each model part, there are two questions. In each part, we consider two principal dimensions of import for us. The first one is that of the use of Box and Cox (1964) transformations of any X_k, written commonly:

$$X_k^{(\lambda_k)} = \begin{cases} \dfrac{X_k^{\lambda_k} - 1}{\lambda_k} & \text{if} \quad \lambda_k \neq 0, \\ \ln X_k & \text{if} \quad \lambda_k \to 0. \end{cases} \qquad (14)$$

This transformation is of interest due to the numerous model improvements obtained through its use during the last 30 years. Of course, gains in theoretical reasonableness and fit are clear. But, as signs in regression depend not just on the variance of regressors considered individually but also on their covariances – and as these are modified by transformations of variables – the very existence of statistical correlation depends on mathematical form (and conversely, as form parameters are jointly estimated). The first question asked of models is, therefore, whether their form was assumed or estimated because it is easy to argue that behavioural reasonableness, numerical best fit and statistical robustness of coefficient estimates require rejection of untested results based on *a priori* fixed forms.

The second dimension can be best defined if we start from the traditional specification of complete demand equation systems where, using Logit terminology for clarity, any utility function V_m of an alternative can in principle include the transport conditions (prices or service levels) of all other alternatives, as in the land freight example of Table 13.5 where such conditions (prices in this case) are laid out in square matrix format (each accompanied by a regression and a form parameter, both here subjected to some equality restrictions).

Table 13.5 Example of full price matrix used in three-mode representative utility function set

Three V_m representative utility functions (m = r, c, f), each dependent on all prices P_r, P_c and P_f, where prices obtain (here constrained) regression and Box-Cox transformation coefficients				
Road by truck V_r	= [β_{01}]	+ $\beta_1 P_r^{(\lambda 1)}$	+ $\beta_3 P_c^{(\lambda 3)}$	+ $\beta_2 P_f^{(\lambda 2)}$ [... + ...]
Intermodal rail V_c	= [β_{02}]	+ $\beta_2 P_r^{(\lambda 2)}$	+ $\beta_1 P_c^{(\lambda 1)}$	+ $\beta_3 P_f^{(\lambda 3)}$ [... + ...]
Traditional rail V_f	= [β_{03}]	+ $\beta_2 P_r^{(\lambda 2)}$	+ $\beta_3 P_c^{(\lambda 3)}$	+ $\beta_1 P_f^{(\lambda 1)}$ [... + ...]

Source of the actual function: Gaudry *et al.* (2007).

Of course, traditional diagonal dominance, defined as strong own elasticities and weak cross elasticities, is expected in demand systems defined by the Table 13.5 format, but what we define as diagonal slavery is something else: when the cross terms are absent from (12)–(13), IIA consistency is per force imposed (Luce, 1959, 1977). The second question asked of models is therefore whether or not such substitution or complementarity "cross" terms have been considered.

Concerning interdependency among markets, we will argue that it is one thing to estimate models *ex post* using IIA-consistent formulations, but that more flexible formulations should be used to account for interdependency; and that the importance of this generalization arises notably through tourism for Generation-Distribution models and more generally in mode or path choice (Split) models where a reference alternative matters to the explanation of consumer choices.

This implicitly means that *ex ante* airport demand forecasts are all the more biased that an incorrect form has been used and real interdependency has not been duly accounted for. Our examples are drawn from passenger and freight markets and their implications for airport demand are briefly stated.

3.2 The Generation-Distribution Gravity-Type Part Determining Trip Length

Transport trip length is increasing rapidly, In France, for instance, average distance to work doubled from 6 to 12 km between 1993 and 2001 and air trip length doubled to 1200 km between 1981–82 and 1993–94, and is expected to have increased further in the current survey (2006–07). During the 1980–95 period, world air tonne-km increased 20 per cent more than air tonnes. Total market size (for all modes) and trip length are central to any airport service demand issue.

A Within IIA-consistent structures: the issue of form – gravity models and the IIA axiom The gravity structure within (8) and (9) is arguably the most successful modelling structure in all of economics, compatible with many theories (e.g. Rossi-Hansberg, 2005), and still formidably robust empirically in both mathematical form and broad specification.

Suggested in strict Newtonian form by Carey (1858–59) to describe human spatial interactions, it was first used in transportation by Lill (1891) on Vienna–Brünn–Prague passenger rail flow data. We write it here in so-called "Generation-Distribution" form, without the double constraints that guarantee that all of what comes out of *i* or comes into *j* is assigned. We exclude such constraints because they bias the estimation of parameters by imposing a very special restriction on error terms and thereby artificially induce a form of accounting substitution among all O-D pairs that is absent from the core behavioural model structure. Indeed, this basic unconstrained Gravity structure may be written simply:

$$(\text{Flow indicator}_{ij}) \leftarrow (\text{Activities}_{i \text{ and } j}; \text{Socioeconomic}_{i \text{ and } j}; \text{Ease of interaction}_{ij}) \quad (15)$$

where the ease, or (dis-) utility, of interaction is an impedance metric that can be of many types (generalized cost, geographical, preference or taste, cultural, legal or regulatory etc.). This specification makes abundantly clear the fact that, because the flow from i to j depends only on variables that have i and j indices, the characteristics of other paths or locations have no influence on the flow from i to j. Consequently, the ratio of the demands for two O-D pairs does not depend on the characteristics of other links or locations than those involved in the ratio, a property consistent with Luce's Independence from Irrelevant Alternatives (IIA) axiom of choice theory. We wish to address the relevance of both mathematical form improvements – first studied on intercity trips in Georgia by Kau and Sirmans (1979), who, contrary to what most other authors have since found, could not reject the multiplicative gravity form – and of IIA axiom avoidance efforts within this structure.

Form and response to transport conditions In (15), trip length depends on the response to the "ease of interaction", i.e. to transport conditions which are strongly correlated with distances in any spatial cross-section. Actual response depends on the true shape of the demand curve: we will note that, as a rule, its elasticity is not constant and that all untested constant elasticity log-log models generate incorrect trip length forecasts and incorrect diagnostics of the changing role over time of distance, used as a proxy for cost.

We start with freight trade models, all still using distance and all less advanced today than intercity transport models were 30 years ago,[10] with both flexible form and rich border effect specifications. In Table 13.6, using Box-Cox transformations on known freight models greatly increases log-likelihood (comparing columns 2 to 1 and 6 to 5), despite estimated Box-Cox power values differing little *numerically* from 0, and shows that the optimal form flow elasticities with respect to distance, evaluated at sample means, are almost identical across continents: at about -1,20 (in detail, -1,21 and -1,13 without heteroskedasticity and -1,16 and -1,25 with it). Clearly, one would not wish to use[11] the fixed form constant elasticity results, except as a first approximation.

But how good is distance at representing true transport conditions? This matters if the simplistic log-log form of Columns 1 and 5 is used to compare impedance effects from samples pertaining to different moments of time (Buch et al., 2004): assumed proportionate changes in costs[12] between the samples will be accounted for in the constant but not in the coefficient of the distance term.

---

10 For instance Gaudry and Wills (1978) demonstrate both the effect of flexible forms on elasticity estimates and the import of (non-dummy) linguistic border effects within Canada in 1972. Concerning the former, the optimal value of the λ used for the U_{ij} utility term is 0,03 but the log-likelihood gains of 22 units of log-likelihood (4 degrees of freedom are involved due to the fact that all other variables are also optimally transformed) are massive even if the estimated value differs *numerically* only slightly from 0. For freight, even regional studies use border effects: in their study of 1989 France–Germany freight flows among 36 regions, Becker et al. (1994) show that the border divides total freight flows by 8,25 (with corresponding modal values of 25 for train, 6,78 for road and 0,33 for river).

11 Results for North America shown in B were presented to McCallum in June 1994 to check his previous results (McCallum, 1993) before their publication as A, but he chose the fixed form model results of Column 1, as did Helliwell (1998) after seeing B results for both North America and Europe in February 2006, despite their quality of fit.

12 This problem, whereby distance is not always a proper surrogate for transport cost, does not arise if a Box-Cox form is used to transform the distance term: changes over time in distance elasticities will then be proper surrogates of changes in price elasticities. Clearly, gains from moving away from the fixed log-log

Table 13.6 Flexible form, heteroskedasticity and spatial correlation in freight trade models

Continent		North America (1988)				Europe (1987)			
YEARLY FREIGHT FLOWS		SHIPMENTS IN VALUE BETWEEN 30 AMERICAN STATES AND 10 CANADIAN PROVINCES, AND AMONG THE LATTER				SHIPMENTS IN TONS BETWEEN 10 EUROPEAN COUNTRIES AND 11 GERMAN PROVINCES, AND AMONG THE LATTER			
Gaudry (2004)		C				C			
Gaudry et al., (1996)		B				B			
McCallum (1995)		A							
	Column	1	2	3	4	5	6	7	8
Variable	Case	Log-log	Box-Cox	As (2) + heterosk.	As (3) + autoc.	Log-log	Box-Cox	As (6) + heterosk.	As (7) + autoc.
Distance	elasticity	-1,42	-1,21	-1,16	-1,37	-1,29	-1,13	-1,25	-2,47
	t-statistic	(-22,66)	(-25,72)	(-24,08)	(-4,94)	(-15,54)	(-19,00)	(-20,37)	(-17,44)
	Box-Cox λ	0,00	-0,03	-0,05	0,01	0,00	-0,13	-0,11	0,03
Border dummy	elasticity	3,09	2,64	2,86	2,86	0,99	0,86	0,94	0,76
	t-statistic	(23,69)	(22,60)	(20,20)	(27,34)	(6,38)	(5,42)	(5,36)	(3,10)
Other variables		[...]	[...]	[...]	[...]	[...]	[...]	[...]	[...]
Spatial autocorrelation ρ					0,65				0,87
Distributed lag π					0,45				0,12
Log-likelihood		-8089	-8043	-8032	-7960	-2356	-2348	-2341	-2318
Degrees of freedom		0	2	3	5	0	2	3	5

Conversely, the use of distance as a proxy for price in time series, say, from 1970 until 1995 or 1999 (Ghosh and Yamarik, 2004; Fratianni and Kang, 2006), is incorrect in a gravity trade model of log-log form, even with period dummy variables: it simply reflects basic constant propensities to trade with more or less distant partners, independently from the desired critical role of changing transport conditions in explaining increased globalization flows. Therefore distance has to be used *in addition to* transport costs if trade flow explanations are tested with time-series data, as some authors do (Giuliano et al., 2006). And, to the extent that more sophisticated measures of impedance than distance are in rough proportion to it, any proper comparison of the roles of distance and of its more sophisticated replacements, *even for the same period*, also requires that the form *not be* logarithmic.

Impedance: from distance to a total utility term U To answer the question of the quality of distance used as a proxy for transport conditions, we turn then to the advanced intercity passenger models accounting for "induction" through the sophisticated U_{ij} utility term in (8) or (9) and using Box-Cox transformations to obtain an adequate fit. In all models of Table 13.7 the log-log form is demonstrably less adequate than the Box-Cox form.

In Parts II and III, distance is replaced by a utility measure derived from the sum of the mode choice denominator terms; obviously, apart from sign which must be reversed, the distance and utility elasticities (indicated in bold on grey backgrounds) are extremely close. Certainly, any sophisticated index of the structure of modal transport conditions (such as the particular Expected Maximum Utility index consisting in the (log of) the denominator of a Logit model) is strongly correlated with distance in any spatial cross-section. Although optimal forms of the impedance variable are not too far, numerically speaking, from the log form of the initial popular gravity model, it makes a difference to use the best fitting form even in a cross-section.

---

form go much beyond those of mere fit – some "distance" effects can only be identified if the log-log form is *not* optimal.

Form and results for other regressors We note in Table 13.6 that the *border effects* decrease by about 10 per cent if Box-Cox forms are used (comparing columns 2 to 1 and 6 to 5). The German database is homogeneous and was not obtained, like the Canadian one, by merging heterogeneous sources (one providing spatial structure and another providing totals): its border dummy variable is therefore interpretable without undue risk.[13]

Concerning other variables, in particular *activity* terms, it is extremely difficult to develop Generation-Distribution models of international passenger flows by inclusion of activity variables other than population, perhaps corrected by income and some measure of employment structure. We do not see any new modelling trends in this area, except the innovation introduced by Last (1998), who successfully weighted the trips by the propensity to travel by age or class. But, a form of aggregation of market segments, this innovation has no definable effect on the structure of the hubs markets. There have also been a number of attempts to use Box-Cox transformations to estimate something more general than a simple multiplicative form of activity variables. Although exponents other than 1 are found, and differ as between emanation and attraction zones, the gains beyond fit do not directly bear on our issue except to weigh heavily against attempts to use system constraints to modify the unconstrained basic form of the Gravity equation.

Form and difference in forecasts A simple and global way to compare two versions of a model differing in form is to compare the elasticities evaluated at the sample means, as is done in Table 13.6 and Table 13.7. However, it is possible to be more systematic in the comparisons of two forecasts. Consider two variants 1 and 2 of a model differing by assumption only in the functional form applied to each model, and assume that we want to see how the forecasts differ with respect to the transport condition variable X_{qn} present in both variants. The two models are, with indices 1 and 2 denoting the models and n an observation subscript:

$$Y_{1n} = \beta_{10} + \sum_{k} \beta_{1k} X_{kn}^{(\lambda_{1k})} + u_{1n} \tag{16}$$

and

$$Y_{2n} = \beta_{20} + \sum_{k} \beta_{2k} X_{kn}^{(\lambda_{2k})} + u_{2n} \tag{17}$$

and their difference of interest, neglecting residuals,

$$\Delta Y_n = \left[\beta_{20} + \sum_{k} \beta_{2k} X_{kn}^{(\lambda_{2k})}\right] - \left[\beta_{10} + \sum_{k} \beta_{1k} X_{kn}^{(\lambda_{1k})}\right], \tag{18}$$

reaches a turning point with respect to changes in X_{qn} when $\partial \Delta Y_n / \partial X_{qn} = 0$, namely in:

---

13   Both McCallum, co-author of B, and Helliwell used only the North America database (supplied to us by McCallum in 1994 to test his specification), containing a mixture of data from two sources within Statistics Canada – whereby errors of scaling, bi-proportional Fratar (1954) or not, of interregional flows would directly modify only the USA–Canada border dummy variable. The German database, also supplied to us by Blum in 1994 to produce the intercontinental comparison found in B, does not suffer from attempts to merge distinct databases on trade total and spatial structure: its homogeneity makes the border dummy variable meaningful and its usual interpretation admissible. Fortunately, the measurement of the role of impedance in the determination of trip length is unaffected by the merging error in the North America data set and straightforward in the European data set. We carried out tests with the databases as supplied.

Table 13.7 Flexible form, heteroskedasticity and spatial correlation in passenger transport models

Part I		Single-country models, all modes and all trip purposes (domestic flows)					
Gaudry et al., (1994)		Canada 1976 (120 observations)*			Germany 1985 (286 observations)†		
	Column	1	2	3	4	5	6
Variable	Case	Log-log	Log-log +1st autoc.	2 Box-Cox + 1st autoc.	Log-log	Log-log +1st autoc.	4 Box-Cox + 1st autoc.
Utility	elasticity	0,63	0,57	0,66	0,40	0,25	0,24
	t-statistic	(42,33)	(25,27)	(21,13)	(9,68)	(5,59)	(5,71)
Other variables		[...]	[...]	[...]	[...]	[...]	[...]
	Box-Cox λ	0,00	0,00	-0,05	0,00	0,00	0,41
	Spatial autocorrelation ρ	0,00	0,75	0,44	0,00	0,80	0,73
	Distributed lag π	--	1,00	1,00	--	0,74	0,49
	Log-likelihood	-1318	-1294	-1279	-3057	-3028	-2995
	Degrees of freedom	0	2	4	0	2	6

Part II		Joint France (1993–1994) and United Kingdom (1991) model, all modes by trip purpose (domestic flows and flows between them or with 8 other European countries)‡					
Last (1998)		Vacation trips (6540 observations)			Private trips (8334 observations)		
	Column	7	8	9	10	11	12
Variable	Case	Log-log	3 Box-Cox	3 Box-Cox	Log-log	2 Box-Cox	2 Box-Cox
Distance	elasticity	-0,23	-0,20		-0,72	-0,57	
	t-statistic	(-8,71)	(-12,95)		(-22,46)	(-30,00)	
Utility	elasticity			0,29			0,52
	t-statistic			(12,00)			(26,99)
Other variables		[...]	[...]	[...]	[...]	[...]	[...]
	Box-Cox λ	0,00	-0,74	0,12	0,00	-0,57	0,31
	Log-likelihood	-65407	-65313	-65390	-80804	-79991	-79995
	Degrees of freedom	0	3	3	0	2	2

Part III		Business trips full set of 5711 observations			Business trips subset of 544 observations		
Gaudry et al. (1994); Last (1998)							
	Column	13	14	15	16	17	18
Variable	Case	Log-log	3 Box-Cox + heterosk.	3 Box-Cox + heterosk.	Log-log	8 Box-Cox + heterosk.	As (17) + 1st autoc.
Distance	elasticity	-0,33	0,20				
	t-statistic	(-8,78)	(-11,02)				
Utility	elasticity			0,22	0,11	0,15	0,16
	t-statistic			(12,50)	(4,36)	(5,66)	(5,62)
Other variables		[...]	[...]	[...]	[...]	[...]	[...]
	Box-Cox λ	0,00	-1,02	0,33	0,00	0,25	0,25
	Spatial autocorrelation ρ						0,18
	Log-likelihood	-50846	-50762	-50737	-4890	-4858	-4852
	Degrees of freedom	0	5	5	0	10	11

*Results drawn from Table B.2, Columns 1, 3 and 5. The Utility term is constructed from a 4-mode Box-Cox Logit model (also found in Liem and Gaudry (1994)) and specified in RATE format (for this, see a discussion below) with distinct transformations applied to its four network variables (Price, Speed, Distance, Frequency) but not its socio-economic variables.

† Results drawn from Table 3, Columns 1, 3 and 5. The utility term is constructed from a 3-mode Box-Cox Logit model specified in RATE format with distinct transformations applied to its four network variables (Price, Speed, Distance, Frequency) but not to its socio-economic variables.

‡ Results are drawn from the Appendices of the source paper except for the autocorrelation tests, which are drawn from Table 11. Note that the Utility term is constructed from 3-mode Box-Cox Logit models using only one transformation for the Fare and In-vehicle time network variables used for each of the three trip purposes (except for the Vacation trip model that only has a Fare term. The LEVEL 1.4 algorithm (Liem et al., 1993) is used for estimation except in the presence of spatial autocorrelation when an earlier version of the LEVEL 2.1 algorithm (Tran and Gaudry, 2008) is used. The log form of the sum (i.e. the log-sum) is always rejected, as in Part I of this table, even when the actual value of the transformation applied to the Utility term is close to 0 numerically.

$$\frac{\partial \Delta Y_n}{\partial X_{qn}} = \beta_{2q} X_{qn}^{\lambda_{2q}-1} - \beta_{1q} X_{qn}^{\lambda_{1q}-1} = 0, \tag{19}$$

where

$$X_{qn}^{**} = \left(\frac{\beta_{1q}}{\beta_{2q}}\right)^{\frac{1}{\lambda_{2q}-\lambda_{1q}}}, \quad (\lambda_{1q} \neq \lambda_{2q} \text{ et } \beta_{1q}/\beta_{2q} > 0) \tag{20}$$

This means that an analytical value can be calculated for the *point of maximum difference* $\partial \Delta Y_n / \partial X_{qn} = 0$ between the two forecasts. The interested reader may consult Appendix 9 of Gaudry et al. (2007), where it is shown that, although the *crossing point* $\partial \Delta Y_n / \partial X_{qn} = 0$ cannot be found analytically but only by solving an equation numerically, $\partial^2 \Delta Y_n / \partial X_{qn}^2 = 0$ the *inflexion point of the difference* (18) can also be obtained analytically by the expression:

$$\tilde{X}_{qn} = \left[\frac{\beta_{1q}(\lambda_{1q}-1)}{\beta_{2q}(\lambda_{2q}-1)}\right]^{\frac{1}{\lambda_{2q}-\lambda_{1q}}}, \quad (\lambda_{1q} \neq \lambda_{2q} \text{ et } \frac{\beta_{1q}(\lambda_{1q}-1)}{\beta_{2q}(\lambda_{2q}-1)} > 0) \tag{21}$$

Form therefore makes quite a difference in elasticities in forecasting, and is it possible to say something systematic about the difference between two models: clearly, it is unavoidable.

B Away from IIA Consistency: The Issue of Spatial Competition

If endogenous forms give better estimates of the impact of transport conditions on demand, they have no effect on the issue of interdependence among flows. Therefore how should it be introduced and tested? IIA consistency is not expected in economics because it does not "take the theory seriously" as theory makes all prices relevant, not just own prices on the diagonal. A difficulty here is that there are N^2 flows, or even $(N^2 - N)/2$ if the diagonal is neglected and symmetry imposed. Another is that theory further allows for both substitutes and complements; intervening opportunities models were a way of dealing with complements along paths. We want to propose a parsimonious approach that does not require the inclusion of all prices, but only for the relevant ones, and allows for possible complements among flows.

All-in but excluding complements So we exclude, as already stated above, the artificial linkage among flows implied by imposed double "additivity" constraints on emissions and attractions, a practice often well documented (e.g. Batten and Boyce, 1986). These constraints simultaneously bring transport conditions *all-in* and *force substitution* on the pattern: modifying a link will readjust all others in the same direction to meet the constraints. Taking theory seriously requires Occam's razor, because not all competing O-D pairs matter to explain the flow for a particular one, and the possibility of complements, as in classical demand systems.

There are many ways to be profligate, practise indiscriminate "all-in" specifications and to exclude complements at the same time. A first representative example is provided by Wei (1996), who introduces an accessibility measure defined as a weighted average of distances to all trading partners, in the old fashion of Trip Generation studies devoid of transport conditions as explanatory variables and desperately trying to bring the network back in somehow. Another example is provided by Anderson and Wincoop (2003), based on Anderson (1979), who introduce in the model "multilateral resistance terms in the form of an atheoretic remoteness variable related to distance from all bilateral partners". The latter is an economic version (based on expenditures) of the traffic flow additivity condition just noted, but imposes its own heroic assumptions about the distribution in space of expenditure shares that never hold in trade models. Why is it not a promising option for us to pursue?

Trade models typically deal simultaneously with final and intermediate goods (not just with final goods) and always exclude service flows and balancing financial flows from the analysis. In the absence of such offsetting items required by balances of payments constraints, it is natural to find in flow matrices that do not respect such accounting constraints: (i) when testing, for instance with Box-Cox transformations, power values of Generation terms (whether based on Population or on Income or Output) systematically different from those of corresponding Attraction terms; (ii) and both of course different from 1. In addition to such pound foolishness, such studies simultaneously practise theoretical penny wisdom by imposing substitution between the own flow of interest and all others.

A better way to be systematic is to introduce just the right number of cross-utility terms by an analysis of the spatial correlation among residuals of, say (8) or (15), allowing the particular error v_t associated with an observation t to be correlated with many values of the same vector v_n, for instance for each of 2 orders:

$$v_t = \sum_{l=1}^{2} \rho_l \left(\sum_{n=1}^{N} \tilde{r}_{l,tn} v_n \right) + w_t \qquad (22)$$

where $\tilde{r}_{l,tn}$ is the typical element of the matrix $\tilde{R}_l$, a notation which may be clearer in matrix form:

$$v = \sum_{l=1}^{2} \rho_l \tilde{R}_l v + w \qquad (23)$$

and $\tilde{R}_l$ is the (row or column) normalized Boolean matrix R. The Boolean matrix expresses a hypothesis concerning the presence of correlation, for instance among first neighbour zones at the origin and destination of the ij flow, or perhaps (based on other assumptions) among some competing or complementary destinations. It is possible to have both substitutes and complements in (23), depending on the nature or source of the potential correlation tested by each $-1 < \rho_l < 1$. One could, for instance, imagine that different regions might be competing for tourists ($\rho_1 > 0$) but that sub-areas within regions could be simultaneously complementary ($\rho_2 < 0$). As only some O-D pairs pertaining to non ij flows are selected when spatial autocorrelation is detected in this way, the resulting system is one of diagonal dominance because the cross terms have, with $-1 < \rho_l < 1$, a smaller role than that of diagonal terms in explaining ij flows, as the reader can readily verify by substituting (22) into (16).

But a *more or less* all-in structure is also possible with an extension, with the data deciding on the role of distant flows. Assume that one wants to test the influence of neighbours of neighbours and so on, in geometrically declining strength, on the relevant flow, in this case we may redefine the $\tilde{R}_l$ matrix as:

$$\tilde{R}_l = \pi_l \left[I - (1 - \pi_l) \bar{R}_l \right]^{-1} \bar{R}_l, \quad (0 \prec \pi_l \leq 1), \qquad (24)$$

where the proximity parameter π_l allows endogenization of the relative importance of near and distant effects considered in the rule of construction of R_l. If π_l equals one, $\tilde{R}_l$ is equal to $\bar{R}_l$, indicating that only the adjacent neighbours have an impact on the correlations among the associated residuals assumed in defining the matrix of neighbours $\tilde{R}_l$: this corresponds exactly to a classical case of spatial correlation. By contrast, as π_l tends towards zero, the near effect is reduced to a minimum in favour of the distant effect. Generally, the π_l weigh the *relative* importance of near and distant effects with a single parameter defining the sharpness or slope of the decline. It is therefore possible, with such a distributed lag, to test for and weigh influences beyond those assumed by the original $\tilde{R}_l$ structure. This is far more acceptable than full all-in systems where the balancing constraints treat all O-D flows in the same way.

We do not discuss here the results of Tables 13.6 and 13.7 obtained with such "long tail" π_l, estimated with the algorithm documented in Tran and Gaudry (2008), but solely those obtained only with first neighbours defined in the Boolean matrix R. In Table 13.6, first neighbours are defined for any given flow as all those other flows having origins or destinations within 400 miles and 300 km, respectively in columns 4 and 8. In Table 13.7, the similar rule is 320 km (columns 2 and 3), or in a range between 100 and 180 km (columns 5 and 6) or between 100 and 300 km (column 18). Naturally, the log-likelihood gains are much greater when all trip purposes are considered (columns 5 and 6) than when only the business trip purpose (column 18) is. In all cases, flows are substitutes because the estimated sign of ρ_l says so, not by assumption.

Parsimony without diagonal slavery depends on reality   What is our message here? Obtaining the right amount of off-diagonality cannot depend on adding up constraints applied automatically, but on the nature of each case. Clearly, theory does not warrant having all prices *in*: only those missing from the regression. Such distinctions could be crucial to the modelling of air tourism. Surely, European tourism flows to any tourist area (say, the Mediterranean) depend on the prices to *some* other zones (say, the Atlantic coast, the Baltic Sea and North Africa) and not on *all* O-D prices and transport conditions in Europe.

3.3 The Split-Type Part Determining Mode/Company/Scheduled Flight and Path

Air transport demand modelling does not depend only on how much travel occurs among origin-destination pairs. It depends critically on mode choice. We therefore proceed to show what difference it makes to air mode demand to use a proper mode choice specification. As the workhorse of mode choice analysis is the Logit model, this has to discuss the specification of that model, which is also the critical model of air path choice and air itinerary modelling. In some sense, the discussion of hub demand is often primarily a discussion of Logit mode choice, if one has an interest in the issue at all.

Contrary to the Gravity model, which has a long history and a reference multiplicative (Logarithmic) form, the classical Logit model has short history and a reference Linear form. To explain, say, the probability of choice (or the market share) of a mode m (1,..., m,..., M), it may be written in multinomial garb as:

$$p\ (m) = \frac{\exp(V_m)}{\sum_{j \in C} \exp(V_j)}, \qquad (25)$$

with all M utility functions defined as:

$$V_i = \beta_{i0} + \sum_n \beta_{in}^i X_n^i + \sum_s \beta_{is} X_s + u_i \qquad (26)$$

where we have changed slightly the usual notation to identify the X_n^i (the upper index denotes the mode and the lower ones the utility function index i and the network variable n), the network characteristics that belong to a particular mode and therefore vary across alternatives, and it is clear that in (26) the X_s denotes socio-economic characteristics of consumers that are common across alternatives.

The issue of the nature of competition among modes, paths and company services is everywhere studied almost exclusively with this classical Linear Logit model and its enrichments. We will see that the explanation of such choices can differ much depending on form, used here both to evaluate better the influence of transport conditions on choices, as above for the Gravity model; but also as a tool to introduce interdependence among alternatives in order to move from diagonal "slavery" to diagonal "dominance" and allow for the possibility of complements.

A. Within IIA-Consistent Structures: The Issue of Form in Logit Choice Modelling

Logit models and the IIA axiom: emergence between 1961 and 1970 Although the logistic curve was discovered and named by the Belgium mathematician Verhulst (1838, 1845), who used it to describe population growth – as did others independently later in the Unites States (e.g. Pearl and Reed, 1929) and elsewhere – it emerged in binomial canonical form (BNL) only relatively recently. It was first promoted in the bioassay literature (Berkson, 1944) and then in transportation (Warner, 1962), an area suffering from severe underreporting in essays on the history of the Logit model (e.g. Cramer, 2003) excessively focused on the two waves of multinomial development of 1968–1970 and 1977.

For instance, the seminal paper by Abraham and Coquand (1961) on a road path choice multinomial model linked to utility, where the Logit is applied (as an approximation to a Probit) ten years before it appears in the United States for the same problem (Dial, 1971), should be used by historians of the Multinomial Logit (MNL) as its inception in transportation. It stands out well ahead of the 1968–1970 "first wave", consisting of McFadden (1968 or 1976) on the choice of road *tracé*, and in Theil (1969), Ellis and Rassam (1970) or Rassam *et al.* (1970) on the choice of mode.

The multinomial format brings out a feature hidden in the binomial version: the compatibility with the IIA axiom, which translates as equal cross-elasticities of demand. As new transport services or modes never draw proportionately from incumbents, some developments of the Multinomial Logit, such as nested hierarchies, were effectively proposed in 1977 largely to mitigate this excessive property. Nesting hierarchies, however, have their own problems: the different tree structures are not special cases of a general specification – this means that they cannot easily be compared by statistical tests; moreover, they will not properly directly address the structural defect itself, the separability assumption – this requires the reintroduction of off-diagonal terms. Yet, at the time of the introduction of hierarchies, many researchers were thinking of structural reform, as demonstrated by background theoretical work on the Universal Logit and on the Generalized Box-Cox Logit, both outlined further below in the chapter: naturally, the reintroduction of off-diagonal terms, often done in *ad hoc* fashion[14] in ordinary Logit applications undermines the hierarchies

---

14 For instance, the utility functions of public transport modes will suddenly contain the natural logarithm of car ownership (*sic*) in a model altogether linear and where modal utilities are assumed to be independent.

precisely by correcting the specification errors on which they are based. Avoiding the unbearable property of equal cross-elasticities of demand means rejecting the *coup d'état* from which they derive.

Among regal classical demand equation systems, a puritanical coup d'état Such a model as the MNL, enriched or not by the flowerings of 1977 to be listed shortly, is indeed a *coup d'état* because the Logit is simply an additive system of demand equations used during this high growth period after 1968, primarily in transportation but also elsewhere, for instance to explain portfolio shares (Uhler and Cragg, 1971). The demand system background from which it distinguished itself uses ordinal (not cardinal) utility specifications where utility is neither separable – because all prices intervene in all functions, except in Houthakker's indirect addilog system (1960), which is effectively a logarithmic case of the Box-Cox Logit – nor constant at the margin (functions are of multiplicative (logarithmic) form).

This background notably included at the time the Linear Expenditure system (Stone, 1954) and the popular Rotterdam system (Barten, 1969), to which the Almost Ideal Demand system (Deaton and Muellbauer, 1980) was later added. This classical tradition, allowing for both complements and substitutes, is still much alive (e.g. Barnett and Seck, 2006). So proposing to measure utility cardinally as precisely as the computer chip will allow, making utility separable and dependent only on own transport conditions (on the diagonal) and assuming linearity of the representative utility functions and constant marginal utility, is iconoclastic.

The year of grace 1977 The year 1977 saw an extraordinary flowering of the MNL as three new streams of work simultaneously extended it by: (i) allowing for nested hierarchies (e.g. Williams, 1977); (ii) treating the coefficients in (26) as random instead of fixed (Johnson, 1977, 1978), an approach later known as the "Mixed" Logit; (iii) applying Box-Cox transformations to the explanatory variables (Gaudry and Wills, 1978), a modification first implemented as the "Standard Box-Cox" Logit and later generalized to include off-diagonal terms under the name "Generalized Box-Cox" Logit, as will be seen below.

As generally applied, the hierarchies of nested linear logits do not change the characteristics of the model for our purposes, essentially because they are still IIA consistent across main branches. Similarly, randomization of regression coefficients poses considerable problems of credibility because the distributions of coefficients are unknown most times – does the distribution of a gender variable follow a particular law? Is it related to testosterone levels? – and at best unclear due to an information matrix[15] that does not have a closed form, which implies an undefined efficiency bound (Cirillo, 2005). It is of course a formal way of dealing with a type of market segmentation, long practised in combinatorial fashion by energetic analysts.

More importantly, Orro et al. (2005) have shown in effect that the recent popularity of the Multinomial Mixed Logit (e.g. McFadden and Train, 2000) may well be due to the fact that the true relationships are not linear and should have their curvature estimated rather than postulated. Figure 13.1 shows the importance of curvature in explaining the response to changes in transport conditions. These authors advocate, for tests of curvature, rewriting (26) in so-called Standard Box-Cox Logit utility terms, namely:

$$V_i = \beta_{i0} + \sum_n \beta_{in}^i X_n^{i(\lambda_{in}^i)} + \sum_s \beta_{is} X_s^{(\lambda_{is})} \qquad (27)$$

---

15 This point was recently brought to our attention by Lasse Fridstrøm.

Form and response to transport conditions in mode choice Allowing for non-constant marginal utility means that the famous sigmoid shape of the Logit response curve, shown as the dark line of Figure 13.1, is no longer symmetric but can have accelerations and decelerations as illustrated. Conversely, asymmetric logistic response curves imply non-linearity of the response to the transport condition X_n graphed on the abscissa. To the best of our knowledge the classical form (26) was rejected, in both freight and passenger applications, in all Box-Cox tests made since their first application (Gaudry and Wills, 1978).

Elasticities, signs and marginal rates of substitution Form makes a huge difference to modal choice elasticities, as was demonstrated at the time with Figure 13.2 and Figure 13.3,[16] among others, where the contrast between classical linear form (on the extreme right of the abscissa) and optimal form (close to the centre, but different from 0 in this case) values are overwhelming. Clearly, modal share forecasts made with an incorrect form have practical implications, as we shall document further and explore presently.

And, as the optimal form is often quite different from the initial linear one, sign inversions that correct for previously unreasonable results (e.g. Fridström and Madslien, 1995), or form corrections that correct for silly value of time results obtained under linearity (Gaudry *et al.*, 1989), are a common experience with the Standard Box-Cox Logit model. In fact, contrary to the Gravity form, where the multiplicative form is often a very good approximation of the true form, we have not found a single case where the Classical Linear Logit stood up under Box-Cox tests, no doubt because constant marginal utility is untenable.

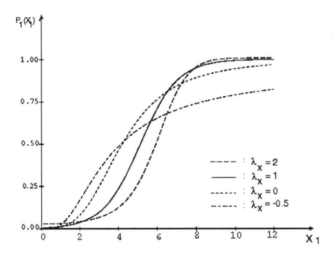

Figure 13.1 Response asymmetry and non-linearity: Classical Linear-Logit vs Standard Box-Cox-Logit

---

16 Figures 13.2 and 13.3 are the same as Figure 6 and Figure 7 in Gaudry and Wills (1978). In these figures the single Box-Cox transformation applied to the three transport condition variables is estimated at -0,19. When variable-specific transformations are applied, estimated values are -0,26 (fare), -0,05 (travel time) and 0,57 (frequency); the log-likelihood increases by 5.69 points with the 2 extra degrees of freedom awarded.

Any Linear Logit result untested for form is therefore suspicious, because all known tested models to date turned out to have form parameters distinct from both 0 and 1. This is true for instance in Europe-wide passenger models, as with the VACLAV model (Schoch, 2000) used in TEN-STAC (2004), or with current freight demand models in Germany (Seltz, 2004) or France (Gaudry et al., 2007). It is true in hierarchical structures applied to study high-speed rail with Canadian intercity passenger flows (Ekbote and Laferrière, 1994) and in European stated preference long-distance freight mode choice studies (e.g. JLR Conseil and Stratec, 2005).

Specification of transport conditions: expenditure or rate format? The first application of the Standard Box-Cox Logit illustrated in Figure 13.2 and Figure 13.3 demonstrated its usefulness with the transport condition variables of (27) specified in the "EXPENDITURE" format, i.e. as [*f* (Fare, Travel Time)], found in classical studies (e.g. Domencich and McFadden, 1975) based on Origin-Destination pair data. But that format is not straightforward: in classical microeconomic demand systems, prices *per unit* are used as explanatory variables, with Income scaling choices. In this case, with variables specified in the "RATE" specification, i.e. as [*g* (Price per km, Speed, Distance)], results will differ from those obtained with the expenditure specification only if the utility function is not exactly logarithmic. If it is, the coefficients of the rate specification can obviously be collected to predict exactly those of the expenditure specification and the two models have identical log-likelihoods.

In practice, as the optimal form is generally no more logarithmic than it is linear, there is a real choice between these two (non-nested) specifications. The rate specification gave better results in every case we have seen. It also makes microeconomic sense in that Distance plays in it the role of Income in demand systems: Distance becomes a money and time budget line. Furthermore, unit (money or time) prices are less correlated than their expenditure form. Colinearity and specification gains are therefore other features of the Standard Box-Cox Logit specification which make it possible to choose between the expenditure and rate formats of transport conditions. We forecast that the rate format will generally dominate in an increasing number of applications and that some silly forms of market segmentation based on trip length, fare or income will disappear because non-linearity is a distinct dimension from segmentation (Algers and Gaudry, 2004).

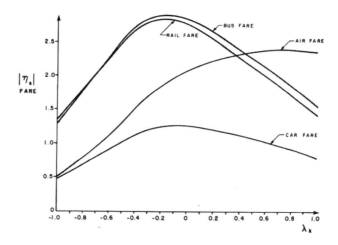

Figure 13.2 Fare elasticities in intercity Box-Cox Logit model (Canada, 92 city-pairs, 1972)

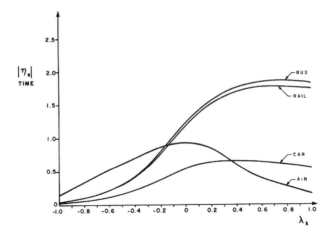

Figure 13.3 Travel time elasticities in intercity Box-Cox Logit model (Canada, 92 city-pairs, 1972)

Impact of non-linearity on forecasts These functions have considerable implications for the behaviour of market shares because the marginal impact of price and service characteristics are no longer constant and, as Figures 13.2 and 13.3 make abundantly clear, depend on the optimal form. We can therefore go beyond obviously different elasticities or marginal rates of substitution (such as values of time) and beyond other specification and estimation gains to ask, as we did above for Generation-Distribution models, whether anything systematic can be said about the difference between the forecasts from two variants 1 and 2 of a model differing by assumption only in the functional form applied to each model. With indices 1 and 2 denoting these variants and n an observation subscript, we focus on the transport condition variable X_{iqn} present in the following multinomial ($i, m = 1,\ldots, M$ alternatives) variants:

$$p_{1in} = \frac{\exp V_{1in}}{\sum_m \exp V_{1mn}} \tag{28}$$

and

$$p_{2in} = \frac{\exp V_{2in}}{\sum_m \exp V_{2mn}} \tag{29}$$

with representative utility components:

$$V_{1in} = \beta_{1io} + \sum_k \beta_{1ik} X_{1ikn}^{(\lambda_{1ik})} \tag{30}$$

and

$$V_{2in} = \beta_{2io} + \sum_k \beta_{2ik} X_{2ikn}^{(\lambda_{2ik})} \tag{31}$$

where $k = 1, K$ denote the independent variables, $(\beta_{1i0}, \beta_{1ik}, \lambda_{1ik})$ and $(\beta_{2i0}, \beta_{2ik}, \lambda_{2ik})$ are the parameters associated with variants 1 and 2, respectively.

Focusing on the difference between forecast shares of the ith alternative Δp_{in} due to changing a variable $X_{1iqn} = X_{2iqn} = X_{iqn}$, the so-called own effect [as we neglect here the cross effect Δp_{jn} $(j \neq i)$] is given by:

$$\Delta p_{in} = p_{2in} - p_{1in} = \frac{\exp V_{2in}}{\sum_m \exp V_{2mn}} - \frac{\exp V_{1in}}{\sum_m \exp V_{1mn}} \tag{32}$$

In contrast with the case previously discussed of Generation-Distribution models, where two out of the three points of interest in the behaviour of this difference could be obtained analytically, here[17] the *crossing point* $\Delta p_{in} = 0$, the *point of maximum difference* $\partial \Delta p_{in} / \partial X_{iqn} = 0$ and the *inflexion point* $\partial^2 \Delta p_{in} / \partial X_{iqn}^2 = 0$ can only be obtained by simulation, unfortunately (and the same holds naturally for cross effects).

Examples of the behaviour of Δp_{in} in passenger and freight models We will see that simulations have indicated that non-linear forms imply higher market share changes in relatively long trips, and lower market share changes in short trips, than those forecast by linear models. This important finding can be demonstrated by calculating, for all observations, Δp_{in} for any mode in a model specified with generic[18] coefficients applied to transport conditions: here we use changes of rail transport speed and price based on results found in Table 13.8 to perform the simulations.

This table presents modal share elasticities and relevant statistics for the two models for which simulations have been made to date. The *ceteris paribus* procedure modifies only one variable in (30) and (31): in the passenger case, train speeds were increased to forecast the effect of the *introduction of ICE trains* in Germany; in the freight case *intermodal train prices were decreased everywhere by 10 per cent* to determine the effect on all trans-Pyreneean flows.

Distinguishing between summary and detailed sample-based forecasts Starting with the discrete choice passenger model for Germany, we first note in columns 1 and 2 of Table 13.8 that the air *time elasticity* evaluated at sample means is 2,16 [= -1,62/-0,75] times larger at the optimum point (λ = 0,24) than at the non-optimal point of linearity[19] (λ = 1,00). But this global statistic hides a structure revealed in Figure 13.4, where differences between the train share forecasts following the introduction of the faster trains depends on O-D distance or trip length and exhibit an S-shape with a crossing point around 150 km (and hypothetical second one outside the range), an inflexion point around 350 km and a maximum difference at about 650 km.

Considering now in Table 13.8 results of the aggregate model of European trans-Pyreneean freight flows, we note in Columns 3 and 5 that the rail price elasticity evaluated at the sample means is 6,30 [= -6,30/-1,34] times larger at the point of optimal form (λ = -1,83) than at the non-optimal point of linearity (λ = 1,00). And we see in Figure 13.5 that, when intermodal container train prices are reduced by 10 per cent on all origin-destination pairs, these overall statistics also

---

17 As the interested reader may verify in Appendix 9 of Gaudry *et al.* (2007).
18 If the coefficients of transport conditions were not generic but specific, the exact shape of the S structure of the differences would depend on the mode considered.
19 In Figure 13.3 pertaining to the aggregate model for Canada, the corresponding ratio is 5,05 [= -0,86/-0,17] with an optimal form point (λ = -0,23) at an equal distance of 0, but negative. Had we considered the comparable ratios for air costs, the ratio in Table 13.8 is 1,84 [= -0,24/-0,13] for Germany and 1,94 [= -2,84/-1,46] for Canada in Figure 13.2.

hide differences among forecasts related to trip length in an S-shaped manner, with crossing points around 600 and 2500 km, an inflexion point around 1,400 km and a maximum difference at about 1700 km. As compared to Figure 13.4, there are two crossing points here because, given parameter values, the range of distances pertaining to trans-Pyreneean freight flows linking origins and destinations within Europe (we excluded the small numbers of flows from Morocco to Poland) is sufficiently large to allow it, in contrast with the case of domestic German passenger flows.

Revenues depend on the product of Generation-Distribution and Mode Choice models Such findings have major implications for revenue generation because short trips are more numerous than long trips, as determined in Generation-Distribution models above, where elasticities with respect to distance are even typically greater than 1 (in absolute value), implying a faster than proportionate decrease. This will be true whether modes or itineraries are considered: non-linear forms seem to increase the market shares effects of changes in transport conditions for relatively long trips and decrease those of relatively short length.

Ekbote and Laferrière (1994) examined such revenue implications with a nested Logit specification applied to stated preference survey data pertaining to high-speed rail (HSR) options in the Quebec–Windsor corridor of Canada and found, after setting revenue-maximizing prices for each of the two trip purposes, that the dominant non-linear forms – with optimal λ values found for price at 0,25 (business) and 0,31 (other purposes) and also duly estimated separately for travel time and access time (but not for frequency of service) – implied lower HSR revenues with optimal forms than those obtained with the (non-optimal) linear form of the same models. We now turn to another study of the same corridor as we try to introduce off-diagonal terms in the utility functions to go beyond the use of own characteristics in diagonal structures.

Table 13.8 Selected own share elasticities of Box-Cox Logit models used in the simulations

Share/probability elasticity evaluated at sample means	Intercity passengers, Germany 1979, all trip purposes (6000 observations)			Trans-Pyrenean freight flows between the Iberian peninsula and 14 European countries 1999, all freight categories (749 observations)			
Column		1	2	3	4	5	
Case	Mandel et al. (1997)	Linear MKI₁	Box-Cox MKBC	Linear Model 58	Box-Cox Model 59	Box-Cox Model 60	Gaudry et al. (2007)
Original reference							
		Fare (per O-D)			Price (per ton-km)		
Elasticity	Plane	-0,99	-0,62	-1,34	-6,30	-6,30	Intermodal train
	Train	-0,13	-0,24	-3,02	-2,08	-1,98	Classic train
	Car	-0,04	-0,04	-0,47	-0,64	-0,65	Truck
	t-statistic of β_k	(-8,39)	(-5,48)	(-7,50)	(-10,94)	(-10,98)	t-statistic of β_k
		Transit time (per O-D)			Speed (km/hr)		
Elasticity	Plane	-0,75	-1,62	0,68	0,57	0,67	Intermodal train
	Train	-0,63	-1,00	0,69	0,64	0,76	Classic train
	Car	-0,08	-0,14	0,16	0,15	0,19	Truck
	t-statistic of β_k	(-14,92)	(-15,14)	(2,54)	(2,35)	(3,41)	t-statistic of β_k
	Other variables (not reported on here) [...]			Other variables (not reported on here) [...]			
Box-Cox λ on Fare and Time		1,00	0,24				
Box-Cox λ on Price				1,00	-1,75	-1,83	
Box-Cox λ on Distance						6,49	
Log-likelihood		-1230	-1189	-2581	-2550	-2535	
Degrees of freedom		0	1	0	1	2	

Used in simulations shown in Figures 13.4 (ICE added) and 13.5 (container trains)

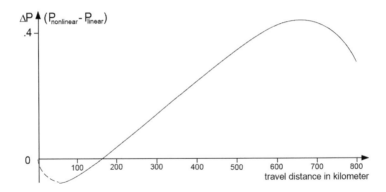

Figure 13.4 Comparing Linear and Box-Cox Logit forecasts of an ICE train scenario in Germany

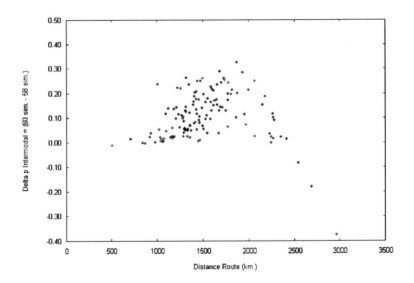

Figure 13.5 Comparing Linear and Box-Cox Logit forecasts of intermodal train gains

B. Away from IIA Consistency in Mode Choice: Diagonal Dominance Instead of Diagonal Slavery

Theory: early but lonely   We indeed wish to show, by outlining alternatives to diagonal slavery and consequent IIA, that it is possible to bring the Logit back into the fold of demand systems and to take interdependence of utility into account by using Box-Cox transformations on the additional terms that, if theory is "taken seriously", should differ from those used on the diagonal: were the optimal forms the same, the enriched model would not be identifiable. As we will see in both Table 13.9 and Table 13.10, identification does not pose undue problems, as the estimated values of Box-Cox transformations shown in bold indicate.

We outlined this strategy in detail early enough (Gaudry, 1978) after the flowering of 1977 had, in our view, shown the unsustainability of linear Logit utility functions and the practical manageability of Box-Cox forms applied to diagonal terms[20] and to socio-economic terms: their possible use to obtain a workable specification of the Universal Logit, originally specified at a high level of generality by McFadden (1975), came naturally.

But adding the characteristics of competing alternatives in own mode utility function did not appeal to everyone: some researchers (e.g. Ben-Akiva, 1974) feared to find unexpected results such as complementary modes or colinearity as in the classical systems. To proceed carefully, we start in Table 13.9 with a partial (passenger) model, where only the car travel time variable is used in all utility functions, and then proceed to further develop a full price matrix model on the basis of the freight example just reported on.

From diagonal slavery to a weak form of diagonal dominance   The results shown in Table 13.9 indicate that:

1. The original linear model yields, for non-business trips, an incorrect sign on the travel cost variable (and consequently negative values of time), as evidenced in the dark braided frames.
2. As one introduces 4 Box-Cox transformations (one for each transport condition and one for Income), the gains in log-likelihood are considerable[21] (from columns 1 to 2 and from columns 4 to 5) and the incorrect sign is corrected. Clearly, it is useless and incorrect to avoid the problem by constructing a generalized cost variable for non-business trips, as was done by those who first analysed this high-quality Via Rail Canada database of 12,000 trips with (nested or not) linear Logit forms (KPGM, 1990), or to ignore this trip purpose entirely (Bhat, 1995).
3. The introduction of car travel time in all utility functions increases considerably the log-likelihood (from Column 2 to 3 and from Column 5 to 6), especially for non-business trips. This completely new addition modifies the error correlation structures which arbitrary nests are meant to address and in fact provides a continuous alternative to non-nested nesting structures. It also certainly modifies the variance of the error terms that may or may not be originally homoskedastic[22] in linear space, an issue beyond our concern here but discussed elsewhere jointly with that of the estimation of Box-Cox transformations (Gaudry and Dagenais, 1979).

Also, the values of time obtained in Table 13.9 (measured in 1987 Canadian dollars), differentiated by mode, make more sense than the unique ones of the linear form. It should also be noted that all off-diagonal term elasticities are strong and their t-statistic results (in greyed cells) very high, which may be due to the addition of only a single off-diagonal term. If only one such term is chosen, time by car is surely the right one because the car, pervasive in that corridor (having almost 90 per cent of the market in the total sample), is "the reference". We also note in passing the presence of non-linearity with respect to Income.

---

20   The estimation algorithm was written and distributed early (Liem and Gaudry, 1987, 1994).
21   It is possible to show that the only transformation which is not significantly different from 1, the linear case, is that of the Frequency variable.
22   Heteroskedascity, as properly spelled (McCullogh, 1985), was studied by Bhat (1995), using only the business trip purpose data, and dropping entirely the bus mode (3264 observations) – an entirely arbitrary stance in modelling.

Table 13.9 Comparing Linear with Standard and Generalized Box-Cox elasticities and values of time

Gaudry and Le Leyzour (1994)		Quebec City–Windsor corridor of Canada 1987 (domestic intercity flows)					
		Business trips (4402 observations)			Other trips (8535 observations)		
Weighted aggregate probability point elasticity*	Column Case	1 Linear	2 Standard Box-Cox	3 Generalized Box-Cox	4 Linear	5 Standard Box-Cox	6 Generalized Box-Cox
	Original reference	Model 3	Model 5	Model 6	Model 9	Model 11	Model 12
Cost (access + in-vehicle)							
- Plane own cost		-0,53	-0,46	-0,12	0,15	-0,05	-0,02
- Train own cost		-0,05	-0,13	-0,07	0,02	-0,08	-0,02
- Bus own cost		-0,03	-0,13	-0,10	0,04	-0,15	-0,05
- Car own cost		-0,05	-0,10	-0,05	0,01	-0,06	-0,01
t-statistic of β_k		(-3,66)	(-7,23)	(-2,85)	(2,45)	(-4,96)	(-2,58)
Associated Box-Cox λ_k		1,00	0,28	-0,22	1,00	-0,29	-0,23
Travel time (access + in-vehicle)							
- Plane own time		-0,32	-0,11	-0,21	-0,07	-0,05	-0,00
cross w.r.t. car travel time		--	--	0,52	--	--	0,29
- Train own time		-0,18	-0,10	-0,12	-0,07	-0,03	-0,05
cross w.r.t. car travel time		--	--	0,09	--	--	0,11
- Bus own time		-0,26	-0,14	-0,14	-0,18	-0,08	-0,42
cross w.r.t. car travel time		--	--	0,13	--	--	0,15
- Car own time		-0,11	-0,05	-0,09	-0,04	-0,02	-0,06
t-statistic of own β_k		(-10,19)	(-5,91)	(-4,31)	(-6,26)	(-2,13)	(-4,10)
t-statistic of cross β_k		--	--	(4,59)	--	--	(13,67)
Associated own generic B-C λ_k		1,00	1,80	0,61	1,00	-0,12	4,96
Associated cross generic B-C λ_k		--	--	4,94	--	--	1,59
Frequency							
- Plane own frequency		0,39	0,46	0,45	0,23	0,30	0,22
- Train own frequency		0,02	0,01	0,01	0,02	0,01	0,02
- Bus own frequency		0,12	0,10	0,11	0,15	0,15	0,15
t-statistic of β_k		(12,52)	(11,67)	(11,88)	(12,14)	(12,78)	(14,05)
Associated generic Box-Cox λ_k		1,00	1,53	1,34	1,00	1,18	0,79
Income (Gross Individual)							
- Plane (reference: car)		0,23	0,18	0,18	0,12	0,08	0,09
t-statistic of specific β_k		(4,99)	(5,00)	(5,05)	(4,09)	(3,10)	(2,83)
- Train (reference: car)		-0,04	-0,04	-0,04	-0,02	-0,02	-0,02
t-statistic of specific β_k		(-2,61)	(-2,99)	(-3,00)	(-6,56)	(-7,82)	(-7,71)
- Bus (reference: car)		-0,10	-0,10	-0,10	-0,13	-0,14	-0,15
t-statistic of specific β_k		(-5,80)	(-6,89)	(-6,93)	(-19,45)	(-21,15)	(-21,02)
Associated generic Box-Cox λ_k		1,00	0,16	0,18	1,00	0,41	0,41
Other variables not reported		(Trip origin in a large city) [...]			(Trip origin in a large city; party size) [...]		
Log-likelihood		-1068	-1058	-1049	-3934	-3835	-3753
Degrees of freedom		0	4	6	0	4	6
Values of time (1987 $Can.)							
- Plane value of time		36,60	15,00	10,20	-19,20	33,00	2,58
- Train value of time		36,60	7,20	16,80	-19,20	3,12	7,20
- Bus value of time		36,60	5,34	7,20	-19,20	2,46	2,34
- Car value of time		36,60	6,60	21,60	-19,20	2,34	9,60

* The probability point elasticity is the usual elasticity of the probability multiplied by the probability of the alternative. For details, see Liem and Gaudry (1987, 1993).

Although our emphasis is on off-diagonal transport condition terms, it should be mentioned that Box-Cox transformations make it possible to solve also in principle the under-identification problem associated with coefficients of socio-economic variables, a problem shared with alternative-specific constants – discussed below with equation (34) – in linear formats. In Table 13.9, alternative-specific effects are estimated (with the car mode as reference) with Income successfully transformed, in contrast with the situation found in Table 13.10, where generic regression coefficients (with a truck reference) are estimated for the Distance variable (also successfully transformed) and for the Market size variable (linear after tests). In no case presented here are alternative specific coefficients estimated for all of the alternatives, although this was tried without success for the Market size variable and is clearly of interest.

We did not have time to systematically introduce car cost and other modal characteristics, and did not wish to do so because the EXPENDITURE format of the utility functions that we inherited from the first analysts were never compared by them to a RATE format. This is normally a second step performed only after it has been demonstrated that the optimal forms are not logarithmic, but one could envisage a comparison of (non-nested) linear forms of the EXPENDITURE and RATE specifications. In the next example, discussed presently, the correct format was determined (the Rate format won easily) and extensive tests were carried out to find the best sequence to introduce off-diagonal terms: most significant additions on a one-to-one basis constitute the "first round" of interdependence terms, and so on for less significant additions, with the resulting sequence defined already in Table 13.5.

Table 13.10 Linear, Standard and Generalized Logit mode choice model of freight across the Pyrenees

Flows between the Iberian peninsula and 14 European countries 1999, all freight categories, by land across the Pyrenees (749 observations) Model numbers 58 to 62 are as in the source: Gaudry et al. (2007)							Progressive inclusion of Price variables (added to Speed, Distance and Total O-D market size)		
Log likelihood	D.F.	Constraints from Table 5 and λ on Distance				V_i utility function	Road	Container	Rail
		λ_1 Own	λ_2 First	λ_3 Second	λ Distance		Own percentage point elasticity* (*)		
A. Linear Logit Model (58)									
-2581	0	1,00	1,00	1,00	1,00	V road r	-0,382		
						V container c		-0,194	
						V classic rail f			-0,137
B. Standard Box-Cox Logit Model with one BCT (59)									
-2550	1	-1,75	1,00	1,00	1,00	V road r	-0,517		
						V container c		-0,933	
						V classic rail f			-0,093
C. Standard Box-Cox Logit Model with two BCT (60)									
-2535	2	-1,83	1,00	1,00	6,49	V road r	-0,523		
						V container c		-0,984	
						V classic rail f			-0,093
D. Generalized Standard Box-Cox Logit Model with first round of substitutes or complements (61)									
-2524	4	-1,709	-6,271	1,000	6,272	V road r	-0,572		0,079
						V container c	0,441	-1,232	
						V classic rail f	0,131		-0,120
E. Generalized Standard Box-Cox Logit Model with all rounds of substitutes or complements (62)									
-2510	6	-1,89	1,05	-0,16	6,92	V road r	-0,489	1,165	0,227
						V container c	0,383	-1,233	-0,145
						V classic rail f	0,106	0,068	-0,082

* Note that, although the first three models A, B and C are the same as those reported on in Columns 3, 4 and 5 of Table 13.8, the elasticities found here differ: in Table 13.8 all elasticities presented are the normal *share elasticities*. The latter, multiplied by the market shares, yield the *percentage point elasticities*, as explained in the SHARE program documentation (Liem et al., 1997) where the interested reader may verify that all derivatives of shares with respect to any variable duly take into account its presence in all utility functions where it appears.

A stronger form of diagonal dominance: a practical form of the Universal Logit Although the model considered has Speed, Distance and a measure of Total O-D market size in the utility functions, our emphasis here is on prices, as documented in Table 13.10. To be formal, the inclusion in the utility functions of the characteristics of other alternatives, in the so-called Generalized Box-Cox Logit form, modifies (27) to:

$$V_i = \beta_{i0} + \sum_n \beta_{in}^i X_n^{i(\lambda_{in}^i)} + \sum_n \beta_{in}^j X_n^{j(\lambda_{in}^j)} + \sum_s \beta_{is} X_s^{(\lambda_{is})} \tag{33}$$

which permits complex pattern of substitution and complementarity, as classical demand systems also allow. We must therefore comment on the evolution of the sizes and signs of price own and cross-elasticities of the market shares.

We have already noted above (in our comments on Table 13.8) that passing from (A) to (B) by putting a Box-Cox transformation λ_1 on all own prices on the diagonal of the price matrix has an important effect on own share elasticities, as reiterated in Table 13.10 where, by contrast, almost no impact on the Price elasticities is found in (C) after the addition of the second Box-Cox transformation λ_4 to the Distance term,[23] despite the fact that this second transformation itself increases the log-likelihood by 15 units and obtains a very strong value.

Can transport modes be gross complements? Concerning specifically the introduction of interdependence in utility, note further that the addition in (D) of the first round of three off-diagonal terms naturally yields massive gains (they were tested one at a time on the basis of their individual effect on the log-likelihood value) but provides no indication of complementarity because all off-diagonal terms are positive. The off-diagonal term effects are, in the majority of cases, much smaller than those of the diagonal terms; in the case of the classic rail utility function, where they are about equal, the t-statistic[24] (-4,72) of the road price coefficient underlying 0,13 is much lower (-11,83) than that of the own price underlying (-0,12). Overall, this suggests that, as in demand systems, off-diagonal substitution effects are weaker than diagonal effects which "dominate" them, as they are expected to do in demand systems: it would be a weird case if off-diagonal term cross-elasticities were stronger than diagonal term own elasticities in a model where many such "first round" terms intervene.

How many off-diagonal terms? Can one go further and add a second round of prices? When the price matrix is full in (E): (i) the elasticity values on the diagonal hardly budge, as compared to those in (D); (ii) all off-diagonal terms are *positive* except for -0,145 – with a t-statistic of (9,55) – of the rail price in the intermodal utility function, which is negative like the own elasticity of -1,233, with a t-statistic of (-8,18). One might then be led to think that, because intermodal rail combines road and rail components, this complementarity (of classic rail to intermodal rail) result is reasonable despite the fact that the converse is apparently not the case —intermodal rail is not a complement of classic rail because its reverse elasticity is positive and equal to 0,068, with a t-statistic of (-0,037). Of course, t-statistics have to be used to weigh all results: the converse complementarity effect is not significant.

These (E) results therefore imply the presence of some complementarity, but the (fragile) asymmetry of cross-effects between container and classic rail poses some questions. Another

---

23   Tests made on modal Speed showed that its effect was approximately linear.
24   The t-statistics pertaining to Table 13.10 are found in Table 7 and Appendix 3 or 6 of the source document.

suspicious result is that the elasticity of road demand with respect to the price of combined intermodal transport is higher in absolute terms – at 1,165, with a t-statistic of (11,05) – than the own elasticity, at -0,489, with a t-statistic of (8,18). Are we at the limit of the technique and encountering the same difficulties as classical demand systems: colinearity? Will constraints like (Slutsky-type) symmetry conditions sadly have to be imposed?

A careful analysis of the best colinearity indices, those of Belsley et al. (1980), indicates a serious deterioration of the indices as one goes from (D)[25] to (E). This means that, despite obvious robustness on the diagonal, some of the cross effects are imprecisely estimated. We therefore consider the (E) results as lacking in robustness and, short of a deeper analysis, will not use them to answer the next question.

Impact of additional non-linear off-diagonal terms on forecasts What does the presence of cross terms imply for systematic differences between linear and non-linear forecasts? If the diagonal effects dominate, one should not expect the additional off-diagonal terms in (D) to modify what was found earlier in Figure 13.5 as we proceeded from linear to Standard Box-Cox forms. Note also that the difference Δp_{in} between the two forecasts is still expressed by (32). But, as the variable X_{qn} now appears in all alternatives, its modal index i is no longer necessary and the own and cross elasticity formulas are per force identical. It is also true here that the crossing point $\Delta p_{in} = 0$, the point of maximum difference $\partial \Delta p_{in} / \partial X_{qn} = 0$ and the inflexion point $\partial^2 \Delta p_{in} / \partial X_{qn}^2 = 0$ can only be obtained by simulation, as held in the Standard case. Figure 13.6 exhibits the same behavioural pattern as Figure 13.5 despite the fact that model (61) replaces model (60).

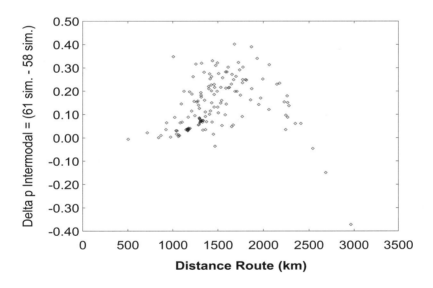

Figure 13.6 Comparing Generalized Box-Cox and Linear Logit forecasts of intermodal train gains

---

25 This Generalized Box-Cox Logit is the model in use since 2005 by the French Ministry of Transport to forecast freight land mode market shares across the Pyrenees. This ministry now frequently requires Box-Cox tests in calls for tenders pertaining to demand models.

The presence of reasonably weak off-diagonal effects does not change the structure of results already obtained with the Standard Box-Cox formulation. Non-linearity therefore makes the strongest difference to our problem – implying a sigmoid behaviour in the difference between the linear and non-linear forecasts – despite specification enrichment. We are back in classical demand systems, where all prices matter and utility is not presumed to be separable, but the diagonal dominates without enslaving us and the full matrix contains a trace of modal complementarity – another entirely new result attributable to an excellent database.

C. Implications for Path, Company or Service Choice in Air Markets

Company, Schedule and Path Choice As a cross-sectional problem, the choice of company, scheduled flight and passenger itinerary, is not different from the mode choice problem just discussed at length. Generally, path choice models are of the Linear Logit kind, with the same properties as in mode choice applications except one: that of the constants, to be discussed shortly. Until as recently as 2004, Lufthansa was using a Linear Logit procedure for path choice problems. As the number of paths used by air passengers tends to increase with distance (from, say, three within Germany up to 16 between Germany and the United States), the modelling procedure is of some import.

The emergence of the non-linear Box-Cox Logit model implies that some paths will incorrectly dominate other paths under a linear form, but these effects have yet to be studied in comparative fashion as they are expected to mimic those found for the mode choice models noted above. They have implications for the appropriate size of planes: linear forms would put larger planes than necessary on relatively short flights and planes of somewhat insufficient size on relatively long flights, as the horizontal S-shaped curve of differences in forecasts between models implies in Figure 13.5 or 13.6.

Efficient stated preference sample design Use of non-linearity of the Box-Cox type should therefore have a strong impact on *marketing studies* and Stated Preference surveys: it implies that experimental data should contain sets of variables that are orthogonal not in linear but in non-linear space. Indeed, the correlation that matters for the reproduction of reality and the establishment of statistical causality is defined among the optimal non-linear forms of the variables of interest, a point particularly well made in Cirillo (2005).

In the analysis of air carrier and path choice, there are currently few exceptions to the use of a Linear Logit and such exceptions as exist, such as the PODS procedure derived from an earlier Boeing procedure (Boeing Co, 1996), works very much like a Linear Logit (Carrier, 2003), no doubt partly because its "diagonal" utility functions are linear, as they are in complex variants of the Multinomial Logit applied to air path choice (Walker and Parker, 2006; Djurisic, 2007). Importantly, PODS also attempts to take into account capacity constraints on links, which artificially generates some complementarity among them, in the spirit of the seminal MAPUM air path choice algorithm (Soumis, 1978; Soumis *et al.*, 1979).

The many ways to avoid interdependence and develop compensating mechanisms This "linear logit-like + overflow and overspill" practice should bias decisions towards high-frequency hubs and stabilise path assignments by motivating the development of complementary block groupings (such as Toronto–Paris, Montreal–Paris and Montreal–Toronto–Paris) in part because of the simplistic linearity assumptions preventing reference paths from affecting the utility of all alternatives. In many of these problems, this "reference" choice (shortest path at desired departure time etc.) is like the car time characteristic of the Quebec–Windsor mode choice model developed to determine the market for high-speed rail in central Canada: it belongs in all utility functions because the utility of any alternative is not separable from that of the reference alternative.

The problem of constants in unimodal path assignments, notably air networks   In this context, a problem with the Logit is that its mode-specific constants are underidentified: consequently one ends up estimating differences in their coefficients with respect to an arbitrarily chosen reference one, as in:

$$\left.\begin{array}{rl} V_i = & (\beta_{i0} - \beta_{r0}) + \sum_n \beta_{in}^i X_n^i + \sum_s (\beta_{is} - \beta_{rs}) X_s \\ V_i = & \beta_{i0}^\nabla + \sum_n \beta_{in}^i X_n^i + \sum_s \beta_{is}^\nabla X_s \end{array}\right\}$$ (34)

where the problem is illustrated for these constants and for the socio-economic variables common to all functions.

This problem is minor in mode choice problems if no new mode is to be introduced, but in company/path choice procedures it matters whether the path constants are equal because the game is precisely to introduce new paths all of the time. The reason why this is a hard problem in air networks is because there is no natural labelling of paths (first, second, third etc.) serving an O-D pair, as there is in mode choice analysis (bus, plane, car, train etc.), because it hardly makes sense to define a reference path as one might define a reference mode: it has to come out of the analysis. Some research is going on in this area. The issue is finessed by some path analysts (Abraham and Coquand, 1961; Dial, 1971) by setting all constants at 0 or by ignoring it while working on the choice of road *tracé* (McFadden, 1968).

Constants and ignorance parameters   One way out of this constant identification quandary is to use an Inverse Power Transformation (IPT) envelope (Gaudry, 1981) applied to a Logit (IPT-L) path choice model, as was done by Laferrière (1987) to identify all his path-specific air constants for a Canada-wide model built with a 100 per cent sample (16 million individual trips) of domestic air trips made on Air Canada and Canadian Pacific Airlines in 1983. If one further accepts a defined ordering of air paths as a testable device, reference alternatives should be identifiable. Moreover, some additional gains can be expected from the IPT-L form, such as the identification of captivity to certain paths (or airlines and airports) and the detection of leftover non-linearity or asymmetry in the response to such ordered path utilities. Captivity, or modeller's ignorance, amounts to non 0 or 1 asymptotes of the response curves in Figure 13.1. It is distinct from the issue of its asymmetry. This new orientation is therefore promising for the proper modelling of air company/path competition that generates airport demand.

4 Conclusion: A TAM and Remedies to Build in Interdependence among Flows

In this chapter, we first defined a new four-part Traffic Accounting Matrix (TAM) to register all spatial flows of interest for air demand forecasting, effectively extending the scope of classical algebraic input-output analysis by doubling up and reinterpreting the intermediate and final transactions components of two-part Input-Output (I-O) matrices. Strictly defined subsets of a TAM can then be matched to, and explained by, the usual procedures pertaining to the distinct generation, distribution, mode choice and assignment steps of traffic demand planning, or to their combinations.

We then focused on the key properties of such demand models in order to evaluate their relevance to the explanation of airport or hub competition and considered, among potential remedies, the estimation of form with Box-Cox transformations, but pointed out that their demonstrated relevance to the measurement of the impact of transport conditions is insufficient to solve the problem at hand. In both Generation-Distribution and Split Choice mode-company-path structural steps, the predominant use of Independence from Irrelevant Alternatives (IIA) consistent cores must be rejected to account properly for competition among destinations in Generation-Distribution models and for the prevailing importance of reference alternatives in Split Choice mode-company-path models.

We provided a first partial literature summary of numerous results obtained with endogenous functional forms in both of these structural steps but argued that, because the issue of non-separability of utility is not directly addressed by standard Box-Cox transformations, their increased explanatory power and realism – as compared to the popular fixed form a priori logarithmic (in Gravity models) and linear (in Logit models) specifications – is more relevant to the proper measurement of the role of transport conditions (distance, level of service or price) than to the necessary representation of interdependence among alternatives, which mandates the abandonment of "diagonal slavery" in utility formulations.

Form corrects mistakes and is consistent with non-constant marginal utility   Of course, the proper role of transport conditions still matters decisively in both Generation-Distribution models of transport or trade and in the Mode or company-path Choice splits. In the former class, proper curvature defines the total market reach and data determined forms rectify the demonstrably incorrect use of distance in the many logarithmic pooled time-series and cross-sectional models. In the latter class, allowing for changing marginal utility profoundly modifies the relative sensitivity of longer over shorter length trips, as compared to their behaviour in prevailing untested linear constant marginal utility forms of the same functions – never theoretically very credible or empirically sustainable. But none of these benefits and remedies to current dominant practice allows for interdependence (non-separability) of utility, the key future demand modelling challenge if *ex ante* forecasts are to be of relevance to the air demand question at hand.

Introduction of non-diagonal terms is necessary   To point to real remedies, we summarized some recent promising attempts to deal with interdependence in manageable ways expected to yield "diagonal dominant" results: through the use of spatial autocorrelation in Box-Cox Generation-Distribution models and of Generalized Box-Cox specifications in Split models. Separable utility is thereby rejected by the data, but without using too many independent off-diagonal terms pertaining to transport conditions: if the denial of any separability has been the scourge of classical demand equation system, its blind imposition has been that of Gravity and Logit demand systems. Considerate and flexible middle ways are now within reach, and they matter most to model new interdependent markets such as tourism.

References

Abraham, C. and P. Coquand. "La répartition du trafic entre itinéraires concurrents: réflexions sur le comportement des usagers, application au calcul des péages". *Revue générale des routes et aérodromes*, 357, 57–76, 1961.

Algers, S. and M. Gaudry, "Are Nonlinearity and Segmentation Substitutes, Complements, or Unrelated in Logit Models? The Swedish Intercity Passenger Case", Publication AJD-28, Agora Jules Dupuit (AJD), Université de Montréal, p. 18, 1994, 2001, 2004.

Anderson, J.E., "A Theoretical Foundation for the Gravity Equation", *American Economic Review* 69, 1, 106–16, 1979.

Anderson, J.E. and E. van Wincoop, "Gravity with Gravitas: A Solution to the Border Puzzle", *American Economic Review* 93, 179–92, 2003.

Barnett, W.A. and Seck, Ousmane, "Rotterdam vs Almost Ideal Models: Will the Best Demand Specification Please Stand Up?", Munich Personal RePEc Archive, MPRA Paper No. 417, p. 42, 2006.

Barten, A., "Maximum Likelihood Estimation of a Complete System of Demand Equations", *European Economic Review* 1, 7–73, 1969.

Batten, D.F. and D.E. Boyce, "Spatial Interaction, Transportation, and Interregional Commodity Flow Models", in Nijkamp, P. (ed.) *Handbook of Regional and Urban Economics, vol. 1 Regional Economics*, pp. 357–406, North Holland, 1986.

Becker, A., Savy, M. and M. Giraud, "Approches régionales des échanges franco-allemands de marchandises", Observatoire économique et statistique des transports, Ministère de l'équipement, des transports et du tourisme, March 1994.

Belsley, D., Kuh, E. and R.E. Welsh, *Regression Diagnostics*, John Wiley, New York, 1980.

Ben-Akiva, M., "Note on the Specification of a Logit Model with Utility Functions that Include Attributes of Competing Alternatives", Working Paper, Department of Civil Engineering, MIT, Cambridge, MA, 1974.

Berkson, J., "Application of the Logistic Function to Bio-assay", *Journal of the American Statistical Association* 39, 357–65, 1944.

Bhat, C., "A Heteroscedastic Extreme Value Model of Intercity Travel Mode Choice", *Transportation Research B* 29, 6, 471–83, 1995.

Blankmeyer, E.C., "A Metro Model", unpublished manuscript, Economics Department, Princeton University, 1971.

Boeing Co, "The Decision Window Path Preference Model": Summary Discussion, Marketing and Business Strategy, Boeing Commercial Airplanes, The Boeing Company, 1996.

Box, G.P. and D.R. Cox, "An Analysis of Transformations", *Journal of the Royal Statistical Society*, Series B 26, 211–43, 1964.

Brems, H., *Pioneering Economic Theory, 1630–1980: A Mathematical Restatement*, The Johns Hopkins University Press, 1986.

Buch, C.M., Kleinert, J. and F. Toubal, "The Distance Puzzle: On the Interpretation of the Distance Coefficient in Gravity Equations", *Economics Letters* 83, 293–8, 2004.

Burghouwt, G. and J. de Wit, "The Temporal Configuration of European Airline Networks", Publication AJD-74, Agora Jules Dupuit, Université de Montréal, July 2003. http://planning.geog.uu.nl/medewerkers/g_burghouwt.htm; www.aaeconomics.com.

Carrier, E., "Modeling Airline Passenger Choice: Passenger Preference for Schedule in the Passenger Origin-Destination Simulator (PODS)", MA thesis, Civil and Environmental Engineering, MIT, June 2003.

Carey, H.C. *Principles of Social Science*, Lippincott, Philadelphia, 1858–59.

Cascetta, E. and M. Di Gangi, "A Multi-Regional Input-Output Model with Elastic Trade Coefficients for the Simulation of Freight Transport Demand in Italy", *Transport Planning Methods*, PTRC Education and Research Services Ltd, vol. P 404–2, 1996.

Cirillo, C. "Optimal Design for Mixed Logit Models", Groupe de recherche sur les transports, Facultés Universitaires Notre-Dame de la Paix, Namur, Association for European Transport and contributors, 2005.

Cramer, J.S., "The Origins and Development of the Logit Model", University of Amsterdam and Tinbergen Institute, Amsterdam, August 2003.

Deaton, A.S. and J. Muellbauer, "An Almost Ideal Demand System," *American Economic Review* 70, 312–26, 1980.

Dial, R.B., "A Probabilistic Multipath Traffic Assignment Model which Obviates Path Enumeration". *Transportation Research* 5, 83–111, 1971.

Djurisic, M., "Modèles de demande et gestion du revenu en transport aérien", Mémoire de maîtrise, Département de mathématiques et de génie industriel, École polytechnique de Montréal, Université de Montréal, May 2007.

Domencich, T. and D. McFadden, *Urban Travel Demand: A Behavioral Analysis*, North Holland, 1975.

Ekbote, D. and R. Laferrière, "Demand Model Developments to Assess High Speed Passenger Train Markets on Windsor to Quebec City Corridor", Proceedings of the 20th Annual Meeting, Canadian Transportation Research Forum/Le Groupe de recherches sur les transports au Canada, 598–609, 1994.

Ellis, R.H. and P.R. Rassam, "National Intercity Travel: Development and Implementation of a Demand Forecasting Framework", US Dept of Transportation Contract T-8 542, Modification No. 1, National Technical Information Service PB 192 455, US Dept of Commerce, March 1970.

Florian, M. and M. Gaudry, "A Conceptual Framework for the Supply Side in Transportation Systems", *Transportation Research B* 14, 1–2, 1–8, 1980.

Florian, M. and M. Gaudry, "Transportation Systems Analysis: Illustrations and Extensions of a Conceptual Framework", *Transportation Research B* 17, 2, 147–54, 1983.

Fratar, T., "Vehicular Trip Generation by Successive Approximations", *Traffic Quarterly* 8, 53–65, 1954.

Fratianni, M. and H. Kang, Heterogeneous Fistance – Elasticities in Trade Gravity Models, *Economics Letters* 90, 68–71, 2006.

Fridström, L. and A. Madslien, "A Stated Preference Analysis of Wholesalers' Freight Choice", Institute of Transport Economics, Oslo, 1995.

Gaudry, M., "A Note on the Economic Interpretation of Delay Functions in Assignment Problems", *Lecture Notes in Economics and Mathematical Systems* 118, 368–81, 1976.

Gaudry, M., "A Generalized Logit mode choice model", publication 98, Centre de recherche sur les transports, Université de Montréal, 1978.

Gaudry, M., "Six Notions of Equilibrium and their Implications for Travel Modelling Examined in an Aggregate Direct Demand Framework", in Hensher, D. and P. Stopher (eds), *Behavioural Travel Modelling*, Croom Helm, 138–63, 1979.

Gaudry, M., "A Study of Aggregate Bi-Modal Urban Travel Supply, Demand, and Network Behavior Using Simultaneous Equations with Autoregressive Residuals", *Transportation Research B* 14, 1–2, 29–58, 1980.

Gaudry, M. "The Inverse Power Transformation Logit and Dogit Mode Choice Models", *Transportation Research B* 15, 2, 97–103, 1981.

Gaudry, M, "Key Substitution-Complementarity Features of Travel Demand Models, with Reference to Studies of High Speed Rail Interactions with Air Services", in Calzada, M. and M. Houée (eds), *Actes du Colloque Déplacements à Longue Distance. Mesures et Analyses*, Ministère de l'Equipement, du Logement et des Transports, et ESA Consultants, Paris, 412–57 1998; and in COST 318, *Interactions between High-Speed Rail and Air Passenger Transport*, Final Report of the Action, European Commission, Directorate General Transport, Brussels, 204–38, www.cordis.lu/cost-transport/src/cost-318.htm, 1998.

Gaudry, M., "The Four Approaches to Origin-Destination Matrix Estimation: Some Considerations for the MYSTIC Research Consortium", Working Paper No. 2000-10, Bureau d'économie théorique et appliquée (BETA), Université Louis Pasteur, September 1999.

Gaudry, M, "The Robust Gravity Form in Transport and Trade Models", *Zeitschrift für Kanada-Studien* 24, 1, Band 44, 144–61, www.ajd.umontreal.ca/Frames/Frameanalysedelademande/frameanalyse123.htm, 2004.

Gaudry, M. "Structure de la modélisation du trafic et théorie économique", Ch. 1 (92 p.), in Maurice, J. and Y. Crozet (eds), *Le calcul économique dans le processus de choix collectif des investissements de transport*, Economica 2007.

Gaudry, M., Blum, U. and J. McCallum, "Fist Gross Measure of Unexploited Single Market Integration", in S. Urban (ed.), *Europe's Challenges*, Gabler Verlag, pp. 449– 61, 1996.

Gaudry, M., Briand, A, Paulmyer, I. and C.-L. Tran, "Choix modal transpyrénéen ferroviaire, intermodal et routier: un modèle Logit Universel de forme Box-Cox généralisée", in Lebacque, P., Boillot, F. and M. Aron (eds), *Modélisation du trafic – Actes du groupe de travail 2004–2005*, Actes INRETS No. 110, pp. 11–119, December 2007.

Gaudry, M. and M. Dagenais, "Heteroscedasticity and the Use of Box-Cox Transformations", *Economics Letters* 2, 3, 225–9, 1979.

Gaudry, M., Jara-Diaz, S.R. and J. de D. Ortuzar, "Value of Time Sensitivity to Model Specification", *Transportation Research B* 23, 2, 151–8, 1989.

Gaudry, M. and A. Le Leyzour, "Improving a Fragile Linear Logit Model Specified for High Speed Rail Demand Analysis in the Quebec-Windsor Corridor of Canada". Publication CRT-988, Centre de recherche sur les transports; Paper 9413, Département de sciences économiques; Publication IEA-94-07, Ecole des Hautes Etudes Commerciales, Université de Montréal, 1994.

Gaudry, M., Mandel, B. and W. Rothengatter, "Introducing Spatial Competition through an Autoregressive Contiguous Distributed (AR-C-D) Process in Intercity Generation-Distribution Models for Canada and Germany", Publication AJD-14, Agora Jules Dupuit, Université de Montréal, 1992.

Gaudry, M. and M. Wills, "Testing the Dogit Model with Aggregate Time Series and Cross Sectional Travel Data", *Transportation Research B* 13, 2, 155–66, 1978.

Ghosh, S. and S. Yamarik, "Does Trade Creation Measure Up? A re-examination of the effects of regional trading arrangements", *Economics Letters* 82, 213–19, 2004.

Giuliano, P., Spilimbergo, A. and G. Tonon, "Genetic, Cultural and Geographical Distances", Discussion Paper No. 2229, Institute for the Study of Labor (IZA), Bonn, July 2006.

Helliwell, J.F., *How Much Do National Borders Matter?* Brookings Institution, Washington DC, 1998.

Houthakker, H.S., "Additive Preferences", *Econometrica* 28, 244–57, 1960.

JLR Conseil and Stratec, "Enquête de préférences déclarées sur le choix modal en transport de marchandises transpyrénéen", Ministère de l'équipement DTT/SI1, Paris, 31 August 2005.

Johnson, L.W., "Stochastic Parameter Estimation: An Annotated Bibliography", *International Statistical Review* 45, 257–72, 1977.

Johnson, L.W., "Regression with Random Coefficients", *Omega* 6, 71–81, 1978.

Kahn, A., *Lessons from Deregulation*, AEI-Brookings Joint Center for Regulatory Studies, 2003.

Kau, J.B. and C.F. Sirmans, "The Functional Form of the Gravity Model", *International Regional Science Review*, 4, 2, 127–36, 1979.

KPGM Peat-Marwick and F.S. Koppelman, "Proposal for Analysis of the Market Demand for High Speed Rail in the Québec/Ontario Corridor", Report presented to the Québec/Windsor Corridor High Speed Rail Task Force, KPGM, 1990.

Laferrière, R., *Une Aggrégation Nouvelle des Itinéraires du Transport Aérien (ANITA)*, Publication AJD-87, Agora Jules Dupuit, Université de Montréal, December 1987.

Last, J., "Reisezweckspezifische Modellierung von Verkehrsverflechtungen", dissertation, Institut für Sirtschaftspolitik und Wirtschaftsforschung, Universität Karlsruhe, 1998.

Leontief, W., *The Structure of the American Economy, 1919–1929*, Cambridge, MA, 1941.

Liem, T.C. and M. Gaudry, "PROBABILITY: The P-2 program for the Standard and Generalized BOX-COX LOGIT models with disaggregate data", Publication CRT-527, Centre de recherche sur les transports, Université de Montréal, 1987, 1993.

Liem, T.C., Dagenais, M. and M. Gaudry, "LEVEL: The L-1.4 program for BC-GAUHESEQ regression – Box-Cox Generalized AUtoregressive HEteroskedastic Single EQuation models", Publication CRT-510, Centre de recherche sur les transports, Université de Montréal, 1993.

Liem, T.C. and M. Gaudry, "QDF: a Quasi-Direct Format Used to Combine Total and Mode Choice Results to Obtain Modal Elasticities and Diverstion Rates", Publication CRT-982, Centre de recherche sur les transports, Université de Montréal, June 1994.

Liem, T.C., Gaudry, M. and R. Laferrière "SHARE: The S-1 to S-5 programs for the Standard and Generalized BOX-COX LOGIT and DOGIT and for the Linear and Box-Tukey INVERSE POWER TRANSFORMATION-LOGIT models with aggregate data", Publication CRT-899, Centre de recherche sur les transports, Université de Montréal, 1997.

Lill, E., *Das Reisegesetz und zeine Anwendung auf den Eisenbahnverkehr*, Vienna, 1891.

Luce, R.D., *Individual Choice Behavior: A Theoretical Analysis*, Wiley, New York, 1959.

Luce, R.D., "The Choice Axiom after Twenty Years", *Journal of Mathematical Psychology* 15, 215–33, 1977.

Mandel, B., "The Interdependency of Airport Choice and Travel Demand", in Gaudry, M. and R.R. Mayes (eds), *Taking Stock of Air Liberalization*, Kluwer Academic Publishers, 189–222, 1999a.

Mandel, B., "Measuring Competition in Air Transport", Publication 99-18, Centre de recherche sur les transports, Université de Montréal, 1999b.

Mandel, B., Gaudry, M. and W. Rothengatter, "A Disaggregate Box-Cox Logit Mode Choice Model of Intercity Passenger Travel in Germany and its Implications for High Speed Rail Demand Forecasts", *The Annals of Regional Science* 31, 2, 99–120, 1997.

McCallum, J., *National Borders Matter: Regional Trade Patterns in North America*, Working Paper in Economics 12/93, Economics department, McGill University, 1993.

McCallum, J., "National Borders Matter: Canada–U.S. Regional Trade Patterns", *American Economic Review* 85, 3, 615–23, 1995.

McCulloch, J.H., "On Heteros*edasticity", *Econometrica* 53, 2, 483, 1985.

McFadden, D., "The Revealed Preferences of a Government Bureaucracy", Technical Report No. 17, Institute of International Studies, University of California at Berkeley, 1968.

McFadden, D., "On independence, structure, and simultaneity in transportation demand analysis", Working Paper 7511, Institute of Transportation and Traffic Engineering, University of California, Berkeley, 1975.

McFadden, D., "The Revealed Preferences of a Government Bureaucracy", *Bell Journal of Economics* 7, 55–72, 1976.

McFadden, D., and K. Train, "Mixed MNL Models for Discrete Response", *Journal of Applied Econometrics* 15, 447–70, 2000.

Moses, L., "The Stability of Interregional Trading Patterns and Input-Output Analysis", *American Economic Review*, 45, 5, 803–32, 1955.

Orro, A., Novales, M. and F.G. Benitez, *Nonlinearity and Taste Heterogeneity Influence on Discrete Choice Model Forecasts*,, Association for European Transport and contributors, 2005.

Pearl, R. and L.J. Reed, "The Population of an Area Around Chicago and the Logistic Curve", *Journal of the American Statistical Association* 24, 65, 66–7, 1929.

Quesnay, F., *Tableau Oeconomique*, Versailles, 1st edition, 1758.

Rassam, P.R., Ellis, R.H. and J.C. Bennett, "The N-Dimensional Logit Model: Development and Application", Peat Marwick and Mitchell & Co, Washington, DC, 1970.

Rossi-Hansberg, E., "A Spatial Theory of Trade", *American Economic Review* 95, 5, 1464–91, 2005.

Schoch, M., *VACLAV-VIA, The IWW-MKmetric Passenger Model*, Institut für Wirtschaftspolitik und Wirtschaftsforschung, Universität Karlsruhe (TH), September 2000.

Seltz, T., "The BVU Nested Box-Cox Logit Freight Mode Choice Model Developed for the German Federal Transport Plan in 2002", Presentation at the MODEV-M Workshop, Fondation Reille, Paris, 22 June 2004.

Soumis, F., "Modèle de comportement des clients du transport aérien", Research Report No. 78–16, Département des méthodes quantitatives, Ecole des Hautes Etudes Commerciales, November 1978.

Soumis, F., Ferland, J.A. and J.M. Rousseau, "MAPUM, A Model for Assigning Passengers to a Flight Schedule", Publication CRT-142, Centre de recherche sur les transports, Université de Montréal, August 1979.

Stone, J.R.N., "Linear Expenditure Systems and Demand Analysis: An Application to the Pattern of British Demand," *Economic Journal* 74, 255, 511–27, 1954.

TEN-STAC: *T*rans-*E*uropean *N*etwork – *S*cenarios, *T*raffic Forecasts and *A*nalysis of *C*orridors, European Commission Contract No. ETU/B5-7000A-SI2.346797, http://www.nea.nl/ten-stac, 2004.

Theil, H., "A Multinomial Extension of the Linear Logit Model", *International Economic Review* 10, 251–9, 1969.

Tran, C.-L. and M. Gaudry, "LEVEL: The L-2.1 estimation procedures for BC-DAUHESEQ (Box-Cox Directed AUtoregressive HEteroskedastic Single EQuation models) regressions", Publication AJD-107, Agora Jules Dupuit, Université de Montréal, April 2008.

Uhler, R.S. and J.G. Cragg, "The Structure of Asset Portfolios of Households", *Review of Economic Studies* 38, 341–57, 1971.

Verhulst, P.-F., "Notice sur la loi que la population. suit dans son accroissement", *Correspondance mathématique et physique*, vol. X, 113–21, 1838.

Verhulst, P.-F., "Recherches mathématiques sur la loi d'accroissement de la population", *Nouveaux mémoires de l'académie royale des sciences, des lettres et des beaux-arts de Belgique*, 18, 1–38, 1845.

Walker, J.L. and Parker, R.A., Estimating Utility of Time-of-Day Demand for Airline Schedules Using Mixed Logit Warner, S.L., *Stochastic Choice of Mode in Urban Travel: A Study in Binary Choice*, Northwestern University Press, 1962.

Model, Transportation Research Board Annual Meeting 2006, Paper No. 06-2434, 2006.

Wei, S.-J., Intra-national versus International Trade: How Stubborn Are Nations in Global Integration? Working Paper 5531, National Bureau of Economic Research, 1996.

Williams, H.C.W.L., "On the Formation of Travel Demand Models and Economic Evaluation Measures of User Benefit", *Environment and Planning* 9a, 3, 285–344, 1977.

PART C:
Case Studies of Airport Competition

Chapter 14
Competition in the German Airport Market: An Empirical Investigation

Robert Malina

Introduction

Much research has been done in the past few years in order to understand and optimize the regulatory framework of airport markets.[1] There are a multitude of approaches to this regulatory framework – sometimes complementary, mostly, however, competing. Economists generally agree that airport regulation of some sort is necessary if one can identify persistent market power within the relevant market. To date, much less emphasis has been placed on the problem of measuring an airport's market power. This is especially true for the German airport market, where much research is still needed.

The extent of the market power of an airport is a result of different factors, such as current and potential intramodal, intermodal and intersectoral competition, and the countervailing power of airlines. In this chapter we present a new approach for measuring the market power of an individual airport in respect to other airports (intramodal competition) in its core business (facilitating air traffic) and apply it to the German airport market. We develop a concentration ratio that accounts for different buyer preferences. We call this ratio the 'substitution coefficient'. It quantifies the quality of the best substitute for a certain airport. The substitution coefficient is defined as the share of residents within the relevant regional market of an airport that would consider another airport, which has been identified as meeting the demands of the airlines, to be a good substitute.

The chapter is structured as follows: the next section gives a short overview on the German airport market and its shareholder structure, outlining the fundamental changes over the last decade. We then present our model for measuring market power and apply it to 35 German airports. In the penultimate section the results and some important restrictions of the model are discussed, while the final section concludes.

Governance and Incentive Structure in the German Airport Market

For many years German airports were regarded as passive providers of infrastructure with no interest in generating profits. The airport's sole purpose was the provision of capacity for the air transport market. During the last decade, however, airports have started perceiving themselves as enterprises competing for passengers and airlines. This increasing profit orientation can be attributed to three major developments. First, and most important, there has been a change of governance structure at many German airports. Although organized as private companies for decades (contrary to other

---

1   Forsyth (1997) and (2001); Kunz (1999); Gillen/Morrison (2001); Starkie (2001); Niemeier (2002) and Wolf (2003).

German transport infrastructure providers that were structured as public agencies), the shareholders were public entities such as the federal or state government, public companies or municipalities. This has changed significantly. Of the 35 airports in Germany offering scheduled or holiday charter flights in the summer of 2004, 11 are now at least partially owned by private investors (Table 14.1), and there are plans to open two more airports (Munich and Cologne) to private capital.[2] In 2005, more than 50 per cent of all passengers in Germany started or ended their journey at an airport with private capital involvement. This figure will rise to about 75 per cent if the federal government sells their stakes in Munich and Cologne.

Table 14.1 Ownership structure of privatized German airports in spring 2008

Airport	Airport operating company	Shareholders	Share
Altenburg-Nobitz (AOC)	Flugplatz Altenburg-Nobitz GmbH	County Altenburger Land	60.00%
		Stadtwerke Altenburg GmbH	19.00%
		Municipality of Nobitz	10.00%
		Altenburger Destillerie & Liqueurfabrik GmbH	3.00%
		Altenburger Brauerei GmbH	3.00%
		Wellpappenwerk Luck GmbH	3.00%
		County Chemnitzer Land	2.00%
Düsseldorf-International (DUS)	Flughafen Düsseldorf GmbH	City of Düsseldorf	50.00%
		Airport Partners GmbH[a)]	50.00%
Düsseldorf-Mönchengladbach (MGL)	Flughafen Gesellschaft Mönchengladbach GmbH	Flughafen Düsseldorf GmbH	70.03%
		Stadtwerke Mönchengladbach GmbH	29.96%
		City of Willich	0.01%
Frankfurt-Hahn (HHN)	Flughafen Frankfurt-Hahn GmbH	Fraport AG	65.00%
		Federal state of Rhineland-Palatinate	17.50%
		Federal state of Hesse	17.50%
Frankfurt/Main (FRA)	Flughafen Frankfurt/Main AG (Fraport AG)	Federal state of Hesse	31.62%
		Stadtwerke Frankfurt Holding	20.19%
		Deutsche Lufthansa AG	9.96%
		Julius Bär Holding AG	5.09%
		The Capital Group Companies, Inc.	4.70%
		Others	28.44%
Friedrichshafen (FDH)	Flughafen Friedrichshafen GmbH	County Bodenseekreis	20.10%
		City of Friedrichshafen	20.10%
		Luftschiffbau-Zeppelin GmbH	10.74%
		ZF Friedrichshafen AG	13.10%
		Technische Werke Friedrichshafen GmbH	7.87%
		Dornier GmbH	2.95%
		MTU Friedrichshafen GmbH	2.95%
		Federal state of Baden-Württemberg	17.38%
		Chamber of Commerce Bodensee	4.80%

---

2   Siegmund (2004), p. 72; Moring/Zamponi (2005), p. 22 and Handelsblatt (2004), p. 4.

Table 14.1   Concluded

Hamburg (HAM)	Flughafen Hamburg GmbH	City of Hamburg	51.00%
		Hochtief Airport GmbH	34.80%
		Hochtief Airport Capital	14.20%
Hanover (HAJ)	Flughafen Hannover Langenhagen GmbH	Hannoversche Beteiligung GmbH[c]	35.00%
		City of Hanover	35.00%
		Fraport AG and Nord LB	30.00%
Lübeck (LBC)	Flughafen Lübeck GmbH	City of Lübeck	10.00 %
		Infratil Airports Europe Ltd	90.00 %
Niederrhein (NRN)	Flughafen Niederrhein GmbH	Airport Niederrhein Holding GmbH[d]	99.93%
		County Kleve	0.04%
		Municipality of Weeze	0.03%
Saarbrücken (SCN)	Flughafen Saarbrücken Betriebsgesellschaft mbH	Fraport AG	51.00%
		Flughafen Saarbrücken Besitz-gesellschaft [e]	48.00%
		City of Saarbrücken	1.00%

Notes: [a] Airport Partners GmbH: Hochtief AirPort GmbH 30 per cent, Hochtief AirPort Capital 30 per cent, Aer Rianta PLC (Dublin Airport Authority PLC) 40 per cent; [b] Artisan Partners Ltd Partnership 3.87 per cent, Taube Hodson Stonex Partners Limited 3.01 per cent, Morgan Stanley 2.90 per cent, free float 18.66 per cent; [c] Sole shareholder: federal state of Lower Saxony; [d] Sole shareholder: Airport Network BV; [e] Sole shareholder: federal state of Saarland.

Source: Own compilation based on airport operators' reports and publications by shareholders.

Second, severe public budget problems lead to increased awareness of costs and profit for public shareholders, who strive for some return on their investment in the airport or at least expect the airport operator to minimize financial losses.

Third, the liberalization of the downstream airline market has led to a change in the airport business environment from 'mutual existence' with airlines to confrontation on service quality and prices. Airports have to cope with airlines' demands for lower charges and better quality and the threat of airlines switching to another airport, thus 'forcing' airports into competition.

With the increasing profit orientation of German airports the question of whether airports in Germany possess market power and therefore need to be regulated is gaining more importance. In essence, an airport with market power is able to restrict the amount supplied and to raise the price in order to increase its profits at the expense of consumers; with the result that social surplus is reduced. Even if the airport can discriminate in pricing and thus charges the marginal consumer prices equal to marginal costs – so that there is no shortage of supply compared to the competitive level – overall social surplus might be lower because of inefficient operations and higher marginal costs than in a competitive environment. The airport could also seek to reduce quality below consumer preferences in order to increase profits.

Therefore, airport regulation is not only a question of market power abuse in the form of capturing consumer surplus and thus a distribution issue, it is also a question of allocative efficiency.

Figure 14.1 shows the spatial distribution, the size – measured in weekly departures – and the ownership structure of all 35 airports in Germany with scheduled and holiday charter traffic in the summer of 2004. Many of these airports do not operate independently, but are part of airport groups with overlapping shareholder structures. Members of airport groups do not compete with each other; therefore, they do not mitigate potential market power of their partners. We will account for this aspect in our analysis of market power.

Measuring the Market Power of Airports

Introduction to the Approach

Airports provide a number of services to their customers:

- aeronautical services (that is, infrastructure provision, rescue, security, fire-fighting, runway and taxiway maintenance);
- aeronautical-related commercial services (that is, baggage, passenger and cargo handling, catering, supply of fuel and lubricants, waste disposal);
- non-aviation services (that is, retailing, car rental, parking, banks, hotels, restaurants).

Aeronautical and aeronautical-related commercial services are often called 'aviation services' in order to emphasize their core role for the functioning of the air transport market and to differentiate them from additional services not needed for facilitating air traffic (non-aviation services). This chapter focuses only on market power in aviation services. Non-aviation services of an airport are generally regarded as being subject to competition from inner-city leasing or retailing.[3]

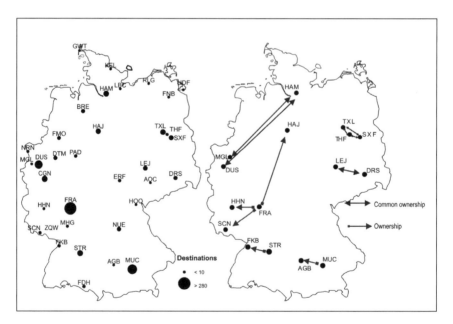

Figure 14.1 Spatial distribution, size and ownership structure of the German airport market

---

3   Productivity Commission (2002), p. 177 and Brunekreeft/Neuscheler (2003), p. 270.

Moreover, we concentrate on market power in enabling passenger transportation. We do not focus on cargo-only flights due to the fact that the market for providing aviation services for cargo-only flights and the cargo market itself are unanimously regarded in the relevant literature as being already highly competitive, contrary to the passenger market.[4]

Because of prohibitively high entry barriers in the German airport market (due to legal, administrative and political restrictions), competitive constraints can derive only from current substitutes.[5] Therefore we do not analyse the potential for future competition, but focus only on current competition.

The standard approach for assessing market power is a two-stage process consisting of the definition of the relevant market followed by an analysis of market power within the relevant market. There are various methods for defining the relevant market. One of the most common methods is the concept of close substitutability, in which the relevant market is described as 'a group of sellers of close-substitute outputs who supply a common group of buyers'.[6] Close substitutability of the output of different suppliers is a result of the fact 'that they are all varieties of the same sort of good or service ... with similarity in form or function and fulfilling the same sort of specific want or need of buyers'.[7]

When we transfer this definition to the airport market, a close substitute of an airport can be every supplier that is able to satisfy the needs of the customers of aviation-services (passengers and airlines) as successfully as the currently used airport. Such suppliers can be other airports (intramodal substitutability), but also other means of transportation such as road and rail (intermodal substitutability), or even firms from other economic sectors (intersectoral substitutability) such as IT companies that offer video-conferencing.[8] In this chapter we exclusively analyse market power due to a lack of intramodal substitutability.

As for the definition of the relevant market, there are a variety of approaches for identifying supplier power within that market.[9] Some of the most important concepts are Lerner's degree of monopoly, Bain's concept of excess profits and concentration ratios such as the Herfindahl-Hirschman Index (HHI).[10] Lerner's and Bain's approach have conceptual disadvantages when being used for identifying the market power of an individual airport, and are therefore not suitable for our analysis.[11] The HHI measures market share concentration under the assumption that the offered products are homogenous goods. This assumption is not appropriate for the airport market because some residents within the relevant market of an airport might perceive another airport to be a good substitute, for example because driving distance is very short from their home, whereas it is not good for others living further away. Therefore we develop a concentration ratio that accounts for different buyer preferences in the intramodal analysis. We call this ratio 'substitution coefficient'. It quantifies the quality of the best substitute for a certain airport. The substitution coefficient is defined as the share of residents within the relevant regional market of an airport that consider another airport, which has been identified as meeting the demands of the airlines, to be a good substitute.

---

4 Gillen/Henriksson/Morrison (2001), p. 47; Tretheway (2001), p. 39.
5 Malina (2005b), p. 132 for a detailed discussion of entry barriers.
6 Bain (1959), p. 6.
7 Bain (1959), p. 211.
8 However studies show that intersectoral substitution of air transportation is very low. Hughes (1995); Stephenson/Bender (1996); Elsasser/Rangosch-du Moulin (1997); Mason (2002); Denstadli (2004).
9 Lerner (1934); Bain (1941), p. 272; Herfindahl (1950).
10 Lerner (1934); Bain (1941), p. 272; Herfindahl (1950).
11 Malina (2005b), p. 32.

Factors Influencing Airport Substitutability

We look at airport substitutability from the position of an airline as the direct customer of aeronautical services. The question we try to answer is: What requirements does an airport have to meet in order to be considered by the airline as a good substitute for a currently used airport?

One can differentiate between three types of demands. *Customer-oriented demands* are requirements derived from the factors determining airport choice by passengers. An air carrier is only able to profitably switch flights to another airport if the passengers are willing to use this airport. A potential substitute has to meet *infrastructural demands* as well. In order to be regarded as a good alternative the airport must be capable of providing the necessary services for the type of air traffic the airline operates at the currently used airport. Third, an alternative airport has to meet the necessary *legal requirements* for enabling the kind of air traffic that the airline operates at the current airport. We look at these demands in detail below.[12]

Customer-oriented demands of an airline Much econometrical work has been done on the determinants of passengers' airport choice. Starting with Kanafani et al. in 1975, numerous studies from various countries have shown that airport choice is influenced by four main factors:[13]

- Flight availability: Is there a flight from airport A to destination B?
- Flight frequency: How often can destination B be reached by flights departing from A?
- Ticket price: How much does it cost to fly from A to B?
- Access time: How long does it take to get to airport A?

The first factor is a precondition for an airport to be shortlisted and, thus, is equally important for all customer segments. Kanafani et al. (1975), Ashford/Benchemann (1987), Harvey (1987), Windle/Dresner (1975), Pels et al. (2000) and Hess/Polak (2003) show, however, that the importance of the other three determinants differs significantly between business and leisure travellers. Whereas frequency of flights and access time are the most important factors in the utility function of a business traveller (the more frequent the flights and the shorter the access time, the higher the utility), leisure travellers' utility is influenced most strongly by the ticket price (the cheaper the ticket, the higher the utility). These findings are consistent with estimates of the price elasticity of business and leisure travel demand, which show that the one of business travellers is much lower than the one of leisure travellers.[14] The results are also supported by passenger surveys on various German airports. The survey results indicate that the airports with a higher percentage of leisure travellers have larger catchment areas than the airports where business travel is more

---

12 We do not look at prices as a factor of airport choice for two main reasons. First, we assume that airports are able to successfully discriminate in prices between new and old customers. This means that airlines that are willing to shift to another airport can be offered competitive prices. Second, as economies of scale of airport operations are generally perceived to flatten out at around 5 million workload units a year, airports can reach a competitive overall cost structure by making airlines switch operations. See, for example, Doganis/Graham (1995); Salazar de la Cruz (1999); Pels (2000) and Gillen/Henriksson/Morrison (2001).

13 Kanafani et al. (1975); Harvey (1987); Ashford/Bencheman (1987); Windle/Dresner (1995); Cohas et al. (1995); Mandel (1999a) and (1999b); Pels et al. (2000) and (2003); Moreno/Mueller (2003); Hess/Polak (2003) and (2006).

14 Gillen/Morrisson/Stewart (2003).

prevalent.[15] These surveys also indicate that leisure customers of low-cost carriers are even more price conscious and less time sensitive than the average leisure traveller.

Ticket prices, flight availability and flight frequency, however, are endogenous for an airline because it is the airline that sets prices, defines its network and sets the flight frequency. Only access time is an exogenous factor that cannot be influenced by an air carrier. Thus, an airline looking for an alternative to the currently used airport has to check carefully to make sure access time at the new airport is adequate for its customers.

Infrastructural demands The airport infrastructure mainly consists of a runway system and facilities for processing passengers or freight. The dominant factor in the decision of a carrier to shift flights to a new airport is the runway system. Depending on the type of aircraft and the intensity of usage, the requirements of an airline regarding the configuration of the runway differ significantly, both qualitatively and quantitatively. If an aircraft is heavier than average, the runway has to be longer and must be able to withstand more load (qualitative dimension). If an airline wants to shift a flight to a new airport, this airport must have the capacity to process this flight (quantitative dimension).

The Aircraft Classification Number/Pavement Classification Number (ACN/PCN) system indicates whether an aircraft is suitable for regular take-off or landing on a specific runway without damaging the runway. An ACN value is assigned to every type of aircraft, depending on the load it exerts on the runway. This value is compared with the PCN of the runway, which shows its strength. If the ACN is lower than the PCN, the aircraft is suitable for regular use of the runway.[16] In general, the ACN of an aircraft rises with its weight.[17] The same is true for the take-off distance required (TODR) and the landing distance required (LDR). TODR and LDR, however, are also influenced by external factors such as runway elevation and weather conditions.[18] As LDR is always shorter than TODR, the required take-off distance is the limiting factor of runway usage.

Table 14.2 shows the TODR of six different aircraft types, from wide-body aircraft to a 50-seater regional aircraft.[19] As departing aircraft are often not at their maximum take-off weight (because of, for example, shorter flight length and thus less fuel), we also display TODR with ¾ load capacity. The table indicates that, even for regional aircraft 1,000 metres, is the minimum required runway length.

Required runway strength and length are quite similar for some aircraft types; therefore we can categorize aircraft with regard to the runway requirements, starting with category 6 for regional aircraft such as ATR 42/72, Bae 146 and going up to category 1 for wide-body aircraft such as B747 or A340.[20]

Apart from runway strength and length, the quality of landing aids at the airport is also important for airport choice. In order to operate reliable regular air transport, an airport has to be equipped with an instrument landing system (ILS) that enables precision approaches under poor visibility. Depending on the quality of the system, an ILS can be divided into three main categories, starting with Cat. I, which allows for landings for suitably equipped aircraft in weather as low as 550 metres visibility and a decision height of not less than 60 metres, and going up to Cat. IIIc that enables zero/zero operation. If visibility is poor and an airport does not possess an ILS, the airline

---

15   Klophaus/Schaper (2004); Valentinelli et al. (2004).
16   The Boeing Company (1998) for an accurate and detailed description of ACN and PCN.
17   Annex 1, ACN values of selected aircrafts.
18   Fraport AG (2003) and ICAO (1993).
19   See Annex 14.3 for data on more aircraft.
20   Annex 14.2.

Table 14.2 Take-off distance required (TODR) for various aircraft with maximum take-off weight (MTOW) and 3/4 load capacity

Aircraft	TODR in metres	
	MTOW	¾ load capacity
B747-400	3,488	2,616
B767-300 ER	3,188	2,000
B757-300	2,780	1,962
B737-800	2,523	1,853
B737-500	1,655	1,363
ATR 42-500	1,165	990

Notes: TODR is calculated in the following conditions: dry and even runway at sea level, 24°C, no wind.
Source: Own calculations based on information from aircraft manufacturers.

has to redirect the flight to another airport. This causes additional costs for staff and passenger for alternate transportation to the scheduled point of arrival. Therefore, it is not surprising that every airport in Germany that is used for scheduled traffic, or even holiday charter traffic, is equipped with an ILS of at least Cat. I.

Airport infrastructure has to meet the demands of an air carrier not only qualitatively but also quantitatively. All of the infrastructure components of an airport (runways, apron, terminals etc.) have individual capacities. In essence, the capacity of the runway system determines the overall capacity of the airport in the long run. This is due to the fact that terminals and aprons can be enlarged more easily than runway capacity because of less restrictive legal and administrative barriers.[21] If the runway is used to its capacity limit, an airline that would like to shift flights to this airport is not able to do so because it cannot supersede the incumbent airlines, even if it is willing to pay higher airport fees. The incumbents can rely on so-called 'grandfather rights', which guarantee incumbents their take-off and landing rights at the airport.[22]

Legal demands An airport has to fulfil numerous legal requirements in order to provide scheduled or leisure traffic. First of all, the facility needs approval as a public airport or public airfield. Some airports in Germany – such as Hamburg-Finkenwerder, Lemwerder and Oberpfaffenhofen – are so-called *Werksflughäfen* (special airfields). They are not open for general public use and thus not suitable for shifting flights to them. In addition, the airport must have the right to allow take-off and landing of the aircraft type the airline wants to use. Some smaller airports in Germany have legal restrictions concerning the maximum permissible weight (MPW) of aircraft using the runway. There are other important restrictions which are not due to runway length or strength, such as environmental concerns, the protection of nearby residents and 'reasons of public interest'.[23] Some airfields, for example, are restricted to an MPW of 14 t or 20 t, although their runway is suitable for heavier aircraft. Such restrictions lead to a limitation of usable aircraft types to small regional

---

21 Wendlik (1995), p. 5; Hüschelrath (1998), p. 47; Wolf (2003), p. 65; Urbatzka/Wilken (2003), p. 7.
22 Ewers et al. (2001).
23 In some federal states, 'public interest' is primarily a euphemism for protecting publicly owned incumbents from competition from other airports; for example the airport policy in the federal states of Brandenburg and Berlin as analysed in Malina (2005b), p. 150.

aircraft such as DO-328 (<14 t) or ATR 42, ATR 72, CRJ 200 (<20 t). These airports, therefore, cannot be used for holiday charter traffic, for most normal feeder-flights to hubs or for point-to-point traffic because the dominant aircraft in these segments are Fokker 100, B 737 and A 320 family or heavier aircraft.

Apart from weight restrictions, authorities sometimes impose limitations on the number of aircraft movements, preventing the airport from fully utilizing its technical runway capacity. This can lead to flights being rejected by an airport although the runway has ample technical capacity. There could also be shorter operating hours set, so that the airport is not suitable for intercontinental flights that often start or end early in the morning or very late in the evening. There is not a single European airport with a relevant number of intercontinental destinations that is closed for more than six hours per day. For holiday traffic the possibility of night-time operations is essential because, in contrast to business passengers, leisure travellers are willing to accept flights at night. Thus, the airline is able to increase the utilization of aircraft, thereby raising the profitability of operations. Some airports are restricted in operations not only at night but also during the day. The legal approval to enable air traffic on an airport's infrastructure might be limited to weekdays or some hours throughout the day. These restrictions, in effect, make the airport unsuitable for the type of traffic we analyse in this chapter.

All the above mentioned restrictions are based on national law. For some airports, however, there are severe limitations on the kind of traffic that can be operated on their infrastructure that result from bilateral agreements: Whereas the air transport market within the European Union (EU) is fully liberalized (each air carrier from any member state is allowed to use every airport in the union),[24] traffic to countries outside the EU is subject to agreements of various regulatory intensity between the country of origin and the country of destination.[25] Irrespective of the regulatory intensity, all bilateral agreements contain so-called 'ownership clauses', which specify that only carriers that are under substantial ownership and effective control of shareholders, either by the country of origin or the destination country, are allowed to fly routes from the one country to the other.[26] This clause prevents airlines from other EU member states from shifting non-EU flights to a German airport and vice versa. This is a problem particularly for European network carriers with a high percentage of flights to destinations outside the common market, such as Lufthansa or Air France/KLM. They are restricted to using airports in their home country as hubs.

Application to the German Airport Market

After analysing the customer-oriented, infrastructural and legal demands of an airline's choice of an airport, we now apply the findings to the German airport market in detail. Our perspective is that of an airline that is currently using one of the 35 German airports with scheduled and holiday charter traffic. This airline is looking for a good substitute airport. In principle, this substitute can be situated in Germany or another country. As the new airport has to be accepted by the passengers, we only include foreign airports with an access time of less than two hours (by private transportation) from the nearest German district. There are over 100 airports within that area with a runway of more than 1,000 metres. Only 67 of them, however, are open to public use. Fifty-nine of these public airports are equipped with an instrument landing system of at least Cat. I. Figure 14.2

---

24 Sinha (2001), p. 73.
25 Jung (1999), p. 33.
26 Lelieur (2003).

shows the spatial distribution of the airports. We will use these sites for the following assessment of the substitutability of one of our 35 German airports.

First of all, we categorize the 35 German airports based on their passenger structure. Depending on the importance of low-cost traffic, traditional leisure traffic and classic feeder or point-to-point traffic by full-service carriers, we divide them into three groups:

- low-cost airports at which this segment has a market share of at least 40 per cent (measured in weekly departures);
- holiday airports with the same 40 per cent market share threshold in the holiday traffic segment;
- standard airports that reach a 40 per cent market share neither in leisure nor in low-cost travel.[27]

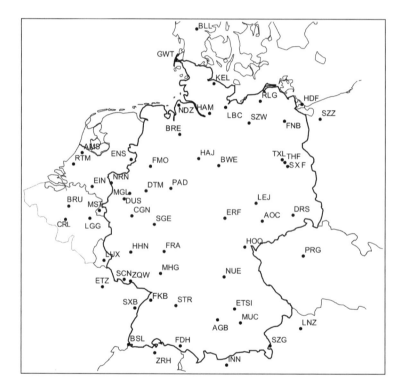

Figure 14.2 Spatial distribution of German and neighbouring foreign airports open to public traffic with a runway of at least 1,000 metres length and ILS

---

27   We use 40 per cent as the relevant threshold because both holiday and low-cost traffic is operated using bigger and heavier aircraft on average than other types of traffic (for example, B 737/800 in comparison to Bae 146 or CRJ 200). As airports usually impose aircraft fees based on MTOW and, additionally, passenger fees, a market share of 40 per cent based on the number of flights leads to a share of this segment of overall revenues of more than 50 per cent. This shows that the segment is pivotal for the airport operator. See Annex 14.4 for details.

Based on the analysis of airport choice of passengers we assume that passengers of a low-cost airport are willing to accept an access time of a maximum of two hours; passengers of a holiday airport 90 minutes; and those of a standard airport 60 minutes. Using route planning software we calculate access time from all German and neighbouring foreign districts to the 35 airports in our analysis to determine core catchment areas. For each district within the catchment area of an airport (we call this the 'base airport' in the following) we then try to find another airport that can be reached within a certain threshold time (= acceptance by passengers) and that also satisfies the legal, infrastructural and ownership demands of the airline. The threshold time we use depends on the type of airport we examine (60, 90 or 120 minutes). The data is again compiled by using route planning software. The airport that fulfils these demands and is accepted by the highest number of residents within the catchment area of our base airport is the best substitute of the base airport. For this airport/airport combination we calculate substitution coefficient (s_A) by dividing the number of residents within the catchment area of the base airport that can reach the substitute within the threshold time, by the population in this area. As a result, we get the share of people in the catchment area that regard an airport as a good substitute that is perceived as a good substitute from the airlines' point of view as well. The value of the substitution coefficient indicates the ability of an airline to switch to another airport without losing its customers or having to change flight operations: s_A values range between 0 and 1. The higher the value, the better the quality of the alternative airport, and therefore the higher the competitive constraints on the base airport.

Infrastructural requirements are now taken into consideration. Analysing airport timetable data for all of our 35 base airports for the summer of 2004, we identify the biggest/heaviest aircraft that regularly use the infrastructure of one airport. Based on runway strength and runway length, we then analyse the biggest aircraft that the 59 airports can cater to.[28] We then compare the current aircraft usage at the base airport with the potential aircraft usage at an alternative airport. If the infrastructure of the alternative airport is capable of handling aircraft of at least the weight category that is currently used at the base airport, then this alternative is investigated in greater detail. We then can determine if the alternative airport has ample capacity to take up traffic from the base airport. Most of the analysed airports indeed have excess capacity; only Frankfurt/Main (FRA), Düsseldorf (DUS) and Berlin-Tegel (TXL) operate more or less permanently at full capacity.[29] This means that these three airports are capable of providing services for only a few more flights. DUS, for example, is still able to additionally dispatch the eight daily flights that used Mönchengladbach (MGL) in 2004, but not to dispatch a relevant share of the average 300 daily flights in Cologne. We include the capacity situation of these three airports in the analysis by categorizing all airports according to the number of weekly departures and by restricting substitutability of other airports by these three 'at capacity' airports to small airports of category 5 and 6 (fewer than 100 flights per week).[30]

---

28   Annex 14.5.

29   DFS (2004). FRA is confronted with a demand that exceeds technical capacity throughout the day. At DUS, the technical capacity cannot be fully used because of aircraft movement restrictions by local authorities. Estimates by DFS (2004), however, show that even if these administrative restrictions were lifted, demand would exceed technical capacity at DUS. The situation at Tegel is different from the other two airports because it is not the runway system that is operating at full capacity, but the terminal and apron. As public shareholders of Tegel have officially decided to only marginally increase terminal capacity in TXL due to the planned upgrade of Berlin-Schönefeld (SXF) and the closure of TXL in 2010, we have to take terminal and apron capacity at TXL as is.

30   Annex 14.4.

We now look at the legal demands. Some of these have already been taken into account when choosing our 59 potential substitute airports, for example the necessity of being an approved public airport. What remains is the implementation of operating hours and ownership clauses. We integrate operating hours into the model by requesting that a potential substitute for a base airport with regular flights at night must be open for traffic during the night as well. Ownership clauses are integrated by eliminating foreign airports from the analysis if the base airport has a significant share (> 20 per cent of departures) of non EU-flights.

Finally, the ownership structure between the base airport and an alternative is taken into account. As previously shown, airports with substantial overlap of ownership do not compete with each other and thus do not reduce each other's market power. Thus we eliminate all airports from the analysis that have a similar shareholder structure to the base airport. Scheme 14.1 gives an overview of the calculation of the substitution coefficient.

When calculating the substitution coefficient we can divide the airports into four groups (Table 14.3). The first group shows high values of more than 0.70, which indicates very high competitive constraints from other airports and thus low market power. The second group encounters lower but still substantial competition from another airport (0.5 to 0.69). The third group has a substitution coefficient lower than 0.5 but higher than 0.2; the last shows very low values, with $s_A = 0$ for 12 airports.

Discussion of the Results

The results show that, rather than assessing market power of German airports in general, each airport has to be analysed individually. Market power in respect to intramodal competition is the result of the spatial and infrastructural configuration of the airport market and of passenger and airline demands regarding airport location, infrastructural configuration and legal requirements. Thus it is not the sheer market share of an airport that leads to market power, but the lack of good substitutes in its relevant market. Frankfurt-Hahn for example – an airport in the federal state of Rhineland-Palatinate belonging to Fraport AG, and a passenger throughput of more than 3 million in 2005 – is confronted with substantial competition from former military airport Zweibrücken with fewer than 30,000 passengers in 2005. Düsseldorf, which is the third biggest German airport, is subject to high intramodal competition from Cologne/Bonn, an airport with a similar catchment area and infrastructure quality but only half the number of Düsseldorf's passengers.

When analysing the results of the model, one should keep in mind some limitations. First of all, a low substitution coefficient does not automatically imply that an airport possesses high market power and thus should be regulated in order to prevent abuse of that position. It only shows that an airport is not confronted with substantial *intramodal* competition. As mentioned before, an airport can also be disciplined by intermodal substitution such as road or rail, intersectoral substitution – which, however, is generally perceived as being very low – and the countervailing power of airlines. Moreover, when analysing hub airports, such as Frankfurt and Munich, one should keep in mind that these airports might also face competition for transfer passengers from other hub airports. As we have not included these competitive dimensions, we are not able to draw a conclusion on the degree of substitution that stems from these factors and, as far as we know, little research has been conducted in order to measure their effect on an individual airport's market power. This is especially the case for countervailing power, which has been analysed only in a more general, somewhat impressionistic way to date.[31] Further research could therefore aim to systematically isolate the factors determining the degree of countervailing power and apply them to specific airport markets.

$$s_{A_i} = \frac{E_{KA_{F_i}}}{E_{GA_{F_i}}}$$

$$T = \left\{ z \mid d_{zF_i} \leq \begin{cases} 60, \text{ for all } F_i \ \forall \ BU_{F_i} > 2 \wedge BLC_{F_i} > 2 \\ 90, \text{ for all } F_i \ \forall \ BU_{F_i} \leq 2 \wedge BLC_{F_i} > 2 \\ 120, \text{ for all } F_i \ \forall \ BLC_{F_i} \leq 2 \end{cases} \right\}$$

$$E_{GA_{F_i}} = \sum_{z=1}^{n} E_z, \forall \ z \in T$$

$$E_{F_j} = \sum_{z=1}^{n} E_z \ \forall \left\{ z \in T \mid d_{zF_k} \leq \begin{cases} 60 \text{ for all } F_i, F_j \ \forall \ FP_{F_j} \leq AF_{F_i} \wedge BU_{F_i} > 2 \wedge BLC_{F_i} > 2 \\ 90 \text{ for all } F_i, F_j \ \forall \ FP_{F_j} \leq AF_{F_i} \wedge BU_{F_i} \leq 2 \wedge BLC_{F_i} > 2 \\ 120 \text{ for all } F_i, F_j \ \forall \ FP_{F_j} \leq AF_{F_i} \wedge BLC_{F_i} \leq 2 \end{cases} \right\}$$

for all $j \in \{F^{-i} \cup L\}$

s.t. $Kap_{F_j} = 0 \vee BG_{F_i} \geq 5$

$\wedge \ NF_{F_iF_j} = 0$

$$E_{KA_{F_i}} = \max E_{F_j}$$

AF: Biggest/heaviest category of aircraft currently using the base airport
BLC: Importance of the low-cost segment for an airport (≤ 2 means market share higher than 40%)
BU: Importance of the holiday traffic segment for an airport (≤ 2 means market share higher than 40%)
d: Access time (in min)
d_{zF_i}: Access time (in min) from district z to the base airport F_i
d_{zF_j}: Access time (in min) from district z, from which the passengers can reach the base airport F_i within the threshold time, to an alternative airport F_j
E_{F_j}: Number of residents that can reach alternative airport F_j and base airport F_i within the threshold time
$E_{GA_{F_i}}$: Number of residents that can reach alternative airport F_i within the threshold time
$E_{KA_{F_i}}$: Number of residents that can reach a substitute of the base airport F_i within the threshold time
E_z: Population of district z
F_i: Base airport, $i \in F, F = \{1,2,...,35\}$
F_j: German or foreign alternative airport, $j \in \{F^{-i} \cup L\}$
FP: Biggest/heaviest aircraft category that can use an airport without relevant weight restrictions
Kap: Capacity utilization of an airport (takes the value 1 when operating at full capacity, 0 in case of no relevant excess demand)
L: Number of analysed foreign airports, $L = \{1,...,20\}$
n: Number of analysed districts z
$NF_{F_iF_j}$: Comparison of night-time operating hours of airports F_i and F_j. Takes the value 1, if the base airport has no evening closure and it is used at night by airlines and the alternative airport F_i has restricted night-time hours. Takes the value 0 in all other cases
SAi: Substitution coefficient of base airport i
T: Assistance set
z: German or foreign district

Scheme 14.1 Formal presentation of calculating s_A

---

31 Malina (2005a), pp. 22–9.

Table 14.3 Best substitutes and substitution coefficients for 35 German airports

Base airport	Best/second best substitute	s_A
Altenburg-Nobitz	Leipzig	0.94
Lübeck-Blankensee	Hamburg	0.93
Niederrhein	Eindhoven/Düsseldorf	0.96/0.93
Kiel	Hamburg	0.79
Paderborn/Lippstadt	Münster/Osnabrück	0.77
Mönchengladbach	Cologne/Bonn	0.76
Münster/Osnabrück	Paderborn	0.76
Düsseldorf	Cologne/Bonn	0.75
Friedrichshafen	Zurich	0.68
Karlsruhe	Frankfurt/Main	0.61
Frankfurt-Hahn	Zweibrücken/Cologne/Bonn	0.61/0.61
Cologne/Bonn	Liege	0.56
Saarbrücken	Zweibrücken/Luxembourg	0.56/0.52
Augsburg	Manching	0.53
Dortmund	Cologne/Bonn	0.52
Mannheim	Frankfurt Main	0.43
Zweibrücken	Frankfurt Main	0.39
Rostock-Laage	Neubrandenburg	0.11
Leipzig	Berlin-Schönefeld	0.08
Bremen	Münster Osnabrück	0.07
Dresden	Altenburg Nobitz	0.03
Berlin Airport System (TXL, THF, SXF)	–	0.00
Erfurt	–	0.00
Frankfurt	–	0.00
Hamburg	–	0.00
Hanover	–	0.00
Heringsdorf	–	0.00
Hof-Plauen	–	0.00
Munich	–	0.00
Neubrandenburg	–	0.00
Nuremberg	–	0.00
Stuttgart	–	0.00
Westerland/Sylt	–	0.00

Furthermore, a low substitution coefficient does not imply that the substitutability of this airport cannot be increased. Of course, a change of airport location to a more competitive airport region is not practical. Market entries of newcomer airports are almost just as unlikely due to ever increasing entry barriers such as legal, administrative and political restrictions. In contrast, a change of ownership structure could lead to a significant increase of competition in some airport markets. This holds true for the metropolitan region of Berlin, where the relevant airports (Tegel, Tempelhof and Schönefeld) are operated as an airport system and owned by a public company of the federal and state governments. Splitting the airport system into three independent airport companies would increase the substitution coefficients to 0.93 for Tegel, 0.91 for Schönefeld and 1.00 for Tempelhof, thus making economic regulation obsolete. Unfortunately, in this example of the Berlin airports, current airport policy is heading in the opposite direction, and their plan is to close Tegel and Tempelhof in order to generate enough traffic for Schönefeld, which will be upgraded to 'Berlin Brandenburg International'.[32]

The caveats for low substitution coefficients, however, do not apply to high values. High coefficients are a sufficient condition for trusting market forces to control the market conduct of an airport, rendering economic regulation superfluous. If an airport is confronted with substantial competition from other sites, misconduct in the form of high prices or poor quality will be responded to by customers switching to another airport, thus forcing the airport to improve its services and lower its prices. This would hold true even if there were no additional intermodal or intersectoral competition.

Conclusion

Much research has been done in the last several years in order to optimize the regulatory framework of airport markets. Much less emphasis has been placed upon the problem of measuring an airport's market power, although economists generally agree that airport regulation is only needed if an airport has persistent market power within its relevant market.

In this chapter we have presented a new approach for measuring the market power of an individual airport with respect to other airports (intramodal competition) in its core business (facilitating air traffic), and have applied it to the German airport market. We have developed a concentration ratio – called substitution coefficient – that accounts for different buyer preferences. It takes into account infrastructural, legal and customer-oriented demands. The results show that nearly half of the 35 German airports analysed have substitution coefficients higher than 0.5, which indicates substantial intramodal competition. Substitution could be higher for some airports if they were independent entities and not owned by the same holding company. A high substitution coefficient indicates that economic regulation of the airport is unnecessary as market forces are strong enough. However, low substitution coefficients do not necessarily imply the need for regulation, as there might be competition from other modes of transportation, intersectoral competition or countervailing power from airlines that are not included in the model. This has not been accounted for in the model. Further research could therefore focus on measuring the effects of these factors on an airport's market power and integrating it into the model.

---

32 Berliner Flughäfen (2004).

Annex 14.1 Aircraft Classification Number (ACN) of selected aircraft

Aircraft	Weight	ACN							
		Flexible surface and subgrade strength category				Rigid surface and subgrade strength category			
		A	B	C	D	A	B	C	D
A380-800	MTOW	71	79	99	136	53	61	76	94
	OWE	29	31	35	48	25	26	30	35
A340-300	MTOW	62	68	79	107	54	62	74	86
	OWE	37	39	44	57	34	36	42	48
B747-400	MTOW	59	66	82	105	54	65	77	88
	OWE	23	24	27	35	20	23	27	31
A330-300	MTOW	55	60	70	94	46	54	64	75
	OWE	41	44	50	66	36	39	46	54
B757-300	MTOW	36	41	51	64	35	42	49	56
	OWE	16	17	20	27	15	17	21	24
A320-200	MTOW	39	40	45	51	42	45	48	50
	OWE	20	21	22	26	22	23	25	26
B737-500	MTOW	33	35	39	43	38	40	42	43
	OWE	16	16	18	21	18	19	20	21
Bae 146-200	MTOW	22	23	26	29	24	26	27	29
	OWE	11	12	13	15	12	13	14	15
CRJ 700	MTOW	18	19	21	24	21	22	23	24
	OWE	10	10	11	13	11	12	12	13
ATR 72	MTOW	11	12	14	15	13	14	14	15
	OWE	6	6	7	8	7	7	8	8
Dash 8-300	MTOW	9	9	11	12	10	11	11	12
	OWE	5	5	6	7	5	6	6	7
ERJ-145	MTOW	12	13	15	16	14	15	15	16
	OWE	5	6	6	7	6	7	7	7

Source: Transport Canada (2001).

Annex 14.2 Aircraft classification

Category	Aircraft type (selection)
1	B747-400, A340-600
2	B767-300, B777-300, A340-300
3	B757-300, A330-300
4	B737-800, A321-200, A320-200, A319-200
5	B737-500, Fokker 100, A318-100, CRJ 200/700, ERJ 145
6	ATR 42-500, ATR 72-500, Dash 8-400, Bae 146-200, Do 328-110

Annex 14.3 TODR and LDR of selected aircraft

	Dry runway					Wet runway	
	15°C, sea level		24°C, sea level	22°C, 500 metres above sea level		sea level	500 metres above sea level
	TODR	LDR	TODR	TODR	LDR	LDR	LDR
B747-400	3,320	2,130	3,619	4,087	2,379	2,450	2,735
A340-300	2,765	1,830	3,014	3,404	2,044	2,105	2,350
B757-300	2,550	1,748	2,780	3,139	1,952	2,010	2,245
B737-800	2,315	1,600	2,523	2,850	1,787	1,840	2,055
A330-300	2,270	1,660	2,474	2,795	1,854	1,909	2,132
A320-200	2,190	1,440	2,387	2,696	1,608	1,656	1,849
CRJ 700	1,565	1,509	1,706	1,927	1,685	1,735	1,938
ERJ-145	1,550	1,290	1,690	1,908	1,441	1,484	1,657
B737-500	1,518	1,362	1,655	1,869	1,521	1,566	1,749
ATR 72-500	1,223	1,048	1,333	1,506	1,170	1,205	1,346
ATR 42-500	1,165	1,030	1,270	1,434	1,150	1,185	1,323

Source: Own compilation based on Fraport AG (2003) and aircraft manufacturer data.

Annex 14.4 Airport classification concerning traffic structure

WG:	Overall number of flights
Cat	Thresholds
1	At least 2,000 flights per week, at least 150 destinations
2	At least 1,000 flights per week, at least 75 destinations
3	At least 400 flights per week, at least 50 destinations
4	At least 100 flights per week, at least 15 destinations
5	At least 14 flights per week, at least 3 destinations
6	Less
BU:	**Importance of holiday traffic on the airport**
Cat	Thresholds
1	At least 60% share of weekly flights
2	At least 40% share of weekly flights
3	At least 15% share of weekly flights
4	Less
BLC:	**Importance of low-cost traffic on the airport**
Cat	Thresholds
1	At least 60% share of weekly flights
2	At least 40% share of weekly flights
3	At least 15% share of weekly flights
4	Less

Annex 14.5 Current and potential runway usage on German and foreign airports

Airport	AF	FP	Airport	AF	FP
German airports			PAD	4	4
AGB	6*	6 a)	RLG	4	4
AOC	5	5	SCN	5	5
BRE	4	4	SGE	–	5
BWE	-	5	STR	3	2
CGN	3	1	SXF	4	2
DRS	4	4	SZW	–	3
DTM	4*)	4*	THF	5	5
DUS	3	2	TXL	3	2
ERF	4	4	ZQW	4	3
ETSI	–	3	**Foreign airports**		
FDH	4	4	AMS	–	1
FKB	4	4	BLL	–	2
FMO	4	4	BRU	–	1
FNB	5	4	BSL	–	2
FRA	1	1	CRL	–	4
GWT	6	5	EIN	–	2
HAJ	3	1	ENS	–	2
HAM	3	1	ETZ	–	4
HDF	6	4	INN	–	5
HHN	4	3	LGG	–	2
HOQ	6	6	LNZ	–	3
KEL	6	6	LUX	–	1
LBC	5	5	MST	–	4
LEJ	3	1	PRG	–	2
MGL	6	6	RTM	–	4
MHG	6	6	SXB	–	4
MUC	1	1	SZG	–	4
NDZ	–	4	SZZ	–	4
NRN	4	4	ZRH	–	1
NUE	4	4			

Notes: Only AF-values for base airports are shown. * Restrictions on maximum permissible tyre pressure.

References

Ashford, N. and Bencheman, M. (1987): Disaggregate behavioural airport choice models, n.p.
Bain, J.S. (1941): Profit rate as a measure of monopoly power, *The Quarterly Journal of Economics*, Vol. 55, No. 2, pp. 271–94.
Bain, J.S. (1959): *Industrial Organization*, New York.
Berliner Flughäfen (2004): Airport Berlin Brandenburg International BBI, Berlin.
Brunekreeft, G. and Neuscheler, T. (2003): Preisregulierung von Flughäfen, in: Knieps, G. and Brunekreeft, G. (eds), *Zwischen Regulierung und Wettbewerb, Netzsektoren in Deutschland*, 2. Auflage, Heidelberg, pp. 251–80.
Cohas, F.J., Belobaba, P.P. and Simpson, R.W. (1995): Competitve fare and frequency effects in airport market share modeling, in: *Journal of Air Transport Management*, Vol. 2, No. 1, pp. 33–45.
Denstadli, J.M. (2004): Impacts of videoconferencing on business travel: the Norwegian experience, in: *Journal of Air Transport Management*, Vol. 10, No. 6, pp. 371–6.
Deutsche Flugsicherung GmbH (DFS) (2004): *Studie zur Kapazität an den Flughäfen mit DFS-Flugplatzkontrolle*, Frankfurt am Main.
Doganis, R. and Graham, A. (1995): 'The economic performance of European Airports', *Research Report 3*, Department of Air Transport, Cranfield University.
Elsasser, H. and Rangosch-du Moulin, S. (1997): Verkehrsreduktion durch Telekommunikation? in: Egli, H.-R. et al. (ed.), *Spuren, Wege und Verkehr*, Festschrift für Klaus Aerni zum Abschied vom geographischen Institut, Bern, pp. 156–69.
Ewers, H.-J. et al. (2001): *Möglichkeiten der besseren Nutzung von Zeitnischen auf Flughäfen [Slots] in Deutschland und der EU, Ein praxisorientierter Ansatz*, Berlin.
Forsyth, P. (1997): Price regulation of airports: principles with Australian applications, in: *Transportation Research E*, Vol. 33, No. 4, pp. 297–309.
Forsyth, P. (2001): Airport Price Regulation: Rationales, Issues and Directions for Reform, Submission to the Productivity Commission Inquiry: Price Regulation of Airport Services, Clayton.
Fraport AG (2003): Flughafen Frankfurt-Hahn, Studie zur Verlängerung der Start- und Landebahn 03/21 – Aktualisierung der Studie von 1998, Frankfurt am Main.
Gillen, D., Henriksson, L. and Morrison, B. (2001): Airport Financing, Costing, Pricing and Performance, Research conducted for the Canada Transportation Act Review, www.reviewcta-examenltc.gc.ca/CTAReview/CTAReview/english/reports/gillen.pdf, access: 4 September 2004.
Gillen, D. and Morrison, W.G. (2001): *Airport Regulation, Airline Competition and Canada's Airport Policy*, Waterloo.
Gillen, D., Morrisson, W.G. and Stewart, C. (2003): Air Travel Demand Elasticities: Concepts, Issues and Measurement, Department of Finance Canada, http://www.fin.gc.ca/consultresp/Airtravel/airtravStdy_e.html, access: 25 October 2004.
Handelsblatt (2004): Kabinett billigt Eichels Rekordverschuldung, Issue No. 195, p. 4.
Harvey, G. (1987): Airport choice in a multiple airport region, in: *Transportation Research A*, Vol. 21, pp. 439–49.
Herfindahl, O.C. (1950): Concentration in the US Steel Industry, PhD thesis, Columbia University, New York.
Hess, S. and Polak, J.W. (2006): Exploring the potential for cross-nesting structures in airport-choice analysis: A case-study of the Greater London area, in: *Transportation Research E*, Vol. 42, No. 2, pp. 63–81.

Hess, S. and Polak, J.W. (2003): Development and application of a model for airport competition in multi-airport regions, paper presented at the annual conference of the Universities Transport Studies Group, Newcastle, January 2004, http://www.cts.cv.imperial.ac.uk/StaffPages/StephaneHess/papers/Hess_and_Polak_airport_choice.pdf, access: 1 March 2004.

Hughes, D. (1995): Desktop video conferencing may offset some air travel, in: *Aviation Week & Space Technology*, 17 July, pp. 37–8.

Hüschelrath, K. (1998): *Infrastrukturengpässe im Verkehr, Die Vergabe von Start- und Landerechten an Flughäfen*, Wiesbaden.

ICAO (1993): Manual of the ICAO Standard Atmosphere (extended to 80 kilometres (262 500 feet)), Doc 7488, 3rd edition.

Jung, C. (1999): Luftverkehrsmärkte im Europäischen Wirtschaftsraum – Staatsverträge, Deregulierung und 'Open Skies', in: Immenga, U., Schwintowski, H.-P. and Weitbrecht, A. (eds), *Airlines und Flughäfen: Liberalisierung und Privatisierung im Luftverkehr*, Baden-Baden, pp. 11–62.

Kanafani, A., Gosling, G. and Thaghave, S. (1975): Studies in the demand for short-haul air transportation, Special report 127, Institute of Transportation and Traffic Engineering, University of California, Berkeley.

Klophaus, R. and Schaper, T. (2004): Was ist ein Low Cost Airport?, Ergebnisse einer vergleichenden Fluggastbefragung an den Flughäfen Frankfurt-Hahn und Bremen, in: *Internationales Verkehrswesen*, Vol. 56, No. 5, pp. 191–6.

Kunz, M. (1999): Airport Regulation: The Policy Framework, in: Pfähler, W., Niemeier, Hans-Martin and Mayer, O. G. (eds), *Airports and Air Traffic Regulation, Privatisation and Competition*, Frankfurt am Main, pp. 11–57.

Lelieur, I. (2003): *Law and Policy of Substantial Ownership and Effective Control of Airlines*, Aldershot.

Lerner, A.P. (1934): The concept of monopoly and the measurement of monopoly power, in: *Review of Economic Studies*, Vol. 1; pp. 157–75.

Malina, R. (2005a): Market power and the need for regulation in the German airport market, Diskussionspapier Nr. 10, Institut für Verkehrswissenschaft, Münster.

Malina, R. (2005b): Potenziale des Flughafenwettbewerbs und staatlicher Regulierungsbedarf von Flughäfen, Dissertation, Münster.

Mandel, B.N. (1999a): The interdependency of airport choice and travel demand, in: Gaudry, M. and Mayes, R.R. (eds), *Taking Stock of Air Liberalization*, Boston, Dordrecht, London, pp. 189–222.

Mandel, B.N. (1999b): Measuring Competiton in Air Transport, in: Pfähler, W., Niemeier, H.-M. and Mayer, O.G. (eds), *Airports and Air Traffic Regulation, Privatisation and Competition*, Frankfurt am Main, pp. 71–92.

Mason, K.J. (2002): Pricing Strategies of Low Cost Airlines, in: ATRS (ed.), Papers of the 6th Air Transport Research Society Conference, Seattle, 14–16 July 2002, CD-ROM.

Monopolies and Mergers Commission (1987): Manchester Airport PLC: A Report on the Economic Regulation of the Airport, London.

Moreno, M.B. and Muller, C. (2003): Airport Choice in Sao Paulo Metropolitan Area: An Application of the Conditional Logit Model, in: ATRS (ed.), Papers of the 7th Air Transport Research Society World Conference, Toulouse Business School, 10–12 July 2003, CD-ROM.

Moring, A. and Zamponi, R. (2005): Große Pläne für Lübecks Flughafen, in: *Hamburger Abendblatt*, No. 87, p. 22.

Niemeier, H.-M. (2002): Regulation of airports: the case of Hamburg Airport – a view from the perspective of regional policy, in: *Journal of Air Transport Management*, Vol. 8, pp. 37–48.

Pels, E. (2000): *Airport Economics and Policy: Efficiency, Competition, and Interaction with Airlines*, Amsterdam.

Pels, E., Nijkamp, P. and Rietveld, P. (2000): Airport and airline competition for passengers departing from a large metropolitan area, in: *Journal of Urban Economics*, Vol. 48, pp. 29–45.

Pels, E., Nijkamp, P. and Rietveld, P. (2003): Access to and competition between airports: a case study for the San Francisco Bay area, in: *Transportation Research Part A*, Vol. 37, No. 1, pp. 71–83.

Productivity Commission (2002): Price regulation of airport services, Report No. 19, Canberra.

Salazar de la Cruz, F. (1999): A DEA approach to the airport production function, in: *International Journal of Transport Economics*, Vol. XXVI, No. 2, S. 255–70.

Siegmund, H. (2004): Down under in Schleswig-Holstein, in: *Aero International*, No. 2, p. 35.

Sinha, D. (2001): *Deregulation and Liberalisation of the Airline Industry, Asia, Europe, North America and Oceania*, Aldershot.

Starkie, D. (2001): Reforming UK airport regulation, in: *Journal of Transport Economics and Policy*, Vol. 35, No. 1, pp 119–35.

Stephenson, F.J. and Bender, A.R. (1996): Watershed: The Future of U.S. Business Travel, in: *Transportation Journal*, Vol. 35, No. 3, pp. 14–32.

The Boeing Company (1998): Precise methods for estimating pavement classification number, Document No. D6-82203, www.boeing.com/assocproducts/ aircompat/faqs/estimatingpcn.pdf, access: 25 July 2004.

Transport Canada (2001): *Aircraft Classification Numbers (ACNs)*, Ottawa.

Tretheway, M. (2001): Airport Ownership, Management and Price Regulation, Report prepared for The Canada Transportation Act Review Panel, 14 March 2001.

Urbatzka, E. and Wilken, D. (2003): Future Airport Capacity Utilsation in Germany: Peaked Congestion and/or Idle Capacity?, in: ATRS (ed.), Papers of the 7th Air Transport Research Society Conference, Toulouse Business School, 10–12 July 2003, CD-ROM.

Valentinelli, N., Liefner, I. and Brandt, O. (2004): Fluggastpotenzial des Flughafens Hannover-Langenhagen, *Hannoversche Geographische Arbeitsmaterialien*, No. 29, Hanover.

Wendlik, H. (1995): *Infrastrukturpolitik im europäischen Luftverkehr unter den Bedingungen des wachsenden Marktes*, Ostfildern-Kemnat.

Windle, R. and Dresner, M. (1995): Airport Choice in Multiple-Airport Regions, in: *Journal of Transportation Engineering*, Vol. 21, No. 4, pp. 332–7.

Wolf, H. (2003): *Privatisierung im Flughafensektor, Eine ordnungspolitische Analyse*, Berlin.

Chapter 15
Competition among Airports and Overlapping Catchment Areas: An Application to the State of Baden-Württemberg

Daniel Strobach

Introduction

The Federal Republic of Germany has one of the densest airport networks in Europe. Against the background of emerging competition within the European airport industry, this means new challenges for the future of German airports. First steps have been done, as the privatization of a few airports shows. But this will not be the only change to come.

Low-cost carriers (LCCs) are emerging in Germany, and this will be a chance for smaller airports with spare capacity. Another way of creating new business is to focus on freight services. The airport of Leipzig is developing as a hub for DHL, a subsidiary of Deutsche Post. This strategy is possible especially for airports that have no operational constraints during the night-time. Here, smaller airports could even compete with large hub airports such as Frankfurt, which are facing operational and capacity constraints.

Competition with hubs is only feasible for regional airports when it comes to certain businesses areas. Regarding the number of services and destinations offered, competition among hubs and regional airports differs. Hub airports are concerned with a wider catchment area. They are competing on a European level. On a regional level, competitive pressure is rising because many former military airfields are waiting to be converted into civil airports. In some cases regional governments are trying to promote these new airports without taking a closer look at costs, existing competitive forces and a modest future for their new airport.

This chapter discusses the meaning of competition among airports on a regional basis. It creates a picture of the current situation of competition for the southwest part of Germany, in particular the state of Baden-Württemberg.

The starting point for considerations is the nature of competition in two-dimensional space. The factors that influence passengers the most when deciding on an airport flow into a ranking system to describe airport significance. This construction is used to compare attraction and competitive strength of different airports in two-dimensional space, and to map the results.

The Development of Catchment Areas

Catchment Areas and Competition

All the classical models[1] employ quite restrictive conditions, which seems very much an ideal case in comparison to the geography and topography of the real world. Proceeding on the assumption that the surface is a plane, where travelling in all directions is possible under unique cost and time conditions, this leads to a distinction between the market areas of different firms. The theoretical result is a network of strictly divided monopolistic areas. But topography causes the emergence of a traffic axis, which means that different travel directions are connected with varying cost and time components. A strict division of market areas and the existence of regularly shaped networks are not likely under these premises.

Catchment areas, as a synonym for market areas, play a vital role in the airport business. The term describes a geographical space, within which the probability of selection is so high that the majority of potential passengers living in the region decide on this particular airport. In other words, an airport gets the bulk of its traffic volume out of this area.

Catchment areas are not static structures. They can vary because of changes in determinant factors. What these factors are is the subject of the next section. Also, catchment areas are overlapping.[2] The definition via the probability of choosing a particular airport shows that alternatives are imposed with positive probabilities as well. A point in space could rarely be assigned to one airport solely.

Competition among airports is based on catchment areas. Irregularity in shape is typical, and overlapping is vital in this process. Location plays an important role in competition, but airports are immobile. Once a location has been chosen, this strategic decision cannot be changed. If a firm cannot be shifted, it has to rely on the flexibility of its customers. Thus passengers' recognition and choice of an airport determine its catchment area, market share, revenue and profit.

This chapter provides a method estimating the size and shape of overlapping catchment areas, based on freely available data. Proceeding from different locations in space, the attractions of a certain number of competing airports are determined and compared with each other. The process is a substitute for the appointment of concrete choice probability when appropriate data is lacking. It also has the advantage of being easily reiterated and compared over time.

Relevant Factors

Certain factors make up an airport's recognition and cause the existence of a positive choice probability. These factors determine to which extent passengers choose a specific airport and, therefore, strongly influence the dimension and shape of catchment areas, as well as competition among airports.

For the discovery and description of factors that are important for a passenger's choice, Multinomial Logit (MNL) or Nested MNL (NMNL) can be used.[3] Because a broad database is necessary for such investigations, many examinations rely on a survey covering the San Francisco Bay area in 1995.[4] Within the scope of most of these investigations, the quality of airport access,

---

1   See, for example, Thünen (1826), p. 2ff; Lösch (1962), p. 71ff.
2   See Zeike (2003), p. 71f.
3   An overview on the different models is in Bondzio (1996), p. 11ff.
4   Study on the San Francisco Bay area: Basar and Bhat (2004), p. 895ff; Hess and Polak (2004), p. 4ff; Pels, Nijkamp and Rietveld (1998), p. 6ff. Study on the extension of Sheffield Airport: Thompson and Caves (1993), p. 138ff. Study on airport access in southern Germany: Bondzio (1996), p. 42ff.

offered frequency of flights and price are emphasized as the most important factors.[5] When analysing an airport, these and additional factors are the base of competition. Passengers are influenced by these parameters, choose a particular airport on this basis and thus determine market share. The most common factors are outlined below.[6]

Airport access quality This factor can be described by values of time, distance and cost. While it is relatively easy to transform distance into monetary dimensions, this is much more controversial when it comes to the valuation of time. Usually the ways of reaching an airport are divided into private and public transport.

Access time is an important, and maybe the central, factor in airport competition. It can be shown that access time reaches higher elasticities concerning choice probability and market share than flight frequency.[7] The importance of access time may vary according to flight length. On short-haul routes, access time acquires a higher weight in relation to total travel time than it does on long-haul or intercontinental routes. Thus it is a more important criterion on short- or medium-haul routes, for instance national or European traffic, than on intercontinental routes.[8]

One explanation for these high values of access time could be risk minimization. Long access times and distances increase the risk of delays and missed flights.

Frequency Airports offer their passengers a variety of flights and destinations. Besides the number of destinations reachable via direct connections, the frequency of flights is the decisive parameter. Flights are originally supplied by airlines. It is their competition parameter to reach or maintain market share on specific routes or airports. Airlines have an incentive to compete by offering a dense network of services, which is of advantage to the airport. These try to attract airlines in order to proffer an attractive supply. From an airport's point of view, an optimal timetable contains a sufficient number of flights. At minimum a pair of flights should be offered per day. Flight frequency has a decreasing marginal utility. Nine flights per day can be seen as a kind of saturation point. As access time and frequency are the most important factors, the tendency is that if two airports offer a similar number of flights, travellers choose on the basis of other factors, most likely the nearest airport.[9]

Air fare Besides the two parameters above, air fare seems to be very important, although not all studies verify this assumption.[10] Air fare causes severe problems in data collection, which is why many authors do not include it in their studies. Airlines offer different service categories and use for price differentiation. Ticket prices may vary with time, capacity and category. Thus accuracy in data collection is hard to achieve.

Experience This is a subjective element in airport choice. Positive experiences with a specific airport increase the probability that the passenger will choose this airport again. But how to measure experience is difficult.

---

5   See Harvey (1987), p. 443ff; Cohas, Belobaba and Simpson (1995), p. 41ff; Valentinelli, Liefner and Brandt (2004), p. 57ff and the sources mentioned in the previous footnote.
6   Essential to this topic are the sources contained in the preceding two footnotes.
7   See Basar and Bhat (2004), p. 899f.
8   See Harvey (1987), p. 448.
9   See Cohas, Belobaba and Simpson (1995), pp. 34, 39; Harvey (1987), p. 446ff; Hess and Polak (2005), p. 63f; Thompson and Caves (1993), p. 143, Zeike (2003), p. 78f.
10   Pro: Thompson and Caves (1993), p. 139, 145; contra: Pels, Nijkamp and Rietveld (2003), p. 79.

Other factors include:

- Tax: Total air fare includes charges and taxes, which may differ between airports.
- Type of aircraft: Despite the efficiency of turboprop aircraft on short-haul routes, many passengers prefer jets to propeller-driven planes.[11]
- Aircraft size: Larger aircraft are linked with greater comfort.
- Delays and punctuality: These points include inconvenience and the risk of missing a flight.

A common distinction separates business and leisure travel. In general, business travellers are more time-sensitive and leisure travellers more cost-sensitive. Business travellers are more likely to accept higher fares in exchange for higher frequencies.[12] Local residents have a broader base of knowledge and experience. They react sensitively to changes in access quality and air fare. Visitors mainly focus on access time.[13]

Description of Competition among Airports

Outlining Goals and Course of Action

Passengers choose their preferred airport based on the attributes described above. They thus influence shape and size of an airport's catchment area, and its market share. Passengers' preferences establish the foundation of competition among airports. These preferences, more exactly the attributes and factors they are based on, are used in the following sections to map competition among airports. This is done in a general form, without falling back on data provided by interviews, using freely available data material. It would be complicated and a long-term process to gather sufficient data by interviews on a large number of airports covering a wider geographic area.

Another advantage of the chosen design, besides its easier handling, is that it allows repetition at any time if the basic facts change, and an analysis of other regions. Thus the results of different mappings are comparable in a spatial and temporal way.

The current study covers Baden-Württemberg. To develop a spatial reference, the state of Baden-Württemberg is subdivided into cells based on existing administrative districts. Within these, the administrative centre of a district usually reflects the centre of economy and population and, therefore, serves as a reference point for the determination of access quality. In special cases, the reference point or divison of districts can vary, if necessary.[14]

A problem concerning the transfer of information relating to one specific point in space automatically arises: the finer the subdivision, the more accurate the information is. But improved accuracy implies higher expenditures. For deeper insights, the administrative districts should be subdivided further.

A ranking method is used to rate all values of a specific factor. The score ranges from 9 for the best to 1 for the worst performer. These score values are weighted and accumulated. At the end, this method leads to one value per cell and airport. This can be interpreted as a measure of an airport's

---

11 For Innes and Doucet (1990), p. 514f this factor is of high relevance.
12 See Hess and Polak (2005), p. 65f; Pels, Nijkamp and Rietveld (2003), p. 79.
13 See Harvey (1987), p. 440; Hess and Polak (2005), p. 66.
14 For a closer look on the division, see Table 15.4 and Figures 15.1 and 15.2.

attraction concerning a particular cell. Conclusions about competition among airports and their catchment areas can be derived from comparing the results.

The factors used in the present study are grouped into a cell-specific and an airport-specific part. The cell-specific part measures the connection between an airport (as a point in space) and a geographically differentiated area. This is done on the basis of attributes concerning access quality.

Airport-specific factors assess the airport itself. On one side is traffic supply, with flight frequency as the most important attribute. Connected to this point are factors which describe the conducting of traffic. On the other side are attributes concerning the convenience of an airport. They measure how stress-free and 'lazy' an airport's design is for travellers. This point is closely related to the question of how positive the traveller's experience is. Such experiences enhance the possibility of reusing a specific airport. At the end, the evaluation of an airport as a point in space is combined with the assessment of airport access. Together, the results have a multidimensional shape, reflecting competition in space.

Passengers' preferences and an airport's attractiveness for travellers are useful indicators of competition, but are not a perfect measure. Ideally, the results should be compared with data reflecting traffic streams on a regional base (that is, passenger's origin at selected airports, as well as the alternatives travellers of a specific region are currently choosing from). The latter information is difficult to obtain, and a detailed verification on the basis of such data is an important topic of further research. The present study is a first approximation to the spatial dimension of airport competition, not as an exact measure, while keeping in mind the overall goals mentioned above.

Selection of Attributes

The following section is an overview on selected factors employed in the present study. Airport access can be measured in quantities of time, distance and cost. In the case of private transport, distance can be translated into cost using a linear function. However, this is difficult because cost of fuel and the use of a particular type of car have to be determined. Therefore, only distance and time are calculated. Cost is not considered. Because of the linear relationship this does not cause any distortion. It is assumed that a passenger leaves a train station at a reference point at eight o'clock in the morning and prefers quick routing. Regarding public transport, ticket prices (regular tariffs) and travel times are calculated. The passenger leaves the mentioned train station around the same time with the fastest connection possible. If an airport has no mainline station, connecting services are added and a transfer time of 10 minutes is included.[15]

One problem in this context concerns the perception of costs by the customer, and the emergence of low-cost carriers. In many cases, travellers tend to evaluate only out of the pocket instead of total cost. This leads to distortion in airport choice, which cannot be modelled. Modelling and mapping have to rely on the ideal of an economically efficient decider.

Flight frequency is the most important factor within the bounds of airport-specific attributes. With overlapping catchment areas, direct connections (point to point) are of main interest. To equalize the conditions for all airports under view, the present study concentrates on destinations in Germany and Europe (Table 15.1).

---

15 Private transport elements are calculated using an actual commercially available digital roadmap. The website of Deutsche Bahn AG (http://www.bahn.de) and municipal transport services are used to calculate access via public transport. Thursday 25 August 2005 is the fixed date.

Table 15.1 Worldwide arrivals at selected European airports, 2004

Destination/Airport		Code	Passengers
Germany			
Berlin	Schönefeld	SXF	3.382.166
	Tegel	TXL	11.048.000
	Tempelhof	THF	441.580
Düsseldorf		DUS	15.256.500
Hamburg		HAM	9.893.700
Hanover		HAJ	5.249.169
Europe			
Amsterdam		AMS	42.541.200
Barcelona		BCN	24.549.600
Brussels		BRU	15.583.700
Budapest		BUD	6.444.700
Graz		GRZ	898.504
Istanbul	Atatürk International	IST	n/a
	Sabiha Gokcen	SAW	n/a
Copenhagen		CPH	18.965.700
London	City	LCY	n/a
	Gatwick	LGW	31.461.500
	Heathrow	LHR	67.344.000
	Luton	LTN	7.500.000
	Stansted	STN	20.908.100
Milan	Bergamo	BGY	3.334.161
	Linate	LIN	8.947.900
	Malpensa	MXP	18.554.000
Palma de Mallorca		PMI	20.410.900
Paris	Orly	ORY	24.032.200
	Charles de Gaulle	CDG	50.860.600
Prague		PRG	9.696.413
Wien		VIE	14.785.500

Source: ACI (2005); websites of the subject airports.

The most frequently served destinations from the centrally located airport of Stuttgart[16] constitute the starting point. The inclusion of the main administrative and economical centres, as well as popular destinations of low-cost carriers, guarantee a balanced choice. Principal destinations of package holidays are excluded. Only northern German cities, where flying is a viable alternative when travelling from the southern part of the country, are included in order to minimize intermodal competition.

As a criterion for flight frequency, the number of flights per week is measured for the summer period 2005.[17] Two flights per weekday is the minimum and nine flights per day is the saturation point. With just six connections per day at the weekend, this leads to a saturation level of 57 flights per week. Fewer than 14 flights per week have the lowest score (1), while 57 flights or more have the highest (9). The range in between is divided evenly, and the remaining scores from 2 to 8 are attached accordingly.

Additional factors concern the handling of traffic. Minimum connecting time (MCT) and the number of gates and check-in-counters describe the ability to process passengers without delays and to ensure a stable process control. MCT defines the time range in which an airport guarantees the transfer of passengers and luggage from one flight to another. The lower the probability of a negative experience for passengers, the more likely they are to make the same choice again.

A second part of the valuation of an airport's quality focuses on convenience for passengers. This also relates to the positive experience passengers should get. Most passengers still reach an airport by car.[18] Price for and number of parking lots may be good indicators to judge whether convenient conditions exist. Although these factors do not concern all passengers, they affect the majority of accessing travellers. Concerning costs, the cheapest price per day and week were measured. The number of parking spaces was compared to the number of passengers in order to make airports comparable. An attachment to the convenience factors in a wider meaning was chosen because of the 'long-term' nature of parking costs. Access is complete when reaching the target, namely the airport. Parking costs are due on return from a certain journey. They reflect how expensive or convenient it is to leave a car at the airport. In contrast, access in a 'hop-on' or 'drop-off' way is usually free of charge.

The same was done with the size of the terminal and the shopping area, as well as with the number of gates and check-in counters. Narrow terminals can be a reflection of insufficient capacity. They may cause delays, stress and a negative experience. Terminal amenities such as shopping, catering and similar services ensure a comfortable and enjoyable stay. In addition, they give an airport the possibility to gain independence from aeronautical businesses. Profits from the non-aeronautical field may enable the airport to offer better conditions on the aeronautical side and to strengthen its competitive position.

The weights used in the study take their pattern from known schemes of travel choice. They are orientated on examinations of passenger surveys over the last two decades.[19] All high-weighted factors were shown as important in these past surveys. Airport access quality and frequency of

---

16   See Flughafen Stuttgart GmbH (2005), p. 28.

17   Timetables published (via the internet) by the airports in question make up the source for the survey. A flight was taken into consideration if it was offered for more than half of the period.

18   See, for instance, data concerning Munich Airport in 2004 (published on the internet homepage of the airport): Car (47 per cent), S-Bahn (31 per cent), taxi (10 per cent), bus (6 per cent), rental car (6 per cent). Data for Hannover: Valentinelli, Liefner and Brandt (2004), pp. 47f. Older data for southern Germany: Bondzio (1996), pp. 49ff. The relationship between public and private transport is also reflected in the construction of weights for measuring access quality (Table 15.2).

19   See, in this context, footnotes 4 and 5.

service, for example, are by far the most important factors in the current study, as shown by investigations of the motives of travellers' airport choice.

As said before, this study has no underlying passenger interviews. Therefore, it has to rely on earlier measurements to imply realistic weights. Even if the chosen weights may seem arbitrary, varying them in a way that takes into account earlier studies of a traveller's choice does not affect the ranking itself. A concertina effect can be observed, changing the value and intervals in between them but leaving the arrangement itself in line. However, the focus of this study of airport competition lies exactly on the ranking. Table 15.2 summarizes the structure and weights of attributes influencing the study.

All airports offering a certain measure of size and traffic are taken into consideration as competitors for passengers living in Baden-Württemberg. A condition is that the airports had to serve at least four of the selected destinations during the summer season 2005. Table 15.3 gives an overview of the included airports.

Table 15.2   Share of individual factors

Factors	Share of final results (%)		
Access quality			50
Private transport		27,5	
Distance	13,75		
Time	13,75		
Public transport		22,5	
Expense	11,25		
Time	11,25		
Airport quality			50
Traffic supply and handling			42,5
Frequency		36,125	
Minimum connecting time		2,125	
Gates		2,125	
Check-in		2,125	
Convenience			7,5
Parking		2,55	
Expense per week	0,765		
Expense per day	0,765		
Total supply of parking	1,02		
Terminal area		2,475	
Area for shopping and services		2,475	

Source: D. Strobach.

Table 15.3   Airports included in the study

Airport	Code	Passengers	Freight (to)	Aircraft movements	Ownership structure
Basel	BSL	2.545.687	88.312	57.915	50% République francaise 50% Schweizerische Eidgenossenschaft
Frankfurt	FRA	51.098.300	1.750.996	477.475	31,9% Land Hessen 20,4% Stadt Frankfurt 18,3% Bundesrepublik Deutschland 29,4% Free Float
Friedrichs-hafen	FDH	500.892	–	7.884	25% Stadt Friedrichshafen 24,75% Bodenseekreis 16,14% ZF Friedrichs-hafen AG 13,23% Luftschiffbau-Zeppelin GmbH 20,88% other shareholders
Innsbruck	INN	728.138	3.957	39.377	49% Innsbrucker Kommunalbetriebe 25,5% Land Tirol 25,5% Stadt Innsbruck
Munich	MUC	26.814.500	309.828	359.568	51% Freistaat Bayern 26% Bundesrepublik Deutschland 23% Stadt München
Nürnberg	NUE	3.648.580	71.000	71.818	50% Freistaat Bayern 50% Stadt Nürnberg
Strasbourg	SXB	1.942.296	–	–	Under the management of CCI Strasbourg & Bas-Rhin
Stuttgart	STR	8.821.533	18.227	125.220	50% Land Baden-Württemberg 50% Stadt Stuttgart
Zurich	ZRH	17.214.500	363.537	231.086	46,76% Kanton Zürich 5,4% Stadt Zürich 47,84% Free Float

Note: Data reflects the year 2004.
Source: Websites and company reports of the studied airports, ACI.

Results

Table 15.4 presents the results of the study. The numbers represent scores valuing an airport's quality of supply and access from a particular point in space. Based on these scores, a ranking for all cells can be established.

Table 15.4 Resulting scores

Reference point	Airport								
	BSL	FRA	FDH	INN	MUC	NUE	SXB	STR	ZRH
Aalen	3,0486	5,5193	5,0759	1,9218	6,4807	5,1408	3,3286	6,8431	4,7576
Baden-Baden	5,0736	6,8818	3,5759	1,7843	4,4557	3,0033	5,6911	6,2306	5,3076
Balingen	4,5736	5,4943	5,2134	1,7843	5,5182	2,3658	4,1286	6,8431	6,4201
Biberach an der Riß	3,4986	4,6943	5,8759	2,8593	6,6182	3,6158	2,8286	6,4556	5,8951
Böblingen	3,1861	6,1568	4,6009	1,7843	5,0932	3,7033	4,6286	6,9556	6,0326
Calw	3,9361	6,6568	4,2384	1,7843	4,4557	3,1408	5,1911	6,9556	5,7826
Emmendingen	6,4611	5,8568	3,7634	1,7843	4,4557	2,7533	5,1911	5,3181	6,4201
Esslingen	3,0736	6,2943	4,7134	1,7843	5,9307	3,5908	4,8536	6,9556	5,0326
Freiburg i. Breisgau	6,5736	5,7193	4,1509	1,7843	4,4557	2,7533	5,3036	5,2056	6,4201
Freudenstadt	5,1236	5,6068	4,1259	1,7843	4,5932	2,9783	5,3286	6,8181	6,1201
Friedrichshafen	5,2986	4,2193	5,8759	3,5843	5,5682	2,4783	2,4661	5,7056	7,0326
Göppingen	3,1861	5,6318	5,1009	1,7843	6,5682	3,7783	4,5786	6,9556	4,6451
Heidelberg	4,5986	7,7693	3,6884	1,7843	4,5682	3,7783	4,8036	6,4556	4,5576
Heidenheim an der Brenz	2,6861	5,5443	5,1884	2,2843	6,5932	5,1408	3,1911	6,8431	4,8701
Heilbronn	4,0736	7,2693	3,8259	1,7843	4,5682	4,6408	4,4161	6,9556	4,9451
Karlsruhe	5,0736	6,9943	3,6009	1,7843	4,5682	3,4158	5,1911	6,8431	4,6701
Künzelsau	3,1861	6,4318	3,9384	1,7843	5,7307	5,3658	3,9661	6,8431	4,8701
Lörrach	6,5736	5,3818	4,4634	2,6718	4,2057	2,2283	4,4661	4,9806	7,0326
Ludwigsburg	3,3236	6,5443	4,3009	1,7843	5,0682	4,3658	4,7411	6,9556	5,0576
Mannheim	4,8236	7,7693	3,5759	2,0593	4,7932	3,5783	4,8036	6,4556	4,6701
Mosbach	4,4611	7,3818	3,4634	2,0093	4,5682	4,4783	4,4661	6,8431	4,4451
Offenburg	5,8486	6,2693	3,9384	1,7843	4,4557	2,7533	5,8036	5,7306	5,4201
Pforzheim	4,8486	6,8818	3,6009	1,7843	4,8182	3,5908	5,1911	6,9556	4,5826
Ravensburg	4,4111	4,3318	5,8759	3,3343	6,1182	2,7283	2,5786	5,9806	6,8951
Reutlingen	3,0736	6,1568	4,9884	1,7843	5,6807	3,4283	4,2661	6,9556	5,7826
Rottweil	4,5736	5,3818	5,2384	1,7843	4,5932	2,9783	4,2161	6,9556	6,5326
Schwäbisch-Hall	3,0736	6,4318	4,0759	1,7843	5,5932	5,3658	3,9661	6,9556	4,8701
Sigmaringen	4,8236	4,7193	5,8759	2,0343	5,8182	2,7283	3,5786	6,1181	6,7576
Singen (Hohentwiel)	5,2986	4,7693	5,7384	2,4718	4,8182	2,3408	3,8036	5,7306	7,3076
Stuttgart	3,0736	6,5443	4,4384	1,7843	5,0682	4,2283	4,7411	6,9556	5,5576
Tauberbischofsheim	3,3236	7,4068	3,6884	1,7843	5,7307	4,9783	3,9661	6,7306	4,6701
Tübingen	3,7986	6,0443	4,8509	1,7843	5,3182	3,0658	4,1536	6,9556	6,2826
Tuttlingen	4,8236	5,0193	5,6259	2,1968	4,6807	2,6158	3,9411	6,4556	6,7826
Ulm	3,0236	4,8568	5,4884	2,6968	6,8432	3,9158	3,1036	6,8431	5,7076
Villingen-Schwenningen	4,8236	5,1568	5,4884	2,0593	4,4557	2,7283	4,4161	6,5931	6,6451
Waiblingen	3,0736	6,5443	4,0509	1,7843	5,2307	4,4783	4,7411	6,9556	5,2826
Waldshut-Tiengen	6,1861	4,7693	4,8759	2,9468	4,6807	2,0908	4,0536	5,0931	7,4201

Source: D. Strobach.

To analyse the competitive situation, a division into different zones is introduced. Referring to the highest value of a particular cell, two limits are set at 90 per cent and 80 per cent of the maximum value. The choice of these limits should create a sensible relation between potential improvements and competitive positions. Extensive investments ought to allow a leap upwards in the ranking and make an upper zone reachable, reflecting growing competitive strength. Incremental improvements should have no drastic effect on an airport's relative position. The limits are selected according to this goal. It should be noted that, under different circumstances, determination of limiting values could change. This categorization generally aims for further differentiation of the competitive situation.

From a district's point of view the closeness of valuation results or scores indicates how competitive the relationship between two airports is. The closer the values, the more similar are the airports' supplies and attractiveness to passengers, and the more likely is equality to prospective passengers; thus they could be interchangeable.

Area A (90–100 per cent) contains all competitors in close proximity to the leading company; area B (80–90 per cent) all those with a looser connection to the best performing airport. The occupation of these areas, especially the first one, shows how competitive a district is and how many airports are considered as stronger or weaker alternatives by passengers. For example, the district of Karlsruhe has a maximum value of 6,994, achieved by Frankfurt as the best performing airport. Stuttgart Airport reaches a score of 6,293 and can be seen as a strong competitor to Frankfurt. With no other airport reaching similar valuations, competition in the district of Karlsruhe tends to be duopolistic. Both airports, Frankfurt and Stuttgart, are attractive to customers located in Karlsruhe, and they seem to be almost interchangeable for them. Table 15.5 gives an overview on the situation in each district, when using the mechanism described above.

Competition now can be described from an airport's point of view. Therefore, five zones are introduced. Zones 1 and 2 mark the areas where a particular airport gets the highest scores and holds, because of that, the leading role. In doing so, zone 1 describes the situation where no close competitors exist, that is no other scores within area A,. Zone 2 represents the competitive case of leadership. In districts categorized as zone 2, a particular airport gets the highest score, but has to cope with at least one close competitor performing on a similar level. Within districts of zone 3, the airport under review does not occupy a leading position. It is situated in area A, performing like another airport in leading position. The distance between it and the leader grows, when ranked in zone 4. The score of the specific airport now ranges within area B. That means its competitive strength and substitutability with the leader is further weakened. The weakest competitive position is represented by zone 5, where an airport is no longer seen as a strong competitor to the leading company.

In the district of Tuttlingen, for example, Stuttgart Airport (6,4556) has a position as strong competitor to Zurich Airport (6,7826), which occupies the leading position. Friedrichshafen Airport (5,6259) falls into a lower category than Stuttgart. In the eyes of Stuttgart Airport, the district of Tuttlingen belongs to zone 3.

The distinction between different areas and zones aims at forming a more differentiated picture of airport competition. It allows us to bring out overlapping catchment areas. The typical pattern consists of a core occupied by the regional market leader and lacking effective competition, as well as areas of stronger rivalry among different airports. How these differentiations form a picture of an airport's catchment area is shown in Figure 15.1 with the example of Stuttgart Airport. Figure 15.2 assigns all districts to the airport with the highest score. This contains zones 1 and 2, where an airport plays a leading role.

Table 15.5 Leader and competitors

Reference Point	Max. value	Leader	Limit A (90%)	Close competitors	Limit B (80%)	Other competitors
Aalen	6,8431	STR	6,1588	MUC	5,4745	FRA
Baden-Baden	6,88175	FRA	6,1936	STR	5,5054	SXB
Balingen	6,8431	STR	6,1588	ZRH	5,4745	MUC FRA
Biberach an der Riß	6,6182	MUC	5,9564	STR	5,2946	ZRH FDH
Böblingen	6,9556	STR	6,26	–	5,5645	FRA ZRH
Calw	6,9556	STR	6,26	FRA	5,5645	ZRH
Emmendingen	6,4611	BSL	5,815	ZRH FRA	5,1689	STR SXB
Esslingen	6,9556	STR	6,26	FRA	5,5645	MUC
Freiburg im Breisgau	6,5736	BSL	5,9162	ZRH	5,2589	FRA SXB
Freudenstadt	6,8181	STR	6,1363	–	5,4545	ZRH FRA
Friedrichshafen	7,0326	ZRH	6,3293	–	5,6261	FDH STR
Göppingen	6,9556	STR	6,26	MUC	5,5645	FRA
Heidelberg	7,76925	FRA	6,9923	–	6,2154	STR
Heidenheim an der Brenz	6,8431	STR	6,1588	MUC	5,4745	FRA
Heilbronn	7,26925	FRA	6,5423	STR	5,8154	–
Karlsruhe	6,99425	FRA	6,2948	STR	5,5954	–
Künzelsau	6,8431	STR	6,1588	FRA	5,4745	MUC
Lörrach	7,0326	ZRH	6,3293	BSL	5,6261	–
Ludwigsburg	6,9556	STR	6,26	FRA	5,5645	–
Mannheim	7,76925	FRA	6,9923	–	6,2154	STR
Mosbach	7,38175	FRA	6,6436	STR	5,9054	–
Offenburg	6,26925	FRA	5,6423	BSL SXB STR	5,0154	ZRH
Pforzheim	6,9556	STR	6,26	FRA	5,5645	–
Ravensburg	6,8951	ZRH	6,2056	–	5,5161	MUC STR FDH
Reutlingen	6,9556	STR	6,26	–	5,5645	FRA ZRH MUC
Rottweil	6,9556	STR	6,26	ZRH	5,5645	–
Schwäbisch-Hall	6,9556	STR	6,26	FRA	5,5645	MUC
Sigmaringen	6,7576	ZRH	6,0818	STR	5,4061	FDH MUC
Singen (Hohentwiel)	7,3076	ZRH	6,5768	–	5,8461	–
Stuttgart	6,9556	STR	6,26	FRA	5,5645	–
Tauberbischofsheim	7,40675	FRA	6,6661	STR	5,9254	–
Tübingen	6,9556	STR	6,26	ZRH	5,5645	FRA
Tuttlingen	6,7826	ZRH	6,1043	STR	5,4261	FDH
Ulm	6,8432	MUC	6,1589	STR	5,4746	ZRH FDH
Villingen-Schwenningen	6,6451	ZRH	5,9806	STR	5,3161	FDH
Waiblingen	6,9556	STR	6,26	FRA	5,5645	–
Waldshut-Tiengen	7,4201	ZRH	6,6781	–	5,9361	BSL

Source: D. Strobach.

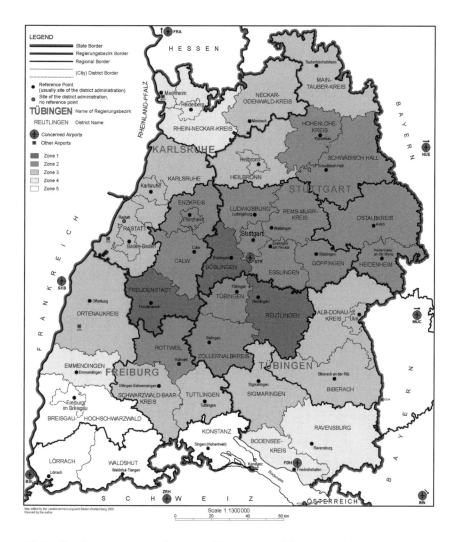

Figure 15.1 Catchment area and competitive zones of Stuttgart Airport
Source: D.Strobach.

When considering the results and the depicted maps, a few things become obvious. One of them is the increased influence of Frankfurt Airport, stretching south along the Rhine Valley. On one hand, the Black Forest creates a kind of natural barrier to the east, with higher travel resistance. But, on the other, the transport infrastructure along the Rhine Valley is excellently developed. Travelling to Frankfurt is becoming more attractive because, despite longer distances, travel times are short. Contrary valuations of access time and cost for Frankfurt Airport in the Rhine Valley districts confirm this assumption.

An existing traffic axis of road or railway infrastructure thus plays a vital role in the moulding of catchment areas. Along these axes, even in more distant regions, the connectivity can be adequate enough so that the higher number of flights and destinations compensate for disadvantages in access quality. Another example is the two districts in the east of Baden-Württemberg, which

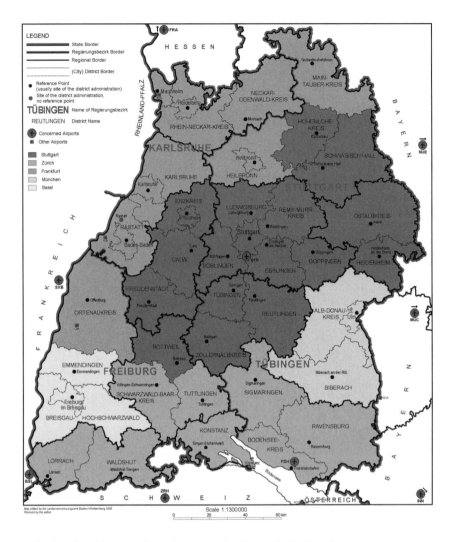

Figure 15.2 Leadership position of airports in Baden-Württemberg

Source: D.Strobach.

belong to the catchment area of Munich (München) Airport. Both have better access to Stuttgart Airport than to Munich. But they are also both very well linked to the highway and/or railway network serving the Bavarian capital. This restricts the disadvantage of poorer access quality in a favourable way, so that it can be compensated by advantages on the airport-specific side.

Therefore, it is no wonder that the most competitive districts are in the Rhine Valley (Emmendingen and Ortenaukreis) and in the east of Baden-Württemberg (Alb-Donau-Kreis and Ulm). These findings support the assumption that competition is most intense in centrally located districts between two airports. Reasons for the two cases are mentioned above.

The Rhine Valley is an especially interesting case because of the high density of competitors. The airports of Basel, Strasbourg and Stuttgart are in close proximity. Furthermore, the more distant hub airports of Frankfurt and Zurich are of great importance. It should be noted that three airports failed to be included into this survey because of their small size (Baden-Airpark Karlsruhe and Mannheim) or limited operating licence (Black Forest Airport Lahr). As potential competitors, they could be of significance in the future.

The lowest intensity of competition appears at the Swiss border (Singen). It is a typical example for the assumption that a district in close proximity to an airport, especially a hub airport, will be 'ruled' by this company in the sense of undisturbed market power (in this case Zurich Airport). Relatively large differences in scores between the leader and followers underline such an outstanding position.

Both assumptions concerning the spatial distribution of competitive forces and market power have to be interpreted as tendencies. According to these tendencies, it must be assumed that Stuttgart airport exerts a similar kind of distinct leading role in the centre of Baden-Württemberg. But the airport has only three districts (Böblingen, Freudenstadt and Reutlingen) assigned to zone 1. Within all other districts in its vicinity, the airport must deal with competitive pressures. Stuttgart airport is in the unlucky situation of facing competition from large hub airports from three directions. Those three main competitors and the natural barrier in the west set the tone for developing a catchment area, in the end. For the hub airports, it is easier to compensate for disadvantages and, therefore, enlarge their catchment area.

Outlook

Concerning the widening of its catchment area, the potential for the development of Stuttgart airport as the central airport of this study seems to be limited. Together with EuroAirport Basel, it fills an intermediate position between hub and regional airports, especially when focusing on airport-specific factors. There is a risk of being stuck in the middle. Yet, even with a difficult starting point, new opportunities are arising. The development of another classical hub does not seem a profitable strategy. The attraction of charter and low-cost airlines could create growth. However, the dangers of too close ties with these types of airlines should be kept in mind. The new congress and exhibition centre (*Landesmesse*) built in the direct neighbourhood of the airport offers the opportunity to generate new traffic. As part of the project 'Stuttgart 21', a new long-distance train station is planned in front of the terminal. The example of Frankfurt airport shows that such investment opens up the opportunity to better exploit more distant catchment areas and strengthen its own competitive position.

Further studies could widen the geographical area examined, focus on several case studies or examine in more depth the construction of weights used in such a study. Another interesting field could be the further development of existing theoretical models, for example including highways or natural barriers – factors of different travel resistance.

References

ACI (Airports Council International), 2005. Airport Traffic Report, December 2004.
Başar, G. and Bhat, C.R., 2004. A parameterized consideration set model for airport choice: an application to the San Francisco Bay Area. *Transportation Research B* 38 (10), 889–904.
Bondzio, L., 1996. Modelle für den Zugang von Passagieren zu Flughäfen. Bochum.
Cohas, F., Belobaba, P. and Simpson, R., 1995. Competitive fare and frequency effects in airport market share modelling. *Journal of Air Transport Management* 2 (1), 33–45.

Flughafen Stuttgart GmbH, 2005. Statistischer Jahresbericht 2004.

Harvey, G., 1987. Airport choice in a multiple airport region. *Transportation Research A* 21 (6), 439–49.

Hess, S. and Polak, J., 2004. On the use of discrete choice models for airport competition with applications to the San Francisco Bay area airports. 10th Triennial World Conference on Transport Research, Istanbul, Turkey, 4–8 July 2004.

Hess, S. and Polak, J., 2005. Mixed logit modelling of airport choice in multi-airport regions. *Journal of Air Transport Management* 11 (2), 59–68.

Innes, J.D. and Doucet, D., 1990. Effects of access distance and level of service on airport choice. *Journal of Transportation Engineering* 116 (4), 507–16.

Lösch, A., 1962. Die räumliche Ordnung der Wirtschaft, 3rd edition, Stuttgart.

Pels, E., Nijkamp, P. and Rietveld, P., 1998. Access to airports: a case study for the San Francisco Bay Area. 38th Congress of the European Regional Science Association, Vienna, Austria, 28 August–1 September 1998.

Pels, E., Nijkamp, P. and Rietveld, P., 2003. Access to and competition between airports: a case study for the San Francisco Bay Area. *Transportation Research A* 37 (1), 71–83.

Thompson, A./Caves, R., 1993. The projected market share for a new small airport in the North of England. *Regional Studies* 27 (2), 137–47.

Thünen, J.H.v., 1826. *Der isolierte Staat in Beziehung auf Landwirtschaft und Nationalökonomie*, Hamburg.

Valentinelli, N., Liefner, I. and Brandt, O., 2004. Fluggastpotenzial des Flughafens Hannover-Langenhagen. Hannover working papers in economic geography, No. 29.

Zeike, O., 2003. *Nachfrageveränderungen im Rahmen von Flughafenkooperationen – Analyse des Verhaltens von Luftverkehrskunden und unternehmen sowie der Maßnahmen zur Förderung von Verkehrsverlagerungen zwischen Flughäfen*, Hamburg.

Websites of the Airports Focused on in this Chapter

Basel:	http://www.euroairport.com
Frankfurt:	http://www.fraport.d; http://www.airportcity-frankfurt.de
Friedrichshafen:	http://www.fly-away.de
Innsbruck:	http://www.innsbruck-airport.com
Munich:	http://www.munich-airport.de
Nürnberg:	http://www.airport-nuernberg.de
Strasbourg:	http://www.strasbourg.aeroport.fr
Stuttgart:	http://www.flughafen-stuttgart.de
Zurich:	http://www.unique.ch; http://www.flughafen-zuerich.ch

Chapter 16
Airport Competition in Greece: Concentration and Structural Asymmetry

Andreas Papatheodorou[1]

Introduction

Situated in south-eastern Europe, Greece has a population of about 11 million people and is nearly 132,000 km² in size. Islands cover almost 20 per cent of this area. The Greek archipelago has more than 6,000 islands and islets, of which 227 are inhabited. Many of the Greek islands attract significant tourism flows, both domestic and inbound, and about 60 per cent of tourist accommodation establishments in Greece are located on the islands (Greek National Tourism Organization, 2006). Tourism is a major business in Greece, accounting directly for about 7 per cent of gross domestic product (GDP); nonetheless, when all indirect and induced effects are considered, the travel and tourism economy contribution exceeds 15 per cent of GDP (World Travel and Tourism Council, 2006). About 14.3 million international tourists visited Greece in 2005 (United Nations World Tourism Organization, 2007). More than 75 per cent of these visitors travelled by air.

In spite of the importance of tourism and air travel for the country's economy, many airports in Greece remain largely underdeveloped. Traffic flows are essentially concentrated into a limited number of airports (especially on the mainland), with important implications for airport competition in both international and domestic passenger markets. This chapter aims at highlighting the structural asymmetries that exist in the Greek airport sector and proposing measures to alleviate the problem. In particular, the next section briefly reviews airport competition and the recent empowerment of regional airports in Europe. The chapter then discusses the profile of Greek airports, showing the evolution of relative size distribution over time. The penultimate section undertakes a correlation and an asymmetry analysis to show the great potential of airports in mainland Greece. Finally, the chapter concludes by suggesting appropriate consultation with all stakeholders involved in airport development leading to a way forward.

Airport Competition and the Empowerment of Regional Airports in Europe

Until the early 1990s, airport competition in Europe was essentially taking place at an international level. The creation and expansion of mega-hubs such as London Heathrow, Frankfurt, Paris Charles de Gaulle and Amsterdam Schiphol aimed at capitalizing on the lucrative connections between North America, Europe and Asia. Hub airports were largely dominated by flag carriers, which also operated domestic services to selected destinations; these were mainly chosen on their market potential within a hub-and-spoke system and their ability to provide feeder traffic to international

---

1 The author would like to express his thanks to Mr Konstantinos Polychroniades for his assistance in data collection.

flights. In some cases, Public Service Obligations (PSO) routes were offered to remote areas subject to specific terms and conditions. On these grounds, regional airports were clearly playing a peripheral role and were very dependent on the network configuration shaped by the centre.

As a result of the intra-European air transport liberalization completed in April 1997, the advance of low-cost carriers (LCCs) empowered the regions: point-to-point services emerged between satellite airports (such as London Luton, Brussels Charleroi and Frankfurt Hahn) and the newly rediscovered small, regional airports across the Continent (such as Carcassonne, Jerez and Pescara). These routes bypassed the main hub airports and rendered widely accessible those areas previously underserved. To a significant degree, the so-called 'Ryanair effect' redefined air transport geography by offering extremely low fares to a wide range of destinations across Europe (Papatheodorou and Lei, 2006). To sustain these operations and low fares, the Irish carrier was, in several cases, subsidized by local airports and/or regional authorities. The rationale behind this aid was based on the direct, indirect and induced effects of airport traffic related to income and employment generation (Debbage, 1999; Debbage and Delk, 2001; Gillen and Hinsch, 2001). Improved accessibility could also strengthen backward and forward linkages with the local economy and act as a propulsive, self-reinforcing mechanism of agglomeration and the realization of external economies of scale (Papatheodorou, 2004). Leisure tourism is a major economic activity, which benefits from this trend as cheap air transport motivates people to travel and consume local goods and services.

In this context, it is important to understand the essence of airport competition. To a major extent, this depends on whether the starting point is the airline or the airport (and surrounding destination). To illustrate the argument, consider five cities, namely A, B, C, D and E. For simplicity, we assume that A and E act as origins (not related to each other), while B, C and D as destinations: as a result, the A→B traffic is essentially related to outgoing passengers from city A, while the B→A traffic refers predominantly to passengers returning back to A. To assess competition between airports B and C, one may focus on airline markets and the substitutability of related routes. Starting from direct routes A→B and A→C, these are good substitutes if the activities of origin A passengers in destinations B and C are not location-specific. In other words, if A passengers fly to B mainly on business, substitutability between A→B and A→C may be low. On the other hand, if both B and C are sea-and-sun leisure destinations, then A passengers can easily switch from B to C (and vice versa) and hence A→B and A→C (and subsequently airports B and C) are in direct competition. In terms of indirect routes, there are two main cases to consider:

1. Assume that there is no A→C service; hence people who wish to travel between A and C have to fly first to B and then reach C either by air or another mode of transport – a possible start of a direct A→C flight will then generate traffic for airport C to the detriment of B, which loses its local connection role.
2. Assume that both airports B and C act as connecting hubs for the services A→B→D and A→C→D respectively; then, a possible fare reduction in the A→B leg may have negative repercussions for the A→C→D traffic, as people may switch away from hub C in favour of B. As a result, airports B and C are again in competition.

Similar conclusions may be drawn for services originating from E. On the other hand, and because origins A and E are not related to each other, services such as E→B and A→C are not regarded as good substitutes, i.e. airports B and C do not appear to be in competition for these routes.

Nonetheless, things may become more complicated if we consider competition not from the airline but from the airport point of view. In the previous example, assume that the authorities of airport B decide to heavily subsidize an airline to fly between E and B (acknowledging the market potential of origin E) to boost local tourism growth. If this effort is successful, then destination B may gain popularity and develop a number of built attractions and tourism infrastructure. This may then raise the interest of people residing in A, who may decide to switch away from resort C in favour of B. As a result, the action taken by airport authorities on the E→B route had positive repercussions on the A→B route and negative on the A→C route, in spite of the fact that E→B and A→C routes are not substitutes. Had airport C (instead of B) decided to take similar action, the converse outcome could have emerged. Similarly, a possible decision of airport B to expand its hub facilities to facilitate indirect flights to D would have a positive effect on E→B→D and A→B→D services and a negative effect on A→C→D. In other words, two airports may be in competition even if some of their related airline routes are not in the same market (in terms of substitution) due to the emerging network externalities. The recent empowerment of regional airports in Europe is not related to hub development but generates airport competition to capture direct flights of leisure passengers from a multitude of origin places.

In contrast to the widely publicized fierce reaction of major carriers against the subsidization of LCCs – which produced serious litigation, e.g. the European Commission (2004) decision on Charleroi Airport – the large hub airport authorities have been rather modest or even welcoming in their views. This is either because they benefit from this trend as proprietors of regional airports (e.g. BAA owns London Stansted and Southampton Airport, while Fraport owns Frankfurt Hahn) or because they feel unaffected by this airport competition, i.e. regional airports have basic facilities and are unable to serve network carriers. On the other hand, independent small airport authorities face the danger of engaging in a zero-sum game of destructive beggar-thy-neighbour policies: more specifically, to attract LCCs, small airports may offer too large a subsidy to reap any wider development benefits even in the longer term. In this case, the clear winners are the LCCs, and possibly their passengers: the LCCs can always play airports against each other due to the inherent flexibility of flight operations as opposed to the sunk nature of airport infrastructure. Ryanair's substitution of Strasbourg Airport in France for Karlsruhe-Baden Airport in Germany is a good example of this case. This is not to say that local authorities should abstain from taking measures to improve the air transport accessibility of their regions: still, they should be able to fully assess the economic impacts of such practices (European Commission, 2005). Similarly, policymakers in countries with limited exposure to LCCs, such as Greece, may learn some useful lessons on how to design their regional airport strategy in the future.

Airport Profiles and Distribution of Traffic in Greece

Thirty-nine airports were open to public access and operated on a commercial basis across Greece in 2007. Eleven of them were located on the mainland and 28 on the islands (Hellenic Civil Aviation Authority, 2007). Although this distribution is clearly asymmetric in favour of the islands (both in area and population terms), it is largely justified on the fragmented spatial structure of the country, the existence of poor transport substitutes (sea fares are usually cheaper but journeys are long) and the popularity of the islands among tourists. The Greek state is the dominant player in the market, owning 34 out of the 39 operating airports. Interestingly, however, Athens International Airport (AIA), the largest airport in Greece, is a joint venture between the state and Hochtief AG on Build-Operate-Transfer (BOT) terms. The role of local authorities has also increased over time:

four operating airports are currently owned by municipalities. In terms of traffic origination, 15 airports are classified as international, 13 as hybrid (i.e. accessible to international flights on an ad hoc basis) and 11 as domestic (Hellenic Civil Aviation Authority, 2007).

About 403,000 aircraft movements were recorded in 2005, almost equally split between domestic and international flights. Interestingly, however, international passenger arrivals exceeded domestic arrivals by more than 120 per cent, i.e. 12.5 million passengers compared to 5.6 million. This is essentially related to the use of larger aircraft for international flights; the latter are dominated by leisure tourism, as 7.3 million international passengers (or about 58.5 per cent) arrived in Greece on charter services in 2005 (Hellenic Civil Aviation Authority, 2007). Based on the European Commission (2005) traffic size classification criteria, in 2005 Greece had one 'A Class' airport (exceeding 10 million passengers per year), accounting for about 39.1 per cent of total traffic (that is AIA); no 'B Class' airport (between 5 and 10 million passengers per year); seven 'C Class' airports (between 1 and 5 million passengers per year), handling about 48.7 per cent of the traffic; and 30 'D Class' airports (fewer than 1 million passengers per year) with the remaining traffic market share of nearly 12.2 per cent (Hellenic Civil Aviation Authority, 2006): Kalymnos Airport, another very small D Class airport, opened in 2006.

In other words, the airport sector in Greece is characterized by high concentration and a notable dualism, where a very large airport coexists with a multitude of very small ones. Many among the latter are largely dependent on PSO. In particular, Greece has designated 25 routes as PSO (22 operating and three idle as their tender process is still pending): 12 between Athens (airport code: ATH) and the islands; five between Thessaloniki (the second largest city, airport code: SKG) and the islands; five inter-island connecting two to five islands; two linking mainland points with the island of Crete and one connecting Thessaloniki with a mainland destination (European Commission, 2007). It seems, therefore, that almost all PSO in Greece have been designed to support the island network due to limited alternative transport modes and lack of sufficient demand to sustain profitable operations out of the tourist season. In spite of the subsidization of the PSO fares, small regional island airports fail to prosper because of a significant lack of infrastructure. In many cases, and due to a lack of appropriate flat field, runways are very short (700–1,500 m), narrow (25–40 m) and built on cliff tops near mountainous terrain exposed to gusty winds: as a result, aircraft performance is significantly limited. In fact, the small aircraft operating these routes (ATR42, ATR72, Dash8, etc.) may occasionally have to divert to other airports due to unfavourable weather conditions or other unforeseen conditions such as power disruptions or communications failure on the ground. In such cases, the passenger is eligible to receive a refund (as the actual service was not delivered), despite the fact that the airline incurs a transport cost (Fragoudaki et. al., 2005).

To elaborate further on the issue of size distribution from a dynamic perspective, passenger traffic in Greek airports for the period 1986–2005 was first classified into nine spatial segments using as criteria the location of the airport (mainland and island) and the type of traffic (domestic and international). In this context, the nine groups refer to domestic traffic originating and departing from mainland airports, island airports and all of the airports in aggregate (three groups); incoming and outgoing international traffic from mainland airports, island airports and all of the airports in aggregate (three groups); and total (i.e. domestic and international) traffic originating and departing from mainland airports, island airports and their sum (the last three groups). Further to the previous discussion on airport competition, it is important to note that these nine segments do not bear any market-related connotations as the concept of substitutability has not been taken into consideration. For example, although the domestic ATH and SKG traffic are added among others to calculate the mainland domestic traffic, it is very unlikely that their associated flights (e.g.

Athens–Chios and Thessaloniki–Mykonos) are substitutes for each other. The only spatial segment where there might be some degree of substitutability is the island international traffic, when focusing on specific tourism origin countries (or cities). Given that the overwhelming majority of international passengers to and from the Greek islands are leisure tourists on charter flights, it can be argued, for example, that the London Gatwick (LGW) to Heraklion (HER) and the LGW to Rhodes (RHO) traffic (jointly considered in the above spatial segment) are very good substitutes for British tourists travelling to the Mediterranean; on the other hand, the LGW–HER and the Düsseldorf (DUS)–RHO traffic are not good substitutes from an airline perspective. As discussed previously, however, infrastructural and other developments at HER may not only have positive repercussions on the LGW–HER route, but may also switch passengers away from DUS–RHO in favour of DUS–HER. Similarly, the granting of international status to a number of island airports in Greece in the early 1980s resulted in the initiation of direct flights between foreign cities and these islands, to the detriment of indirect flights through ATH or SKG. Airport competition steps beyond the boundaries of airline markets.

On these grounds, and despite the divergence from market-based criteria, the nine segments classification can be useful in revealing certain aspects of airport competition and in showing how the distribution of airport traffic in Greek mainland and island airports has evolved over time. From a quantitative perspective, this can be done using concentration and asymmetry indexes such as the Herfindahl-Hirschman Index (HHI), the coefficient of variation, the Gini index and Theil's entropy measure. Reynolds-Feighan (2001) argues that the Gini index is the most appropriate when considering a single point in time. Nonetheless, the estimation of both Gini and entropy indexes heavily depends on the number of participating units (operating airports in our case). If this changes over time, then the different values of the Gini or the entropy indexes may not necessarily reflect a change in market concentration or asymmetry rendering comparisons over time problematic. This can be an important issue in the present context. Illustratively, there were 22 operating international airports at a country level in 1986 and 31 in 2005, i.e. an increase of over 40 per cent – changes in subcategories (e.g. mainland international) may be more acute. To address this problem, it was decided to focus on the HHI, which is defined as (Papatheodorou, 2003):

$$HHI = \sum_{i=1}^{N} s_i^2 \qquad (1)$$

where s_i is the traffic market share of airport i in a total of N. The HHI does not depend on the number of participating airports (all that matters is the share), and ranges between zero (infinitesimal firms – totally fragmented market structure) and 10,000 (the case of monopoly, with one firm having a 100 per cent market share). Due to the square power, the HHI puts more emphasis on the share of larger compared to smaller units, acknowledging in this way their power in shaping the market configuration and conduct. Although this may not be directly relevant in our case (as a result of the loose definition of the airport market), it is important to understand the weight of major airports at a system-wide level especially if infrastructure issues (e.g. size of runway, ground handling facilities etc.) and economies of scale are of importance. Values of HHI over 1,800 provide good evidence of concentrated markets.

According to Figure 16.1, traffic concentration among airports in mainland Greece was very high for the whole period under consideration, with HHI values exceeding 5,000. In contrast, if all 11 mainland airports shared the traffic equally, i.e. each with a share of 9.09 per cent, then the HHI would be 909. International traffic (shown by Mainland I in Figure 16.1) was extremely concentrated in the mid-1980s, with HHI values close to 8,000 as a result of restricted bilateral agreements on scheduled services (based on a duopoly of national carriers flying between country capitals) and a limited number of charter flights. Athens Airport (ATH) accounted for 89 per cent of international traffic in mainland Greek airports in 1986. Since then the pattern has exhibited a downward trend, stabilizing around a HHI value of 6,000 from the mid-1990s onwards. On the other hand, concentration in domestic traffic (shown by Mainland D in Figure 16.1) has been stable over time. As a result, total traffic in mainland airports (shown by Mainland T in Figure 16.1) lies somewhere in between, with values ranging between 7,000 (in 1987) and 5,900 (in 2005).

Traffic concentration at island airports is in sharp contrast with that of the mainland airports. The absence of large urban centres – Athens (ATH) and Thessaloniki (SKG) are both on the mainland – and the dispersion of population and airports over a large number of islands has resulted in a fragmented structure for domestic traffic, with stable HHI values in the area of 1,000. On the other hand, international traffic has been slightly more concentrated due to the large number of charter services operating on a seasonal basis to the most popular island tourist destinations (i.e. Crete, Rhodes and Corfu). Therefore, total traffic exhibits remarkable stability, around an HHI value of 1,500. For benchmarking reasons, it should be noted that if all 27 island airports shared the market equally, i.e. each with a market share of 3.70 per cent, then the value of HHI would be 370.

With respect to air traffic concentration for the country as a whole, its pattern lies between the heavily concentrated mainland and the fragmented islands in most cases. If all 38 airports shared the market equally, the HHI value would be 263. In 1986, mainland airports handled

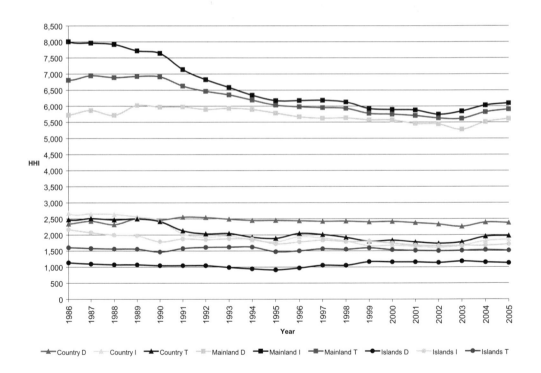

Figure 16.1 Evolution of the HHI, 1986–2005

Note: Calculations are based on relevant traffic data downloaded from the internet database of the Hellenic Civil Aviation Authority, http://www.hcaa.gr/home/index.asp.

52 per cent of total international traffic, while the island airports handled 48 per cent. In 2005, the figures were 48 per cent for the mainland and 52 per cent for the islands. As a result, the weight of island airports in the determination of the international traffic concentration at a country level has increased. Given the larger number of airports taken into consideration at a country level (which inevitably puts downward pressure on the HHI value), it is not surprising that since the early 1990s, the HHI value for overall international traffic has almost coincided with the HHI value for island international traffic. On the other hand, in 1986 mainland airports accounted for 61 per cent of total domestic traffic, while island airports accounted for 39 per cent. In 2005 mainland airports handled 63 per cent of domestic traffic and the islands accounted for 37 per cent. This is roughly the same. Domestic traffic concentration at a country level is also relatively stable, around a HHI value of 2,500. Finally, in 1986 mainland airports accounted for 56 per cent of total traffic, while island airports accounted for 44 per cent of total traffic. In 2005 the figures were 53 per cent for the mainland and 47 per cent for the islands. Total traffic concentration at a country level has also exhibited a downward trend over time, with values of HHI converging to those of overall international traffic.

Structural Asymmetries in the Greek Airport Sector

The analysis undertaken in the previous section highlighted the significant levels of traffic concentration that exist in the Greek airport sector, especially on the mainland. Still, this concentration may well be explained by the prevailing market and spatial forces: for example, if market potential (in terms of population and purchasing power) is concentrated in a few urban centres then traffic concentration is inevitable. To validate this argument, the results of an interesting correlation and asymmetry analysis is presented in this section.

As a first step, data on gross domestic product (GDP), population and area size were collected for the 51 Greek regions at NUTS3 level. The Nomenclature of Units for Territorial Statistics (NUTS) provides a hierarchical classification of all EU economic areas at different spatial scales – level 3 is equivalent to a county or a prefecture. The most recent available census data refer to years the 2000 and 2001. Subsequently, the 38 operating airports (in 2000 and 2001) were assigned to specific NUTS3 regions according to their actual location. The important issue here is the administrative classification rather than the identification of airport catchment areas: although the latter is an important task, it would complicate the analysis to a very significant degree. Catchment areas are often endogenously determined, i.e. lack of a local airport may artificially expand the catchment area of a larger, yet remoter, area. Following this data preparation, a set of correlations were calculated, as shown in Table 16.1. A value of -1 (1) detects perfect negative (positive) linear correlation; a value of zero shows no linear correlation.

The first correlation is very supportive of the 'market potential argument' as the high coefficient value suggests that prefectures with large purchasing power are associated with airport presence and significant traffic. Interestingly, however, the second correlation reveals that the previous one is somewhat spurious, in terms of being significantly affected by the inclusion of the two largest urban centres in Greece, i.e. Athens and Thessaloniki. Illustratively, the two prefectures jointly account for 50 per cent of GDP and 47 per cent of passenger traffic. When excluded, the correlation between GDP and traffic remains positive, albeit significantly weaker. Similar results hold for the correlation between population and traffic. On the other hand, the results show no relation between the size of a NUTS3 region and its level of passenger traffic. This may be related to major differences in population density observed across regions in Greece.

Table 16.1   Results of correlation analysis

Correlation	Coefficient value
$GDP_0 - PAX_0$	0.88
$GDP^*_0 - PAX^*_0$	0.37
$POP_1 - PAX_1$	0.87
$POP^*_1 - PAX^*_1$	0.29
$SIZE - PAX_1$	0.09
$SIZE^* - PAX^*_1$	-0.10

Note: GDP is adjusted into purchasing power units, PAX is passengers, POP is population and SIZE refers to NUTS3 areas. The subscript 0 refers to the year 2000 and the subscript 1 to 2001. Series with asterisks exclude Athens and Thessaloniki (the two largest urban centres in Greece). All variables are expressed in percentage terms (market shares at a country level).

Source: GDP, POP and SIZE data, Petrakos and Psycharis (2004).

It seems, therefore, that the market potential of NUTS3 regions in Greece is not as strongly represented as it should be in airport passenger traffic. Geographical fragmentation and the over-representation of the islands in airport infrastructure and leisure tourism traffic are major factors. To elaborate further on this issue graphically, an asymmetry analysis is performed, based on the conceptual framework provided by income distribution inequality studies (Papatheodorou, 2003). All Greek NUTS3 regions have been ranked in descending order to express GDP, population (POP) and passengers (PAX) in cumulative percentage terms. Figures 16.2 to 16.5 show scatters related to the first four correlation pairs in Table 16.1. Points on the diagonal line correspond to perfectly proportional representation (e.g. 50 per cent of POP to 50 per cent of PAX); similarly, points above the diagonal are related to over-representation and points below to under-representation in passenger traffic.

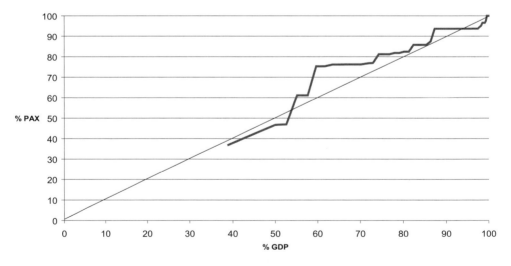

Figure 16.2   GDP and passenger traffic in NUTS3 regions, 2000

All four graphs are characterized by the existence of linear segments parallel to the X axis, which represent NUTS3 regions with no airport traffic. Illustratively, 26 out of the 29 NUTS3 regions with no traffic in Greece are situated on the mainland. As a good example of this asymmetric structure, we may consider the Larissa prefecture (located in central Greece): although it accounts for 2.51 per cent of PPP-adjusted GDP, 2.55 per cent of the population and 4.09 per cent of the country's size, the prefecture has no access to commercial airport facilities. On the other hand, Zakynthos prefecture (a popular tourist destination situated in the Ionian Sea) handled 2.75 per cent of total airport traffic in Greece in 2001, in spite of having only 0.27 per cent of PPP-adjusted GDP, 0.36 per cent of the population and 0.31 per cent of the country's size. Therefore, though the improvement of airport infrastructure on the islands will be beneficial to their accessibility, policy priority should be given to the largely under-represented mainland.

The 'Ryanair effect' of cheap, scheduled, point-to-point services is more likely to appear on the mainland than on the islands if the appropriate airport infrastructure becomes available and accessible to commercial airline services. First, urban facilities and functions are better developed on the mainland and operate throughout the year. This is in sharp contrast with the islands, where the majority of tourist establishments and ancillary services usually operate only on a seasonal basis, i.e. between April and October. Inbound mainland tourists may engage in various activities while staying in Larissa compared to Zakynthos or even Rhodes during the winter: a visit to the site of Meteora, where monasteries are built on steep rocks situated in the middle of the Thessaly plain, is a good example of tourist activity in this case. Second, stage distance and/or block times between western/eastern European states and mainland Greece are in most cases shorter than journeys to the islands; this is also associated with weather conditions, which can be unsettled on the islands throughout the year.

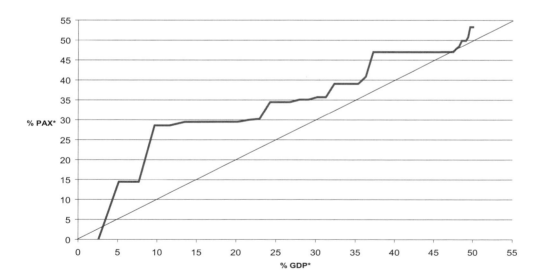

Figure 16.3 GDP and passenger traffic in NUTS3 regions, 2000 (excl. Athens and Thessaloniki)

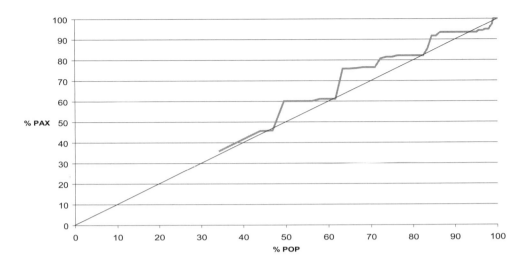

Figure 16.4 Population and passenger traffic in NUTS3 regions, 2000

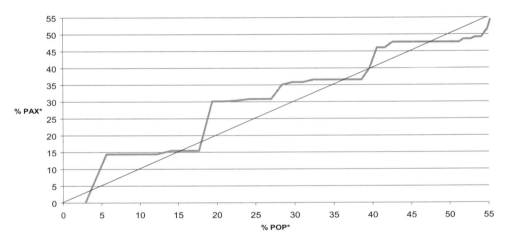

Figure 16.5 Population and passenger traffic in NUTS3 regions, 2000 (excl. Athens and Thessaloniki)

In essence, and by securing direct scheduled flights from several European cities, mainland regional airports and their surrounding destinations may enter the international leisure traffic market, offering a valid substitute to the product supplied by other mainland and/or island airports (and destinations) in Greece and elsewhere. As a result, some scope for airport competition in specific airline markets is generated. Moreover, mainland airports are often located in areas of adequate or at least promising market potential (in terms of purchasing power and population) to financially sustain at least some outbound and inbound leisure and business traffic. Consequently, these direct flights to and from Europe may deny part of the hub role currently played by ATH (and SKG to a lesser extent) or reduce the need for extensive intermodal considerations. For example, instead of flying from London Heathrow to Athens Airport and then domestically to Alexandroupolis (a major city situated in north-eastern Greece close to the land border with Turkey), a passenger may save time by flying directly from London Stansted to Alexandroupolis; or instead of flying from Frankfurt to Athens and then spending about four hours on the motorway to reach Volos (a major city near Larissa), a passenger may decide to fly directly from Hahn to Nea Aghialos Airport (situated between Volos and Larissa) and spend less than 30 minutes to reach their final destination. In this way, Nea Aghialos and Alexandroupolis airports can exercise some competitive pressure at least on ATH. To quantify this, econometric techniques such as discrete choice (Greene, 2003) and stated preference (Louviere et. al., 2000) can be used to analyse results from structured questionnaires distributed to potential inbound and outbound travellers. In addition to Likert-scale questions, potential travellers should express their opinion on the total additional cost they would be prepared to incur to fly directly from their local airport, and hence avoid driving, going by train or even flying first to ATH.

In any case, however, it should be clearly understood that it would not be sustainable from a financial perspective to upgrade all existing under-utilized airport infrastructure, or even to build new airports throughout mainland Greece. A strategic choice should be made to maximize the emerging benefits of regional and tourism development and the generation of healthy competition among airports. In this context, it might be advisable to focus first on the development of a limited number of under-utilized airports whose location bears significant advantages. The airports of Nea Aghialos (to satisfy the needs of both Volos and Larissa), Ioannina (capital of the prefecture of Epirus and gateway to Albania), Araxos (situated in western Peloponnesus close to Patras – the third largest city in Greece – and the emerging tourist resorts of the prefecture of Ilia) and Alexandroupolis are very good candidates to promote international scheduled flights, at least in the beginning. Still, any policy to upgrade airports in mainland Greece should be accompanied by an appropriate consultation process, as now argued in the concluding section of this chapter.

Conclusions and the Way Forward

Passenger air traffic in Greece is highly concentrated on the mainland (due to the predominance of two large urban centres, namely Athens and Thessaloniki), yet asymmetrically distributed in favour of the islands due to insufficient or nonexistent airport facilities in many NUTS3 mainland regions. Moreover, and with the exception of Athens International Airport, all other commercial airports in the country are owned and/or managed by the state. As a result, airport competition in Greece is far from being effective at present.

To capitalize on the good market potential of the mainland regions, co-ordinated action of all related stakeholders is required, and for this reason opinion questionnaires and open interviews should be conducted with tourism and aviation authorities, hospitality service providers, elected representatives of local and regional administration, environmentalists and members of local communities. Although clashing interests will inevitably emerge, it is important to reach a consensus and design a sustainable regional airport strategy that faces Nimby (not in my back yard) problems among others (Thomas and Lever, 2003).

Securing private funds to finance an active aviation policy is very important. Although the Greek state will be expected to play an important role in channelling money from the European Union in this direction, the establishment of public-private partnerships (PPP) will also be required to meet the European Commission (EC) 'private investor's principle' (European Commission, 2005), especially if issues of airport subsidies to carriers emerge. In this context, airport competition within Greece could become a reality as the government may decide to attract investors by privatizing (partly at least) the small regional state-owned airports on preferential terms. Similarly, the government may exercise pressure on the Greek Air Force to consolidate its facilities and open up a number of well-situated military airports to commercial operations. Nonetheless, such moves should be handled carefully not only to secure approval from the political opposition, but most importantly to avoid selling public infrastructure at nominal value or having airports become engaged in mutually destructive bidding wars for airline services.

References

Debbage, K. G. (1999) Air Transportation and Urban Economic Restructuring: Competitive Advantage in the US Carolinas. *Journal of Air Transport Management* 5: 211–21.
Debbage, K. G. and Delk, D. (2001) The Geography of Air Passenger Volume and Local Employment Patterns by US Metropolitan Core Areas: 1973–1996. *Journal of Air Transport Management* 7: 159–67.
European Commission (2004) *The Commission's Decision on Charleroi Airport Promotes the Activities of Low-cost Airlines and Regional Development*. IP/04/157. Brussels: European Commission.
European Commission (2005) *Community Guidelines on Financing of Airports and Start-up Aid to Airlines Departing From Regional Airports*. Brussels: European Commission.
European Commission (2007) *Public Service Obligations – List of Routes Concerned (30th October)*. Brussels: European Commission.
Fragoudaki, A., Keramianakis, M. and Yiankovich, S. (2005) The Greek PSO Experience, 4th International Forum on Air Transport in Remoter Regions. Stockholm, 24–26 May.
Gillen, D. and Hinsch, H. (2001) Measuring the Economic Impact of Liberalization of International Aviation on Hamburg Airport. *Journal of Air Transport Management*, 7: 25–34.
Greek National Tourism Organization (2006) *Information on Greece*. Website: www.gnto.org (accessed on 21 March 2006).
Greene, W. (2003) *Econometric Analysis*. 5th edition, Harlow: Prentice Hall.
Hellenic Civil Aviation Authority (2007) *Information on Greek Airports*. Website: www.ypa.gr (accessed on 15 November 2007).
Louviere, J. J., Hensher, D. A. and Swait, J. D. (2000) *Stated Choice Methods: Analysis and Application*. Cambridge: Cambridge University Press.
Papatheodorou, A. (2003) Corporate Strategies of British Tour Operators in the Mediterranean Region: An Economic Geography Approach. *Tourism Geographies*, 5(3): 280–304.
Papatheodorou, A. (2004) Exploring the Evolution of Tourist Resorts. *Annals of Tourism Research*, 31(1): 219–37.
Papatheodorou, A. and Lei, Z. (2006) Leisure Travel in Europe and Airline Business Models: A Study of Regional Airports in Great Britain. *Journal of Air Transport Management*, 12: 47–52.
Petrakos, G. and Psycharis, Y. (2004) *Regional Development in Greece*. Athens: Kritiki (in Greek).
Reynolds-Feighan, A. (2001) Traffic Distribution in Low-cost and Full-service Carrier Networks in the US Air Transportation Market. *Journal of Air Transport Management*, 7: 265–75.
Thomas, C. and Lever, M. (2003) Aircraft Noise, Community Relations and Stakeholder Involvement, in Upham, P., Maughan, J., Raper, D. and Thomas, C. (eds). *Towards Sustainable Aviation*. London: Earthscan Publications.

United Nations World Tourism Organization (2007) *UNWTO World Tourism Barometer 5(2)*. Madrid: UNWTO.

World Travel and Tourism Council (2006) *Greece: The Impact of Travel and Tourism on Jobs and the Economy: Preliminary Findings*. Website: www.wttc.org.

Chapter 17
The Airport Industry in a Competitive Environment: A United Kingdom Perspective[1]

David Starkie

Introduction

This chapter provides an overview of UK airports from the perspective of a business enterprise. Its object is to show, through the medium of the UK industry, that effective competition between airports is possible and that a competitive industry can be financially viable. In the UK case viability is achieved at all levels of output, thus refuting the suggestion that high fixed costs are a significant barrier to positive returns, particularly for airports of limited output. This viable industry operates for the most part in the private sector of the economy and it has evolved without the imposition of a strategic plan. It is competition that has driven the dynamics of the industry, an industry that in its symbiotic relationship with the airline industry has been an economic success story, helping to produce strong economic growth in the service sector of the UK economy.

The structure of the paper is as follows: the following section provides a snapshot of the UK airports industry, from which it is evident that economies of scope are important.[2] This is followed by a section on the ownership structure of the industry, drawing attention therein to the market for corporate control (with its implications for productive efficiency). The chapter then outlines the relationship between airports and airlines, stressing the recent development of long-term contracts which are important for understanding the current nature of competition. The following section places competition in a spatial setting by summarising recent analyses of hinterlands or catchments from which the individual airports attract traffic; it is shown that most of these catchments overlap to form a chain of substitution. The penultimate section analyses the financial performance of the industry; its conclusion is that airports at all levels can operate without subsidy and, overall, the industry's profitability is similar to that for the non-financial sector of the UK economy. The chapter concludes by arguing that governments should encourage competition in the market for basic airport services, if necessary by restructuring and by unbundling concentrated ownership. Economic regulation, even when well designed, is very much a second-best solution.

Size and Diversity

In spite of its relatively small area, the UK in general, and England in particular, has a surprisingly large number of airports with scheduled passenger services; in 2007 there were about 40 such

---

1   First published as OECD/International Transport Forum Discussion Paper No. 2008-15, prepared for the Round Table of 2–3 October 2008 on Airline Competition, Systems of Airports and Intermodal Connections. Published here by kind permission of the OECD.

2   For a recent excellent general overview of the UK industry, see Graham (2008).

airports. Their size distribution, with reference to passenger numbers, is bimodal with four large airports with numbers in excess of 20 million and the remainder with fewer than 10 million passengers. There is one major transfer (connecting) hub, London Heathrow, which, of course, is a supreme example of its genre. I could continue to cut and slice the data on UK airports countless ways and possibly bore you in the process. Instead, I can refer you to the comprehensive airport data on the UK Civil Aviation Authority's (CAA) website.[3] I am more interested in the 'airport' as a business entity, or enterprise, operating as the UK airports do within a competitive market economy with economic regulation limited to three co-owned London airports; this gives the analysis of the data a particular slant.

I have already referred to passenger numbers and it is a common practice to use this as the defining characteristic of an airport's size, but to do so tends to diminish the fact that airports are in most cases multi-product entities supplying to the market a bundled group of services (Office of Fair Trading, 2006). Apart from handling passenger traffic, other activities include shipping airfreight (including mail); providing for air-taxi services and general aviation; acting as a base for flying training; aircraft maintenance; flight testing and corporate jet activity; and providing for a large number of other specialist aviation services. Further complexity is added because the activities of the airport company can extend beyond the supply of airport services per se. The property assets within the airfield boundary might also serve non-aviation related activities. At the smaller airfields, it is not unusual to find former hangars and similar obsolete or stranded assets used for storage or as units for light industry.

Looked at in business terms, a more appropriate measure of firm size is turnover. Table 17.1 ranks selected UK airports by turnover in 2005/6. (There are probably another dozen relatively small airports exceeding the minimum turnover threshold shown, £5 million, for which financial data are not readily available.) The range of turnover is huge, as one might expect, and, although there is a high correlation between this statistic and passenger numbers for medium and large airports, there are, nevertheless, cases where financial turnover relative to passengers is disproportionately high; Nottingham East Midlands, Cardiff and London City are examples. For small airports the relationship between passenger numbers and turnover is less close.[4] A further indication of this is the relative proportion of Air Transport Movements (ATMs), those with a passenger focus, to total aircraft movements, which for these airports is often less than half (see Table 17.1, columns 3 and 4).

But small airports are of greater significance than their size suggests. In a competitive airport environment (about which I will say more later) they are a source of competition for the larger airports. The industry is dynamic and rapid growth from a small base is not uncommon. Liverpool, for example has experienced exceptional growth during the last decade, which has had repercussions for other airports in its region. A competitive challenge has also come from newly established civil operations at former military bases using the stranded assets of the defence 'industry' (Doncaster, Newquay, Manston) and at airfields located alongside aircraft manufacturing plants (Belfast City), as well as from the occasional new airport on a greenfield site, such as London City and Sheffield City. But there has been exit from the industry as well and the latter example is a case in point.

---

3   At www.caa.co.uk/airportstatistics.
4   For UK airports with under £30m turnover (n=13), total passenger numbers explained two-thirds of the variance in turnover.

Table 17.1 Selected financial and operating data for UK airports, 2005–06

	Turnover (£000)	ATMs*	Other movements†
London Heathrow	1,195,400	472,954	5,981
London Gatwick	361,500	254,004	9,058
Manchester	290,553	217,396	16,421
London Stansted	176,500	180,729	15,465
Birmingham	111,109	113,668	9,731
Glasgow	82,615	97,610	13,296
Edinburgh	77,381	117,312	9,808
London Luton	77,021	87,690	20,203
Newcastle	51,360	55,164	23,798
Nottingham East Midlands	50,566	56,224	24,490
Bristol	49,619	59,854	20,670
London City	40,180	61,179	9,733
Aberdeen	33,954	94,665	17,851
Belfast International	31,206	43,780	37,093
Liverpool	28,799	43,312	37,347
Cardiff	22,103	20,689	22,337
Southampton	22,022	45,109	13,351
Leeds Bradford	21,023	36,330	31,641
Exeter	17,707	14,481	40,572
Bournemouth	14,440	14,041	69,600
Coventry	14,123	13,951	54,134
Norwich	12,089	20,894	30,145
Humberside	10,934	11,342	25,996
Durham Tees Valley	10,834	53,532	52
London Biggin Hill	6,892	4,834	62,666
Blackpool	6,333	13,028	61,985
Southend	4,973	1,548	47,798

Source: Centre for Regulated Industries, *Airport Statistics 2005/6*, Appendices D1 and B2.

Note: * Movements of aircraft engaged in the transport of passengers, cargo or mail on commercial terms.
† Includes test and training flights, aero club movements, military movements and private flights.

Ownership and Capital Markets

Until little more than 20 years ago, virtually all runway and terminal assets at UK airports were owned by the public sector (although the private sector often played a major role, through concession agreements in the running of the airports or, more typically, parts of them). The transfer in 1987 to the private sector of all the share capital of the British Airports Authority, a corporate enterprise owned by central government, was the first important change of ownership in the UK industry. This transfer, by flotation of shares on the London stock exchange, established BAA plc (confusingly referred to as a public quoted company) with a substantial capitalization. Between 1993 and 1999 many local government-owned airport assets were also sold. This was a period when strict controls were imposed on local government spending on airport assets so that, to expand such airports, private capital was needed; but further privatizations have occurred since removal of the capital spending constraints (see Table 17.2). Unlike the public flotation of BAA, disposals by local governments to the private sector took the form of trade sales, that is, sales to existing trading entities.

Table 17.2   Ownership patterns at main airports in the United Kingdom, 2007

	Present ownership	Private interest (%)	Privatization	
			Date	Resales*
Aberdeen	ADI (BAA)	100	1987	1
Belfast City	Ferrovial	100	n.a.	1
Belfast International	ACDL	100	1994	2
Birmingham	Local authorities/Dublin Airport Authority/Macquarie Airports/employees	51	1997	
Bristol	Ferrovial/Macquarie Airports	100	1997	
Cardiff	ACDL	100	1995	1
Edinburgh	ADI (BAA)	100	1987	
Glasgow	ADI (BAA)	100	1987	2
Leeds Bradford	Bridgepoint	100	2007	
Liverpool	Peel Holdings	100	1990	
London City	AIG/GE/Credit Suisse	100	n.a.	2
London Gatwick	ADI (BAA)	100	1987	1
London Heathrow	ADI (BAA)	100	1987	1
London Luton†	ACDL	100	1998	1
London Stansted	ADI (BAA)	100	1987	1
Manchester	Local authorities	0	n.a.	
Newcastle	Copenhagen Airport	49	2001	
Nottingham East Midlands	Manchester Airport Group	0	1993	1
Prestwick	Infratil Ltd	100	1987	2
Southampton	ADI (BAA)	100	1961	2

Source: Adapted from Graham, 2008. All airports in the United Kingdom with more than 1 million annual passengers in 2005 are included.

Note: n.a. = not applicable. * 'Resales' indicates the number of changes of owner since the first privatization, or initial sale in the case of Belfast City and London City. † 30-year concession contract; ownership remains with the local authorities.

The majority of the financial transactions have been outright sales to the private sector, but with some exceptions. Local government retains a majority share in Newcastle upon Tyne airport, a minority share in Birmingham airport and a tiny share in Blackpool airport; whilst London Luton airport is a 30-year concession agreement. The latter commenced in 1998 and recent events suggest that this approach is not without its problems. The concession holder, Airport Concessions and Development Ltd (ACDL), has now decided not to pursue earlier plans for major investment, citing as a reason the limited period remaining before the end of the concession agreement (although central government's support for, arguably, premature expansion of nearby London Stansted has probably complicated matters). Not all airports have been sold to the private sector, or introduced private equity capital. Manchester (the UK's fourth largest), which belongs to a consortium of local governments in north-west England, is a significant exception. The UK airport industry is thus a mixed private-public sector industry, but one currently dominated by the private ownership of assets.

An important feature of the market in UK airport assets (and indeed the assets of the privatized utility industries in general in the UK) is that it is a market with a global reach, the final impediments to which disappeared in 2006: when BAA was privatized the government capped the number of shares that any one shareholder could hold to 15 per cent but, following a ruling by the European Court of Justice that this restriction impeded the free movement of capital in the European Union, it was removed. The take-over of BAA soon followed when a consortium led by Ferrovial, the Spanish construction, infrastructure and services group,[5] outbid Goldman Sachs, the US investment bank. Another Spanish-led consortium has an interest in Belfast International and Cardiff airports, as well as holding the Luton airport concession; Macquarie Airports, an investment trust which is part of the Australian bank of that name, owns Bristol airport; a New Zealand investment group has an interest in two smaller airports; Copenhagen airport (in which Macquarie also has an interest) holds the minority stake in Newcastle airport; and private equity groups have been involved in the two most recent privatizations of local government airports, Leeds Bradford and Exeter. Both these latter sales took place at a price of about 30 times earnings (before allowing for interest, tax, depreciation and amortization). This suggests high expectations by the purchaser that substantial cost efficiencies can be achieved and/or strong market growth is attainable.

The market for corporate control generally in the UK is a very active one and this characteristic applies equally to the market in airport assets. Ownership by unquoted private companies (BAA plc was the exception) has not prevented several of the airports changing hands since they were first transferred to the private sector, some several times (see Table 17.2). It would be reasonable to suppose, therefore, that as a consequence much of the industry is subject to capital market disciplines, which bear in particular upon its productive efficiency: the acquiring firm will aim to increase the profitability of the airport taken over, securing a good investment return by improving the airport's operational efficiency. And, in so far as the remaining (corporatized) public sector airports find themselves competing in the market for air transport services with (for-profit) private sector airports, competition for private sector airport assets in the global capital markets will have had the effect of increasing the productive efficiency of the sector as a whole.

---

5   BAA is owned by Airport Development and Investment Limited (ADI), in turn a wholly owned subsidiary of SGP Topco Limited, in which Grupo Ferrrovial SA holds 61.06 per cent of the ordinary shares through two of its subsidiaries. The other two shareholders are Airport Infrastructure Fund LP, which is managed by Caisse de Dépôt et Placement du Québec, which has 28.9 per cent of the ordinary shares; and Baker Street Investment PTE Limited, a subsidiary of GIC Special Investments PTE Limited, which holds the remaining 10 per cent.

Competition for Contracts

It was expected that liberalization of European air transport, the final phase of which was completed in 1997, would lead to a much more competitive air transport market throughout Europe. Not anticipated was the role that the low-cost carrier (LCC) would play in driving these market reforms, or the profound effect the carriers would have on the airport industry. The consequence has been to greatly increase competition between airports and to increase the bargaining power of the airlines. The catalyst in this transforming process has been the introduction of formal, specific (long-term) contracts between the airport and downstream airline customers. These vertical supply contracts, arguably, should rank alongside the use of on-line internet booking systems and the introduction of cheap one-way fares (which undermine the ability of the 'legacy' carriers to price discriminate) as a major innovation in contemporary civil aviation. The privatized UK airport industry has played a key role in their introduction.

Vertical supply contracts between airport and airline have long been a familiar feature of civil aviation in other parts of the world, such as in Australia (long-term leases on terminals) and, especially, in the United States (gate leases and 'majority-in-interest' clauses giving airlines some control over capital expenditure). But the focus of contractual developments in Europe since liberalization, initially in the UK and then more generally, is novel; it has been a focus on negotiated *charges* for the long-term use of basic airport infrastructure.

The traditional relationship between airport and airline user has had at its core a posted tariff of charges (the most important of which are generally structured to reflect aircraft weight), together with associated 'conditions of use'. The interesting feature of this traditional approach is its informality: users do not need a contract with the airport but, in paying the published tariff, they also accept the 'conditions of use' (Condie, 2004; Graham, 2001). Under this arrangement the airport is, in effect, assuming the long-term traffic risk. This was not of concern to airport owners when air services were subject to general regulatory controls on route entry, and thus operated in a less competitive, stable environment. But liberalization of aviation has increased the risk of airport assets being stranded by the opportunistic behaviour of airlines that are now free to change routes and switch airports at will. Consequently, there is now an incentive for the airport to establish with its downstream airline customers negotiated long-term[6] contracts for supply that achieve a better balance of risks. These contracts are not dissimilar to those that exist in other industrial sectors faced with similar economic circumstances – the shipping and ports industry for example.[7]

Besides specifying charges, the negotiated contract usually covers issues such as the quality of service the airport is to provide, for example minimum turnaround times; the amount of marketing support the airline is to receive; and a commitment by the airport to future investment, the nature of which is sometimes specified in detail. Conversely, as part of the agreement the airline commits to basing a certain number of aircraft at the airport; to roll out, per schedule, a route network; and sometimes to guarantee a minimum level of traffic, effectively take-or-pay contracts.[8] The average charge paid by the airline in these contracts is usually much less than the average that would result from the use of the published tariff. Payments are also structured in such a way that traffic risks are shared, for example by using a per passenger charge only. The published tariff is, of course, still used for charging those airlines for which a negotiated contract is less suitable or inappropriate.

---

6   Some have been written with 20-year terms.
7   For a review of similar arrangements in the electricity supply industry, see Littlechild (2007).
8   This description of contract terms is based on those in two contracts, details of which are known to the author.

The negotiated contract has led to a fundamental change in the nature and intensity of competition between most UK airports. Although airports still compete to attract linking air services provided by airlines based at other airports, the prime competitive focus has shifted to encouraging airlines (and associated entities such as express freight carriers) to establish an operating base at which aircraft would be positioned overnight, and to develop from this base a route network. The effect of this has been to greatly increase the bargaining power of many airlines vis-à-vis the airports.

Prior to liberalization, those airlines now commonly referred to as legacy airlines tended to focus their base operations on a specific geographical market. This was especially true of the so-called flag carrying airlines, with their capital-city focus. If they had been required to negotiate commercial contracts with their base airport, and generally the issue did not arise because of the symbiotic relationship between what were usually two public sector entities, the airline(s) would have had little countervailing power unless the city happened to be served by multiple airports with different owners. In contrast, LCCs have no specific interest in a particular geographical market; their objective is to choose locations across Europe that maximize the return on their capital (aircraft) assets. The effect is to increase considerably the countervailing power of such airlines in negotiations; the airline can credibly threaten to take their 'capital on wings' to a different location. The point is exemplified by Ryanair's frequent practice of announcing a shortlist of airports at which it might base its next tranche of aircraft and then to hold a 'beauty parade' in order to secure best terms. Thus, competition between airports is no longer simply and only a matter of competition between spatially adjacent airports; competition in the new regime takes place over a very wide geographic market, which reflects, in particular, the willingness of the LCCs to open new bases throughout Europe.

Once such a base has been established, the airline will have sunk a certain amount of costs but, until at least until the end of its contract period, it will be protected from the airport behaving opportunistically and, as already mentioned, these contracts are usually of long duration. It is also likely that the airport will wish to compete against rivals to attract the basing of future increments of capacity (not already prescribed in an existing contract) and this too will constrain its behaviour. Of course, at the end of the contract period the airline will be in a different position and will face switching costs, but it will have a stable and known environment prior to the contract termination date in which to negotiate a replacement contract or to make other arrangements; this in itself should reduce transitional costs.[9] Equally, at the end of the contract the airport also faces losing a chunk, possibly a large chunk, of its business in circumstances where most likely it is faced with a level of fixed costs much higher than the (location-specific) fixed costs faced by the airline. There will be incentives for both parties, therefore, to negotiate new contract terms and it is not immediately obvious that either the airport or the airline will have the upper hand; the most likely outcome is that the bargaining positions will be reasonably balanced.

In contrast, airlines that established operating bases at a time when negotiated contracts were not available, and operate essentially by reference to the published tariff and associated conditions of use, appear more vulnerable to increases in posted charges or other forms of opportunistic behaviour on the part of the airport; and there has recently been much discussion between airlines and regulators concerning the size of the switching costs that might be involved in moving flight operations between airports. But what is to be emphasized is that it is the *net* cost of switching

---

9 Note that the airline is not necessarily dependent upon the existence of a potentially competing airport in the same region. It is to be stressed again that the airline will be looking for the best return on its airline capital across a wider European market.

that is the important factor and financial inducements, such as marketing support by competing airports which reduce this net cost, should be taken into account. There is no reason to suppose that base airlines currently paying the published tariff, and thus supposedly stranded by their switching costs, would not be subject to inducements from rival airports. Thus, in so far as those airports strongly associated with legacy airlines find themselves competing with (low-cost) airports willing to enter into long-term contracts with airlines, they too have to respond to competition by adjusting prices for their established airline customers.

Manchester airport provides a very good example of this. It is an airport now of very similar size to London Stansted, having been in the past much larger than the latter. It also used to be head and shoulders above other airports in its north-west region, though that is no longer the case. For a time, Manchester shunned the LCCs and declined to enter into negotiated contracts with them. It preferred to continue to focus on serving the legacy airlines (British Airways had a base there) and, in this capacity, it served some long-haul routes as well as many European and domestic services (the latter including an important shuttle service to and from London Heathrow). It was also an important base for inclusive tour (charter) traffic.[10] But in the last 10 years or so it has faced increasing competitive pressures from, first, Liverpool and then more recently from Leeds Bradford and the new Robin Hood Doncaster Sheffield airport, so that its growth rate slowed and its overall share of UK passenger traffic stagnated (at around 10 per cent). It was forced to respond and it did so by selectively supporting a large number of airlines in the form of either a reduction in airport charges (or rebates), or by making large contributions to joint marketing campaigns. Between 1998 and 2003, for example, about 75 different airlines received support, although this was highly skewed – with 20 airlines receiving over 90 per cent of the expenditure. Importantly, to prevent reductions in services, it provided support (to the extent of nearly one-quarter of the total support budget in 2002–03) to airlines (and charter carriers) 'that would otherwise have ceased or reduced services'.[11]

Because the UK was at the vanguard of air services liberalization and the synergistic LCC revolution, competition between airports to attract airline operating bases soon followed, and today there are a large number of such bases in the UK. Table 17.3 shows the UK operating bases for four non-legacy airlines,[12] easyJet, Ryanair, Flybe and Jet2, in the summer of 2008. These four airlines have, between them, 32 bases in total, spread across 19 airports. It is not an exhaustive list of operating bases in the UK: bmibaby, another quasi-LCC, currently has four; VLM bases aircraft at London City airport; Aer Lingus has just opened a base at Belfast International; and, of course, British Airways has a number of bases. But Table 17.3 shows those bases most likely subject to negotiated long-term contracts along the lines described above; Gatwick and Stansted are known exceptions as BAA currently declines to enter into such contracts. The list also indicates that the use of long-term negotiated contracts now extends beyond the type of airline that essentially pioneered the approach.

---

10 It is also one of the UK's major centres for air freight.
11 Competition Commission (2002), Appendix 7.5
12 I have referred to this group as 'non-legacy' airlines rather than LCCs because only easyJet and Ryanair maintain the essential characteristics that define the original LCC brand.

Table 17.3 UK operating bases for four non-legacy airlines, summer 2008

	easyJet	Flybe	Ryanair	Jet 2
Belfast				
- Belfast City		•	•	
- Belfast International	•			•
Birmingham		•	•	
Blackpool				•
Bournemouth			•	
Bristol	•		•	
Edinburgh	•		•	•
Exeter		•		
Glasgow				
- Prestwick			•	
- Renfrew	•			
Leeds Bradford				•
Liverpool	•		•	
London				
- Gatwick	•			
- Luton	•		•	
- Stansted	•		•	
Manchester	•	•		•
Newcastle	•			•
Nottingham East Midlands	•		•	
Southampton		•		

Competitive Hinterlands

The attractiveness of an airport to an airline – and thus the contractual terms an airline is prepared to accept in order to establish at it an operating base (or serve it as an end point on a network, the hub of which is located elsewhere) – depends on a number of factors, including those that could affect operating performance and those that bear upon the anticipated average fare yield. The airline will be cognizant of the airport's infrastructure (its runway length, the standard ('category') of its instrument landing system (ILS), terminal facilities); how much spare capacity the airport has; its potential for future expansion and its freedom from operating restrictions. The fare yield will depend upon the presence of potential competitors already at the airport or at a nearby airport, and, of course, upon its perceived attractiveness to potential passengers. This latter, in turn, depends foremost upon the airport's location in relation to a market demand, the extent and depth of which is determined by factors such as population density, income levels, business activity, international trade links, tourism potential and the quality of the transport links – particularly of the regional road network – which will determine airport access times.

Access times are important in determining the overall size of the regional market, and the UK CAA has suggested that a significant number of leisure passengers are, in general, willing to tolerate access times of around 2.0 hours to reach a chosen airport (CAA, 2006, 22.17); although, for business travel, 1.0 to 1.5 hours is thought to be more appropriate. Its analysis indicates that a 2-hour drive time accounts for around 80 to 90 per cent of passengers using an airport.[13] These statistics are derived from data for the larger (and more leisure-oriented) airports in the UK, each of which serves a large number of destinations. Consequently, they might draw from a larger than average area of 'catchment'. On the other hand, airports with a smaller volume of passenger traffic might draw most of it from a more restricted catchment, perhaps within 1.0 to 1.5 hours' drive time of the airport.

From a competitive viewpoint, the issue is whether and to what extent the catchments of different airports overlap. I have previously pointed out (Starkie, 2002) that airports (and airlines) cannot segment their customer base by residential/business location; that is to say that they do not have the ability to price discriminate between customers according to where the latter are located with respect to the airport. The consequence of this is that even a small degree of catchment overlap might have a potent effect on prices. The point is illustrated in Figure 17.1. This shows stylized catchments for two airports (A and B), that overlap in the shaded area. Within this latter area, air services from the two airports compete directly for customers. However, passengers located at point Z (well outside the catchment of airport A and thus captive to airport B) are potential beneficiaries of the price set for customers located in the area of direct competition. Unless it is possible to separate passengers at Z into a market separate from those in the overlapping zone, the former passengers when using airport B will also benefit from the competitive price offered to passengers in the overlapping catchment.

The CAA has undertaken detailed analysis of the degree to which airport catchments overlap in both the London and north-west (Manchester) regions of the UK. The analysis, based on driving times, reveals that there are extensive overlaps between the various London airports for both the

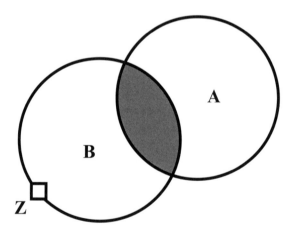

Figure 17.1 Competition and catchment areas

---

13 For London Stansted the figure was about 80 per cent, but closer to 90 percent in the case of London Luton and London Gatwick. This difference between Stansted and the two other airports might reflect the fact that Stansted is dominated by Ryanair, which has a lower average fare yield than the low-cost airlines that are relatively more important at Gatwick and Luton; thus passengers might be driving longer distances to benefit from lower fares.

1- and 2-hour 'isochrones' and, for the latter, overlaps with airports outside the London region, particularly with airports located to the north and west. In the case of the north-west region, again there is a strong overlap. Manchester is by far the largest airport in this region, with a turnover about tenfold that of nearby Liverpool (see Table 17.1); but the overlap of its catchment with those of other airports is considerable, and this remained generally the case when the analysis was repeated for different market segments (domestic, long-haul and so on).

The UK Competition Commission (CC) has also conducted a broadly similar analysis of the London and Scottish airports in relation to its *Emerging Thinking* Report (Competition Commission, 2009).[14] However it made use of a different methodology and, in particular, it did not depend upon the analysis of driving times, but used instead the results of passenger surveys. The CC's starting point was to assume that if a significant percentage of passengers originating from a district used a particular airport, then it could be inferred that all passengers in that district were potential customers of airlines operating from that airport. For the initial analysis the threshold of significance was set at 20 per cent. Therefore, the CC measured the degree of catchment area overlap by measuring the percentage of an airport's passengers that came from local authority areas where another airport accounted for 20 per cent or more of its passengers. In terms of Figure 17.1, one would calculate for Airport A the percentage of its passengers from the shaded area when Airport B attracted at least 20 per cent of the shaded area's passengers.

Subject to the availability of survey data, this analysis can be refined further. For example, it can be repeated for different market segments – leisure, business etc. – and the CC did this (although when judging whether such a disaggregated approach is really necessary again one should be mindful of constraints on the ability of airports and airlines to segment and thus price discriminate in spatial sub-markets). The CC also carried out a yet more stringent examination, deriving catchment areas only for those passengers travelling on air routes served from more than one airport. However, my own view is that this approach is too stringent. For (short-haul) leisure passengers it ignores the issue that many such passengers are purchasing a commodity with certain attributes, so that an airline route to destination X is a substitute for a route to destination Y. There are also market dynamics to consider: that in a liberalized, competitive airline market airlines are free to enter in response to perceived opportunities so that the picture regarding parallel route competition is fluid. And, in the longer term, improvements in transport infrastructure also change the shape and size of catchment areas.

The initial conclusions of the CC following its analysis of the London region airports match, in large measure, those of the CAA: that in the regions analysed there should be significant potential for airport competition (para. 167), although the common ownership of three London airports by BAA is likely to adversely affect competition between them. With respect to the Scottish airports, the CC's current view is that there is potential for competition between Glasgow and Edinburgh (para. 274) and probably also between them and Aberdeen; but again joint ownership of these three airports by BAA is a feature that adversely affects competition.

---

14   One of the UK's two competition agencies, the Office of Fair Trading (OFT), conducted in 2005/6 an investigation of UK airports (OFT, 2006). This led to a referral to the second agency, the Competition Commission, with the request that it carry out a market investigation into the supply of airport services by BAA, the dominant airport operator in two regions of the UK, with a view to increasing competition in this part of the market. This inquiry is ongoing but in April 2008 the Commission published its *Emerging Thinking* report. The Commission planned to publish its interim conclusions in August 2008 and its final report in March 2009. Unfortunately, the stipulated timetable for completion of this written paper preceded the publication of the CC interim conclusions.

The analyses of both the CAA and the CC have a particular regional focus (London, north-west England and Scotland) because each was serving a particular purpose. In order to examine the potential for competition between airports more *generally*, I have carried out an analysis of driving times between significant airports across England and Wales as a whole (Starkie, 2008). Table 17.4 shows driving times between proximate English/Welsh airports.[15] Included in the database are all those airports with scheduled passenger services and with more than 400,000 passengers in 2005/6: a total of 21. The entries in the table show times between those airports that are within 2.0 hours' drive of each other (unless the nearest neighbouring airport exceeded 2.0 hours drive time). For this purpose, driving times are taken from the RAC's Route Planner and are based on assumed speeds of 60 mph (96 kph) – for motorways the UK national speed limit is 70 mph (112 kph) and 30 mph (48 kph) for all other roads; times, therefore, are derived from very conservative speed estimates.

In spite of this conservative estimate of travel times, many airports are in surprisingly close proximity to at least one other airport (although, as noted, in south-east England as well as in Scotland, most proximate airports are owned by BAA); bear in mind that, for example, a driving time of up to 2.0 hours between two airports implies that residents located halfway (in terms of driving time) can get by car to either airport within 1.0 hour. There is, in fact, only one airport, Norwich, lying more than 2.0 hours from its closest neighbour; all of the remaining 20 airports lie within 1.5 hours of at least one other airport (and, in some cases, several airports). The average driving time beween airports is slightly more than 1.0 hour, implying an average journey time of about 0.5 hours for passengers located at the half way point on the fastest routes. This, of course, is well within the time criterion set by the CAA.

Overall, the results of these three separate analyses of airport hinterlands suggest that in the UK there is a large degree of overlap of general catchments for passenger traffic, and in this context the airport industry appears to have a potentially competitive structure (which will limit the average fare yield that the airlines can expect and, in turn, will influence negotiated contact prices). There will of course be a degree of differentiation in the market: not all airports have the infrastructure to serve long-haul destinations (although a surprising number do have this capability); for environmental reasons some have restrictions on their hours of operations; and the market for freighter operations is concentrated on only a few airports.[16] But, on the whole, the market for the provision of airport services in the UK appears to be strongly, if imperfectly, competitive; the one possible exception is the sub-market for international connecting traffics in which London Heathrow has carved out for itself a national dominance.[17]

6. Financial Performance

If the structure of the UK's airport industry is strongly competitive (at least outside the London region and Scotland where BAA is dominant), individual airports will have limited market power and are more likely to be price takers, especially when it comes to negotiating contracts to attract new business. But if, as received wisdom will have us believe, airports are subject to very high fixed costs, and thus pronounced economies of density (as well as supposed economies of scale),

---

15 Cardiff International is the only significant airport in Wales.

16 The sunk costs associated with the specialized facilities required for freight operations are also protected in a number of cases by long-term contracts.

17 In this part of the market, London Heathrow competes with mainland European hubs.

Table 17.4 Driving times between adjacent airports (hours.minutes)

	BHX	BLK	BOH	BRS	CWL	DSA	EMA	EXT	HUY	LBA	LCY	LGW	LHR	LPL	LTN	MAN	MME	NCL	NWI	SOU	STN
BHX							0.48														
BLK										1.44											
BOH																				0.42	
BRS					1.23			1.17													
CWL				1.23																	
DSA							1.22		0.48	1.20						1.44					
EMA	0.48					1.22															
EXT				1.17																	
HUY						0.48				1.32											
LBA		1.44				1.20			1.32							1.06	1.29				
LCY												1.01	0.44								0.47
LGW											1.01		0.44		1.14					1.28	1.19
LHR											0.44	0.44		1.14	0.40					1.08	1.09
LPL													1.14			0.44					
LTN	1.26											1.14	0.40							1.37	1.01
MAN	1.34	1.01				1.44				1.06				0.44							
MME										1.29								1.04			
NCL																	1.04				
NWI																		2.12			
SOU			0.42									1.28	1.08		1.37						
STN											0.47	1.19	1.09		1.01						

BHX: Birmingham BLK: Blackpool BOH: Bournemouth BRS: Bristol CWL: Cardiff DSA: Doncaster EMA: Nottingham
EXT: Exeter HUY: Humberside LBA: Leeds Bradford LCY: London City LGW: Gatwick LHR: Heathrow LPL: Liverpool
LTN: Luton MAN: Manchester MME: Durham Tees Valley NCL: Newcastle NWI: Norwich SOU: Southampton STN: Stansted

such competition can be expected to result in airports with small traffic volumes – and perhaps even not-so-small airports – generating financial losses. In part of mainland Europe it is this belief, that competition will lead to average prices below average costs for many airports, that has encouraged a planned rather than market-led approach to the development of the airport industry.

To examine whether airports in a competitive environment are generally loss-making (or whether airports are so differentiated and competition so imperfect that profits are excessive), I have used summary statistics on the financial performance of UK airports compiled by the Centre for Regulated Industries at the University of Bath. These data have the great advantage that they are subject to consistent (UK) accounting standards, but, nevertheless, their use is not without its problems. First, in spite of the data being compiled on an annual basis, comparison between years is difficult because of changes in accounting standards.[18] Second, and more importantly, airports have often reported year-by-year results covering different periods of time: either 9 months, 12 months or 15 months. Third, different airports have different depreciation policies. Finally, there are two sets of accounts available, one based on Company House returns and the other based on returns to the CAA for regulatory purposes; the two sets are, for the most part, the same but there are a few differences. The following analysis focuses on Company House data for 2005/6, which has the virtue that all airports are reporting 12 months' results.

There are 27 *individual* airports reporting financial data in the series, ranging from Southend to the east of London, with a turnover of just less than £5m at one extreme, to London Heathrow, with an excess of £1bn of annual sales.[19] But there is a discontinuity in the size range: the four airports, Heathrow, Gatwick, Stansted and Manchester, are very much larger than the remaining 23. Because of this, the fact that the financial performance of medium-sized and small airports is of more interest – and especially because in 2005/6 all four of these airports were subject to price controls[20] – they have been excluded from the following analysis. This gives a range of turnover for the remaining airports of between £5m and £111m.

Pertinent data for the 23 airports are shown in Table 17.5. Listed are: turnover, operating profit/loss (after allowing for depreciation), net profit/loss (after allowing additionally for tax and interest), operating profit as a percentage of turnover, and operating profit as a percentage of fixed assets (except for Coventry, which was excluded because of anomalies in the data). These data refer to all the activities engaged in by the respective airports, including what the economic regulatory accounts refer to as non-operational activities.

The data show that, in 2005/6, of the 23 airports nearly all were profitable; only two, Blackpool, with a turnover of £6.3m, and Durham Tees Valley, with a turnover of £10.8m, made an operating loss *and* a net loss overall.[21] Coventry, with a turnover of £14.1m, also made an operating loss but recorded a net profit, whilst Cardiff, with a turnover of £22.1m, made an operating profit, large in relation to turnover, but an overall net loss.[22] Humberside also recorded a net loss on a more modest turnover of £10.9m.[23]

---

18   Recent examples include changes in the financial reporting of employee pension liabilities.
19   Also included in the series, but in aggregate form only, are the results for the Highland and Island group of airports controlled by the Scottish Executive. Because of the aggregation, these are excluded from this analysis.
20   Manchester was de-designated in 2008, thus removing price controls.
21   Blackpool's operating loss and net loss were virtually identical, recording no movement on the tax account and virtually zero movement on the interest account.
22   Cardiff's net loss is the result of an exceptionally large tax charge.
23   Humberside's net loss is the result of a large interest payment.

Table 17.5   Financial data for the smaller UK airports, 2005–06

	Turnover (£000)	Operating profit/loss (£000)	Net profit/loss (£000)	Operating profit as % of turnover	Operating profit as % of fixed assets
Birmingham	111,109	35,477	19,458	31.9	9.9
Glasgow	82,615	25,789	15,153	31.2	10.0
Edinburgh	77,381	31,381	18,335	40.6	12.1
London Luton	77,021	12,878	5,643	16.7	13.5
Newcastle	51,360	19,072	15,309	37.1	10.9
Nottingham East Midlands	50,566	15,804	7,433	31.3	25.8
Bristol	49,619	25,344	23,465	51.1	33.7
London City	40,180	7,587	6,024	18.9	164.8
Aberdeen	33,954	10,944	8,715	32.2	11.1
Belfast International	31,206	9,436	4,700	30.2	7.9
Liverpool	28,799	18,336	20,606	63.7	17.7
Cardiff	22,103	5,953	-2,188	26.9	7.8
Southampton	22,022	8,791	5,941	39.9	9.6
Leeds Bradford	21,023	1,357	571	6.5	2.9
Exeter	17,707	1,019	32	5.8	6.1
Bournemouth	14,440	2,951	1,513	20.4	5.6
Coventry	14,123	-1,739	1,415	-12.3	N.A.
Norwich	12,089	563	71	4.7	2.3
Humberside	10,934	642	-751	5.9	2.2
Durham Tees Valley	10,834	-2,715	-1,242	-25.1	-9.8
London Biggin Hill	6,892	391	246	5.7	32.1
Blackpool	6,333	-2,953	-2,952	-46.6	-46.4
Southend	4,973	137	118	2.8	7.1

Source: Adapted from Centre for Regulated Industries, *Airport Statistics 2005/6*, Appendix D.
N.A. = not available.
Note: There is some variability in depreciation policies which might have an effect on the figures for operating profits as a percentage of fixed assets.

Although the few airports recording losses of one sort or another are among the smaller airports in the group examined, there are seven other airports falling within a similar range of turnover (up to £22m) that made both an operating profit and a net profit. These include the smallest airport, Southend; a small turnover per se does not appear to be an impediment to profitability. On the other hand, the margin of profit does appear to increase with turnover, but so too does the ratio of fixed assets to turnover; consequently operating profits expressed as a percentage of fixed assets, a broad indication of the return on capital employed, do not show such a strong association with turnover (see Figure 17.2).

Generally speaking, the better return on fixed assets were produced by airports occupying the middle range of turnover but, overall, the performance measure in Figure 17.2 does not suggest an inability of small- to medium-sized airports to make a decent return on fixed assets; the ratio of operating profits to fixed assets was over 32 per cent for the relatively small Biggin Hill for example. Most probably the ability of such airports to perform well is assisted by the multi-product nature of the industry and associated economies of scope.

Nor do the performance measures suggest, as do proponents of the 'airports are natural monopolies' school, that if no price controls are imposed the industry will make excessive returns. The average return (operating profits as a percentage of fixed assets) for the 22 airports is 15.3 (or 10.9 if we exclude the single most positive and negative outlier). This is very similar to the overall return for the non-financial sector in the UK in 2005 and 2006. To extend the comparison, non-financial *service* sector companies in the UK made an average net return of 17.9 per cent in 2005 and 19.5 per cent in 2006, whilst the corresponding returns for the *manufacturing* sector were 9.1 and 7.8; the return in the UK airports' industry falls neatly between the two (Table 17.6). Competition appears to be a most effective regulator.[24]

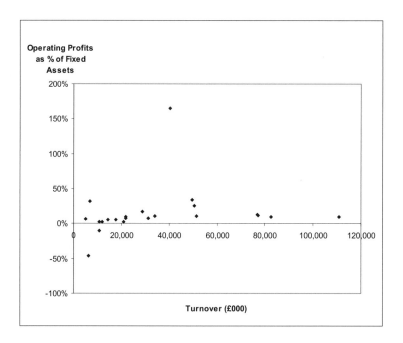

Figure 17.2 Operating profit as a percentage of fixed assets v turnover (£000)

---

24 It is perhaps useful to note that a survey of 50 European airports, most of which were in the public sector, by SH&E (2006), found that the average return on capital was 4.6 per cent.

Table 17.6   Net return (%), airports and UK private non-financial sector, 2005–06

Airports*	2005–06 15.2 (10.9†)	
	2005	2006
Non-financial service sector	17.9	19.5
Manufacturing sector	9.1	7.8
All private non-financial corporations	14.0	14.5

Source: National statistics and author's calculations.

* Airports listed in Table 17.5;

† Excluding outliers.

Conclusions

On the basis of the foregoing evidence, I would argue that a competitive framework is an achievable objective for a national airports policy. It is by no means evident that the industry is inherently a natural monopoly industry and thus requires regulation of prices or financial returns. On the contrary, the UK illustrates the ability of an airports industry to evolve a competitive structure whereby competition is an effective regulator of what the airport can charge the airline. Where there have been problems it is because of the failure to break up the state enterprise, the British Airports Authority, when it was privatized in the mid-1980s so that proximate airports in two UK regions, London and Scotland, continue in common ownership. The lesson to be drawn is clearly apparent.

And yet, in those countries where privatization or corporatization of the airports industry is on the policy agenda, as is currently the case in Spain and Portugal, it appears that economic regulation is considered the natural adjunct of such a policy, a necessary appendage.[25] The alternative approach, of restructuring ownership to provide a less concentrated, more competitive industry structure and then allowing competition to drive the industry forward, does not appear on the radar screen. I suspect that one of the reasons for this is that unless the national administrations are familiar with the processes, the fundamental problems associated with even well-designed economic regulation[26] will not be fully appreciated (and there will be an army of advisers with

---

25   For a very good overview of the European industry see Gillen and Niemeier (2008).

26   In the case of the UK utility industries this has taken the form of the price control model 'RPI-/+X' developed during the 1980s and, because this model provides incentives for economic efficiency (particularly because of its forward looking approach), it is generally considered to be superior to rate base (rate of return) regulation which preceded it. The generic model has a number of key features: a periodic review process; a building blocks approach focused on a Regulatory Assets Base (RAB) which integrates (depreciated) past and planned investments; and a process for deciding upon an allowable return on the RAB. The RAB, allowable return and efficient operating costs then form the basis for determining future allowable prices, usually for a period of five years, benchmarked against the Retail (Consumer) Price Index – hence the RPI-/+X formula. There are different variants of the approach based on this core, with a different emphasis given to different components at different times and according to the industry concerned (for example, applied to the airports industry there can be commercial revenues from retailing to take into account).

a vested interest in a regulated solution; the consulting industry, for example, does well out of offering solutions to regulatory problems).

The most important of these perceived problems is that the price control approach could discourage investment, the so-called 'hold-up' problem, and encourage the undersupply of service quality. In the former case, because the regulator can commit to a regulatory settlement for only a limited period of time (usually five years) but investments are amortized over much longer periods, the regulated firm is faced with the risk that (future) regulators might renege on the regulatory settlement, thus reducing incentives to invest.[27] In the case of service quality the price-regulated firm can save costs and increase its return by skimping on quality, but the regulator finds the problem difficult to address because of the difficulty of judging an optimal level of service quality.

And yet the irony of this situation is that, as we have seen, the unregulated airports industry reaches its own solution to these problems: it establishes long-term vertical supply contracts with its airline customers. The long-term nature of the contract provides the security that the airport needs to sink costs in additional infrastructure, thus avoiding the hold-up problem, and the terms of the contract stipulate the quality of service that the airline expects from the airport. It is, after all, the way in which similar issues are resolved in much of the market economy.[28] In contrast, the effect of regulation can be to crowd out the efficient solution.[29]

It would be better, therefore, if policymakers when undertaking industry reviews, instead of reaching first for the regulatory tool-box, pose the question: is the structure of the industry such that a reasonably competitive outcome is likely? If not, can the industry be restructured to make it more competitive? For the UK's airport industry, as I have tried to show, competition appears to have worked well and led to a dynamic industry, free of subsidy[30] yet profitable for both small and large airports, with an overall level of profit similar to that for the non-financial sector of the UK economy: a most satisfactory state of affairs.

References

Civil Aviation Authority (2006) Airport Regulation: Price Control Reviews – Initial Proposals for Heathrow, Gatwick and Stansted Airports, December, CAA, London.

---

27 There are counter-arguments that suggest that price-controlled airports might overinvest. Furthermore, there is also the issue of a suitable cost of capital. It is argued that the standard approach, deriving the weighted average cost of capital (WACC) with reference to a partly debt-financed RAB and an equity-based capital expenditure programme, does not provide enough incentive for equity capital (particularly in circumstances where there are, as is probably the case with large airports like Heathrow, decreasing returns to scale).

28 By coincidence at the time I was writing this chapter, the following example appeared in the *Financial Times* (3 July 2008): 'Scottish Coal has agreed to sell about 2 million tonnes of coal a year – half its current output – to Scottish Power to feed its two coal-fired power stations ... The coal will be sold at an undisclosed fixed price, which Scottish coal said gave both parties the certainty they need to invest in new mines and power generation equipment.'

29 London Luton airport provides an interesting footnote on this point. EasyJet, an important customer of Luton, attempted in its earlier days at the airport to get the airport subject to price control. It failed to do so but it then reached a negotiated long-term contract with the airport, which, when judged against the published tariff, was on terms most favourable to the airline.

30 The exceptions are the Highland and Island airports subsidized by the Scottish Executive for social reasons.

Competition Commission (2002) Manchester Airport PLC: A Report on the Economic Regulation of Manchester Airport PLC, CC, London.

Competition Commission (2008) BAA Market Investigation: Emerging Thinking (April), CC, London.

Condie, S., (2004) Powerful Customers: Working with the Airlines, in Vass, P. (ed.), *The Development of Airports Regulation – A Collection of Reviews*, CRI, School of Management, University of Bath.

CRI, (2006) Statistical Series: The UK Airports Industry, Airport Statistics 2005/6, CRI, School of Management, University of Bath.

Gillen, D. and Niemeier, H.-M. (2008) The European Union: Evolution of Privatisation, Regulation, and Slot Reform, in C. Winston and G. De Rus, *Aviation Infrastructure Performance: A Study in Comparative Political Economy*, Brookings Institute, Washington DC.

Graham, A., (2001) *Managing Airports: An International Perspective*, Butterworth Heinemann, Oxford.

Graham, A., (2008) Airport Planning and Regulation in the United Kingdom, in C. Winston and G. De Rus, *Aviation Infrastructure Performance: A Study in Comparative Political Economy*, Brookings Institute, Washington DC.

Littlechild, S.C., (2007) Beyond Regulation, in C. Robinson (ed.), *Utility Regulation and Competitive Markets*, Edward Elgar, Cheltenham.

Office of Fair Trading (2006) UK Airports: Report on the Market Study and Proposed Decision to Make a Market Investigation Reference, OFT 882, London.

SH&E (2006) Capital Needs and Regulatory Oversight Arrangements: A Survey of European Airports. ACI Europe.

Starkie, D. (2002) Airport Regulation and Competition, *Journal of Air Transport Management*, 8, 63–72.

Starkie, D. (2008) *Aviation Markets: Studies in Competition and Regulatory Reform*, Ashgate, Aldershot/IEA, London.

Chapter 18

The Effect of Low-Cost Carriers on Regional Airports' Revenue: Evidence from the UK

Zheng Lei, Andreas Papatheodorou and Edith Szivas

Introduction

There are approximately 200 airports in Europe which are classified as underutilized, with fewer than 1 million passengers per annum: the majority of these 'Class D' airports – to use the European Commission (EC) nomenclature – are loss-making, publicly owned and subsidized by central, regional or local governments (Caves and Gosling, 1999; European Commission, 2005; Scheers, 2001). This phenomenon is also apparent in Great Britain, where the lack of a national airport policy resulted in the development of a large number of regional airports without consideration of what was offered at nearby competing airports (Graham, 2004). Hence, several neighbouring regions have airports with overlapping catchment areas (e.g. Liverpool – Manchester, Cardiff – Bristol, Birmingham – East Midlands, Newcastle – Teesside). Illustratively, the majority of the population in the United Kingdom has more than one regional airport within a 90-minute drive (Graham, 2004). Consequently, competitive pressure on airports has led airport managers to seek a passenger throughput increase to reach critical mass for the sustainable financial operation of their facilities.

Nonetheless, a generic strategy to increase the number of passengers may prove ineffective unless complemented by an analysis to assess the financial impact of alternative airline business models (i.e. full-service, charter and low-cost) on the airports' revenue structure (i.e. aeronautical and non-aeronautical). In fact, one of the most critical questions to ask is whether regional airports should trade off a reduction of their aeronautical revenue for an increase in their non-aeronautical income streams. This is especially important as the recent rise of low-cost carriers (LCCs) in Europe has relied heavily on the assumption that such a trade-off is mutually profitable (i.e. for both the airline and the airport). In this context, the present chapter sets out to measure, compare and contrast the impact of LCCs, full-service and charter carriers on the revenue of regional airports. The next section briefly discusses the importance of LCCs for airport competition. We then present the empirical study based on the British experience, as this is the largest, oldest and most competitive market for LCCs in Europe. The final section summarizes and concludes.

Low-Cost Carriers and Regional Airports

Over the past decade, the nature of air travel in Europe has changed dramatically. The deregulation and gradual liberalization of the intra-European aviation market, completed in April 1997, encouraged new no-frills, low-cost carriers to establish scheduled services across Europe. Their low fares, high frequency and flexibility stimulated significant demand for short-haul air travel and led to a phenomenal growth of traffic at regional airports (Papatheodorou and Lei, 2006).

The rise of LCCs may be essentially attributed to their cost leadership strategy. This is a challenging task to pursue in every competitive environment as each decision made by the management team must be guided by the principle of cost control (Sorenson, 1991). Achieving cost leadership in the airline industry is even more difficult as there is strong evidence against the existence of increasing returns to scale in airline networks (Caves, 1962; Strazheim, 1969; White, 1979; Gillen et al., 1985). As cost minimization is their top priority, LCCs argue that, having pushed all other cost inputs to low levels, the only cost variable that they can further reduce is airport charges (Doganis, 2002).

Led by Ryanair, LCCs always aim at securing concessions for airport charges, as the widely publicized case of Brussels Charleroi Airport has revealed (European Commission, 2004). This is because LCCs are in a strong bargaining position to threaten flying elsewhere unless reductions in charges or commercial incentives are granted by the airport. Such threats are credible because unlike network airlines, LCCs do not rely on transfer passengers (Starkie, 2002). Moreover, LCC passengers are usually price-sensitive and willing to travel longer distances to the airport if fares are low enough. Therefore, LCCs have incentives to seek out airports that minimize their operating and station costs and then create a market around these airports (Starkie, 2002). Consequently, they have more scope for switching operations between different airports to reduce costs (Papatheodorou, 2003).

Obviously, not every regional airport is in a vulnerable position when dealing with LCCs. London Luton is an interesting case of a satellite airport serving a large population, and an important business and tourism destination: thus, the airport has much stronger bargaining power over LCCs than smaller regional airports in peripheral areas of Europe. After its initial five-year contract ended, easyJet saw a near four-fold increase in its landing charges at Luton, i.e. from £1.57 per departing passenger to £5.5: still, this is considerably lower than normal charges (Button et al., 2002).

On the one hand, attracting LCCs is an appealing attempt for airport managers to improve financial performance, as LCCs could bring in passengers from a much wider catchment area (Barrett, 2004; Francis, et al., 2003). On the other hand, it is observed that LCCs sometimes take advantage of their bargaining power and drive the airport charge below marginal cost (Francis et al., 2003). To provide a rationale for their aggressiveness, LCCs claim that because they do not offer complimentary in-flight catering or other services, their passengers tend to spend more at airports than travellers on full-service or charter flights; hence, the aeronautical revenue lost by the airport can be compensated from increased non-aeronautical revenues. Despite the importance of this issue, few empirical studies have been carried out so far (Francis et al., 2003).

Empirical Study

Twenty regional airports in the UK enter the sample over a nine-year period (1995–2004). The airports considered differ substantially in size: London Gatwick (the largest in the sample) recorded 32.1 million passengers in 2000–01, while Blackpool (the smallest) handled only 46,820 passengers in 2002–03. For this reason, the data set is subdivided into two groups – large and small airports – using 3 million passengers per year as the cutting threshold. Of the 20 airports in the study, ten are classified as large-size: namely, Belfast, Birmingham, Bristol, Edinburgh, Glasgow, London Gatwick, London Luton, London Stansted, Manchester and Newcastle (in alphabetical order). The remaining ten airports are classified as small-size, namely, Aberdeen, Blackpool, Bournemouth, Cardiff, East Midlands, Exeter, Leeds Bradford, Liverpool, Southampton and

Teesside. Airline passenger data are obtained from the UK Civil Aviation Authority (CAA) and adjusted to the UK financial year (1 April to 31 March). Airport revenue data are collected from the Annual Airport Statistics published by the Centre for the Studies of Regulated Industries and deflated using the UK Consumer Price Index (CPI). It would be interesting to distinguish different types of non-aeronautical revenue, particularly revenue from car parking. As compared to airport charges, car parking charges could be quite high, especially for holiday travellers. This could be a more important source of non-aeronautical revenue for airports with a predominance of origin traffic than for those that mainly serve as destinations. However, as data for car parking are not available, only aggregate data for non-aeronautical revenue are used in this analysis. Table 18.1 shows relevant descriptive statistics.

Although airport personnel can affect the magnitude of non-aeronautical revenue (*NAR*), the primary drive is passengers (Gillen and Hinsch, 2001). Aeronautical revenue (*AR*) in the UK is based on charges levied according to passenger numbers and aircraft size. Aircraft size data is difficult to obtain, hence only passenger numbers are used in this analysis. Two empirical models are estimated in linear form. The number of passengers carried by LCCs, charter and full-service carriers are used as explanatory variables in both *NAR* and *AR* models. The following stochastic equations demonstrate the models estimated:

$$NAR_{it} = \alpha_i + \beta_1 Lccpax_{it} + \beta_2 Charterpax_{it} + \beta_3 Fullservicepax_{it} + \varepsilon_{it} \quad \text{(Model 1)}$$
$$AR_{it} = \delta_i + \gamma_1 Lccpax_{it} + \gamma_2 Charterpax_{it} + \gamma_3 Fullservicepax_{it} + u_{it} \quad \text{(Model 2)}$$

where the subscripts *i* and *t* indicate the airport and time period (year) respectively; α and δ are coefficients that vary across airports but are constant over time; β and γ are coefficients which are constant across airports and time; ε and u are the error terms, which are identical and independently distributed for all *i* and *t*; *NAR* and *AR* are non-aeronautical and aeronautical revenues, respectively, in airport *i* in year *t*; *Lccpax, Charterpax* and *Fullservicepax* are the total number of terminal passengers carried by LCCs, charter and full-service carriers, respectively, in airport *i* and year *t*. All β and γ are expected to have positive signs.

Models 1 and 2 have been estimated using panel data techniques to account for differences across airports and time. In principle, estimation can be done in three ways, depending on whether the individual cross-section effects are considered to be constant, fixed or random. Following Greene (2003), Breusch-Pagan Lagrange multiplier (LM) and Hausman tests are applied to choose the appropriate model formulation. The LM test is used first to compare the random

Table 18.1 Descriptive statistics

Variables	All airports (N=20)		Large airports (N=10)		Small airports (N=10)	
	Mean	St. Dev.	Mean	St. Dev.	Mean	St. Dev.
Total pax	5,020,429	6,963,668	8,885,861	8,149,590	1,154,998	946,312
Non-aeronautical revenue	2.65E+07	4.46E+07	4.77E+07	5.55E+07	5,245,582	3,301,733
Aeronautical revenue	2.76E+07	3.26E+07	4.68E+07	3.69E+07	8,510,971	6,063,746
LCC pax	872,499	2,078,598	1,509,491	2,752,844	235,507	539,347
Charter pax	1,754,613	2,812,517	3,058,631	3,504,760	450,596	432,327
Full-service pax	2,393,317	3,899,565	4,317,739	4,780,763	468,895	488,227

effects specification against the simple linear model. Acceptance of the null hypothesis means that the classical regression model with a single constant term is appropriate, i.e. the model can be estimated by Pooled Ordinary Least Squares (POLS). Conversely, rejection of the null hypothesis is in favour of the random effects specification. However, even under this circumstance, we cannot jump to the conclusion that the model has random effects as there is another competing model, i.e. fixed effects model. The Hausman test has been subsequently used to determine the choice between fixed and random effects models. Under the null hypothesis of the Hausman test, both models are consistent but the random effects specification is more efficient; acceptance of the alternative hypothesis suggests that the random effects model is inconsistent and the fixed effects model is more appropriate.

Table 18.2 reports the results of Model 1. The significant LM test statistics suggest that a common slope for the different panel groups in the pooled regression cannot be assumed: a simple OLS regression of a straightforward pooling of all observations without considering heterogeneity would report biased results. Therefore, the choice would be either fixed or random effects models.

One particular issue in the panel data analysis is that the classical error component disturbances in the panel model assume that the only correlation over time is due to the presence of the same individual across the panel (Baltagi, 2005). This may be a restrictive assumption for the economic relationship between airline traffic and airport revenue, as an unobserved shock in period t may affect the behavioural relationship for at least the next two periods. This type of serial correlation cannot be embedded into the simple error component model. Nonetheless, ignoring the presence of serial correlation may result in consistent but inefficient estimates of the regression coefficients

Table 18.2   Impact on regional airport's non-aeronautical revenue

Dependent variable: NAR (non-aeronautical revenue)			
	All airports	Large airports	Small airports
	AR(1) Fixed effects	AR(1) Random effects	AR(1) Random effects
Constant	1.45E+07	-4604311	1945806
	(6.51)***	(1.11)	(2.90)***
Lccpax	2.84	3.59	3.62
	(4.25)***	(5.63)***	(14.19)***
Charterpax	0.73	8.73	3.10
	(0.26)	(5.94)***	(3.56)***
Fullservicepax	3.39	4.70	2.27
	(2.67)***	(4.54)***	(3.09)***
Overall R^2	0.90	0.93	0.69
rho	0.64	0.64	0.61
LM Test	116.42	26.88	199.76
Hausman Test	9.83	5.39	5.00

Notes: Figures in parentheses are t values; *, ** and *** indicate significance at the 10 per cent, 5 per cent and 1 per cent levels respectively.

and biased standard errors (Baltagi, 2005). In Model 1, the estimated autocorrelation coefficient (*rho*) varies between 0.61 and 0.64, which is fairly large, showing strong evidence of serial autocorrelation. Following Greene (2003), estimates of the fixed effects and random effects models were corrected for first order serial correlation – *AR(1)* – using the Cochrane-Orcutt procedure.

The results of the Hausman test suggest that the fixed effects model is more appropriate for the whole sample and the large airport subgroup, while the random effects model is more efficient for the small airport subgroup. Only the preferred models are shown in Table 18.2. All explanatory variables have the expected signs, and most of them are highly significant across the three samples. Given the linear form, the coefficients may be interpreted directly. In the pooled sample, full-service carriers have a larger positive impact on sample airports than LCCs. On average, every additional full-service carrier passenger contributes £3.39 to an airport's commercial revenue, compared to £2.84 for LCC passengers. The effect of charter carrier passengers on commercial revenue is insignificant. Interestingly, the results for the two subgroups differ. In the large airport group, each additional charter carrier passenger has the largest contribution to airport commercial revenue (£8.73), followed by full-service and low-cost carriers passengers (£4.70 and £3.59, respectively). The results are not surprising as charter passengers usually travel for leisure purposes and have more desire to purchase goods and/or services at airports. Moreover, they usually stay longer at airports, thus providing more commercial revenue opportunities for airports. Similarly, full-service passengers may belong to higher expenditure groups than their low-cost counterparts; alternatively, they may travel on business and wish to make some last-minute purchases (for personal use or gifts) from the airport.

Nevertheless, in the small airports group, LCC passengers turn out to be the most significant contributor to airport commercial revenue. On average, each additional LCC passenger spends £3.62 on airports, which is higher than charter carrier and full-service carrier passengers (£3.10 and £2.27, respectively). A possible explanation for these results is that specialist stores selling luxury products are usually absent in small regional airports, while those stores are important revenue providers for large airports. Moreover, large airports are very likely to derive far more commercial revenue from duty-free sales from charter and full-service carrier passengers than small airports. In contrast, restaurants, newsagents, currency exchange, car rental and small shops are the main commercial revenue sources of small airports. As LCCs usually do not provide free food on board, it is plausible that their passengers tend to spend more on catering than full-service and charter passengers.

With respect to the aeronautical revenue model, the Cochrane-Orcutt procedure is again adopted to correct for first order serial correlation. Based on the same econometric procedure, the fixed effects model is preferred for the whole sample and the large airport subgroup, while the random effects model is more efficient for the small airport subgroup. Only the preferred models are reported in Table 18.3 for parsimony.

Table 18.3 shows that all explanatory variables are highly significant and have the expected signs. In all cases, airports generate more aeronautical revenue from charter airline passengers than full-service and LCC passengers. In the pooled sample, on average, each additional charter carrier passenger contributes £4.74 to the airport's aeronautical revenue, compared to £3.23 and £1.71 generated by each additional full-service and LCC passenger, respectively. These results are close to those obtained in the large airport subgroup, in which it shows the contribution to aeronautical revenue by each additional passenger carried by charter carriers, full-service carriers and LCCs is £4.28, £3.25 and £1.67, respectively. These results are plausible as LCCs are able to negotiate lower aeronautical fees with airports; hence aeronautical revenues from LCC passengers are the lowest in both cases. The reason charter carrier passengers attract more aeronautical charges than their full-service counterpart is most likely because more baggage handling is needed for the former.

Table 18.3 Impact on regional airport's aeronautical revenue

Dependent variable: AR (aeronautical revenue)			
	All airports	Large airports	Small airports
	AR(1) Fixed Effects	AR(1) Fixed Effects	AR(1) Random Effects
Constant	1.07E+07	1.79E+07	1216259
	(10.96)***	(7.30)***	(3.68)***
Lccpax	1.71	1.67	2.34
	(4.98)***	(3.38)***	(10.15)***
Charterpax	4.74	4.28	10.58
	(3.46)***	(2.17)**	(20.75)***
Fullservicepax	3.23	3.25	4.25
	(5.27)***	(3.75)***	(9.54)***
Overall R^2	0.89	0.84	0.97
rho_	0.68	0.68	0.60
LM Test	454.38	164.92	59.73
Hausman Test	10.58	8.23	5.88

Notes: Figures in parentheses are t values; *, ** and *** indicate significance at the 10 per cent, 5 per cent and 1 per cent levels respectively.

Interestingly, in the small airport subgroup, there is a marked difference for aeronautical revenue generated by charter carrier passengers compared to the other two categories of airline passengers. Each additional charter carrier passenger contributes £10.58 to the airport's aeronautical revenue, more than twice as much as that generated by an additional full-service carrier passenger, i.e. £4.25. As for LCCs, on average, each additional passenger contributes only £2.34 to the airport's aeronautical revenue. This can be related to the fees structure adopted by the majority of British airports – international passengers pay the highest charges, followed by intra-EU and domestic passengers. At small regional airports, the majority of charter carrier passengers are long- to middle-haul holidaymakers, while for full-service carriers, regional airports are usually used to feed passengers into their main hubs, hence dominated by short-haul operations. As for LCCs, they mainly fly to short-haul destinations as well. Therefore, this explains the highest aeronautical charge imposed on charter carrier passengers. Again, as LCCs are able to negotiate lower aeronautical charges than full-service carriers, the LCC passengers, on average, end up paying the least fees.

Discussion and Concluding Remarks

This chapter attempts to measure the impact of alternative airline business models on regional airports' revenue. According to the findings of the UK study, the impact of LCC passengers on airport commercial revenue is much smaller compared to charter and full-service carriers in the large airport subgroup. Nevertheless, LCCs emerge as the largest contributor to commercial revenue in the small airports subgroup. In terms of aeronautical income streams, results are clearer: LCCs' contribution is smaller compared to full-service carriers and charter airlines in both the large and small airport subgroups.

Table 18.4 summarizes these results by presenting the impact of the three airline business models on aggregate airport revenue. The numbers reported here have not been calculated by a separate econometric model, but reflect the summation of commercial and aeronautical revenue in each case. Full-service airlines (FSA) seem to offer the best alternative for the pooled sample, as each of their passengers generates a total of £6.62 for the airport on an almost equal income basis – i.e. 51.21 per cent for commercial (NAR) as opposed to 48.79 per cent for aeronautical (AR) revenue. On the other hand, in both sub-samples (large and small airports) charter carriers take the aggregate lead even with converse revenue structures – i.e. a clear focus on NAR (67.10 per cent) for large, and on AR (22.66 per cent) for small airports.

Does it really make commercial sense to adopt an airport strategy aimed at attracting LCCs? Although Table 18.4 would rather suggest a negative answer to this question, the reality seems to be more complicated. In addition to the obvious need to replicate the econometric exercise with a different sample, there are three other issues to consider. First, the present study focused entirely on the revenue side; what really matters, however, is the assessment of profitability, which also requires a thorough study of airport operating costs. For example, Table 18.4 suggests that over 60 per cent of airport revenue from LCCs is derived from commercial activities such as catering facilities etc. If the provision of the relevant airport infrastructure costs less (on a unit basis) compared to the infrastructure related to aeronautical business and/or other types of commercial activities, then the end result on airport profitability may be better.

Second, results in Tables 18.2–4 refer to the incremental (and also to the average, since we assume a linear specification) passenger. LCCs, however, are predominantly characterized by their massive scale of passenger flows resulting from their low fares. For example, if on average an LCC is able to generate 13.68/5.96 = 2.29 times the number of passengers of a charter airline in a small airport, then it is preferable to attract the LCC instead of the charter airline, at least from a revenue point of view.

Finally, and from an airport's competition perspective, a focus of all airports solely on full-service or charter carriers based on the results of Tables 18.2–4 could prove futile as these airlines could then exercise their bargaining power to secure increased concessions, to the detriment of airport profitability. In this context, a differentiation strategy – where some regional airports decide to focus predominantly on full-service airlines, whereas others focus on charter carriers and the remaining ones on LCCs – could be a way forward. Such a strategy would also make sense from a supply side point of view given the different infrastructural requirements (i.e. in terms of boarding bridges, ramp operations, baggage handling etc.) associated with the various airline business models.

Table 18.4   Impact on regional airports' aggregate revenue

	All	%NAR	%AR	Large	%NAR	%AR	Small	%NAR	%AR
LCC	4.55	62.42	37.58	5.26	68.25	31.75	5.96	60.74	39.26
Charter	5.47	13.35	86.65	13.01	67.10	32.90	13.68	22.66	77.34
FSA	6.62	51.21	48.79	7.95	59.12	40.88	6.52	34.82	65.18

References

Baltagi, B.H., 2005, *Econometric Analysis of Panel Data* (3rd edn), Chichester: John Wiley & Sons.
Barrett, S.D., 2004, How do the demands for airport services differ between full-service carriers and low-cost carriers? *Journal of Air Transport Management*, 12, 33–9.
Button, K., Haynes, K. and Stough, R., 2002, *Towards an Efficient European Air Transport System*, Brussels: Association of European Airlines.
Caves, R., 1962, *Air Transport and Its Regulation*, Knoxville: University of Tennessee Press.
Caves, R. and Gosling, G., 1999, *Strategic Airport Planning*, Oxford: Pergamon.
Doganis, R., 2002, *Consultancy Advice on Aviation Issues for the Department of the Taoiseach, Ireland*, London: Rigas Doganis & Associates.
European Commission, 2004, *The Commission's Decision on Charleroi Airport Promotes The Activities of Low-cost Airlines and Regional Development*. IP/04/157, Brussels: European Commission.
European Commission, 2005, *Community Guidelines on Financing of Airports and Start-up Aid to Airlines Departing From Regional Airports*. Brussels: European Commission.
Francis, G., Fidato, A. and Humphreys, I., 2003, Airport–airline interaction: the impact of low-cost carriers on two European airports, *Journal of Air Transport Management*, 9, 267–73.
Gillen, D. and Hinsch, H., 2001, Measuring the economic impact of liberalisation of international aviation on Hamburg airport, *Journal of Air Transport Management*, 7, 25–34.
Gillen D.W., Oum, T.H. and Tretheway, M., 1985, *Airline Costs and Performance: Implications for Public and Industry Policies*, Centre for Transportation Studies, University of British Columbia, Vancouver.
Graham, A., 2004, Financial performance of smaller airports, Westminster Seminar on Regional and Low-cost Air Transport: Opportunities and Challenges, University of Westminster, June.
Greene, W., 2003, *Econometric Analysis* (5th edn), Upper Saddle River NJ: Prentice Hall.
Papatheodorou, A., 2003, Do we Need Airport Regulation? *Utilities Journal*, 6(10): 35–7.
Papatheodorou, A. and Lei, Z., 2006, Leisure travel in Europe and airline business models: A study of regional airports in Great Britain, *Journal of Air Transport Management*, 12: 47–52.
Scheers, J., 2001, Attracting investors to European regional airports. What are the prerequisites? *International Airport Review*, 5(4), 55–63.
Sorenson, N., 1991, The impact of geographic scale and traffic density on airline production costs: the decline of the no-frills airlines, *Economic Geography*, 67, 333–45.
Starkie, D., 2002, Airport regulation and competition, *Journal of Air Transport Management*, 8, 63–72.
Strazheim, M., 1969, *The International Airline Industry*, Washington DC: The Brookings Institution.
White, L.J., 1979, Economies of scale and the question of natural monopoly in the airline industry, *Journal of Air Law and Commerce*, 44, 545–73.

PART D:
Policy Issues

Chapter 19
Competition and the London Airports: How Effective Will It Be?

Peter Forsyth and Hans-Martin Niemeier[1]

Introduction

The performance of the London airports has long been controversial. The major airports, Heathrow, Gatwick and Stansted – all owned by the British Airports Authority (BAA) – have been criticised for providing inadequate capacity, being too crowded, offering poor quality of service and having excessively high costs and charges. Criticism in the press has been longstanding (see the *Economist*, 1995). After several crises in recent years, with the takeover of BAA by Grupo Ferrovial in 2006 and culminating in the problems associated with the opening of Heathrow's Terminal 5, the performance of the London airports has become a live public policy issue. Breaking up BAA, and allowing competition between the London airports, is seen as either the primary way, or at least a major way in which improved performance of the airports can be stimulated (the *Economist*, 2006a and 2006b, reflects these views). BAA also owns three airports in Scotland, including Edinburgh and Glasgow airports, which may be able to compete. The UK Competition Commission has provisionally recommended BAA be required to sell two of its London airports and either Edinburgh or Glasgow airport (Competition Commission, 2008b). Its final report was published in March 2009. BAA has announced it will be selling Gatwick airport.

While BAA's performance has been seen lately as being poor, there are several factors beyond its control which have impacted on it. BAA is subject to regulation which influences the incentives it faces to invest and deliver on quality. Its ability to invest is also constrained by environmental considerations, which have resulted in detailed planning procedures and explicit government intervention. Capacity at London's Heathrow and Gatwick airports is very tight, and this has resulted in congestion and lower-quality services. BAA has not been free to invest where and when it wished. When the London airports are separated, it is probable that their ability to compete will continue to be limited by these factors. Thus the question arises of how effective competition between the airports will be, both in the short and long term.

In this chapter, we concentrate on the competition issue in London. Separating the London airports involves more than this. Separation means ownership changes, and these could impact on performance. Indeed, it could be that new and separate ownership might have a larger impact than competition, per se. It is also possible that separate ownership could lead to less prescriptive regulation, which could impact on performance. It also means that any advantages from coordinating investments between airports, by a common owner, would be lost (though, as noted later, actual investment programmes are very much influenced by regulators, planning controls and government policies).

---

1   We would like to thank Mike Toms and David Starkie for their comments on this chapter. Furthermore, we thank Karsten Fröhlich for the valuable research assistance.

The historical setting is obviously important to understand and evaluate the demand for breaking up BAA. Therefore we start with a historical overview of how BAA managed its efficiency problems and how decisions concerning the regulatory framework and ownership were taken. In the next section we define the efficiency issues more precisely. Thereafter we focus on the following research questions:

- Does separation increase competition so much that regulation is not necessary?
- If regulation is still necessary, how effective is separation in promoting competition and does it increase efficiency?

Thereafter we summarise our results, addressing the question what separation achieves and does not.

Background

The British Airports Authority was established in 1965 as the owner of the major London airports, Heathrow and Gatwick, and, in addition, Prestwick in Scotland (see Table 19.1). Later on it acquired Aberdeen, Edinburgh and Glasgow airports in Scotland. The airports were run as public utilities. The stated objective of BAA was to maximise social welfare. In 1972, like other public utilities, it ostensibly adopted the recommend pricing scheme of Long Run Marginal Cost Pricing (Little and McLeod, 1972). Efficient pricing is linked to the need to provide optimal capacity at the right time (Forsyth, 1972). From the 1970s onwards Heathrow, and later on Gatwick, faced excess demand at certain hours a day. Peak pricing was strongly opposed by airlines (Toms, 1994), but according to Starkie and Thompson (1985) the level of charges was below long-run marginal costs and the peak charges did not clear the market. As with other airports, slots were used to clear the market.

In 1985 the government announced its policy on capacity and ownership in its White Paper on Airport Policy and proposed privatisation of BAA as a group. Starkie and Thompson (1985) criticised this because the disadvantages of divided ownership 'seem to be more apparent than real' (p. 81). They found hardly any empirical evidence for economies of scale in an airport system, and no operational cost savings from common overhead costs (pp. 50ff.). They criticised the cross-subsidisation of Stansted as mainly due to providing excessive quality and thereby potentially distorting competition with Luton. The central message was that effective competition between separated airports was possible because 'Stansted and Gatwick have the potential to compete strongly with each other (and with Luton) in the large and 'foot-loose' inclusive-tour and intercontinental discount fare market. … In addition, Gatwick is shaping up as a promising competitor to Heathrow' (p 81).

This competition should be combined with regulation as 'individual airports would still retain a degree of market power'(p 83). Regulation with separate ownership would be more effective as it will generate competing sources of information (p 83 ff). However the government privatised BAA as a group for a price of £ 1.5 billion. Price cap regulation was established on 1 April 1987. The cap was based on the single till principle and on average revenues per passenger. Vickers and Yarrow (1988) were critical of privatisation and regulation because 'privatisation of BAA was simply the transfer to private hands of a monopoly with valuable property assets' (p 366), which would not 'improve the economic efficiency of airport operations: it was primarily a financial operation designed to serve other objectives' (p. 366).

Table 19.1 Overview of regulatory milestones of BAA

Year	Major events
1965	Airports Act sets up British Airport Authority (BAA) with Heathrow, Gatwick, Stansted and Prestwick
1971 and 1975	Acquires Edinburgh, Aberdeen and Glasgow
1972	BAA applies long run marginal cost pricing with peak pricing
1977	Government establishes traffic distribution rules
1983	Memorandum of understanding between the US and the UK on airport user charges with single till principle
1985	Department of Transport White Paper on airport policy on capacity, ownership and regulation
1985	Study of Starkie and Thompson on privatising London's airports recommends a break-up
1986	Airports Act announcing the sale of BAA as a single entity in 1987
1987	BAA plc was floated
1991	Traffic distribution rules are abolished
1991	The Monopolies and Mergers Commission rejected a break-up
1996	Transport Committee recommends separating Heathrow from Gatwick and Stansted but the government prefers joint ownership
1999	Government investigation rejects break-up
2002	CAA recommends move to dual till and investment incentive mechanism but Competition Commission disagrees. In the end the single till prevails and price caps are adjusted for cost of capital expenditure
2003	White Paper 'The Future of Aviation' proposed a second runway at Stansted (2011/12) and a third runway at Heathrow (2015–20) under environmental standards, or if these are not met a second runway at Gatwick (after 2019)
2006	Office of Fair Trading enquires whether the market structure works well for the consumer
2006	Ferrovial buys BAA
2007	Competition Commission starts investigation into the break-up of BAA
2008	Terminal 5 starts operating, causing substantial delays
2008	Competition Commission recommends break-up of BAA

Sources: Graham (2008), House of Commons Transport Committee (1996), Starkie and Thompson (1985), Toms (2004, Vickers and Yarrow (1988).

Table 19.2 Passenger numbers at London airports (millions)

	1987	1990	2000	2005	2007
Heathrow	34.7	42.6	64.3	67.7	67.8
Gatwick	19.4	21.0	31.9	32.7	35.1
Stansted	0.7	1.2	11.9	22	23.8
Luton	2.6	2.7	6.2	9.1	9.9
Manchester	8.6	10.1	18.4	22.1	21.9
All airports (UK)	80.5	102.4	179.9	228.2	240.7

Source: CAA (2003, 2008).

The break-up issue was considered from time to time, and in 1999 the government investigated the issue again, but concluded that a break-up would not enhance competition because 'the scope for such competition was currently constrained by the lack of unused capacity and the planning system which means that decisions on whether there should be substantial new airport infrastructure in South East will in practice be a matter of government' (Department of the Environment, 2000, para. 220).

Price cap regulation reduced the level of charges substantially at the London airports (see Table 19.3). This was largely due to the single till principle, the duty-free tax exemption and the innovative non-aviation strategy of BAA management, which led to high revenues and profits, which in turn raised the X-factor and lowered the level of charges. As excess demand is rationed by slot controls, this created substantial rents and a situation in which airport charges at Heathrow were lower than at other London airports which have less scarce capacity. The Civil Aviation Authority (CAA) made several attempts to suggest a pricing structure reflecting these scarcities, but BAA was reluctant to follow them given the opposition of airlines. In 2001 the CAA sought to deal with the investment in new terminal capacity. It found that historic accounting values of land were below the opportunity costs. The move to the dual till principle was backed by study of Starkie and Yarrow (2000) on the advantages and disadvantages of the single till principle. The Competition Commission found the 'arguments and current evidence for moving to the dual till … not persuasive' (CAA, 2002a). In the end the CAA compromised with a price cap on a single till but with similar high charges as under a dual till (CAA, 2003, Hendriks and Andrew, 2004).

In 2006 two parallel events happened. In February Ferrovial, a Spanish construction and infrastructure company, announced it was to bid for BAA (*Economist*, 2006a) and succeeded; in June BAA was sold for about £10.3 billion (*Economist*, 2006b). In May 2006 the Office of Fair Trading announced its decision to start an inquiry in the market structure of airports, and informed the bidders that they might recommend a break-up. The political mood had changed and the demands for a break-up were gaining weight (*Economist*, 2006c). As a result of a reference from the Office of Fair Trading, the Competition Commission commenced an investigation into breaking up BAA in London (Competition Commission, 2006). On 20August 2008 the Competition Commission (2008b) provisionally recommended the break-up and announced that it would enquire into how the two the London airports might be sold and how regulation might be reformed. The CAA has supported this recommendation.

Table 19.3   Regulation X-factor

Airport	X-value						
Period	1987–91	1992–93	1994	1995–96	1997–2002	2003–08	2009–13
Heathrow	1	8	4	1	3	-6.5	-7.5*
Gatwick	1	8	4	1	3	0	-2*
Stansted	1	8	4	1	-1	0	

Source: Graham (2008); CAA (2003, 2008). *Additionally the base of the price cap was raised.

Efficiency and Ownership of the London Airports

Efficiency and Distributional Aspects

The London airports are not regarded as particularly efficient providers of airport services, and there are several distinct aspects of efficiency which the introduction of competition, with or without deregulation or regulatory reform, might impact on. In addition, the distribution of the benefits from airport services will alter with structural or regulatory change. Some key aspects of efficiency are discussed below.

Price levels In some markets, price levels are often an important determinant of allocative efficiency – high prices create a deadweight loss. However, if price discrimination is present, as is the case, this deadweight loss is lessened. In the case of the London airports, overall price levels for Heathrow and Gatwick are below market clearing levels, thus there is little deadweight loss arising from prices being too high. If the airports were deregulated, price levels could be much higher. Thus, while price levels will have a big impact on the distribution of the gains from airport service provision, they are not likely to have a significant impact on allocative efficiency.

Congestion and slot availability The London airports are slot constrained and the level of congestion, especially on the runway side, will be determined by the number of slots that are made available, along with other factors such as weather. There is an issue about whether the number of slots made available at the London airports is optimal (see Forsyth and Niemeier, 2008). Slot availability is not something which emerges out of the competitive or price regulatory processes – rather there is a separate regulatory process. Thus we do not discuss this here.

Provision of service quality Choice of the service quality to provide is potentially an important quality issue for airports, especially the London airports. It could be maintained – and indeed the Competition Commission (2008a and 2008b) makes the point strongly – that the BAA airports have provided too low a quality of service, and that users would be willing to pay for higher quality levels. Increasing quality requires the airports to incur higher operating costs and to invest more in facilities. It is well recognised that firms which are subject to price caps have an incentive to undersupply quality (Rovizzi and Thompson, 1992). The London airport regulator does recognise this, and offers incentives for increased quality; though whether these incentives have been set at the right level is an issue (see, for example, CAA 2001a). If the airports were separately owned, they *might* compete on quality (depending on the regulation to which they were subject), though whether this would be so, and the extent to which competition would solve the quality problem, requires investigation.

Productive efficiency Two aspects of productive efficiency are relevant in the case of airports. Firstly, there is productive efficiency in operations, which deals with whether operating costs are being minimised. Secondly, there is productive efficiency in the provision of capacity, which deals with whether the costs of additions to capacity are being minimised. There is some evidence that the productivity of the London airports is relatively low (ATRS, 2008), and there are questions as to whether the costs of building new capacity have been minimised. Where competition is strong, firms are forced to be productively efficient: those which are not will not be able to cover their costs. Thus whether separation of the London airports will lead to pressure for improved productive efficiency is a question worth investigating.

Price structures and capacity utilisation   When facilities are in high demand, the effectiveness with which existing facilities are used is an important efficiency issue. Price structures can have an impact on capacity utilisation. The current price structure at the London airports is one which embodies a high per passenger charge, with a low per movement charge. This is not one which would optimise utilisation. Runway capacity is in short supply at Heathrow and Gatwick, and till recently, terminal capacity at Heathrow has been very constrained. It may remain somewhat constrained for some time as old terminals are redeveloped. A higher per movement charge, coupled with a lower per passenger charge, would induce better utilisation of the airports, as long as there is some scope to handle the extra passengers. Price caps which are based on average revenue per passenger *should* give the airports an incentive to structure their charges so as to increase passenger numbers, by putting greater weight on movement charges (Forsyth and Niemeier, 2008). It is possible that terminal constraints have prevented BAA from moving in this direction in the case of Heathrow, though terminals are not a constraint for Gatwick. Given the high premiums that passengers are prepared to pay to use Heathrow and Gatwick airports, there are potential efficiency gains from increasing passenger numbers. Competition between the airports might stimulate a move to more efficient price structures.

Optimisation of commercial revenues   It is desirable that the airports make effective use of their opportunities to generate commercial revenues (though to an extent, they may be exercising market power when maximising these). The form of regulation will affect incentives to pursue commercial revenues, with dual till regulation creating stronger incentives than the single till regulation which is currently in place (Starkie and Yarrow, 2000). Competition between airports has the potential to influence performance, especially through limiting the use of any market power the airports might possess.

Investment and provision of capacity   Airports are capital intensive facilities, and their efficiency depends critically on the level of capacity provided. With Heathrow and Gatwick both being subject to strong excess demand, there is evidence that capacity provided is too low. It is well recognised that a primary constraint on capacity provision at the London airports, as with other airports worldwide, has been the presence of environmental externalities which make expansion difficult, and the reliance on slow planning procedures. Nevertheless there is a question of whether the owners have a strong enough incentive to push for additions to capacity, and whether competition between them might strengthen this incentive (as the Competition Commission, 2008b, argues).

Slot allocation and rents   Both Heathrow and Gatwick are currently slot constrained, and slot values for Heathrow are very high. One efficiency issue concerns the allocation of slots: the current system of allocation, though grandfathering, does not guarantee an efficient allocation, especially when secondary trading is not extensive. In addition, even if slots were allocated efficiently, they may not be being used efficiently – slot rents might be being dissipated through higher costs or inefficiency (Starkie, 1998). Airlines using the London airports gain very large slot rents, and it is not evident that these are all translated into higher profits. Dissipation of slot rents could be affected by regulation or competition between airports. Under deregulation, high airport prices would eliminate slot rents, since profit-maximising airports would charge at least market clearing prices. Investment in additional capacity would lower, though not eliminate slot rents, thus minimising any efficiency losses from poor slot allocation or slot rent dissipation.

Distributional aspects The sharing of the gains from provision of airport services, as well as the size of these gains, will depend on price levels and structures, slot arrangements and the extent of provision of capacity. Distributional aspects may well influence public policy, which may not be directed solely towards maximisation of efficiency. In the London situation, regulation has sought to keep overall price levels close to cost – or even below long-run marginal costs, below market clearing levels – which, in combination with the slot system, ensures that rents accrue to airlines rather than airports or their passengers. Differing competitive and regulatory arrangements will have implications for distribution, which need to be recognised; though there is no attempt made here to distinguish between the relative desirability of different distributional outcomes.

Ownership and Incentives

Issues of ownership, per se, and the incentives associated with them, are not the primary focus of this chapter; however they are relevant for determining how competition works. An airport which does not seek to maximise profits will respond differently to competition than one which does seek profits.

The traditional approach to modelling private firm behaviour is to assume profit maximisation. BAA is a private firm which has recently been taken over, and thus the starting point would normally be to make this assumption. However, in the past, it has behaved in many instances in ways which appear contrary to this assumption. It has arguably cross-subsidised the development of Stansted, lessening its overall profits by doing so. It has shown reluctance to increase capacity; this may be because it believes investment will be blocked by planning controls, or it may be because it is a profit maximiser which is subject to a very cost-based form of regulation. Nevertheless, one might expect a profit-maximising airport company, which is free to maximise profits subject to a price cap, subject to excess demand to press harder for a regulatory environment under which it can profit from providing more capacity (Competition Commission, 2008b). It has been unwilling to allow other parties to build or operate terminals at its airports – a profit-maximising firm would be open to the possibility since, if others can provide or operate terminal capacity more efficiently, it can add to profits by creaming off some of the rents. Finally, its attitude to the break-up question itself seems more consistent with a non-profit-maximising firm than a profit-maximising firm. Break-up could well increase the profitability of the current BAA company – if BAA could secure more light-handed regulation for itself if it were to agree to sell off some of the London airports once light-handed regulation comes into effect, it could become a smaller, though more profitable firm. Perhaps BAA should have argued that the London airports would be very competitive if separately owned, and thus there is no need for anything beyond light-handed regulation (which would enable the airports to increase prices and profits – as has happened in Australia and New Zealand. BAA seems to have opposed a break-up no matter what the upside for it might be.

It is possible that the new owners are more focused on profits than the previous management. Separation of the London airports could result in some owners being more profit focused than the current owners of BAA. This would influence their behaviour – they would be more aggressive towards opportunities and would be more reluctant to give priority to investments which do not appear to be highly profitable, such as additional runway capacity at Stansted. If one or two of the owners of London airports were strongly profit oriented, this would put pressure on the others to sharpen their focus on profit.

In a situation of excess demand and slot controls, if capacity is inadequate, investment to increase capacity will increase overall efficiency, and it will be in the interest of passengers. Airlines will be unambiguous losers from it through their loss of slot rents, and airport owners may gain somewhat, though by how much depends on how they are regulated.

This poses a conundrum in interpreting public positions on London airport expansion. BAA, which should have an interest in expansion, appears relatively unwilling to press hard for it. This could be because it is not maximising profits, or could be because it sees itself as being subjected to a very cost-based form of regulation which gives it little scope to profit from handling more output. Airlines have been the most vocal in calling for expansion, even though this would not appear to be in their interest. It may be in the interests of some airlines to support expansion, especially those without large slot endowments. It is possible that airlines are advocating capacity expansion to be used solely to reduce congestion rather than to increase output, or that they do not perceive the full implications of capacity expansion for their yields. If BAA is being reluctant to invest so as to protect the position of the airlines, why should it do this, given that it does not gain from this action? It could be that BAA has acquiesced in the inadequate capacity situation because, with excess demand, competition between the London airports is not going to be very effective, and thus the case for the break-up of BAA is weakened.

Separation and Competition under Deregulation

If the London airports operate under separate ownership, the market will be oligopolistic. In the short run, there will be excess demand for Heathrow and Gatwick, though Stansted will have free capacity for a time. Prices can be set at market clearing levels at the very least, and airlines' slot rents will be eliminated – most likely, prices will be set higher, depending on the competitive strategies of the firms. With low demand elasticities, prices are likely to be much higher than at present, though below monopoly price levels. The airports will make use of their ability to price discriminate, though they may not have the same scope to do this as a monopoly. The airports will have the ability to charge more for a higher quality product, and so they will seek to provide the quality level that users are willing to pay for. If the airports are strongly profit oriented, they will keep their costs down. If they are not, they will have the scope to allow their costs to be higher than the feasible minimum – in short, productive inefficiency is a possibility. However, if one airport is achieving lower costs, and using lower charges to gain market share, this puts pressure on other airports to respond through lowering their own costs. However, if airports fear that high profits could provoke re-regulation, they may not minimise costs.

In the long run, the airports will invest and this will increase capacity. However, since they are using their market power and keeping prices high and output down, they will invest less than the optimal level. It is unlikely that there will be free entry, which would force the incumbents to reduce their prices. While with three or four competitors there could be strong competition, with excess capacity and price wars in the longer run it is not likely that this will happen. They could also be tacitly collusive, restrict investment and keep prices and profits high.

In terms of efficiency, the outcome could be quite good. Prices are likely to be high, but the deadweight loss from this will be small because the airports are able to use price discrimination effectively to minimise the loss of output when capacity is adequate. The airports will under invest unless strong competition breaks out between them. Oligopolistic airports should have a strong incentive to invest to provide quality, since they will be able to enjoy higher prices for higher quality service. If the owners are not highly profit oriented, productive efficiency is not guaranteed – the airports will have high prices, which gives them scope to allow their costs to be excessive.

They are not forced by their competitors to keep costs low. Excessive investment and too high a quality level are probably not much of a risk as they could be in a regulated setting.

Thus, apart from the concerns about productive efficiency, a deregulated airport market could deliver well in terms of efficiency. However, it would lead to very substantial shifts in stakeholders' positions. Passengers will lose because of the higher prices. Airlines will lose from higher prices and from the elimination of their slot rents. In the short run, with limited capacity, the airports will gain much higher profits. These effects will be lessened in the long run if the airports compete and add to capacity. There is some chance of capacity and price wars, which would improve the position of the passengers at the expense of the airports, with the airlines being relatively unaffected.

Whether these distributional outcomes are acceptable is a matter for judgement. However, governments have not been comfortable with industries with strong market power earning high profits, even when they are operating efficiently. Governments in the past have been willing to regulate prices such that they deliver modest profits, even when this simply involves shifting rents from airports to airlines. Interestingly the Competition Commission (2008a) does not recommend a full deregulation, but recommends separate ownership under regulatory regime.

Separation and Competition under Regulation

The incentives facing the London airports to compete, keep costs down, deliver efficient levels of quality and investment will all depend on the form of regulation they are subject to. For present purposes we distinguish between cost-based regulation and incentive regulation, in the form of price caps. With cost-based regulation there is rapid adjustment of prices to costs – if costs fall, regulated prices also fall soon after. With price caps, the maximum allowable price that the airports can charge is set for a number of years, and the airports can keep any profits they make during the price period. With a pure form of price cap, the airports' costs would not be used during the setting of prices for the next price period – rather, prices would be set according to external benchmarks, which would give rise to the strongest incentives. In reality, regulators usually employ a hybrid form of regulation (see Baldwin and Cave, 1999, pp. 233–8), relying on price caps for a period, but revising these so that prices are just sufficient to cover expected future costs in the new price period. It is also worth noting that different aspects of the regulatory task may make for different reliance on cost-based and price cap forms of regulation. In the short term price caps might be employed, and the regulator may pay little or no attention to the airports' operating costs. However, in the longer term, the regulator may shift price caps up or down according to the expected actual cost of capacity expansion. The regulator may rely on the airports' own estimates of the cost of this expansion when setting the price caps. To this extent, the regulator may be employing an essentially cost-based approach when it comes to investment, but a more incentive-based approach when it comes to current costs. The airports would have a moderately strong incentive to keep operating costs down, but would not face a strong incentive to keep the costs of capacity expansion as low as possible. The same regulator could be more cost based when it comes to investment than when it comes to current costs.

If regulation is closely based on costs, the incentive to compete between airports is likely to be quite weak. Firms compete if they think they can gain an advantage, through gaining more market share and adding to profits, if they cut prices, increase quality or invest in additional capacity. However, if profits are determined by the regulator, and there is little scope to add to profits by adding to output, there will be little point in competing. Firms which seek to maximise size rather than profit might compete, however; and firms which are subjected to close cost-based regulation may transform themselves in to size maximisers given the constraints they face on adding to profits.

In the discussion below, it will be taken that regulation takes the form of a fairly pure price cap in the short run, since it is under this form of regulation that the possibility of competition becomes more real.

Competition in the Short and Long Run

For firms to have an incentive to compete, they need to be able to increase their profits by increasing their output. By selling more, with (at least perceived) prices above marginal cost, profits are increased. As has been noted, this poses problems for airport competition in London in the short run. Runway capacity is fixed and fully or nearly fully used at Heathrow and Gatwick, and terminal capacity has been in short supply at Heathrow. While the opening of Terminal 5 at Heathrow has eased the constraint, the redevelopment of other terminals will take capacity out of the system, so that the increase in effective capacity will not be as great. Hence there is little point in the airports competing on price levels since, if they gain additional demand, they cannot handle it. The same would be true if they competed on quality – improved quality would be costly and would increase demand, but the airports would not be able to profit from it by handling greater output.

The one form of competition which seems feasible in the short run would be that based on price structures. As noted before, the capacity-constrained airports are not making the best use of their price structures in optimising the use of their capacity. It is possible that separate airports will have a stronger incentive than at present to structure their prices so as to increase utilisation of their facilities. In particular, if Heathrow and Gatwick can handle more passengers, if not more flights, there will be pressure on them to reform their price structures. In the long run, capacity can become greater, and the airports will have the scope to win traffic from one another through competing on price or quality. It would be unlikely that the London airports would have ample capacity throughout the whole day or year. Efficient investment in airport capacity implies that capacity will not be sufficient to handle demand all of the time – it will need to be rationed at some periods, in peak hours and peak seasons. Airports involve lumpy investments, so it is possible that there will be some years in which capacity is ample, followed by years in which capacity is short. Furthermore, as is well known in the London context, investment in airport capacity lags the growth of demand, because of environmental and planning factors. Thus, except after large additions to capacity have been made, airports at any point in time are likely to be partly capacity constrained at peaks. They will be able to compete to attract traffic from one another to the off-peak, though not the peak. The airports will also be able to compete in investing so as to be the one with the greatest scope to attract the peak traffic. They will be able to compete for the peak traffic, but once capacity has been sunk, the scope for competition will be over.

Competition through Price Structures in the Short Run

The price structures of the busy London airports do not optimise their utilisation (Forsyth and Niemeier, 2008). While there is a fixed element in the price schedule, charges are primarily per passenger based. This is in spite of runways being the critical constraint at Gatwick, and one of the two critical constraints at Heathrow. Before the recent addition to terminal capacity at Heathrow, the passenger constraint may have been binding; however, with additional capacity, this should not be the case (though terminal redevelopment will take away from effective capacity). British Airways will be able to handle more passengers in Terminal 5. Thus there are potential gains to be made from altering the price structure to encourage airlines to increase the number of passengers per aircraft (mainly though using larger aircraft). This can be encouraged by increasing the weight of the per flight component in the price schedule to be at least half the average charge (and, arguably, more in the case of Gatwick). The Competition Commission sees 'substantial scope for competition' (2008b, p. 244) by rebalancing the structure and increasing ratio of passengers per movement.

Granted that BAA is subject to a price cap which takes the form of average revenue per passenger, it already faces an incentive to increase passenger numbers by altering its price schedule. However, to some extent, the gains that one airport makes from encouraging more passengers are lost because some of these passengers switch from other BAA airports – the net gain to BAA in passengers and revenue is less than the net gain to any individual airport if airports are separated and competitive. This is moderated by the fact that per passenger charges, and possibly profit margins, at Heathrow are higher than other airports such as Stansted, and thus BAA gains from shifting passengers to Heathrow. In short, the incentive for airports to win more passengers by reforming their price structures is already present, but will be considerably stronger if the airports are separated.

In the short to medium term, this could be the strongest positive for break-up. BAA has had some incentive, and the discretion, to improve the utilisation of its airports by reforming its price schedules; however it has not been effective in exercising this discretion. As the values of slots at the London airports indicate, the premium that passengers are prepared to pay to use the airports, especially Heathrow, are high, and thus even a small gain in passengers per flight would produce a useful efficiency gain. At a premium of £10 per passenger (Forsyth and Niemeier, 2008, based on £10m per daily slot pair), a 5 per cent increase in passengers per flight at Heathrow would produce a gain of around £35m per annum. With recent slot prices of £25m per pair, this would rise to an efficiency gain of around £85–90m.

The Scope for Price Competition in the Long Run

By long run we mean a period sufficiently long for an airport to invest and alter capacity. Airports can choose to expand capacity and have the ability to handle more traffic. For purposes here, it is assumed that they have the flexibility to invest if they wish. It should be recognised that at any point in time airports might have either spare capacity or inadequate capacity – this is inevitable given the lumpy nature of airport capacity investments. Competition in the long run does not necessarily means that competition will be feasible at every point in time: at times when capacity is scarce, the scope for competition will be limited.

In a situation where the regulator sets a price cap, price competition is feasible, though unlikely. An airport could reduce the price it charges below the level which it is permitted to charge. This would be worthwhile if the price reduction increases profits (with other airports maintaining price, the price reduction will have to increase the airport's output by enough such that the additional profit on the output increase outweighs the cost of the price reduction). If there were identical airports, and infinitesimal price reductions, this would result in the price-cutting airport gaining the whole market – and this could result in higher profits, at least temporarily. However the London airports are far from identical and the response to a price cut by one would not be so dramatic – with a smaller quantity response, the likelihood of a profitable price cut is lower. In the past, Stansted airport has charged less than it has been permitted to do so, given the price cap. It has done this because it has ample capacity and it has been able to win traffic from other airports in the region, such as Luton, by doing this. It may not choose to act in this way if and when demand pushes against capacity.

Even if price cutting is temporarily profitable, it does not follow that this strategy is profitable in the longer run. Airports would be well aware of the longer term implications of their actions, and would seek to avoid price wars. They are players in a repeated game, and should find it relatively easy to maintain prices close to the regulated level. This will be more the case granted that airport capacity will normally be constrained, at least at the peak.

If the airports are judged likely to use their market power in the absence of regulation, they are likely to price up to the price cap when regulated. Even when there is spare capacity, airports are unlikely to gain by competing on price.

Competition and Quality in the Long Run

It is often suggested that separated airports would have an interest in competing on quality. An airport can compete and increase output by offering a higher quality at a constant price. Quality competition is similar to price competition – the gain in profits from the increased output has to be balanced against the cost of the increased quality. In principle, competition could result in a quality war, with quality being bid up to the extent that costs equalled the regulated price. Depending on the price that is set by the regulator, this could lead to excessive or inadequate quality. Again, the likelihood of such strong competition in an oligopolistic setting, with fixed capacity in the short run, is not very great. There is a well-recognised problem of price-capped firms not having an incentive to supply the right quality. Profits are higher if they cut quality and thereby cut costs. Suppose a quality improvement which users are willing to pay for. The firm's demand curve shifts upwards. However, with price being regulated, the firm can gain only a small share of the overall welfare improvement. In Figure 19.1, the firm faces a regulated price of P, and its average costs are c_1. If it offers a higher quality level, its costs rise to c_2. This results in a demand shift from D_1 to D_2. The firm's output increases from X_1 to X_2. The gain from this is small relative to the cost of supplying the higher quality, and thus it does not offer the higher quality. This is an inefficient outcome, since the upward shift in willingness to pay exceeds the upward shift in costs – users would be willing to pay for the quality improvement.

Introducing competition between the airports will give them a stronger incentive to supply higher quality. If BAA is an integrated company, it can increase its output through offering higher quality by increasing the use its existing passengers make of its airports, and by winning passengers from other, non-BAA airports in the region. However, if the airports are separated, an individual airport such as Gatwick can also attract traffic from the other ex-BAA airports, Heathrow and Stansted. The demand curve for the individual airports will be more elastic than that for BAA London airports as a whole. A quality improvement will induce the individual airport's demand curve to shift from D_3 to D_4. The output increase is greater, as is the impact on profit. It will still be the case that supplying higher quality will normally be unprofitable for the airport; however, the incentive to increase quality is less weak than before. If quality improvements are highly valued by users, and if they are cheap to supply (i.e. quality has been seriously undersupplied in the past), an airport may find it profitable to supply the higher quality. Ultimately, for the airport to increase quality, it will need to be given an incentive by the regulator to do so, through the regulator allowing a higher price conditional on higher quality.

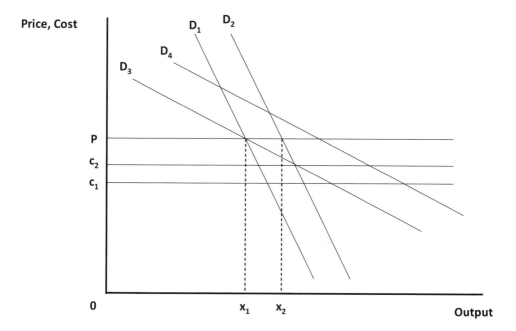

Figure 19.1 Price and quality

In summary, separating the airports will give them a stronger incentive to supply quality; but this incentive is still quite weak, and it will be necessary to adapt regulation so as to create stronger incentives to supply quality. As with price competition, quality competition between the airports will not be very strong in a regulated setting, and if price competition is not regarded as strong enough to enable regulation to be dispensed with, the existence of quality competition will not alter this judgement.

Productive Efficiency and Yardstick Competition in the Long Run

If airports are subject to price caps they will have the incentive to be productively efficient, but they will have some scope to be productively inefficient and allow costs to rise to the level of the price cap. Separation, per se, will not change this. If one airport is more productively efficient, it will enjoy higher profits. Unlike the situation of unregulated competition, this will not necessarily put any pressure on other airports to perform better. However, if the regulator employs yardstick regulation, lower costs of one airport could induce the regulator to set lower price caps for the other airports. This would put pressure on them to increase their productive efficiency. Separation of the London airports is not necessary for yardstick regulation, since the airports can already be benchmarked against other airports in the UK and other countries. In practice, regulators have found it difficult to implement yardstick regulation, but separation does give them a few more benchmarks.

Competition in Capacity Provision in the Long Run

Investment in capacity enables airports to increase their output. If the regulated price cap is above the incremental cost of output, a profit-maximising airport will invest. As with the incentives to make more effective use of existing capacity, a profit-maximising airport will find it worthwhile to invest whether or not it faces competition from other airports. However separation will increase the market open to the firm, since separated airports will be able to win traffic from one another. Underinvestment also leads to the risk that an airport will lose some of its market if the airports are separated. Thus, if anything, separated airports will have a stronger incentive to invest in additional capacity than horizontally integrated airports.

As noted above, the existence of spare capacity is needed for strong competition. It is significant that all of the examples of airport competition noted by the Competition Commission (2008b) involved smaller airports with spare capacity. If the London airports were not subject to constraints on capacity expansion, there would be periods of spare capacity and periods of tight capacity. Efficient capacity expansion in a context of indivisibilities would give rise to this pattern. However, environmental sensitivities are not likely to go away, and airports are not likely to have a free hand in investing. This will result in less rather than more investment, and thus will increase the duration of periods of inadequate capacity and of periods in which competition is limited. Better environmental regulation, under which airports are faced with more effective pricing of the externalities they generate, could result in fewer hold-ups of capacity expansion, and thus more scope for competition.

Separation, under well-designed regulation, will make the airports keener to increase output, and to invest to enable this to happen, than single ownership. Airports have an incentive to press more strongly against planning constraints, and to seek out smaller, less constrained investments which can make short- to medium-term increases in capacity and output feasible. It is not likely that separation would result in capacity wars and excess capacity in London in a context where regulation and planning controls are likely to continue to have a major influence on the amount of investment incapacity which is undertaken.

The Role of Regulation

Since competition is unlikely to be seen as strong enough to dispense with regulation, the airports will be competing in an environment of regulation. This regulation will affect their ability to compete. Price regulation will often rule out price competition and will make it difficult for airports to compete on quality, since they will not be able to increase prices to cover the cost of quality improvements which users are willing to pay for. Regulation also is a key determinant of investment, and thus the availability of capacity in the future which the airports will need to have to compete. Most price regulation focuses on monopolies rather than capacity-constrained oligopolies. Facilitating competition in this environment poses now problems for regulators.

Current approaches to regulation seem inconsistent with competition, especially in the provision of investment. The trigger approach to regulation employed currently by the CAA (whereby it increases the price cap conditional on investment being undertaken) gives the regulator very considerable discretion over investment. If airports are going to compete in investment, they will need to be given more flexibility over investment. With the separation of the airports, a more light-handed form of regulation might be employed. This would give the airports more flexibility to negotiate on quality or investment with airlines.

Since increased competition can only make a modest contribution to improving performance, much of the task will fall on improved regulation. This has been recognised, though it is not clear that the difficulties in designing such regulation have been. Getting regulation right in complex utilities such as airports is very difficult. The UK government, in 2008, began a review of airport regulation. The Competition Commission (2008b) recommends improving current regulation. However, the list of recommendations, such as 'change of control provisions and conditions for revoking the licence, more powers to intervene between quinquennial reviews, a greater role in facilitating agreement between BAA and airlines on service and investment' does not seem to address the real issues of incentive regulation. Intervening in the price cap period contradicts the principle of price cap regulation and might lead to gaming. The CAA (2002a) had proposed a number of substantive measures to reform regulation, such as different price path conditional on investment, a dual till approach, separated price caps and benchmarking. These were all dismissed by the Competition Commission (2002). The peculiarity of rival regulators with different approaches has not only been criticised by the regulated firm (Toms, 2004), and does tend to make regulation more costly and less reliable (Hendriks and Andrew, 2004).

Conclusions: How Effective Will Competition Be?

In this chapter we have explored the ways in which breaking up BAA, to enable separate ownership of the London airports and competition between them, can impact on their performance. Separation will give rise to the scope for more competition between the airports in the long run, especially if they are permitted to invest as they wish and choose their output levels. However, environmental and planning controls will restrain investment and the scope for competition. In addition, efficient capacity expansion would lead to only occasional periods of spare capacity, and hence the scope for moderate competition.

Separation with deregulation would lead to a tight oligopoly market, and the most likely consequence would be desirable from a pure efficiency perspective; but airport prices and profits would be very high. Granted past public policy preferences, this would not be acceptable and price regulation would be seen as necessary. This regulation will impact significantly on how competition works – regulation, in itself, lessens the ability of the airports to compete. Separation is likely to lead to competition in several dimensions:

- The only real form of competition in the short run is likely to be competition in price structures, which can increase the utilisation of the airports' fixed capacity.
- Price competition is unlikely under price caps, even in the long run.
- Quality competition is likely to be weak in the long run, but it is likely to be slightly stronger under separate rather than single ownership.
- Under separation, the airports will have a stronger incentive to invest in additional capacity, though they will still be under significant constraints, and the form of regulation will have a major influence on investment choices.
- It is possible that it may be feasible to move to more light-handed forms of regulation if separation takes place, and this may increase the scope for competition.

Thus, overall, separation can have some positive impacts on performance through opening up the possibility of competition, even though this competition will not be strong, especially in the short run. These benefits need to be compared to any costs in breaking up the single owner, BAA; these costs have not been considered here. For this competition to be effective regulation will need to be very well designed, since cost-based regulation will dampen or eliminate the pressures discussed here. While separation will enable some competition and will encourage some improvement in performance, it is not the magic bullet that many commentators believe it is.

References

Air Transport Research Society (ATRS) (2008), *Global Airports Performance Benchmark Report on World's Leading Airports*, Vancouver, University of British Columbia.

Baldwin, R. and M. Cave (1999), *Understanding Regulation: Theory, Strategy and Practice*, Oxford: Oxford University Press.

Competition Commission (2002), *BAA plc: A Report on the Economic Regulation of the London Airports Companies*, London: The Stationery Office.

Competition Commission (2006), *BAA Airports Market Investigation – Issues Statement*, www.competition-commission.org.uk.

Competition Commission (2008a), *BAA Market Investigation Emerging Thinking*, April, www.competition-commission.org.uk.

Competition Commission (2008b), *BAA Market Investigation Provisional Findings Report*, August, www.competition-commission.org.uk

CAA (2001a), *Heathrow, Gatwick, Stansted and Manchester Airports Price Caps – 2003–2008: CAA Preliminary Proposals – Consultation Paper*, London, www.caa.co.uk.

CAA (2001b), *Peak Pricing and Economic Regulation*, Annex to CAA (2001), *Heathrow, Gatwick, Stansted and Manchester Airports Price Caps – 2003–2008: CAA Preliminary Proposals – Consultation Paper*, London, www.caa.co.uk.

CAA (2002a), *Heathrow, Gatwick and Stansted Price Caps 2003–2008, CAA Recommendations to the Competition Commission*, www.caa.co.uk.

CAA (2002b), *Competition Commission Current Thinking on Dual Till – CAA Statement on Process Press Release*, 13 August 2002, London, www.caa.co.uk.

CAA (2003), *Heathrow, Gatwick, Stansted and Manchester Airports Price Caps – 2003–2008: CAA Decision*, London, www.caa.co.uk.

CAA (2008), *The Competition Commission's Market Investigation of BAA Ltd The Civil Aviation Authority's Response to the Provisional Findings and Remedies Notification*, www.caa.co.uk.

Department of the Environment (2000), *Transport and the Regions*, London.

Economist (1995), 'Why Heathrow is hell', 26 August.

Economist (2006a), 'Tilting at windsocks', 9 February.

Economist (2006b), 'On the runway', 8 June.

Economist (2006c), 'Holding pattern', 24 August.

Forsyth, P (1976), 'The Timing of Investments in Airport Capacity', *Journal of Transport Economics and Policy*, January, 51–68.

Forsyth, P. and H.-M. Niemeier (2008), 'Setting the Slot Limits at Congested Airports', in A. Czerny, P. Forsyth, D. Gillen and H.-M. Niemeier (eds), *Airport Slots International Experiences and Options for Reform*, Aldershot, Ashgate, pp. 63–83.

Hendriks, N. and Andrew, D. (2004), 'Airport Regulation in the UK', in P. Forsyth, D. Gillen, A. Knorr, O. Mayer, H.-M. Niemeier and D. Starkie, *The Economic Regulation of Airports: Recent Developments in Australasia, North America and Europe*, Aldershot, Ashgate, pp. 101–11.

House of Commons Transport Committee (1996), *Second Report from the Transport Committee*, UK Airport Capacity, Session 1995–96 (HC 67).

Little, I.M.D. and K.M. McLeod (1972), The New Pricing Policy of the British Airports Authority, *Journal of Transport Economics and Policy*, 6, 101–15.

Office of Fair Trading (OFT) (2006), *UK Airports Report on the Market Study and Proposed Decision to make a Market Investigation Reference*, December, www.oft.gov.uk.

Office of Fair Trading (OFT) (2007), *BAA The OFT's Reference to the Competition Commission*, April, www.oft.gov.uk.

Rovizzi, L. and D. Thompson (1992), 'The Regulation of Product Quality in the Public Utilities and the Citizen's Charter', *Fiscal Studies*, 13 3 74–95.

Starkie, D. (1998) Allocating Airport Slots: A Role for the Market, *Journal of Air Transport Management*, 4, 111–16.

Starkie, D. (2001) 'Reforming UK Airport Regulation', *Journal of Transport Economics and Policy*, 35, 119–35.

Starkie, D. (2002), 'Airport Regulation and Competition', *Journal of Air Transport Management*, 8, 63–72.

Starkie, D. and D. Thompson (1985), *Privatising London's Airports*, IFS Reports Series No. 16, The Institute for Fiscal Studies, London.

Starkie, D. and G. Yarrow (2000), *The Single Till Approach to the Price Regulation of Airports*, London, www.caa.co.uk.

Toms, M. (1994), 'Charging for Airports', *Journal of Air Transport Management*, 1, 77–82.

Toms, M. (2004), 'UK-Regulation from the Perspective of the BAA plc', in P. Forsyth, D. Gillen, A. Knorr, W. Mayer, H-M. Niemeier and D. Starkie (eds), *The Economic Regulation of Airports: Recent Developments in Australasia, North America and Europe*, Aldershot, Ashgate, pp. 117–24.

Vickers, J. and G. Yarrow (1988), *Privatization: An Economics Analysis*, MIT Press, Cambridge MA.

Chapter 20
Airport Alliances and Multi-Airport Companies: Implications for Airport Competition

Peter Forsyth, Hans-Martin Niemeier and Hartmut Wolf[1]

Introduction

Alliances and mergers among airlines are common. Will they become so for airports as well? Some, such as Doganis, believe so. Doganis (1999, 2001) has forecast that there will be a few, large airport groups in the next decade or so. In this chapter we examine factors which might induce airports to enter alliances or form groups of multiple airports under one owner. We also examine some of the policy problems which could arise should these alliances or groups develop. Several possibilities are present, but two broad types of arrangements are likely to be of most relevance:

- strategic alliances between separately owned airports;
- groups of airports under common ownership.

Many airports are still publicly owned and if there are several airports in a country, it is possible that they will all be owned by the same owner, the central government. They may or may not be operated as a group, as the airports of Spain are in AENA. Such government-owned groups of airports have come about more or less by accident, but if they are to be privatized, the possibility of restructuring has to address those issues raised in this chapter. Here we are more concerned about groups of commonly owned airports which have been consciously built up through merger or takeover; though a company intentionally buying several airports from a government at a time of takeover; or as the result of the privatization of several airports as a group. As with airline alliances, airport alliances can be extensive, covering the whole or many of the airports' functions and product range; or they be targeted to particular functions or products, such as ground handling or retail.

As with airline alliances and mergers, there can be both benefits and costs from a public policy perspective. An alliance or merger may lead to operational cost savings, though better coordination of functions. However, it could also lead to a reduction in competition, especially if the airports are close to each other and could be effective competitors.

Thus it has been suggested that global air transport will soon be governed by a highly concentrated structure of airline alliances and also of airport alliances or companies (though other observers, such as Graham, 2001, take a more sceptical view). However, our analysis leads to a different view. Unlike airlines, airports do not gain substantial economies through mergers and alliances. Therefore we do not expect a strong integration process; but a few, though important,

---

1   We would like to thank Jürgen Müller for his comments on this chapter. We are also indebted to Karsten Fröhlich and Falko Weiser for excellent research assistance.

cases of integration might increase market power. In these cases airport and competition policy will need to assess carefully the benefits and costs of any integration proposals.

To begin with, we shall compare airport alliances with airline alliances, taking account of the view of Doganis (1999) that the formation of airline alliances, as well as airport alliances and companies, is driven by a parallel process, only with a time lag, with airport alliances simply copying airline alliances. Later in the chapter, we examine airport alliances, concentrating on three cases to find out what motivates alliances and what they deliver. Then we look at the motives and patterns of multiple-airport companies. We consider briefly the attempted takeover of Bratislava airport by Vienna airport. Thereafter we shall analyse the benefits and costs of airport alliances and finally we summarize our findings.

Horizontal Integration: Similarities and Differences between Airlines and Airports

Horizontal integration, in any industry, will come about as a result of the benefits that can be reaped from integration, or drivers for integration, which need to be balanced against the costs of integration. It will also be conditional on the regulatory environment – various forms of integration, such as alliances or companies, may be either permitted or prohibited. One reason for the popularity of cross-country airline alliances may be the associated economies of scale on the one hand, but the prohibitions on cross-country mergers on the other hand; this prohibition may not be as strong for airports, and this may induce more reliance on common ownership rather than strategic alliances in airports.

Table 20.1 shows a list of potential economic drivers for horizontal integration within air transport. At the supply side airline alliances and mergers are driven by the network effects resulting from economies of scope and density. Cost savings at the route level and know-how transfer are of less importance. This is different with airports. There are some important know-how transfers, for instance by national airport groups such as AENA, and by international airport companies such as Macquarie Airports, HOCHTIEF AirPort (HTA) or Fraport: by using their management and financing concepts, they may gain some operational and especially financing advantages. Especially this financing aspect should not be overlooked. Both Macquarie and HTA are able to obtain favourable long-term financing from pension funds because they have, at the same time, operational know-how to exercise management control of the airport with limited amount of their own equity

The management of airports in third world countries by major European airports is another example of this kind of know-how transfer. There may also be some network and coordination effects in some airport service activities, like ground handling, parking, retailing and other non-aviation activities, even if most of their costs are local and the cost-reduction potential seems to be limited. For example, international ground handling companies such as Swissport, Menzies, or Groundforce can offer international airlines groups one-stop shopping arrangements, thus avoiding multiple local contract negotiations. But these activities can easily be outsourced.

Still, contrary to airlines, the cost reduction potential for airport groups is limited since most costs are local (Graham, 2001; Niemeier and Wolf, 2002). Therefore, the kind of cost reductions with the magnitude found in airline integration cannot be expected from airport integration, but they may nevertheless be large enough to explain the move towards multi-airport groups.[2]

---

2   According to Oum, A. Zang and Y. Zhang (1996) there are some positive network effects of hub-and-spoke systems. However, private airport alliances have so far not formed such systems, while the state-

Table 20.1 Horizontal integration of airlines and of airports

	Airline alliances and mergers	Airport alliances and mergers
Network effects	++ (Economies of scale, scope and density)	– (Local costs dominate)
Route effects	+ (Only limited economies of scale at route level)	– (Local costs dominate)
Know-how transfer	+ (Some importance in less developed markets)	++ (Potentially relevant for former state-owned and third world airports)
Marketing	+ (Development of local markets)	+ (Some success)
Demand side	++ (Consumer preference for networks and quality)	– (Only mild preference)
Regulation	++ (Restrictive ASA with foreign ownership restrictions)	– (Relatively open airport industry, though constrained by foreign investment policies)
Avoid double marginalization in vertically related markets	+ (High yield routes)	+ (Between an unregulated hub and spoke airport)
Limit competition	++ (Strong incentives in oligopoly)	++ (Potential to extend regional monopoly)

++ highly important + less important – unimportant
Source: Niemeier and Wolf, 2002.

The demand side of merged and aligned airlines shows that marketing and seamless travel are also driving forces. In addition, regulation can be an important driver of airline alliances to overcome foreign ownership restrictions of air service agreements (ASAs). Neither driver is strong in the case of airports. Airports have, at best, only a mild preference for forming alliances, as Galaxy shows (see p. 343). Regulatory barriers to cross-country mergers and the formation of multiple-country airport countries are not as strong as for airlines. Foreign investment rules and specific airport policies may limit foreign investment in airports in some countries, though these are not as pervasive or restrictive as the ownership rules for international airlines.

However, incentives for forming alliances may exist in the airline as well as in the airport industry to overcome market imperfections. The avoidance of double marginalization – a problem within a vertically integrated or jointly provided service relationship of firms – might play a role. Two airlines which each operate a monopoly route that is complementary to the route of the other carrier might earn more profit by coordinating their pricing strategies, thus avoiding a double price

---

owned central systems have failed to price efficiently such systems. Furthermore, these systems show X-inefficiencies.

mark-up on costs (Spengler, 1950). The same might be true for a hub and its feeder airport that also are jointly providing a service. However, in both cases such strategic behaviour depends on the existence of market power of the alliance partners. If markets are contestable, or if efficient regulation exists, this double marginalization problem does not exist.

However, gaining market power can be an effective driver for both airline and airport integration. Parallel airline alliances and mergers can reduce price competition in oligopolistic markets, and airports can secure and extend their monopoly in a region by acquiring nearby airports.

Airport Alliances

Consolidation in the airport industry may take place either by forming strategic alliances or entering into mergers. In the case of an alliance partners coordinate their strategies while staying independent. Thus, decision-making within an alliance is decentralized, in the sense that all partners decide on their operations at their own will but promise each other to take the effects of their own decisions on the alliance into account. In contrast, a full merger of firms generates a centralized ('unified') governance structure within which central ('hierarchical') decision-making for all of the merged firms takes place.

In general, whether an alliance is the preferred mode of consolidation or whether it is a merger depends on the transaction costs associated with each mode. Transaction costs may come either in form of market transaction costs – i.e. the costs of writing and enforcing market contracts that define rules for the behaviour of each independent partner – or they may come in form of managerial transaction costs, i.e. the costs of arranging and managing transactions that take place within a single firm (Williamson, 1985). The latter mainly comprises the costs of setting efficient incentives for those working within the firm. Although the definition of common rules that define each partner's obligations may be costly in the case of complex transactions, the enforcement of market contracts may incur little costs if no legal or economic obstacles prevent partners leaving the agreement and entering into a market contract with another partner. In case one partner violates their obligations written in the contract the other may easily step out of the bilateral (or multilateral) relationship. This exit option creates pressure on each of the partners to fulfil their obligations efficiently. Thus, if partners may leave their mutual relationship at little cost, the forming of a strategic alliance instead of entering a merger saves on managerial transaction costs. To the contrary, irreversible investments may prevent partners leaving an alliance, thus making the threat of leaving an alliance incredible. Instead, they may create a hold up-situation, thus making it more attractive for partners to agree on a unified governance structure to set incentives for (or even to enforce) and protect irreversible investments against opportunistic behaviour by each other, i.e. to enter into a merger. However, it is much harder to set efficient incentives by internal rules for those working within the merged firm than to set efficient incentives for independent firms that together join forces in competitive markets. To sum up, as a general rule alliances may be the preferred mode of consolidation in the case of low barriers to change transaction partners; that is, if the need for irreversible relation-specific investments is low. To the contrary, full mergers or joint ownership of a single firm may be the preferred option in cases where considerable irreversible investment in the transaction is needed.

With regard to airports the provision of infrastructure services needs huge irreversible site-specific investments in aprons, runways and terminals, while the need for such investments is low in areas of ground handling and non-aviation services. Thus, given the reasoning above, the formation of airport alliances is unlikely to occur with respect to the provision of infrastructure services, but to join forces may be an option for independent service providers in the ground handling and non-aviation businesses.

In fact, the only alliances between airports that coordinate major functions of their services that we were aware of at the time of writing were the Galaxy International Cargo Alliance that was established in the year 1999, Aviation Handling Services (AHS), and the Pantares Alliance, both established in 2000.[3] While Galaxy and AHS both provide ground handling services, Pantares mainly covers ground handling services and non-aviation business. In order to understand what drove the formation of these alliances it may be useful to refer to the stated objectives of these cooperations.

Galaxy International Cargo Alliance

Galaxy, which was established by Washington Dulles International Airport and Chateauroux-Doels Airport in 1999, quickly grew to 21 members all over the world in 2001 (Sunnucks, 2002).[4] It still operates today, although its intellectual property has been transferred to The International Air Cargo Association (TIACA Times, 2005). Galaxy stated its goals as to 'create a worldwide organization of airports joining together for the purpose of promoting and developing the air cargo and logistics business'. The concept of the alliance was 'to create an international organization with the common goal of offering both existing and potential customers, such as airlines, forwarders and shippers, an environment that encourages air cargo development ... [by the] ... establishment of service and cost parameters for member airports.' Alliance members should also be 'encouraged to participate in joint marketing initiatives to expand each other's global reach' (Gateway, 2000). Referring to more detail, Galaxy's stated goals are:

1. the promotion of a global brand for cargo;
2. an improvement of the airports' image in the cargo business;
3. joint international marketing;
4. the raising of the attractiveness of member airports for freight alliances;
5. the establishment of an agreement on common operational standards;
6. the operation of a common internet site;
7. the establishment of a common internship programme.

As can be seen the alliance was formed mainly to cooperate on joint marketing by establishing a common brand for member airports. However, it failed to prove any significant value to the allied airports (Tretheway, 2001).

Aviation Handling Services (AHS)

Members of the AHS alliance, which was established in the year 2000 as a joint venture of three German airports and grew in terms of the number of members in the following years,[5] provide

---

3   Beside these alliances there exist many loose agreements among airports to cooperate in other than direct operational issues (e.g. in benchmarking their operations).

4   Among the members of the Galaxy alliance were the airports of Accra, Avalon, Bucharest, Casablanca, Cologne-Bonn, Dublin, Fairbanks, Hamilton, Houston, Kuala Lumpur, Manchester, Milan, Sharjah, Shenzen and Stockholm (Sunnucks, 2002).

5   The founding members of AHS were the airports of Bremen, Hamburg and Hannover. Later on the airports of Muenster/Osnabrueck, Nuremberg and Stuttgart joined the alliance (AHS, 2008).

common quality standards for ground handling services at different airports, enabling customers one-stop shopping. One might conclude that the rationale for establishing the alliance was – as in the case of Galaxy – to improve on the marketing side. In addition, another rationale might be potential benefits from intra-alliance transfer of know-how on best operational practices within the alliance.

The Pantares Alliance

Pantares, which was established in the year 2000 by Fraport and the Schiphol Group,[6] represents an effort of two airport companies to join forces on a number of major airport operations. According to its own statement this alliance aimed to establish cooperation among both members in the areas of (i) terminal and retail management, (ii) handling of aircraft and cargo, (iii) facility management, (iv) information technology and (v) international joint ventures. Financial analysts expected Pantares to reap significant value by cooperation because of complementary skill sets of both companies: While Fraport was very successful in the ground handling market but was rather weak in the area of retailing, Schiphol had the opposite strengths and weaknesses (Morgan Stanley, 2001, p. 35). Therefore one might come to the conclusion that the formation of Pantares was mainly driven by an attempt to exchange know-how between partners with regard to improving each partner's performance in complementary business areas. However, although Pantares stated ambiguous plans at the time of its establishment which attracted much attention in the industry, the performance of this alliance failed to fulfill its expectations. Pantares was never mentioned again by Morgan Stanley Reports and the initial plan to partner up with Aeroporti di Roma and the operator of Milan's Linate and Malpensa airports did not materialize.

To sum up, it is unlikely that possible consolidation of the airport industry will take place by airport companies forming strategic alliances in order to cooperate in the provision of infrastructure services. There might be limited scope to enter into alliance agreements in the areas of ground handling and the non-aviation business. However, if such moves did occur, they would probably be driven mainly by efforts to join forces on marketing in order to create a common branding for members' services; to allow customers one-stop-shopping; and to transfer know-how between members in order to improve performance in specific market segments.

Multiple-Airport Companies

By multiple-airport companies we mean companies which involve common ownership of a number of airports, or cases in which an airport has a majority holding, or at least a strategic minority holding, in other airports. There can be several different types of airport companies:

- Government majority owned national airport corporations, such as Spain's AENA, or regional corporations, such as France's Aéroports de Paris. Some of these, such as Aéroports de Paris, may have some private shareholdings.
- Companies which have been formed as a result of the sale of a group of airports by the government, such as BAA, which owns several London and Scottish airports.
- Major airports which own regional subsidiaries, such as Fraport, which owns the Frankfurt

---

6 Fraport operates, among others, Frankfurt Airport. The Schiphol Group operates, among others, Amsterdam Airport.

hub airport and the smaller Hahn airport and operates a number of other airports around the world.
- Corporations with a diverse portfolio of airports, such as Macquarie Airports, which owns airports such as Sydney, Copenhagen, Brussels and Bristol.
- Airports which hold strategic, though minority interests in other major airports, such as Amsterdam Schiphol's interest in Brisbane airport and Fraport's in Hannover airport,
- Specialist investors, such as infrastructure funds (e.g. Colonial First State).
- Facility owners and operators, such as companies which own terminals in a range of airports.

Patterns of Airport Companies

In Table 20.2 we present some details of airport companies. The list is selective, but it includes most of the groups that are functioning as companies which own and are investing in a number of airports. Most of the main private airport groups are included, and some government-owned groups, such as Amsterdam Schiphol, are listed. There are several government-owned groups of airports which are not included, since for these common ownership is due to government ownership, national borders and geography. The many situations where large airports have investments in other small airports are also excluded.

All of the groups listed in Table 20.1 (other than Dublin) hold majority stakes in more than one airport. Most have strategic stakes in other airports and an influence on management, though Macquarie Airports and HTA are moving to concentrate on majority ownership. Several groups, particularly those which are largely owned by governments, have investments in airports which are potentially competitive.

Table 20.2   Airport companies: Stakeholdings and competition

Airport company	Majority interest in more than one airport?	Strategic holdings in airports?	Ownership of potentially competitive airports?	Operating concessions for other airports?
Abertis (TBI)	Y	Y	N	Y
AENA	Y	Y	Y	Y
Aéroports de Paris	Y	Y	Y	Y
Dublin Airport Authority	N	Y	N	N
Ferrovial/BAA	Y	Y	Y	N
Fraport	Y	Y	Y	Y
Hochtief (HTA)	Y	Y	N	N
Infratil	Y	Y	N	N
Macquarie	Y	Y	N	N
Peel	Y	Y	Y	N
Schiphol	Y	Y	Y	Y

Key: Y – Yes; N – No

Source: Carter and Ezard (2007); company websites.

These groups can come about in several ways. Government-owned airports may be consolidated into a single corporation; companies may buy several airports from a government at time of privatization (BAA); or groups may be formed by corporations bidding for airports, or strategic holdings in them, as they come on to the market (Macquarie Airports). There are several distinct types of owners. Some are government owners. A type of corporation which has developed recently is that of a specialist airport operator – such corporations may seek to develop expertise in operating complete airports, as does Macquarie or HTA, or in some facets of airport operation, including ground handling and terminals. General investors, such as infrastructure funds, may be important in that they may have a say in airport management. Finally, another model is where a successful major airport seeks to extend its expertise to manage other airports.

Rationales for Multiple-Airport Corporations

Multiple-airport groups may come into being for several reasons. Amongst these will be:

- reducing airport competition and gaining market dominance;
- achieving operational and investment coordination;
- utilizing the expertise of specialized managers and investors.
- We consider each of these in turn.

Reducing airport competition and gaining market dominance   If a single company owns two or more airports in a region, it can eliminate competition between them. This rationale is only relevant when the airports are sufficiently close to compete – for example, it is not applicable to the case of Macquarie Airports owning Sydney and Brussels, or Schiphol having a strategic stake in Brisbane. There are a number of examples of where common ownership inhibits competition. Common ownership of the Paris airports by Aéroports de Paris means that the airports do not compete, and the London airports of BAA do not compete against one another. There is some potential for competition between the London airports, though there is some disagreement about how strong this competition might be (see Starkie and Thompson, 1985; Competition Commission, 2008). BAA's ownership of three Scottish airports means that they do not compete, and Frankfurt's ownership of Hahn removes the possibility of competition. Vienna sought to buy the nearby Bratislava airport when it came on the market – competition fears were a strong factor in opposition to the move. If an airport company can eliminate competition between airports it may be able to increase their profitability. Whether it can do so will also depend on regulation – BAA's London airports are all subject to price caps, which limit its ability to raise prices and profits.

Operational and investment coordination   An airport group may be able to achieve economies in operation through coordination of several airports. This is most likely to be the case for close-by airports, serving the same or overlapping markets. Some gains from coordination might be achieved on the supply side – for example, in the purchasing of inputs. However, airports are more likely to coordinate services on the traffic side if they are near each other. Thus an airport could allocate one form of traffic (such as full-service carriers) to one airport and other forms of traffic (such as charter or low-cost carriers) to another. When BAA was government owned, it directed charter traffic to Gatwick and away from Heathrow. Privately owned airport groups may not wish to refuse service to an airline at an airport, but they can develop specific airports to appeal to specific traffic types. Thus, Hahn airport targets low-cost carriers, while Frankfurt seeks to attract full-service carriers. Such coordination of traffic is easiest if a single group owns all the airports in a region. However it is not necessarily the most efficient way to go and, if desirable, it can still be achieved with separately owned airports. Single ownership of nearby airports may facilitate more effective market segmentation, which could be desirable or undesirable on welfare grounds.

Single owned airport groups can also coordinate investment in the different airports more effectively than separately owned airports. This is an argument put strongly by BAA. A single owner can design a programme of investments which can best handle growth in traffic for a city. As a privately owned group it will seek to make investments which maximize its profits, and this need not coincide with the programme which best achieves overall efficiency. This is especially so if it faces price regulation which will affect investment incentives at the different airports. A single owner could avoid duplication of capacity increases and subsequent underutilization of new capacity. Investment in airport facilities is a slow process and it is not likely that separate airports would over- or underinvest because they are unaware of the investments planned for their competitor airports. Decisions about investment in major airport facilities, such as terminals and runways, at large cities are not likely to be left to the owners of the airports. Rather, they will require approval by regulatory authorities, planning bodies and by the government. Thus it is not clear that investment decisions made for separately owned airports would be uncoordinated and that wasteful duplication would take place.

Specialized managers and investors Airport companies may develop expertise in operating airports, and these skills may be transferable. Thus BAA established itself as an innovator in retail, and it has been able to apply its skills where it has invested in airports outside the UK. Macquarie Airports may have originally been intended as a simple infrastructure investor owned by an investment bank, but, as an owner, it has developed expertise which it can apply to other airports it might invest in. It is now seeking to invest in other airports and to use its approach to improve their financial performance. Another airport company, Infratil, has developed expertise in running smaller airports.

Since the privatization of airports began, specialist airport investors have developed. Some infrastructure funds have developed expertise in investing in airports. They have become familiar with the operational and financial demands on airports, and have taken strategic stakes in a number of airports.

There may be other rationales for airports to form groups – Doganis (1999) lists marketing. It is not clear that this will be a major driver for consolidation. On the supply side, large groups are not likely to get significantly better terms from marketing suppliers than smaller groups. On the demand side, it would not seem that there are many advantages to marketing disparate airports such as Sydney and Copenhagen airports together.

Case Study: Vienna's Bid for Bratislava

The competition issue arises when mergers or takeovers are being proposed. Until recently, this has not been much of an issue, since there have been few cases of separately owned airports operating in the same region which were capable of being bought by the same company. As privatization proceeds, more and more cases of close airports which are available for purchase will occur. Thus Vienna airport sought to purchase the nearby Bratislava airport when it was privatized. Concerns about reduced competition were instrumental in blocking the proposal.

On 1 February 2006 the Slovak Conservative government announced the majority privatization of Bratislava and Kosice[7] to a consortium led by Vienna airport. The offer for a share of 66 per cent amounted for the two airports to about €780 million, including major investments at Bratislava (Wieninternational.at, 2006). Vienna Airport declared that it intended to build an airport system like the ones in London or Paris, strengthening the competitive advantage of this region relatively to others. In 2004 Vienna Airport had 14.7 million passengers and Bratislava 1 million passengers, mostly from low-cost carriers (LCCs). Both airports expect strong growth for the whole region, doubling passenger throughput by 2015. The airports are located in the same region, about 60 kilometers away from each other. According to Vienna Airport the benefits of a Vienna-Bratislava airport system lie in the postponement, for three to four years, of an investment of about €400 million in a third runway. Vienna Airport announced its willingness to invest in extending the City Airport train to Bratislava (WirtschaftsBlatt, 2004).

The sale of Bratislava Airport was bound to the condition that it would be approved by the competition commissions of both states by 2006. It was criticized by the opposition party as a sale of assets with strategic importance.

The decisions of both cartel offices are of particular interest. The Austrian cartel office found out that competition in the LCC market would be reduced (Bundeswettbewerbsbehörde, 2006a and b). The takeover would create a monopoly as no other airports exist in the relevant market. Although for full-service airlines (FSA) Bratislava is not a perfect substitute for Vienna, the pressure to reduce charges at Vienna airport to prevent Bratislava winning FSA from Vienna would be reduced in the future with growth of Bratislava. The Austrian cartel office allowed the merger only on the condition that Vienna's airport charges would not rise faster than an index of charges of a group of European airports, including two neighboring hubs – Frankfurt and Munich. The condition lasts for the period 2006/07 to 2011/12, with a phase-out for another five years. It is independent of the sliding scale charges regulation. The decision of the cartel office clearly does prevent an abusive rise of airport charges but it will not lead to a reduction in charges, which would be the normal effect of a market entry. Most of the benchmarking airports have a monopoly and are lightly regulated. While it is too early to judge this decision, the reaction of the capital markets may make one sceptical. Erste Bank commented that profits from Bratislava/Kosice will be low but 'it is still better to pay a high price and receive low contributions than let a strong competitor grow next door' (2006, p. 1).

While the Austrian competitive commission approved the takeover the Slovakian Antimonopoly Commission rejected it on the grounds that 'the only efficient competitor would be eliminated, and thus both current and also potential competition pressure in the relevant market of providing infrastructure to regular regional flights would be eliminated. Regarding the high entry barriers to market and non-existence of potential competition after concentration, the regional monopoly will be created' (Antimonopoly Office of the Slovak Republic, 2006). Rumours that the decision was taken on political grounds were rejected. In the meantime the opposition had won the election, and on 18 October 2006 it announced its final decision to cancel the sale of Bratislava to Vienna Airport and stop the privatization (Reuters, 2006).

---

7   Kosice is a small regional airport in east Slovakia.

Public Policy and Airport Integration

Airport companies and alliances raise a number of questions for competition policy. What are the potential benefits and costs of airport alliances? Will cross-ownership restrictions be established, or should mergers be assessed case by case by general competition policy? As is usually the case, it is a matter of comparing cost savings and service quality improvements against the anti-competitive effects.

Airport alliances might provide benefits by speeding up the process of know-how transfer. This might be especially important in light of the current global trend of privatization and commercialization of airports. Airports which have operated (or still operate) as public utilities can more easily adjust their management to the challenges of the capital markets and the dynamics of the airline industry. However the transfer of knowledge seems to be more a temporary source of cost reduction than a permanent driver of the long-term evolution of the industry. Furthermore, this knowledge can be drawn in from other industries, such as the shopping mall industry in the case of retail business at an airport. Further, the coordination of operations and investment could be a source of gain.

On the negative side, integration of airports might generate social costs if airports buy other airports which could act as a substitute, thereby reducing the competitive pressures in the airport services markets. From the perspective of competition policy both effects have to be balanced against each other. We think that there is a strong case for recommending policymakers to allow alliances or mergers of airports serving different markets. In light of these recommendations it must be seriously doubted if the overall welfare effects of integrating airports which from the perspective of their users can be potential substitutes are positive. In the case of airports operating in the same or overlapping markets, policymakers should be much more cautious about approving alliances or mergers. Privatizing the London airports – Heathrow, Gatwick and Stansted – as a group within BAA has been criticized by Starkie and Thompson (1985) (for further discussion see Forsyth and Niemeier, Chapter 19, this volume). Integrating Frankfurt with its neighbouring Hahn occurred at a time when the rise of low-cost carriers was not foreseen. While the negative effects of integration are currently limited because Frankfurt is capacity constrained and not attractive for low-cost carriers, this might change in the future and will raise the question whether the welfare gains from management know-how transfer outweigh the losses from increased market power. Integrating Bratislava and Vienna led two different assessments by the competition authorities and in the end the merger was stopped. Given that the cost savings stemming from horizontal integration are very limited, but the potential welfare losses might occur and are not always easy to foresee for competition authorities, it might be argued that ownership restrictions are preferable. However such general rules might have potential negative effects as they are hard to design. The substitutability of airports is not a fixed constant easy to define. Distance and time between airports and their catchment areas might change with ground transport. Furthermore, LCCs have changed consumer habits so that airports are now becoming substitutes, which they had not been at the time FSAs dominated the market.

In the case of airports, the balance between the costs of the use of market power and the cost and service advantages arising from integration will be affected by regulation. Many airports are price regulated, and this will lessen the use of market power, and the inefficiencies arising from it. On the other hand, regulation is not costless, as it dampens incentives for productive efficiency, and particular forms of regulation can impact on investment and quality. Thus competition authorities may be willing to allow a merger to go ahead, or a group of airports in a region to be owned by a single company, if they believe that the gains from coordination of operations and investment outweigh the (reduced) costs arising from the use of market power and the efficiency costs of regulation.

While ownership restrictions are superior to a case by case approach of competition law it seems also to be necessary to complement this policy by a privatization policy that increases airport competition rather than hinders it. The case of BAA and Vienna Bratislava show that very often the government tries to increase the revenue proceedings from the sale by strengthening the monopoly position of the airport. Fraport begged by the state government of Hessen and the Federal government proposed to from a German airport system of all major German airports and then to privatise it. However, this was not accepted by the German federal states and would not have been approved by the German cartel office (see Kunz, 1999). In these cases airport policy claims to increase the countervailing power[8] against big airlines. However, the countervailing power of airlines is limited if no attractive substitutes are available. Furthermore this doctrine neglects the danger that the airline and the airport markets become fully monopolized. If unregulated, the integrated airports will increase price above marginal costs up to the monopoly price. Hence, countervailing power of airports against airline alliances might lead to a social welfare loss.

In short, an airport policy trying to increase competition through separated ownership could lead to a more competing airports and competition policy could evaluate on a case by case rule if mergers reduce the substitutability of airports to an anticompetitive result.

Over time, as more airports are privatised, private airport groups will gain control of more airports in Europe and around the world. It is possible, though not inevitable, that major groups will grow up and control significant proportions (say, 20 per cent) of the major city airports. Such groups would only pose a problem for competition if they were permitted to buy airports which are close to one another and which can compete effectively. A group might be large, but this need not mean that it is dominant in any particular city market. McDonalds possesses a significant share of the world hamburger market, but it does not pose problems for competition since it faces many competitors. More like airports, Westfield is a significant owner of shopping centres in many countries, but it is not dominant in the markets it serves. The size of the company is not as important as the degree of its local dominance – currently, BAA and Aéroport de Paris are more dominant in their local markets than the private airport companies are likely to be allowed to become.

Conclusions

While some consolidation amongst airports around the world is taking place, and is likely to continue, we expect a different pattern of consolidation from that which has been observed with airlines were the benefits of consolidation seems so much larger. In the airline sector, a large consolidation has already taking place. Because of the peculiar legal environment of international traffic rights, alliances have been the key form of consolidation, though mergers have also taken place, especially within countries. Regulatory constraints, especially on cross border mergers, partly explain this reliance on alliances – alliances may be the only form of consolidation permitted. By contrast, with airports, controls on ownership are not as restrictive, and cross country ownership

---

8 This concept was developed by Gailbraith in the fifties. See Gailbraith (1980) and for airports Button, Chapter 5.

is easier to achieve. Patterns of consolidation amongst airports have been constrained by public ownership – many airports remain owned by national or regional governments, and are not free to choose to merge with other airports or to be taken over by them. As privatization continues, this constraint on consolidation is being relaxed.

The key motivations for airline alliances or mergers are to obtain network benefits or economies of scope and density, and to increase market power. The latter is relevant in the case of horizontal integration of airports, but the former is much less relevant. Apart from gaining market power, the main motivations for airport integration appear to gain economies through coordination, and the transfer of expertise. These may be easier to achieve under an ownership than an alliance framework. It is notable that several airport companies have formed, yet there are very few airport alliances, even though there are fewer regulatory constraints on forming alliances than multiple airport companies. It has yet to be determined how large these benefits from coordination and knowledge transfer are – they may not be as significant as the network coordination benefits of air services, and thus the move to consolidation in airports may not be as pronounced as in airlines.

The main public policy concern with airport integration is that it may lead to increased market power in a region. This is only a concern when a single airport corporation owns two or more airports in a region. Many airport corporations own airports in diverse locations, so that this problem does not arise. However, competition policy problems have been arising in several cases. Several airports in a region, such as the London airports, have been privatized as a group, thus precluding the possibility of competition between them. This was criticized at the time of privatization, and it continues to be an issue – BAA may well be forced by competition authorities to sell some of is airports to enable competition to take place between them. In other cases, airports have sought to merge with close by airports, which are competing with them, or have the potential to do so. In some cases competition authorities have permitted these mergers (as with Frankfurt's purchase of Hahn) and in others they have prohibited the mergers (as with Vienna's aborted takeover of Bratislava). As in other industries, competition authorities must evaluate the trade-off between promised cost reductions and service improvements, against the losses from greater use of market power. This trade off has been influenced by the presence of regulation, which has sought to lessen the use of market power that airports have – this regulation has made it easier for groups of airports in a region to be owned by a single company.

References

Antimonopoly Office of the Slovak Republic (2006), Press Issue on the Decision on Bratislava and Vienna Airport, Bratislava 14 September 2006, http://www.antimon.gov.sk/eng/article.aspx?c=387&a=2365.
Aviation Handling Systems (AHS) (2008), http://www.ahs-de.com/en/about_us/company_structure.html.
Bundeswettbewerbsbehörde (2006a), Weitere Prüfung im Zusammenschlussfall Flughafen Wien–Bratislava, http://www.bwb.gv.at/BWB/Aktuell/Archiv2006/flughafen_vie_bts.htm.
Bundeswettbewerbsbehörde (2006b), Bundeswettbewerbsbehörde gibt Flughafen Wien 'grünes Licht', http://www.bwb.gv.at/BWB/Aktuell/Archiv2006/bwb_gibt_flughafen_wien_%84gruenes_licht%93.htm.
Carter, C. and Ezard, K. (2007), Global Airport Groupings, *Airline Business*, December, 58–9.
Competition Commission UK (2008) *BAA Market Investigation Emerging Thinking*, April, http://www.competition-commission.org.uk.

Doganis, R. (1999), The Future Shape of the Airline and Airport Industries, in Pfähler, W., Niemeier, H.-M. and Mayer, O. (eds), *Airports and Air Traffic – Regulation, Privatization and Competition*, Frankfurt/New York: Peter Lang Verlag, 103–13.

Doganis, R. (2001), *The Airline Business in the 21st Century*, London: Routledge.

Erste Bank (2006), Flughafen Wien, Company Report, 11 April 2006, http://www.erstebank.at.

Galbraith, J.A. (1980), *American Capitalism – The Concept of Countervailing Power*, revised edition, Oxford: Basil Blackwell.

Gateway (2000), A Publication of the Metropolitan Washington Airports Authority, Vol. 3, Spring, http://www.metwashairports.com.

Graham, A. (2001), *Managing Airports: An International Perspective*, Oxford: Butterworth Heinemann.

Kunz, M. (1999), Airport Regulation: The Policy Framework, in Pfähler, W., Niemeier, H.-M. and Mayer, O. (eds), *Airports and Air Traffic – Regulation, Privatisation and Competition*, Frankfurt/New York: Peter Lang Verlag, 11–55.

Morgan Stanley (2001), Fraport, A Good F-rap(p)ort, *Equity Research*, 20 August, London.

Niemeier, H.-M. and Wolf, H. (2002), Strategische Allianzen zwischen Flughäfen – notwendiges Gegengewicht zur Blockbildung der Fluggesellschaften? in Knorr, A., *Europäischer Luftverkehr – wem nutzen die strategischen Allianzen?*, Schriftenreihe der DVWG, B 246 Bergisch-Gladbach, 187–215.

Oum, T.H., Zhang, A. and Zhang, Y. (1996), A Note on Optimal Airport Pricing in a Hub-And-Spoke System, *Transportation Research B*, Vol. 30, 11–18.

Reuters (2006), Slovakia Stops Bratislava Airport Sale, 18 October, http://news.airwise.com/story/view/1161212078.html.

Spengler, J. (1950), Vertical Integration and Antitrust Policy, *Journal of Political Economy*, Vol. 58, 347–52.

Starkie, D. and Thompson, D. (1985), *Privatising London's Airports, IFS Report Series*, No 16, London: Institute for Fiscal Studies.

Sunnucks, M. (2002), Dulles looks to load up on cargo, *Washington Business Journal*, 10 June 2002, http://www.washington/bizjournals.com.

TIACA Times (2005), Galaxy International Cargo Airports Alliance Will Encourage Improved Standards, Service and Cooperation, Spring, 1–11, www.tiaca.org.

Tretheway, M. (2001), *Alliances and Partnerships Among Airports*, 4th Hamburg Aviation Conference, 14–16 February.

Wieninternational.at (2006), *Bratislava Airport Finally Sold to Vienna Best Bidder TwoOne*, http://www.wieninternational.at/en/node/456.

Williamson, O.E. (1985), *The Economic Institutions of Capitalism. Firms, Markets, Relational Contracting*, New York: The Free Press.

WirtschaftsBlatt, 2004, *Flughafen Wien unterstreicht Vorteil des gemeinsamen Airport-Systems mit Bratislava*, Vienna, 25 November 2004, www.wirtschaftsblatt.at/.

Wolf, H. (2003), *Privatisierung im Flughafensektor. Eine ordnungspolitische Analyse*, Berlin: Springer.

Chapter 21
Airport Competing Terminals: Recent Developments at Dublin Airport

Aisling Reynolds-Feighan

Introduction

This chapter examines the recent history of Dublin Airport, Ireland, and chronicles the Irish government's evolving policy in relation to the development of additional terminal facilities at the airport. McLay and Reynolds-Feighan (2006) reviewed the development of Dublin Airport up until 2005, when the possibility of competing terminals at the airport was an alternative that the government was considering. McLay and Reynolds-Feighan (2006) set out the economic and regulatory issues surrounding such an innovative policy initiative, along with the background to such a policy stance being considered. Ireland's airports are recognised as having major significance for regional and national economic development. The network of airports facilitates access to Ireland's regions and, accordingly, the structure of the Irish airport industry is of interest from a regional economic perspective.

In this chapter, we briefly review the main arguments in McLay and Reynolds-Feighan (2006) and give some background to recent developments at Dublin Airport. The next section gives an overview of Dublin Airport's development and regulatory history. We then review the recent traffic profile of the airport and rehearse the arguments for greater independent private sector involvement in various aspects of the airport's operation. In the final section the decision not to proceed with a competing independent terminal operator at Dublin Airport is outlined and the most recent events are briefly reviewed. We also demonstrate how the Dublin Airport case has been cited internationally as evidence against consideration of within-airport competition.

Development of Dublin Airport – An Overview

Dublin Airport is the key gateway for the island of Ireland, with 23 million passengers passing through the airport in 2007. The airport handles roughly four-fifths of all air passenger traffic in the Republic of Ireland. McLay and Reynolds-Feighan (2006) set out the history and regulatory development of Dublin Airport up to 2005. Table 21.1 gives a summary of the main legislative and regulatory instruments that guided the development of Dublin Airport for the period 1937–2006. The government holding company, Aer Rianta, was initially established as the holding company for the new Irish airline Aer Lingus. Aer Rianta took over the management of the newly constructed Dublin Airport in the 1940s, with charges being regulated by the Minister for Transport. In line with best international practice, the Irish government sought to separate the ownership and regulatory roles by (i) formally transferring ownership of the three airport assets and lands to Aer Rianta (1998); and (ii) establishing an independent regulator to determine airport and terminal charges for the airport and for air navigational services (2001).

The Commission for Aviation Regulation (CAR) was established under the Aviation Regulation Act, 2001. This Act required the CAR to make a determination specifying the maximum levels of airport charges that may be levied by an airport authority at any Irish airport with more than 1 million passengers in the previous year. The CAR imposed a price cap on Dublin, Shannon and Cork airports in its initial determination. The price cap was designed to reflect the company's capital expenditure requirements, and to increase annually in accordance with an 'RPI-X' formula. While this form of regulation allows for charges to rise to cover the increases in the operator's cost base, it also requires the operator to make improvements in productivity of approximately X per cent per annum. Similar regulatory regimes are in place at several UK and German airports. The CAR applied a 'single-till' approach,[1] essentially taking into account both aeronautical and non-aeronautical airport revenues in the setting of maximum airport charges. In order to reduce the ability of Aer Rianta to use revenues derived from Dublin Airport to cross-subsidise its other airports, a specific price cap for Dublin Airport was set below the level of the general cap set for the average revenue per passenger at all Aer Rianta airports.[2]

In 2004, the decision was made to begin separating out the three state airports and, under the State Airports Act, Dublin Airport Authority (DAA) was established. As of 2008, the three state airports have remained under DAA control and ownership, though the 2004 Act set out a process for the establishment of autonomous competing airports. The long-term future and financial viability of Shannon Airport is uncertain as the transatlantic gateway 'stopover' requirement is phased out in 2008/09, and Aer Lingus, the main operator at Shannon, cut its services to the key London Heathrow hub in 2008. Only with an agreed and robust financial strategy for Shannon

Table 21.1 Key legislative and regulatory instruments governing operation of Dublin Airport

Year	Event
1937	Incorporation of Aer Rianta as holding company for Aer Lingus (Irish Airlines)
1940–41	Establishment of Dublin Airport at Collinstown, Co. Dublin; inaugural flight January 1940
1950	Air Navigation and Transport Act gave statutory responsibility for Dublin Airport to Aer Rianta
1966	Aer Rianta and Aer Lingus separated, with Aer Rianta retaining management role for Dublin Airport
1969	Aer Rianta management extended to include Shannon (SNN) and Cork (ORK) airports
1998	Air Navigation and Transport (Amendment) Act granted ownership and management rights to Aer Rianta
2001	Aviation Regulation Act 2001 established the Commission for Aviation Regulation and transferred regulatory responsibility for airport and terminal charges from the Minister for Transport to the new Commission
2004	State Airports Act renamed Aer Rianta as 'Dublin Airport Authority'; established separate boards of management for Shannon and Cork airports and set out the basis for phased distribution of assets, subject to conditions relating to the operational and financial readiness of the three airports to operate as independent and competing entities Also, regulatory remit of the Commission for Aviation Regulation limited to Dublin Airport charges only

---

1 See Reynolds-Feighan and Feighan (1997) for a detailed discussion on 'single-till' versus 'dual-till' regulatory approaches.

2 See the discussion of this issue at p. 20 of CAR (2001a).

and Cork will the terms of the Airports Act permit the establishment of three independent airport authorities.

The 2004 Act also limited the CAR's regulatory regime to Dublin Airport's charges. The passage of the State Airports Act, 2004 may be viewed as a policy response to the Irish commercial aviation industry's widespread dissatisfaction with its experience of the price cap regulation of airport services, and a cautious willingness to explore competition as a market-based alternative to active regulation.[3]

Aer Rianta and, more recently, DAA have had troubled relationships with the airlines operating from Dublin Airport in the last couple of decades. In the mid-1980s, the airport was typical of many European capital city airports of the time, with traffic dominated by the nationally owned flag carrier, operating a network of cross-channel, European and transatlantic routes. Traffic volumes grew marginally in the period 1980 to 1986. Dublin Airport's traffic profile changed significantly during the 1990s with the growth and expansion of Ryanair services. The Ryanair 'low-cost operating model' was imposed in the 1991/92 reorganisation which saw Michael O'Leary take over as Chief Executive. O'Leary has campaigned continuously for lower airport charges at Dublin.

In the next section, the traffic profile of Dublin Airport is set out in more detail to highlight the growing impact of low-cost traffic at this major European airport and the need for development of facilities to cater for this segment of the airline market.

Traffic Profile at Dublin Airport 1996–2006

Table 21.2 illustrates the dominant role of Dublin Airport in Irish aviation – it handles approximately 80 per cent of all of the passenger traffic in the Republic of Ireland. Table 21.2 also illustrates the rapid growth in traffic at the three state-owned airports during the late 1990s, which accompanied Ireland's rapid economic growth during the period. Rising consumer incomes and increased demand for intra-European and North Atlantic passenger and freight services for business activities and tourism fuelled the increase in traffic. Passenger traffic seating capacity increased by 142 per cent at Dublin Airport between 1996 and 2006. The number of airlines operating to and from Dublin Airport increased from 24 in 1996 to 28 in 2000 and to 46 in 2006.

Figure 21.1 shows the regional breakdown of traffic at Dublin, Shannon and Cork airports in 2000 and in 2006. The 2000 data shows the impact of the designated role of Shannon Airport in Irish aviation policy as the key transatlantic gateway for the west and south of Ireland. In transatlantic bilateral air service agreements, all scheduled flights coming to and from Ireland were required to stop at Shannon, whether or not this was the final destination. This policy is now being phased out as part of the new US–EU bilateral 'open skies' agreement. Figure 21.1 illustrates the growing importance of intra-European air services at all three state airports, but particularly at Dublin. The proportion of UK passengers has dropped from 63 per cent of all traffic at Dublin in 2000 to 40 per cent in 2006, while the European passenger share has risen from 33 per cent to 58 per cent. This reflects the strategy by both Aer Lingus and Ryanair to greatly expand the number of intra-European air routes to and from Dublin. Aer Lingus adopted a low-cost/low fares strategy for its intra-European and UK air services after 2001, offering point-to-point single cabin service.

---

3   See McLay and Reynolds-Feighan (2006) for a lengthy discussion of the turbulent and protracted legal battles between Aer Rianta, the CAR and Ryanair during the 2001–05 period.

Table 21.2 Passenger traffic and traffic growth rates at Cork, Shannon and Dublin airports, 1995–2006

Year	Cork Airport		Shannon Airport		Dublin Airport	
	Passenger traffic	Annual traffic growth rate (%)	Passenger traffic	Annual traffic growth rate (%)	Passenger traffic	Annual traffic growth rate (%)
1995	971,319		1,573,770		8,024,894	
1996	1,124,320	16	1,700,174	8	9,091,296	13
1997	1,191,261	6	1,822,089	7	10,333,202	14
1998	1,315,224	10	1,840,008	1	11,641,100	13
1999	1,501,974	14	2,188,154	19	12,802,031	10
2000	1,680,160	12	2,408,252	10	13,843,528	8
2001	1,775,817	6	2,404,658	0	14,333,555	4
2002	1,874,447	6	2,353,530	-2	15,084,667	5
2003	2,182,157	16	2,400,677	2	15,856,084	5
2004	2,254,251	3	2,395,116	0	17,138,373	8
2005	2,729,906	21	3,302,424	37	18,450,439	8
2006	3,010,575	10	3,639,046	10	21,196,382	15
2007	3,180,259	6	3,620,623	-1	23,287,438	10

Source: Aer Rianta/DAA annual reports.

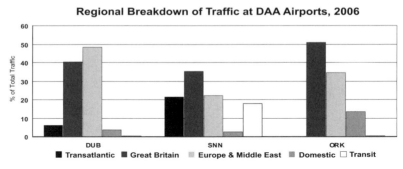

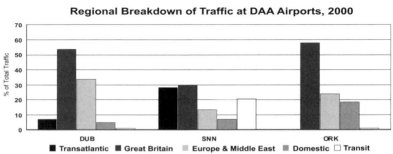

Figure 21.1 Regional breakdown of traffic at the three Irish state airports, 2000 and 2006
Source: Aer Rianta /DAA Annual Reports.

Table 21.3 shows the traffic shares at Dublin Airport operated by the top six airlines in the years 1996, 2000 and 2006. Aer Lingus and Ryanair are the top ranked carriers in all years, but Ryanair takes over the top ranked position in 2006. These two carriers increased their market share from 75 per cent of all traffic in 1996 to 79 per cent of all traffic at Dublin in 2006. Ryanair in this period was greatly expanding its European network not only from Ireland, but from several other European operational bases as well. Dublin Airport's traffic mix has thus become heavily focused on low-cost carrier operations.

Competing Independent Terminals at Dublin Airport – A New Public Policy Response?

In their paper in 2006, McLay and Reynolds-Feighan argued that several features of Dublin Airport's site, situation and traffic profile had given rise to a need to consider new options in an effort to introduce competition in the provision of airport services in Ireland. The airport site at Dublin is unconstrained, and in the long run should be able to cater for greatly increased traffic volumes once appropriate air-side facilities are put in place. Elsewhere in mainland Europe or in the United States, the airports network is sufficiently dense to give rise to competition between neighbouring large and medium airports. Only in less populated regions or on islands does spatial competition between airports tend to be restricted. Dublin Airport is relatively isolated, and terminal competition may be an effective means of introducing competition.

Terminal competition in this context can mean that on a single airport site, competing terminal operators build, own and operate separate terminals. The airlines thus have a choice in the selection of this component of the 'airport turnaround supply chain'. Airlines would use single or multiple runways that might be owned and operated by either one of the terminal companies, or by a third party. In situations where there are no constraints on the development or expansion of runway capacity (as is the case at Dublin Airport), it may be argued that duplication of runways through competing airports rather than competing terminals is inefficient, and that concentration of air traffic at a single location in a region allows for economies of density and scope for all airlines using the single airport.

The traffic mix at Dublin airport is unusual among large European capital city airports in that there are two large airlines dominating: one low-cost carrier and one former flag carrier, that operates point-to-point intra-European air services and 'hubbed' long-haul transatlantic services. As was seen in the previous section, the rapid growth of low-cost services at Dublin Airport gave

Table 21.3 Top six carriers and market shares at Dublin airport in 1996, 2000 and 2006

1996		2000		2006	
Aer Lingus	44.50%	Aer Lingus	47.90%	Ryanair	41.50%
Ryanair	30.60%	Ryanair	28.00%	Aer Lingus	37.60%
Bmi British Midland	7.90%	Bmi British Midland	8.00%	Bmi British Midland	3.30%
Alitalia	5.50%	BA	3.30%	AF	3.00%
Lufthansa	1.90%	Delta	1.70%	SK	1.40%
VS	1.60%	SAS	1.60%	BA	1.20%

Source: Official Airline Guide Databases, 1996–2006.

rise to a need for new fast turnaround airport facilities that allowed for less restrictive movement of passengers and permitted higher aircraft utilisation rates for intra-European travel in the growing single European air transport space that has been evolving from 1992 onwards.

As Dublin Airport's traffic continued to grow rapidly during the 1990s and 2000s, the pressure on Aer Rianta/DAA to provide facilities deemed suitable by the airline customers grew. The discussion was often heated and very much conducted in the public domain. The response by Aer Rianta to the independent economic regulator, the CAR, contributed to a general sense in the Irish aviation industry that regulation had not provided the change in outlook or corporate approach that was warranted in meeting Ireland's long-term aviation needs and challenges. Thus new approaches to introducing competition were explored.

The main economic policy issues identified by McLay and Reynolds-Feighan in their 2006 paper related to:

1. competition compared to regulation in delivering efficient and appropriate services;
2. the possibilities of competition improving productive and allocative efficiency but perhaps decreasing the extent of other efficiencies;
3. the provision of the appropriate infrastructure given the presence of multiple users with heterogeneous demands;
4. operational aspects of delivering the 'airport turnaround supply chain' and the multiplicity of factor markets involved in components of this supply chain conferring varying degrees of market power upon the producer;
5. the implications for the airport business model of the unbundling of the suite of airport services;
6. issues surrounding the allocation of runway services and the need to separate out or regulate runway and terminal services currently provided by a single operator;
7. international examples of multiple terminal operators at a single airport site and the competitive impacts;
8. longer-term investment implications for runway and terminal capacity increases.

McLay and Reynolds-Feighan (2006) concluded that due to the complexity of the industry and its productive processes, considerable further work would be necessary in order to substantiate claims that terminal competition would deliver net benefits in the provision of airport services.

Irish Government Process and Response

There was very little public debate about the merits or demerits of competing terminals in Ireland in the period 2002–05. Table 21.4 sets out the milestones in the evolution of Irish government policy relating to the construction and operation of a second terminal at Dublin Airport. The notion of competing terminals first became a potential policy option when it was included as a possibility to be explored and examined by the incoming government of 2002. The new Minister for Transport quickly began seeking proposals from interested parties to set out their vision of how an independently owned and operated terminal might function at Dublin Airport.

Table 21.4 The road to Terminal 2 at Dublin Airport – key milestones[4]

Date	Event/Decision
June 2002	New government partners come to power in June 2002. Programme for Government presents the possibility of a competing/privately owned terminal at Dublin Airport; Minister Seamus Brennan appointed to Transport Brief
August 2002	Minister Brennan announces decision to seek proposals for second terminal at Dublin Airport
March 2003	Minister Brennan announces public consultation phase on second terminal at Dublin Airport and outlines policy developments in relation to three Aer Rianta airports following publication of 'Expert Review Panel' study of the independent terminal proposals
July 2003	Minister Brennan announces the setting up of new airport boards at Shannon and Cork in advance of the State Airports Act; Minister announces 'I have for some time advocated a policy of introducing greater dynamism to the aviation sector through increased competition and choice. I see the establishing of Dublin, Shannon and Cork as independent airport authorities as heralding the beginning of a challenging and exciting new era in Irish aviation. Under the new arrangements the three airports will compete with each other and vigorously pursue new business, free from central control.'
September 2004	Seamus Brennan removed from Department of Transport and replaced by Martin Cullen in Cabinet reshuffle
May 2005	Government awards development of Terminal 2 to DAA; the possibility of a private Terminal 3 in the future is muted in the announcement
August 2005	Transport Minister directs Commission for Aviation Regulation to take into consideration in determining airport charges the costs of infrastructural development at Dublin Airport
August 2007	Planning permission approved for DAA's Terminal 2 design

In response to a call for 'Expressions of Interest', 13 proposals were received from national and international companies and consortia. The proposals varied in scope and level of detail, ranging from highly developed proposals with layout plans etc. to letters of interest in a subsequent tendering process. An Expert Panel was established by the Minister for Transport to assess the responses and advise the Minister on the viability of an independent terminal operation at Dublin Airport. The Expert Panel study is the only published report setting out in any detail the issues surrounding such a new policy direction.[5]

The review panel considered the traffic profile at Dublin Airport in 2001. At that time, the low-cost traffic component amounted to 26 per cent of total passenger traffic. The review panel suggested that in order for this share to increase at Dublin, a number of key requirements would need to be in place, including (a) competitive landing charges; (b) sufficient facilities to facilitate fast equipment turnarounds; (c) sufficient demand, particularly to new destinations. The review

---

4 Minister Brennan's quote: Irish Department of Transport, 'Government to establish three fully independent and autonomous airport authorities for Dublin, Cork and Shannon', Press Release, 10 July 2003, http://www.transport.ie/upload/general/3769-0.pdf.

5 The report of the Expert Panel may be viewed at http://www.transport.ie/upload/general/3273-0.pdf.

panel noted the changes in the Aer Lingus operational model during late 2001, which saw an increasing focus on low-cost, point-to-point intra-European operations.

The review panel briefly reviewed the idea of an independently owned second terminal at Dublin Airport and concluded that 'international precedents were not favourable because of operational difficulties and that high levels of aeronautical charges would have to be levied to generate an adequate commercial return' (Expert Review Panel Report, 2003, p. 6).

In their review of airport charges in 2003, the review panel concluded that 'airport charges, at their current level, are not a barrier to airlines developing new routes out of Dublin' (ibid.). The review panel also considered additional runway capacity as an essential element for 'a step change in the competitive environment at Dublin' (ibid.).

The panel concluded that 'none of the issues identified are insurmountable and that an independent terminal is a viable strategic option for the development of Dublin Airport and would elicit considerable market interest. If this concept is to be progressed to the next stage of the tender process, it will require detailed consideration in the first instance of all of the issues ...' (ibid., p. 15).

The review panel was favourably disposed to the idea of an independent terminal on operational and technical feasibility grounds. The estimated 'lead-in time' to set up legislative, regulatory and operational aspects of a new independently operated and owned terminal was envisaged to be four to five years by the experts. The Expert Panel set out a number of requirements concerning the legal and operational aspects of the new terminal that would determine its financial viability. These related to contractual relations between airlines and terminal operators for minimum service durations; pricing of services provided; opportunities to develop significant retail and other property revenue streams; and opportunities to have a competitive cost base.

Finally, the review group concluded that the construction and operation of an independent terminal at Dublin Airport would have a profound and significant impact on Aer Rianta and require a re-evaluation of many aspects of its short-, medium- and long-term operational and investment strategies.

The review panel study was published by the Department of Transport in March 2003. The Minister proceeded in the following 12 months to pass legislation to break up Aer Rianta and establish separate boards at the three airports. These changes met with strong opposition and very poor relations between Aer Rianta, officials at the Irish Department of Transport and the Minister (*Sunday Independent*, 2004). There was mounting opposition from unions also. In September 2004, following a cabinet reshuffle, Minister Brennan was moved from the Department of Transport and succeeded by Martin Cullen. Minister Cullen quickly decided against an independent terminal at Dublin Airport and in May 2005 announced the awarding of a new Pier D and the second terminal to DAA.

DAA moved swiftly to build Pier D and in August 2006 submitted plans for Terminal 2. These were duly approved following an independent review of the cost estimates by financial consultants to the Department of Transport. The Minister issued a directive to the CAR to take the costs of the new infrastructural investments into account in determining its new airport charges for Dublin Airport in 2006 and subsequent years.[6] Thus the debate and discussion on competing terminals at the airport ended abruptly in a political decision to go with the status quo.

Planning permission for Terminal 2 was granted in August 2007, subject to a lengthy series of conditions relating to noise, aircraft movements and access transport levies. The Commission

---

6   Ministerial Directive issued to the Commission for Aviation Regulation, 18 August 2005 (see http://www.aviationreg.ie/_fileupload/Image/ABOUT_PR3_2005_AIRPORT_CHARGE.pdf.

for Aviation Regulation issued a new airport charges directive in 2007, which permitted a modest increase in the maximum airport passenger charges for the four years from 2008 to 2011. DAA quickly increased the non-regulated, non-aeronautical charges. In November 2007, DAA increased its car parking charges at Dublin Airport by 50 per cent; in February 2008, check-in desk rental charges were also increased by 50 per cent, along with increases in rental for the floor space required for self-service kiosks. In announcing the car park charge increases, DAA justified the increases on the following grounds: 'The DAA plans to continue levying benchmarked commercial charges for its car parking services in order to maximise the revenues available to the company in a tightly regulated environment' (DAA press statement, November 2007).

UK Government Process and Response

The UK Civil Aviation Authority (CAA) examined the issue of competing terminals as part of its airports policy review conducted in 2005 and 2006 as a lead up to their airport price determination for regulated airports in April 2008. In the consultation document of December 2005 (UK CAA, 2005), the CAA sought contributions from interested parties on the potential merits and case for introduction of competing terminals at regulated airports. The CAA had in an earlier paper (UK CAA, 2001) set out some of the potential benefits of within-airport competition as being:

1. the provision of a better value service at reduced costs to users (since each terminal would price down to marginal cost);
2. more innovative use of existing assets to increase capacity;
3. improving service quality in line with users' needs;
4. the allowance of specialisation of service provision.

In response to the December 2005 CAA Consultation Paper, the British Airports Authority argued at length against the legal and economic rationale for within-airport competition (BAA, 2006). The BAA response suggested that within-airport competition would generate a legal complication with regard to the ability to levy airport charges. It would require the designation of more than one airport operator operating from a single airport site. This in turn would require redefinition by the regulator of the 'regulatory till', as the price cap might not apply to the same bundle of airport activities between the two operators. BAA suggested that this process would give rise to greater regulatory involvement in order to moderate issues of ownership, compensation and allocation of assets.

In terms of the economic case against within-airport competition, BAA challenged the CAA to demonstrate that competition would lead to lower costs and that effective competition could actually exist. BAA suggested that the monopoly provider would most likely have lower marginal costs than multiplier suppliers. BAA also suggested that, for competition to be effective, a competing terminal would have to build sufficient capacity to allow airlines to change location. Such excess capacity at a single airport site would be wasteful and be in contrast to the incremental manner in which most airport operators add terminal capacity. BAA raised issues as to the extent of transactions costs of introducing competing terminals, as well as the reduction in scale economies for the existing single operator. They pointed out that at most airports with multiple terminals, the terminal services are not homogeneous. Airline movements between competing terminals would lead to remodelling costs and limit the extent to which competition could actually take place. BAA went on to cite the experience in Ireland: '*The* experience in Ireland suggests that attempts to introduce competition into terminal construction can significantly lengthen the process of capacity provision and may actually lead to higher costs and delayed opening of terminals' (BAA, 2006, p. 9).

Finally BAA suggested that airlines could use terminal competition to increase their market power by blocking entry bids by new entrant carriers through the uptake of long-term contractual arrangements with terminal operators.

In response to the submissions received and in preparing their price determination for 2008–13, the CAA concluded that:

> the CAA's view at this stage on the case for regulatory intervention to stimulate competition within airports is that the balance of evidence presented so far is against taking such regulatory action. The CAA remains open to further evidence, though, particularly on the issues identified in this chapter, and it notes that several airlines requested that this topic be kept open as the review progresses, in order to allow them to submit views at a later stage. (UK CAA, 2006)

Implications and Conclusions

The key objective of air transport policy must be the provision of safe, reliable and inexpensive air services to maximise the potential economic benefit to regions, and to facilitate access to international transport and distribution networks for all regions. Where current structures and institutional arrangements hinder the achievement of these aims, more flexible structures should be established once these have been analysed and evaluated. The case for competing terminals at Dublin Airport was put forward in the context of general dissatisfaction with the state airport operator's behaviour, even under an independent regulator. The case for such a policy intervention, however, has received little attention and detailed study in Ireland or elsewhere. In the limited number of situations where any form of independent terminal operator has provided services, such as Toronto or Brussels airports, many political and local factors resulted in decisions to discontinue with multiple terminal operators after a short period of time. The Irish case is now added to the shortlist of examples cited by monopoly airport operators claiming that terminal competition is unworkable. However terminal competition, as proposed in the Irish circumstance, remains an untested policy option to date.

References

BAA (2006) 'BAA Response to CAA Policy Issues Consultation Paper – Chapter 7 BAA/Q5/139' (http://www.caa.co.uk/docs/5/ergdocs/dec05/baa_comp.pdf).
CAR (2001a) 'CP2/2001: *Economic Regulation of Airport Charges in Ireland*', Commission for Aviation Regulation, Dublin.
CAR (2001b) 'CP7/2001: *Determination in Respect of the Maximum Levels of Airport Charges that May Be Levied by an Airport Authority in Respect of Dublin, Shannon and Cork Airports in Accordance with Section 32 of the Aviation Regulation Act, 2001*', Commission for Aviation Regulation, 26 August 2001, Dublin.
CAR, (2001c) 'CP8/2001: *Report on the Determination of Maximum Levels of Airport Charges – Part 1: Report on the Reasons for the Determination*', Commission for Aviation Regulation, 26 August, 2001.

Expert Panel Review (2003) '*Dublin Airport – Review of Expressions of Interest for an Independent Terminal*', Expert Panel Report to Minister for Transport, Irish Department of Transport, February (http://www.transport.ie/upload/general/3273-0.pdf).

Fianna Fáil and the Progressive Democrats (2002) An Agreed Programme for Government between Fianna Fáil and the Progressive Democrats, Dublin.

McLay, P.C. and Reynolds-Feighan, A.J. (2006) 'Competition Between Airport Terminals: The Issues Facing Dublin Airport', *Transportation Research A*, 40(2), pp. 181–203.

Reynolds-Feighan, A.J. and K.J. Feighan, (1997) 'Airport Services and Airport Charging Systems: A Critical Review of the EU Common Framework', *Transportation Research E: Logistics and Transport Review*, 33(4), pp. 311–20.

Sunday Independent, 14 November 2004.

UK CAA (2001), 'Competitive Provision of Infrastructure and Services within Airports', Consultation Paper, February (http://www.caa.co.uk/docs/5/ergdocs/competitionwithinairportsfeb01.pdf).

UK CAA (2005) 'Airports review – policy issues – Consultation Paper', December 2005, UK CAA, London.

UK CAA (2006) 'Airports review – policy update', 15 May 2006, UK CAA, London.

Chapter 22
Competition, State Aids and Low-Cost Carriers: A Legal Perspective

Hans Kristoferitsch[1]

Introduction

Competition between airports can be distorted for several reasons. One reason, the misuse of a dominant market position, was dealt with in a separate lecture at the German Aviation Research Society (GARS) conference in Vienna. Another reason for the distortion of competition is financial intervention by public authorities. The Treaty establishing the European Communities (TEC) deals with the permissibility of State aid in its Article 87. While there is a fairly voluminous body of case law on State aid to airlines and State aid for airports, especially in the field of aid for the construction of airport infrastructure, the European Commission was confronted with a new phenomenon in its recent Brussels Charleroi decision: State aid for low-cost carriers serving regional airports.

Since the 1990s, low-cost carriers have become an increasingly important factor in the competition between regional airports and in some cases even in the competition between regional airports and established city airports. Regional airports were putting great effort into attracting budget airlines. Studies, mainly from Great Britain, suggested that these airlines could substantially increase passenger levels at regional airports:

- In 1988 London Stansted provided service to 1 million passengers. In 2002, 16 million passengers used this airport, 52 per cent of which were Ryanair customers. In 2006, the total number of passengers using the airport was 23.7 million.
- In 1997, the year prior to the arrival of Ryanair, Frankfurt Hahn provided service to fewer than 20,000 passengers. In 2005 it was used by 3.1 million passengers. The share of Ryanair customers amounted to 95 per cent. Frankfurt Hahn plans to serve 8 million passengers in 2012 and will soon overtake Dublin and become Ryanair's second largest base.
- Brussels Charleroi, which will be discussed in greater detail later, had fewer than 20,000 customers in 1997, serving only 54 passengers per day. In 2002 1.3 million passengers were counted, 98 per cent of which used Ryanair.[2] In 2004 the 2 million passenger threshold was passed, at 5.479 passengers per day. This is a 10,000 per cent increase in daily passengers between 1997 and 2004.

---

1   Written version of a presentation at the GARS conference on Airport Competition and the Role of Airport Benchmarking, 25 November 2005, Vienna. The author is grateful to the organizers of the conference for the invitation to give a presentation and to publish this chapter and to an anonymous reviewer who provided valuable comments on an earlier draft of this chapter.

2   Data provided by Ryanair, see European Commission 12.2.2004, *Brussels South Charleroi Airport*, O.J. 2004, L 137/1, p. 47. If not indicated otherwise, all paragraphs cited in this article refer to this decision.

Public authorities, who own the majority of regional airports in Europe, were thrilled by the prospect that growing passenger volumes would lead to the profitability[3] of regional airports and eventually to their subsequent privatization.[4] Because there are only a few successful low-cost carriers but many regional airports in Europe, the low-cost airlines are in a very favourable negotiating position when deciding which new routes to open. Therefore in many cases public funds are used to attract low-cost carriers. In some cases public funds were granted to low-cost carriers directly by public authorities. In other cases money was first given to the publicly owned airports, which then used subsidies to attract low-cost carriers.

This practice was fiercely contested by established network carriers. They argued that using public funds to attract low-cost carriers would amount to illegal State aid and would violate Article 87 (TEC). They felt that public funds should be spent to generate long-term growth, but not by giving it away to airlines which lose money with every flight. The network carriers argued that the new routes would lead to a transfer of passenger volumes from their carriers to the low-cost carriers ('market takers'),[5] but that it would not increase passenger volumes.

The low-cost airlines replied that their business model was new. They were receiving public funds not for transporting passengers, but for stimulating tourism in the regions surrounding their destinations and for linking these regions to larger cities. They saw themselves as 'market makers' because their low prices attracted passengers who otherwise would not have travelled. But even if the reason for their success was that they mainly diverted traffic from other suppliers, the low-cost airlines argued that this was due to the lower prices and hence nothing other than standard competition by delivering gains to customers.[6] It was only understandable that they demanded

---

3   According to studies by the Commission an airport needs between 500,000 and 1.5 million passengers to make a profit: 'there are no absolute figures with regard to the break-even point ... there are variations according to the country and the way in which the airports are organised.' European Commission, 6.9.2005, Community Guidelines on financing of airports and start-up aid to airlines departing from regional airports, O.J. 2005, C 312/1, p. 72.

4   Cases such as London Stansted, London Luton, Liverpool and Glasgow Prestwick can be mentioned as examples for privatization after a 'critical mass' of passengers has been attracted by budget airlines (Callaghan, 2005). It is submitted, however, that the profitability threshold may be substantially higher than the Commission estimates if an airport is frequented almost exclusively by low-cost airlines: Frankfurt Hahn achieved a positive *earnings before interest, taxes, depreciation and amortization* (EBITDA) for the first time in 2006 (3.7 million passengers). In 2005, the EBITDA was negative (by EUR 2.4 million), even though 3.1 million passengers had used Hahn.

5   For the distinction between 'market makers' and 'market takers' see Stadlmeier/Rumersdorfer (2006), 300f. A recent study confirms that incumbents are economically hurt by the subsidization of low-cost carriers but suggests that they 'may be more affected by competition than by the subsidy' (Barbot, 2004).

6   For this argument see the comment of John Vickers in his speech to a conference on State aid ('State aid and Distortion of Competition', 14 July 2005, http://www.oft.gov.uk/shared_oft/speeches/spe0605.pdf, p. 3): 'Two principles to guide where the Commission should (and should not) intervene against State aid flow from the points above. The first is whether the aid is likely to have a significant anti-competitive effect in the market. It is more likely to do so if the aid is in favour of a recipient with a substantial market presence. The second, having regard to the lobbying power of vested interests, is whether the recipient is a long-established incumbent operator. If such principles had been established in the past, there might have been no need for intervention against the financial arrangements between Ryanair and Charleroi airport. For historical "national champion" reasons, airline and airport markets are often still characterized by entrenched incumbents with market power and close links to each other and to governments. Low-cost entrants such as Ryanair have brought, and continue to bring, strong pro-competitive challenges into this historically uncompetitive arena, typically bypassing dominant airports by making deals with airports like Charleroi. Large benefits have

money for starting new routes, given the risks inherent in offering flights to airports 'where before the sheep had grazed on the runway'.[7] Ultimately the low-cost carriers insisted that not only public authorities but also private investors were convinced that attracting passengers by offering flights at very low prices would lead to breaking even in the long run.

In retrospect, it took a surprisingly long time[8] before the dispute was fought out at the European Union level. The question of whether or not subsidies for low-cost carriers should be considered as illegal State aid remained unresolved until February 2004, when the Commission decided the Charleroi case.[9] The underlying rationale of this decision was amended and further developed in the 'Community Guidelines on financing of airports and start-up aid to airlines departing from regional airports',[10] published in September 2005. Although these two documents have answered some of the questions on whether or not subsidies for low-cost carriers violate European Commission law, a substantial amount of uncertainty remains. This is due to the fact that the new Commission approach has not been confirmed by the European Courts[11] and the documents unfortunately do not provide an exhaustive response to the underlying legal questions.

The goal of this chapter is to describe the emerging legal framework for State aid granted to regional airports and low-cost carriers and to clarify the unanswered questions. I will first give a brief overview of Article 87 TEC, the most important rule governing State aid (II). Then I will present the Charleroi decision (III) and the new Community Guidelines (IV). After an analysis of these two documents (V) I will conclude by outlining the status quo of legal rules for State aid to low-cost carriers (VI).

Article 87 TEC – The Rule Governing State Aid

Article 87 of the Treaty of the European Community provides that 'all aid granted by a Member State or through State resources in any form whatsoever which distorts or threatens to distort competition by favouring certain undertakings or the production of certain goods shall, in so far as it affects trade between Member States, be incompatible with the common market'.

This statement contains a number of elements:

1. Article 87 TEC only applies if public funds are used or if the state grants an exemption from expenses (i.e. taxes) that an undertaking would have to bear otherwise. As a consequence of

---

resulted for travellers, incumbent airlines have been spurred to make their own offerings more attractive to consumers, and single market goals have been advanced. That the State aid process led to intervention in such a case at least raises questions about policy priorities.'

7   As a representative of Ryanair put it to describe the situation at Charleroi airport before the arrival of Ryanair.
8   This, of course, was very favourable for low-cost airlines because it enabled them to operate free from the legal restraints of the EC regime on State aid. The reluctance of the EU to step in might be due to the popularity of low-cost carriers among European citizens (Soltész, 2003).
9   See n. 1.
10   See n. 2.
11   Quite the contrary: the European Court of First Instance rejected part of the Commission's reasoning in the Charleroi case and declared the decision void in December 2008 (CFI 17.12.2008, T-196/04 *Ryanair Ltd/Commission*, n.y.r.).

this criterion, subsidies to low-cost carriers granted by private airports are not covered by the EC regime on State aid.[12]
2. The aid must affect trade between Member States and must lead to a distortion of competition. The European Court of Justice (ECJ) is very strict in the interpretation of this criterion. Slight distortions of competition suffice to make Article 87 TEC applicable.
3. Ultimately, the measure must amount to an advantage for a particular recipient (selectivity). The function of this criterion is to distinguish general economic measures from measures intending to favour one specific undertaking or one specific branch of the economy over others.[13]

These three criteria can be summarized in a general principle that represents the easiest way to distinguish legitimate use of public funds from prohibited State aid, and is frequently applied by the Commission. This 'principle of private investor' test consists of asking the following question: Would a private investor in a market economy have invested in the economic situation? If there is foreseeable profitability for an investment in the long run, if the State acts like a private investor, Article 87 TEC is not violated.

The underlying rationale of Article 87 TEC is to prevent the State from distorting competition. The prohibition to grant unlawful State aid is monitored by the European Commission. Measures that might violate Article 87 TEC must be reported ('notified') to the Commission in advance. The Commission then decides if the measure is compatible with the Common Market. The TEC prohibits implementation of measures containing State aid prior to EC notification and the outcome of the Commission's assessment (Article 88 (3) TEC). If this obligation is neglected and measures containing State aid are implemented, the Commission can mandate the reversal of the transaction. It is possible for competitors to bring suit both against the recipients of State aid and against the public authority granting aid.

An important exception to the applicability of the EC rules on State aid is established by the ECJ in the Altmark-Trans case.[14] Article 87 TEC does not apply if the State transfers funds to an undertaking performing 'services of general economic interest', subject to the following four conditions:

1. The obligation to fulfil services of general economic interest must actually exist and has to be clearly defined.
2. The parameters for calculating the amount of compensation have to be established in advance.

---

12 There have been cases of circumvention of the EC rules on State aid in which public authorities made large investments in airports prior to their privatization. The European Court of Justice (ECJ) held that a similar case of obvious circumvention (albeit of the EC rules on public procurement) should be resolved by applying the rules which the public entity tried to circumvent. ECJ 10.11 2005, C-29/04 *Stadt Mödling*, ECR 2005, I-9705.

13 To give an example: constructing a highway may have positive effects on companies located close to the new highway. But public investments in general infrastructure are not selective and are therefore not considered as State aid. If the State, on the other hand, funds infrastructure on the site of a particular company, Article 87 TEC is violated.

14 ECJ 24.7.2003, C-280/00 *Altmark-Trans und Regierungspräsidium Magdeburg gegen Nahverkehrsgesellschaft Altmark GmbH*, ECR 2003 I-7747. See also Leibenath, 2003; Werner/Köster, 2003; Michaels/Kühschelm, 2003; Louis/Vallery, 2004; Bartosch, 2004; Meyer, 2005, 193.

3. The compensation must not exceed the actual costs of performing the service (plus a reasonable profit for the undertaking).
4. If the undertaking performing the service is not chosen by way of a public procurement procedure, the level of compensation has to be calculated on the basis of a typical undertaking.

This exception is of particular interest for airports, because the Charleroi decision ascertained that regional airports perform services of general economic interest, and that the exemption from the application of Article 87 TEC can therefore also apply to the funding of regional airports under certain conditions. The Charleroi decision:

Facts of the Charleroi Case

The Charleroi case demonstrates how agreements between public authorities and low-cost carriers have become very elaborate. The case involved three main participants: a) Ryanair; b) the Walloon region, a regional public authority in Belgium, as the owner of the airport infrastructure; and c) Brussels South Charleroi Airport (BSCA), a company controlled by the Walloon region with a concession to manage the airport for a period of 50 years. BSCA is authorized to collect air traffic fees as well as fees for services provided. Sixty-five per cent of air traffic fees remain with BSCA, 35 per cent are passed on to the Walloon region.

In November 2001, Ryanair negotiated two different contracts within four days,[15] one with BSCA and one with the Walloon region. The contract between the Walloon region and Ryanair included a reduction of 50 per cent of landing charges for a period of 15 years. Fees were not calculated based on the weight of the aircraft, but on the number of passengers. In addition, the Walloon region guaranteed to cover losses resulting from any changes in landing fees or hours of operation.

The contract between BSCA and Ryanair included the following benefits for Ryanair:

- ground handling fees of 1 €/passenger instead of 10 €/passenger;
- 250,000 € for hotel costs and staff subsistence;
- 160,000 € for each new route opened by Ryanair, up to a total of 1,920,000 €;
- 768,000 € for the costs of recruiting and training pilots and crew;
- 4,000 € for office equipment;
- free rental of office space and free hangar use.

Ryanair and BSCA further agreed on forming a joint marketing company. The marketing company financed all publicity in relation to the airline's activities at Charleroi. BSCA and Ryanair contributed in the same proportion to the operation of the company (4 €/passenger to the annual budget). Ryanair's obligations consisted of permanently basing two to four aircraft at Charleroi and in operating at least three rotations per departing aircraft over a 15-year period. The airline was obliged to repay BSCA expenditures connected with the opening of the base, in case it decided to abandon the base.

---

15 Not all details of the agreements can be reproduced and discussed here. For further details see pp. 7–12.

The European Commission Decision in the Charleroi Case

There was no doubt that the Walloon region had used substantial funds in order to attract Ryanair to Charleroi airport. The Commission now had to decide to what extent the subsidies violated Article 87 TEC and would therefore have to be recovered from the airline. In its decision, the Commission distinguished between 1) aid granted by the Walloon region, and 2) aid granted by BSCA.

Aid Granted by the Walloon Region

The Commission expressly stated that aid to low-cost carriers by public authorities is not generally prohibited.[16] It claimed that public authorities can set the level of airport taxes as low as they wish, as long as fees are calculated in a transparent and non-preferential manner.[17] Incentive schemes to attract low-cost carriers have to be established on a legal basis, executed in a non-discriminatory manner and should be limited in time.

In the Commission's view the Charleroi case did not follow these rules. The Walloon region did not grant a general reduction of fees, but gave preferential treatment to one company.[18] It did not act as a public authority, but entered into a private law contract. Finally, this contract was not concluded for a short term, but for 15 years. The Commission therefore opined that this selective reduction of landing fees constituted illegal[19] State aid within the rules of Article 87 TEC.[20] Applying a similar argument, the Commission required the guarantee of the Walloon region to compensate Ryanair for its losses from illegal State aid.

Aid Granted by BSCA

The Commission applied the private investor test to BSCA and examined whether a private investor in a market economy would have invested under similar circumstances. It concluded that BSCA, due to its public ownership structure, was able to take risks that a private investor would not have taken.[21] BSCA overestimated its future profit by a total of EUR 33–56 million,[22] while at the same time substantially underestimating the risks involved.

While repeating that reductions of ground handling tariffs for airlines in order to encourage the launching of new routes and increasing frequency may be a necessary tool for the development

---

16 P. 139.

17 The Commission did not apply the principle of a private investor test. It argued that this test was not applicable in the Charleroi case because prescribing airport taxes formed a part of the legislative and regulatory competences of the Walloon region (p. 144); see also *Stadlmeier/Rumersdorfer*, 2006, 296f.

18 Ryanair argued that other airlines could have concluded similar agreements, but in the Commission's view the non-transparent manner in which the deal was concluded proved the opposite.

19 State aid can be compatible with the Common Market for a number of reasons. The Walloon region argued that the measures were compatible with the Common Market according to Article 87 (3) c TEC – funds used to develop the poor or disadvantaged regions of the EU. The region in which Charleroi is situated is a disadvantaged region. In the case at hand, however, the Commission held that Article 87 (3) c TEC was not applicable because a) the aid in the Charleroi case was operating aid; and b) there was no general scheme in place, but only ad hoc measures favouring only one specific company.

20 P. 160.

21 Ryanair claimed that it had concluded similar agreements with other, privately owned airports throughout Europe. But the Commission did not follow this line of argument and was of the opinion that even though the airports were privately owned, public money was involved in order to attract low-cost airlines.

22 Pp. 186, 237.

of regional airports,[23] the Commission insisted that incentive schemes must be proportional to the number of passengers, linked to a start-up period, open to all airlines, limited in time and that there must be sanction mechanisms for misuse of subventions. The Commission opposes all forms of operational aid[24] that is not related to the launch of new routes, such as reduction of fuel prices or catering costs. Aid should be used to make new routes economically viable but not to sustain routes that are not profitable in the long run.

Applying these principles, the Commission decided that the selective reduction of ground handling fees,[25] and operational aid for Ryanair only, plus the fact that it was not published in advance, was incompatible with the Common Market. The Commission mandated the Walloon region to recover the aid provided to Ryanair. Start-up aid, including marketing contributions, one-shot incentives and the provision of office space, were deemed compatible with the Common Market. The Common Market conditions were that the duration of the measures not exceed five years and that the marketing contributions be linked to new routes and to the number of passengers on these routes. The sum total of aid for new routes must not exceed 50 per cent of the cumulative start-up costs and, finally, Belgium was obliged to set up a transparent and non-discriminatory aid regime aimed at ensuring equality of treatment for airlines intending to develop new airline services.

Ryanair protested against the decision and filed an appeal with the European Court of First Instance, but complied with the decision by placing EUR 4 million in an escrow account pending the Court's decision. As a further reaction, Ryanair ceased its connection between Charleroi and Stansted, reduced frequencies on other flights and threatened to abandon its Belgian base. The airport faced an 8 per cent drop in passengers in 2005, but managed in extensive negotiations to sign a new 10-year contract with Ryanair. In 2006, the airline launched five new routes from Charleroi, and passenger numbers increased to 2.2 million.

Even though it had been expected that the Court of First Instance would adhere to the Commission's economic assessment of the facts and that revisions of the decision would be restricted to the amount of aid to be paid back,[26] the Court of First Instance declared the Commission's decision void in December 2008.[27]

European Community Guidelines on Financing of Airports and Start-up Aid to Airlines Departing from Regional Airports

The principles laid out in the Charleroi decision were further refined in the European Commission's 2005 Community Guidelines. Guidelines are not legally binding for the Member States or individual parties, but can be considered as self-binding for the Commission. In these documents the Commission indicates how it will decide future cases in order to enhance predictability. The 2005 Guidelines are complementary to the aviation sector guidelines (1994).[28] While these

---

23 P. 279.

24 P. 279.

25 Because at the time the case was decided Charleroi airport had fewer than 2 million annual passengers, the ground handling directive 96/67/EC was not applicable.

26 *Jaeger*, 2004, 305.

27 CFI 17.12.2008, T-196/04 *Ryanair Ltd/Commission*, n.y.r.; see also *Giesberts/Kleve*, 2009; *Rumersdorfer*, 2009, 263–5.

28 Commission Guidelines on the Application of Articles 92 and 93 of the EC Treaty and Article 61 of the EEA agreement to State aid in the aviation sector (94/C 350/07).

guidelines had been directed mainly at the issue of State aid for airlines, the developments in the budget airline sector brought about a shift of attention towards State aid not only for airlines, but also for airports.[29]

The 2005 Guidelines discern different levels of competition between different types of airports. Generally speaking, major international hubs compete with one another, whereas small regional airports do not compete with other airports – except, in some cases, with neighbouring airports of a similar size whose markets overlap.[30] The Commission distinguishes four different types of airports:

- Category A, large Community airports with more than 10 million passengers per year;
- Category B, national airports, with 5–10 million passengers per year;
- Category C, large regional airports with 1–5 million passengers per year;
- Category D, airports with a passenger volume of less than 1 million.

Public financing granted to category A and B airports will normally be considered to distort competition and to affect trade between Member States. Conversely, funding granted to category D airports is unlikely to distort competition in an extent contrary to the Common Market.[31]

According to the Guidelines, airport management is considered an economic activity.[32] However, not all activities performed at an airport are economic. There are also activities that fall under the responsibility of the State in the exercise of its official powers as a public authority. Some of these activities, such as safety, air traffic control, police and customs, are not of an economic nature and therefore do not fall within the scope of the rules on State aid. Financing of these non-economic activities must not be used to fund other economic activities.

The Commission differentiates four types of airport activities:

1. the construction of airport infrastructure and equipment;
2. the operation of airport infrastructure;
3. the provision of airport services ancillary to air transport, e.g. ground handling;
4. the pursuit of other commercial activities.

The Guidelines apply to all four types of airport activities. If it cannot be ruled out that public funding of one of these activities amounts to State aid, the measure has to be reported. As far as ground handling activities are concerned, the Commission will adhere to the 2 million passenger threshold mentioned in the ground handling directive[33]. If the number of passengers exceeds this threshold, cross-subsidization between different airport activities is no longer permitted, and activities covered by the directive have to be opened to competitors. For example, different levels of ground handling charges may be invoiced if they reflect cost differences linked to the scale of services provided.

---

29 Bartosch, 2005a, 622; 2005b, 1124.
30 2005 Guidelines, p. 11.
31 2005 Guidelines, p. 39. At the time the Charleroi case was decided, Brussels Charleroi had fewer than 1 million passengers annually (now: category D). It is certainly ironic that, had the guidelines been already applicable then, the outcome would probably have differed.
32 Following the ECJ's ruling in the Aéroports de Paris case.
33 Council Directive 96/67/EC of 15 October 1996 on access to the ground handling market at Community airports, *O.J. 1996 L 272/36*.

The Commission further clarifies that airport activities can be considered as services of general economic interest.[34] As mentioned in the first section, public authorities may fund these activities if compensation is granted according to the criteria established by the ECJ in the Altmark-Trans case.[35] The Commission Guidelines provide that compensation for services of general economic interest must be notified, except in the case of category D airports. If a public authority decides to fund the operation of an airport, or to cover the losses incurred, it must officially announce its willingness to do so well in advance.

The Commission acknowledges the necessity for small airports to attract airlines by offering start-up aid and will accept these subsidies under the following conditions:[36]

- The recipient has a valid operating licence.
- Regional airports of category C or D are linked to other EU airports. Subventions granted to category B airports may be acceptable only in exceptional cases, i.e. when these airports are located in disadvantaged or remote regions of the European Union.
- A new route is funded, which leads to a net increase in the volume of passengers. The subvention should not lead to a transfer of passengers from one airline to another.
- The route must be economically viable in the long run. Aid should phase out in time.
- Compensation for costs linked to the opening of new routes, such as marketing costs, recruiting of staff etc., is permissible. The Commission disapproves of public funding for standard operating costs such as the costs of fuel, salaries or catering.
- Start-up aid is permitted only for a maximum period of three years.[37] The amount of aid granted in any one year may not exceed 50 per cent of total eligible costs for that year and total aid may not exceed an average of 30 per cent of eligible costs. For disadvantaged and remote regions these rules are somewhat relaxed.
- Payments have to be linked to a net increase in passengers transported.
- Public bodies planning to grant start-up aid must make their plans public in advance and allocate aid on a non-discriminatory basis. The rules for public procurement and public concessions must be respected where applicable.
- Airlines applying for start-up aid must submit a business plan demonstrating the long-term economic viability of the route.
- The Member States must provide for appeal procedures and penalty mechanisms in case of abuse.
- Start-up aid may not be cumulated with other types of aid granted for the operation of routes.

Start-up aids and start-up schemes must be reported to the Commission without exception. Member States should amend their existing schemes relating to State aid covered by the Guidelines by 1 June 2007 at the latest.

---

34 2005 Guidelines, p. 34.

35 Subventions can also be justified under one of the exemptions mentioned in the treaty (Article 87 (2) and (3), e.g. subventions for regional development). Once more it is of crucial importance to avoid cross-subsidization between activities that lie within the general economic interest and activities that lie outside.

36 2005 Guidelines, pp. 77–81.

37 In contrast, the Charleroi decision had stated that a five-year period would be acceptable, see Conclusions, Article 4 (1).

Analysis of the Charleroi Decision and the European Commission Guidelines

The EC decision and the Guidelines have been heavily criticized by the parties involved in the case. The parties involved in the case include Ryanair, the Walloon region, the Federation of European Regional Airports and the European Low Fares Airline association, and the Committee of the Regions and the Assembly of European Regions (AER).[38] On the other hand, the EC decision and the Guidelines were carefully welcomed by the network carriers and Airports Council International (ACI).[39]

Critics argued that the Commission had failed to take the low-cost carriers' new business model adequately into account.[40] A related argument, which was one of the main points in Ryanair's appeal, was that the Commission had failed to apply the principle of private investor (PPI) test correctly. Ryanair claimed that private investors would have offered similar subsidies in order to ensure profitability in the long run. The EC Guidelines have left the question open on the application of the PPI test to start-up aid.[41] It is to be hoped that the Court of First Instance will take a clear stance on this point.

Some critics further argued that national airlines received State aid in the past and some airlines such as Olympic Airways and Alitalia still receive state aid.[42] Others stressed the importance of the development of regional airports. Many large European hubs are already facing or will soon be facing massive capacity problems.[43] Low-cost carriers operating from regional airports and offering point-to-point connections may contribute to relieving the predominant hub-and-spokes architecture of European air traffic.[44] Other commentators were bewildered by the fact that the Commission will not accept start-up aid for routes which are already served by other airlines or by high-speed trains. This could lead to the perpetuation of existing monopolies.[45]

Overall the legal framework seems quite formalistic and strict: it is certainly striking 'that the airport sector where the Commission has so far had an approach much more liberal than the one pursued by the Directorate General Competition will in future be subjected to a much harsher control of State aid'.[46] The three-year limit for granting start-up aid seems too narrow. In this respect the Charleroi decision had been more generous, allowing for a duration of five years. Further, the duty to notify all start-up aids appears to be overly bureaucratic. If regions choose to compensate airports for services of general interest, they have to stick to the rather complicated Altmark-Trans procedure. Even though compensations for category D airports do not have to be

---

38   Main Conclusions, AER hearing on the possible impact of the co-operation of regional airports and low-cost carriers, Bari, 13.5.2004; *Callaghan*, 2005, 442.

39   *Stadlmeier/Rumersdorfer*, 309.

40   *Gröteke/Kerber*, 2004; *Steinrücken/Jaenichen*, 2004; *Haucap/Hartwich*, 2006; *Callaghan*, 2005 (the author works for Ryanair).

41   *Bartosch*, 2005b, 1134; cf. also *Stadlmeier/Rumersdorfer*, 309. The relationship between the PPI test and the necessity to provide a business plan is utterly unclear. If the guidelines require airports to prove the economic viability of new routes by means of a business plan, why is it necessary to apply the PPI test?

42   *Callaghan*, 2005, 440, esp. n. 12. The argument that distortions of competition in the past should justify further distortions of competition is, however, not convincing.

43   Today the seven largest EU airports attract 33 per cent of all EU air traffic. The 23 largest airports account for more than two-thirds of all EU traffic. Almost half of Europe's 50 largest airports have reached or are close to reaching limits of ground capacity. 2005 Guidelines, p. 8.

44   P. 290.

45   *Bartosch*, 2005a, 627.

46   *Bartosch*, 2005a, 628; see also *Bartosch*, 2005b, 1128.

notified, it is still advisable for regions funding category D airports to stick to the Altmark-Trans procedure, because the Commission can examine subsidies ex officio (irrespective of a duty to notify) and because competitors can bring suit in cases of violations of Article 87 TEC.[47]

The Charleroi decision and the Guidelines are to be welcomed as far as transparency is concerned. Non-disclosure clauses in contracts between airports and low-cost airlines have aggravated the airports' dominant position in negotiations. In extreme cases the behaviour of budget airlines was very undesirable. Cases of budget airlines aggressively stipulating subsidies and immediately forsaking routes as soon as subsidies expired are some examples. It is therefore a sound step that incentive schemes and schemes for compensation of services of general interest have to be published well in advance. The allocation of start-up aid on a transparent basis might lead to increased competition between low-cost carriers.

Besides the pros and cons of the Commission decision and the Guidelines, it should be noted that the documents may have wider ramifications.[48] Because the rules on State aid only apply to subsidies granted by the State or by bodies controlled by the State, they indirectly favour private ownership of airports over public ownership of airports.[49] The indirect and probably unintended consequence of the application of the rules on State aid to regional airports might be an increase in airport privatization.[50]

Summary

The Commission's decision in the Charleroi case and the Guidelines published in September 2005 are cornerstones of an emerging legal framework for State aid for low-cost carriers and regional airports. This framework has since been amended by further Commission decisions.[51] While clarifying the application of the private investor test to airports operated by public and semi-public entities, the CFI's decision to declare void the Commission's Charleroi decision left the European Commission with the task of coming up with a new (and at least partly different) decision on the question of the conformity of subsidies to low-cost carriers with EC State aid law.

---

47 The Guidelines state that subsidies to Category D airports will most likely not distort competition, but it remains to be seen how the Commission will apply this assertion.
48 See also *Callaghan*, 2005, 440.
49 The pressure to privatize goes as far as the Commission ordering Member States to privatize undertakings as a condition of approval to measures potentially including State aid. For authors critical of the pressure to privatize see *Koenig* et al., 1998, 10f.; *Lübbig*, 2001, 517; *Schröder*, 2002, 174; *Devroe*, 1997, 267. *Kruse*, 2005, 66. The pressure to privatize creates a certain tension with respect to Art 295 TEC, which states that 'This Treaty shall in no way prejudice the rules in Member States governing the system of property ownership.'
50 For a critical view on this process cf. *Wimmer/Kahl*, 1999, 161.
51 I.a. Commission cases N 640/04 – Air Route Developing Scheme for Malta; C 48/06 – State aid to DHL and Leipzig Airport, C 24/07 – State sid to Flughafen Lübeck GmbH and Ryanair; C 25/07 – Finland, Tampere Pirkkala Airport; C 26/07 – NERES – Dortmund Airport; C 27/07 – Berlin Schönefeld Airport; and N 156/2007 – Antwerp Airport.

At the national level there are a number of cases, some of them concerning misleading names,[52] for regional airports frequented by low-cost-carriers, and others regarding subsidies for low-cost carriers.[53]

Nevertheless, the legal framework for State aid for low-cost carriers remains fragmentary. By far not all questions were addressed by the two existing Commission documents. The documents are, moreover, not legally binding for the Member States. It is further possible that the Court of First Instance, and maybe subsequently the ECJ, will overrule or modify the rules established in the Charleroi case and the Guidelines. Leaving these uncertainties aside, the status quo of the legal framework for State aid for low-cost carriers can be described as follows:

- The most general rule to identify State aid is to apply the principle of private investor test. As far as aid to airlines is concerned, the Commission does not generally oppose reductions of landing fees. These reductions are permissible, as long as they are part of a general scheme, published in advance and open for all airlines.
- Start-up aid for airlines is acceptable for airlines connecting category C and D airports with other airports if a new route leading to a net increase of passengers is established; and if the aid is limited in time to three years, degressive and linked to the number of passengers per route. Start-up aid has to be notified.

As far as aid to airports is concerned, the Commission asserts that the operation of an airport can be considered as a service of general economic interest. Public authorities can compensate airlines for the shortfall incurred in these activities if the four criteria of the Altmark-Trans case are respected. The Commission further states that there is no obligation to notify compensation for services of general economic interest in cases of category D airports. The Commission will accept operating aid only in very exceptional cases. Subsidies to airports of category A and B will very likely distort competition, whereas this is highly unlikely in cases involving category D airports.

The Charleroi case demonstrates well how local authorities and airlines are creative in order to circumvent the EC rules on State aid. The rather rigid regime which these subsidies will be subjected to will lead to new manoeuvres to evade these rules. Attractive options include the funding of airlines not through public bodies, but through semi-public entities such as associations of the hotel business, recapitalizations and refurbishment of public airports which subsequently are privatized. New business models will further change the pattern of competition. Even though Ryanair's plans to launch in-flight mobile phones have been delayed due to difficulties in obtaining regulatory approvals, Ryanair still plans to outfit its entire 737 fleet with onboard mobile communications equipment. Ryanair expects to eventually make more money by offering gambling, phone connections and advertisements than through airline ticket sales. Even if the decision by the European Court of First Instance did clarify all uncertainty surrounding the permissibility of State aid to low-cost carriers, the debate will continue.

---

52 In its decision 6 U 107/03 of 5.12.2003 a German court (Oberlandesgericht Köln) decided that it was misleading to name the airport Weeze Düsseldorf (located 70–80 km from Düsseldorf). See *Wettbewerb in Recht und Praxis* (*WRP*), 2004, 401; *Kommunikation & Recht*, 2004, 85; *WRP*, 2005, 385. Similarly a court prohibited Ryanair from using the name Hamburg for Lübeck airport (Wettbewerbsrecht Aktuell-Infobrief 2002, 25–6). In yet another case it was decided, on the contrary, that the name Frankfurt-Hahn was not misleading.

53 One French case involved direct payments by the Chamber of Commerce of Strasbourg to Ryanair. A French court stopped the subsidies, Ryanair abandoned the route and opened a new route to the nearby German regional airport of Baden-Baden. *Steinrücken/Jaenichen*, 2003; *Steinrücken/Jaenichen*, 2004.

References

Barbot, L., 2004. Low Cost Carriers, Secondary Airports and State Aid: An Economic Assessment of the Charleroi Affair. Porto University, FEP working paper 159.

Bartosch, A., 2004. Die Kommissionspraxis nach dem Urteil des EuGH in der Rechtssache Altmark – Worin liegt das Neue? *Europäische Zeitschrift für Wirtschaftsrecht*, 295.

Bartosch, A., 2005a. Distortions of Competition on the Markets for the Operation of Airport Infrastructures: The Commission's new Guidelines. *European State Aid Law*, 621.

Bartosch, A., 2005b. Wettbewerbsverzerrungen auf den Märkten für den Betrieb und die Nutzung von Flughafeninfrastrukturen. *Wirtschaft und Wettbewerb*, 422.

Callaghan, J., 2005. Implications of the Charleroi Case for the Competitiveness of EU Air Transport. *European Competition Law Review* 2005, 439.

Devroe, W., 1997. Privatisation and Community Law. Common Market Law Review 34, 267.

European Commission, 2004, *Brussels South Charleroi Airport*, O.J., L 137/1, 12 February.

European Commission, 2005, Community Guidelines on Financing of Airports and Start-up Aid to Airlines Departing from Regional Airports, O.J., C 312/1, 6 September.

Giesberts, L. and Kleve, G., 2009. Private Investor Test. In: EG-Beihilfenrecht – Das Ryanair-Urteil des EuG, *Europäische Zeitschrift für Wirtschaftsrecht*, 287.

Gröteke, F./Kerber, W., 2004. The Case of Ryanair – EU State Aid Policy on the Wrong Runway. ORDO 55, 313.

Haucap, J. and Hartwich, T., 2006. Fördert oder behindert die Beihilfenkontrolle der Europäischen Union den (System-)Wettbewerb? In: Schäfer, W. (ed.), Wettbewerbspolitik im Systemwettbewerb. Duncker & Humblot, Berlin, 93.

Jaeger, T., 2004. Beihilferechtliche Bedeutung von Sonderkonditionen auf Regionalflughäfen anhand der Kommissionsentscheidung, 'Charleroi'. *Wirtschaftsrechtliche Blätter*, 305.

Koenig, C. et al., 1998. Erfüllen die Einstandspflichten des Bundes für die betrieblichen Altersversorgungssysteme der privatisierten Bahn- und Postunternehmen den Beihilfentatbestand des Artikel 92 EGV. *Europäische Zeitschrift für Wirtschaftsrecht*, 5.

Kruse, E. 2005. Privatisierung für notleidende öffentliche Unternehmen? *Europäisches Wirtschafts- und Steuerrecht*, 66.

Leibenath, L., 2003. Gemeinschaftsrechtliches Beihilfenrecht und mitgliedstaatliche Finanzierung von Leistungen der Daseinsfürsorge. *Europarecht*, 1052.

Louis, F. and Vallery, A., 2004. Ferring revisited: The Altmark Case and State Financing of Public Service Obligations. *World Competition* 27, 53.

Lübbig, T., 2001. Die Bedeutung der europäischen Beihilfenaufsicht für die privatrechtliche Transaktions- und Privatisierungspraxis. *Europäisches Wirtschafts- und Steuerrecht*, 517.

Meyer, D., 2005. Dienste von allgemeinem Interesse im Spannungsfeld zwischen Selbstbestimmungsrecht der Mitgliedstaaten und EU-Beihilfenkontrolle, *Europäisches Wirtschafts- und Steuerrecht*, 2, 193–281.

Michaels, S. and Kühschelm, S., 2003. Europäische EPNV-Systeme im Lichte der gemeinschaftlichen Reformtendenzen. *Europäische Zeitschrift für Wirtschaftsrecht*, 520.

Rumersdorfer, B., 2009. Privatisierungen von Airlines, Flughafenbeihilfen und der Privatinvestortest. In: Jaeger, T. (ed.), Jahrbuch Beihilferecht. NWV, Vienna, 243.

Schröder, W., 2002. Europarechtliche Rahmenbedingungen der Daseinsvorsorge. *Europäisches Wirtschafts- und Steuerrecht*, 174.

Soltész, U., 2003. 'Billigflieger' im Konflikt mit dem Gemeinschaftsrecht. Niedrige Flughafengebühren und das Europäische Beihilfenregime. *Wirtschaft und Wettbewerb*, 1039.

Soltész, U., 2006. The New Commission Guidelines on State Aid for Airports – A Step too Far .... *European State Aid Law*, 719.

Stadlmeier, S. and Rumersdorfer, B., 2006. Sonderkonditionen für Low Cost Carriers als Beihilfen. In: Karollus, M. et al. (eds), *Gegenwärtiger Stand und zukünftige Entwicklungen des EU-Binnenmarktes*. Trauner, Linz, 300.

Steinrücken, T. and Jaenichen, S., 2003. Europäische Beihilfenkontrolle und Public Utilities – Eine Analyse am Beispiel öffentlicher Vorleistungen für den Luftverkehr. *TU Ilmenau Diskussionspapier* 35, 3.

Steinrücken, T. and Jaenichen, S., 2004. Towards the Conformity of Infrastructure Policy with European Laws. The Case of Government Aid for Ryanair. *Intereconomics* 39, 97.

Werner, M. and Köster, T., 2003. Zur Beihilfenkontrolle bei öffentlichen Zuschüssen zum öffentlichen Personennahverkehr. *Europäische Zeitschrift für Wirtschaftsrecht*, 496.

Wimmer, N. and Kahl, A., 1999. Öffentliche Verantwortung versus Privatisierung am Beispiel des Betriebes der Bundesländerflughäfen. *Österreichische Juristenzeitung*, 161.

Chapter 23
Subsidies and Competition: An Economic Perspective

Dan Elliott

Introduction

Public subsidies have been a common feature of the airline industry for many years. Historically such aid has usually taken the form of support from governments to national 'flag carriers' to finance restructuring. However, with the liberalisation of air transport, such aid is largely a thing of the past. This chapter looks to the 'future' of public subsidies in the air transport sector by considering the economic aspects of subsidies received by airlines for the use of regional airports. The key precedent in this area is the European Commission's findings with regard to the arrangements between Charleroi Airport and Ryanair, and the subsequent guidelines published by the Commission on the provision of public aid to airports and airlines. Although Ryanair appealed and won the case (European Court of First Instance, 2008) on legal grounds that the Commission did not treat the region with its airport as one entity the economic reasoning of the European Commission on airport subsidies remained largely the same.

In particular, this chapter considers the specific ways in which these public subsidies could be expected to impact on inter-airline or inter-airport competition. The next section summarises the key points in Charleroi case and the Commission's decisions and the guidelines. The chapter then presents a summary of the economic issues that arise with regard to State Aid at regional airports. The final section examines whether the Commission's position is consistent with the promotion of economic efficiency, or whether the Commission may be prohibiting pricing behaviour that could, in principle, be efficient.

Charleroi and the Commission's Guidelines

Article 87 of the European Treaty governs the restrictions on State Aid to private companies within the European Union. This article prohibits aid that favours certain undertakings or economic activities in so far as it affects trade between Member States. However, it is accepted by the Commission that some form of public subsidy to regional airports could be justified under the rules governing 'services of general economic interest'.

In 2004 the Commission ruled that financial arrangements signed in 2001 between Ryanair and Charleroi Airport[1] violated Article 87.[2] Subsequently, in 2005, the Commission issued revised guidance on the application of State Aid rules to airports and airlines.[3] The key elements are very briefly summarised in this section, so that the likely impact of the arrangements on airline and

---

1 Strictly there were separate agreements between Ryanair and both the Walloon region of Belgium and the publicly owned operator BSCA.
2 European Commission (2004).
3 European Commission (2005).

airport competition and the appropriateness of the Commission's Guidelines in this field can be subsequently analysed.

Ryanair's Deal with Charleroi

There were several elements to the deal signed between Ryanair and Charleroi. In exchange for a guaranteed minimum number of based aircraft and aircraft rotations per day, over a 15-year period, Ryanair was granted benefits that included:

- a 50 per cent reduction in landing charges;
- protection against future increases in landing charges;
- a reduction in ground handling charges from €10 per passenger to €1 per passenger;
- lump sums for staff costs such as recruitment, training, temporary accommodation and subsistence;
- a lump sum for each new route opened (up to a fixed maximum sum);
- a 50 per cent contribution to the marketing of all Ryanair's services at Charleroi.

The Commission's Decision

In 2004 the Commission found that the arrangements at Charleroi were in violation of Article 87. The Commission made a distinction between aid provided by the Walloon region directly and aid provided by the airport operator Brussels South Charleroi Airport (BSCA).

As regards the aid from Walloon, this was found to be incompatible with Article 87 because the aid was granted on a preferential basis to Ryanair and was not available to other airlines. It also considered that the period of the arrangement, 15 years, was excessive. However, the Commission was not concerned with the level of the landing charges per se, provided that these charges were set on a non-preferential basis.

As regards the support from BSCA the Commission found that much of this too was in violation of Article 87. The Commission found that the assistance from the airport on ground handling was excessively large given the likely commercial value to the airport – underestimating the risks and overestimating the benefits to Charleroi airport.

Start-up lump sums were found to be consistent with acceptable practice, subject to their being proportional to the number of passengers anticipated on a route, and limited in size and duration.

The Commission's Guidelines

Following the Charleroi decision the Commission issued revised guidelines regarding public subsidy to airports and for the provision of airport services to airlines. These guidelines highlight the role that small regional airports can play in economic development, especially in the most disadvantaged regions of the EU. The guidelines cover aid relating both to the construction and operation of airports and to airlines operating services from those airports.

States are free to choose to invest as they see fit in airport infrastructure at state-owned airport facilities. This is explicitly justified by reference to the example of a State choosing to alleviate congestion at a city airport by developing a new airport adjacent to the city. Implicitly, the state construction of airports is also sanctioned given the view that regional airports provide services of general economic interest and have positive benefits for the regions in terms of acting as catalysts for economic development.

Public funding of the *operation* of airports is discouraged, unless this aid relates to the conduct of some public service obligation. In any event the aid should be proportionate to the costs of the relevant service.

As regards support to airlines operating at regional airports, it is recognised that such aid could be acceptable, but only in specific circumstances. In particular:

- Aid should only be available at the smaller regional airports. Where it is offered it should be offered on equal terms to all operators at that airport.
- It should only be for new routes and additional frequencies, not for services that replace or compete with other services (where other services include those provided by high-speed rail).
- The routes supported must have the prospect to be profitable and self-sustaining in the long run. With this in mind start-up aid should be limited to no more than three years (five years for the most disadvantaged regions).
- The aid must be proportional, which is taken to include being calculated per passenger, relate only to start-up costs not to ongoing costs of operation, and be limited to not more than 50 per cent of the relevant start-up costs.

Economic Issues with the Assessment of State Aid at Regional Airports

This section looks at some of the economic issues raised by the assessment of State Aid to airlines at regional airports. As identified previously, the test as to whether these arrangements constituted prohibited State Aid depends on whether the terms were exclusive and on whether the scale of the discounts received were proportionate to the value of the service provided in return.

The first issue, one of exclusivity is more complex than simply the fact that only one airline is in receipt of these terms in practice. I will return to this point below. The second issue – proportionality – requires the authorities to take a view of the counterfactual: what is the relevant cost, or rather price of the service against which the agreement with the airline should be compared? This is a key test the Commission seeks to apply in State Aid cases: namely the test as to whether a private undertaking would have entered into the same arrangement as the state-owned airport.

But in the case of the pricing of airport services the answer to this question is far from straightforward. Hence the private investor test may be extremely difficult in practice. In particular the price that an unregulated private airport would be able or willing to charge for its services at any point in time will be determined by capacity and competition issues rather than the level of costs. This price will not be constant over time and should not at all times be limited to the marginal cost benchmark favoured by regulators and competition authorities.

Moreover, airport services are also subject to other factors which economists know will lead to efficient prices diverging from direct measures of costs. These factors include externalities (both positive and negative) which mean that there may be economic and social impacts of air traffic movements that are not reflected in direct costs of the aeronautical services provided by airports. In addition, there may be 'network effects' within airports, which mean that changes in the use of an airport by one airline affects the value of the airports services to *other* carriers. Such factors can also justify a divergence between cost and price in order to achieve an efficient economic outcome.

Capacity and Competition

Airports compete with each other to provide services to airlines, which will in turn seek to generate traffic at that airport as a point of departure, a destination or both (including hub airports). Arguably small regional airports distant from existing major airports may be subject to relatively little competition as a point of departure.[4]

Given the spatial nature of competition between airports,[5] the key factors which are likely to determine the intensity of competition between airports will tend to be the physical proximity of potentially competing airports, spare capacity (of an equivalent quality) at those rivals and switching costs, which give rise to lock-in effects on airlines and give airports a degree of market power.

Of these factors the presence of spare capacity at rival airports is likely to be the most significant variable over time that may affect the competitive level of airport charges.

If capacity is a fixed in the short to medium term and sufficient to meet any reasonable level of demand, then the level of prices will be determined by the physical proximity of competing airports, the extent to which *passengers* views these airports as viable alternative points or origin or destination and the costs airlines would incur in switching between airports.

In such a world, if the airports are close substitutes for airlines and passengers and switching costs are low then airport charges will be pushed down towards, but not below, short-run avoidable (or marginal) cost.[6]

However, charges at such a low level cannot persist indefinitely because they would not permit the recovery of the costs of the airport infrastructure. Consequently for airports to be self-sustaining it is necessary either that they are located sufficiently far apart so as to reduce the number of passengers that consider the airports to be close substitutes, or that they restrict capacity to ensure that the scarcity of capacity allows the recovery of capital costs.

Given the importance of airport capacity in determining short-run pricing, it is also worth considering how unregulated airport prices would be expected to be set in the event that some or all airports in a relevant market became capacity constrained.

The first point to note is that if an individual airport runs out of capacity (either in total or during peak times) then the prices charged at that airport will have to rise so as to ration demand to the level of capacity. The airport can price in this way because the spatial nature of competition will ensure that passengers located sufficiently close to the airport would still prefer to use that airport, even if it is more expensive than its rivals. The effect of the constrained airport raising its prices is simply to create for itself a slightly more geographically concentrated catchment.

However, the consequence of an airport behaving in this way is that it reduces the competitive pressure that the constrained airport places on its rivals, which will also be able to raise its prices to some extent. Thus capacity constraints will cause all airport prices to rise. Thus the equilibrium level of prices will be determined by the intensity of competition between the airports; how hard the capacity constraint bites; and whether capacity constraints are ubiquitous, or present only in one or a small number of competitors.

---

4   In a recent report by Frontier Economics on Stansted Airport, econometric analysis of easyJet's passenger sales data was used to demonstrate the extremely strong preference that passengers have for using their closest airport as a point of departure if possible. See Frontier Economics (2007).

5   Especially as regards competition between non-hub airports.

6   Such a model could be described within the differentiated Bertrand framework or similar. An example of a model of this type that has these general properties is given in Barbot (2004).

It is possible that equilibrium prices could be significantly above short-run avoidable cost. However, the actual cost of fixed capacity does not enter into the determination of prices; capacity could be cheap or expensive. It will be the intensity of competition and the nature of the capacity constraint that will determine prices.

A number of conclusions arise when we consider airport pricing in the presence of capacity constraints. First, as airport capacity is typically developed in discrete, relatively large increments it is unlikely that actual airport charges observed at any given point in time will be established at exactly the 'right' level from the perspective of competition authorities, whether that be viewed as short-run or long-run marginal cost.

Rather, at any point in time capacity will almost certainly either be in too plentiful supply to allow the recovery of capital costs or in such short supply, in which case prices may well exceed long-run marginal cost, which would signal the need for further capacity investment at the airport. But the existence of 'high' prices in this situation need not indicate that the airport is pricing excessively over the life of its investment; merely that given the lumpy nature of investment the airport is now in a period in which the costs of capacity can be recovered, offsetting all those periods before when capacity was so plentiful that the airport could only charge to recover short-run avoidable costs.

The interpretation of these factors is essential for the application of the private investor benchmark because the use of a snapshot of prices and costs can give a misleading impression of the long-run implications of a particular pricing policy.

Externalities

The existence of externalities represents the second challenge to the interpretation of airport pricing information in the context of State aid. Specifically, it is only efficient to price goods and services at cost if there are no 'knock-on' effects (or externalities) on other markets.

If there is a negative externality, i.e. activity in the market in question damages economic activity in another market, then this is a reason for prices to be set above cost to offset the externality. On the other hand if the activity question boosts damages economic activity in another market then there is a positive externality, which would justify pricing below cost to encourage the externality.

In practice externalities are commonplace in the operation of airports, although the implications for airport prices vary depending on whether or not the airport can internalise these effects.

The most significant externality created by airport landing services is the knock-on effect of passenger movements on the value to the airport operator of ancillary services offered at the airport, including retail services, hotels and business centres.

An airport can internalise this effect and as a consequence would be expected, in a competitive environment, to reduce landing charges to reflect the value that passenger movements create for other activities at the airport. This is because the airport should calculate the value that additional passengers will bring to it in terms of profits from other commercial activities and would be prepared to compete away these profits to win custom by setting lower landing charges.

The implication for the assessment of State Aid is clear: if the authorities use a private investor benchmark that disregards the value of commercial revenues to the airport they will overestimate the magnitude of any aid deemed to have been provided.

However, airports are also implicated in a range of externalities that the airport operator cannot be expected to internalise in its prices, because the costs or benefits of the externality accrue to other parties. These effects may be positive or negative, including the impact of the airport on regional economic development, the environment or congestion at another airport.

The benefits to regional development of the growth in air services represent a public externality that is one of the key rationales for allowing State aid to regional airports. Left to its own devices the airport will not reflect this benefit in its charges and so there may be a justification for landing charges to be subsidised for the benefit of the regional economy. The key assessment, however, will be whether the extent of any support for this reason is really proportionate to the wider economic benefits created by the airport.

While economic development objectives could be used to justify subsidising landing charges, the impact of airport operations on the local environment acts in the opposite direction.

Although noise and air pollution issues at airports are subject to stringent quality regulation, the existence of these factors nevertheless would justify raising landing charges above cost (via some form of taxation) to reflect the social cost imposed on the local area by air traffic movements.

It is notable, however, that while the Commission's guidelines make much of the positive externalities of air travel providing a rationale for State aid, the countervailing arguments regarding the negative externalities of air travel receive little attention.

A third externality created by airport operations can be the knock-on effect that growth at a regional airport can have on congestion at existing major airports. This effect is frequently cited as a justification for the provision of discounts at regional airports. But it is unclear why it is thought appropriate for an uncongested publicly owned airport to reduce its landing charges in response to this problem.

If the major airport is congested at some time of day or time of year then there is an externality within that airport such that one flight crowds out the ability of other airlines to operate at the same time. As with other negative externalities, the most appropriate response to this should be for the congested airport to *raise* landing charges during the congested period to reflect the cost of congestion. By contrast economic efficiency is not promoted by *reducing* landing charges at another airport, which implies (other things being equal) promoting an inefficiently high level of air traffic.

Network Effects

As airports form part of a network facilitating air transport, it is reasonable also to consider the role of 'network effects' in influencing efficient airport prices.

The simplified assumption in economics textbooks is that the value that a buyer or seller receives from a transaction is independent of the actions of third parties not directly part of that transaction. However, this assumption may not be true with respect to airport services.

This concept is common in the pricing of telecommunications networks. In that case, the value to a subscriber of joining a telephone network depends on how many other people are also connected to that network. As subscribers join the network the value of being on that network increases for *other* subscribers. This is the network effect.[7] In the presence of such network effects it will be efficient for the network operator to discount the price it charges each subscriber for joining its network, to reflect the benefits it expects to receive from other subscribers as a result.

Similar network effects may also apply to airport services. Provided the airport is not congested, adding services may increase the value of using that airport for other carriers, who will then have a higher willingness to pay. In an efficient outcome we would expect the effect that adding a service has on the willingness to pay for airport services of other carriers to be reflected in landing charges. With network effects, landing charges should be discounted relative to the level that would exist without this effect.

---

7   Sometimes referred to as a 'bandwagon' effect. See Rohlfs (2001).

There are various reasons why network effects could apply to airports, especially ones that are initially very underutilised. First, the demand for outbound air services depends on the airport's catchment. But this catchment is not limited to those passengers originating in the region close to the airport. It also potentially includes all passengers arriving on services into that airport, some of whom may use that airport to take connecting flights to other destinations, using the same or a different carrier. This may be an explicit part of the airline's operating model, or merely a consequence of its choices of point-to-point routes. Thus, by providing inbound passengers, the airline increases to some extent the pool of passengers that other carriers can draw on to offer their services from that airport.

Secondly, I have already noted above in the section on externalities that increased passenger numbers will increase the value to the airport operator of other commercial services that they are offering at the airport. An efficient airport operator would reflect the value of this externality in its aeronautical charges. But there may also be a dynamic network effect in this area. This is because the willingness of some airlines to locate at a particular airport may depend on the extent to which other ancillary services of the airport are well developed, including transport links to surrounding towns and cities, and also services at the airport such as hotels, business centres and retail outlets. The viability of such services depends on the airport being able to generate a critical volume of traffic. Hence the addition of new services increases the likelihood of ancillary services being added or improved, which in turn increases the attractiveness of the airport to additional carriers. Again in an efficient outcome we would expect the value of this network effect to be reflected in a reduction in charges for airport services.

The two network effects identified can be internalised by the airport in its pricing decisions, but can be expected to lead to a reduction in landing charges relative to cost. Again this complicates the application of the private investor benchmark by breaking the direct link between costs and prices.

Clearly in each of the cases above, as airports become congested the scope for adding additional services is limited, which reduces the scope for such network effects and consequently reduces the argument for discounting of prices, whether provided by the airport operator or by some public body. But when airports are very empty each of these network effects could be very powerful indeed, leading to a strong argument in favour of the efficiency of below-cost pricing.

Is the Commission's Approach Consistent with Promoting Economic Efficiency?

The Commission's guidelines formalise the principles the Commission wishes to see followed by Member States as regards State Aid given to airlines for operating at regional airports. But in the absence of any subsequent decisions, Charleroi remains the only decision to cast light on the interpretation of these guidelines. In this section I use the issues identified above to highlight the areas of the both guidelines and the Charleroi decision that seem to be contentious from the point of view of economic efficiency.

The purpose here is not to assess whether the Commission arrived at the right or wrong decision in the Charleroi case, but rather to identify whether the Commission has arrived at a framework that is likely to have a positive impact on economic efficiency on a forward-looking basis.

Key Aspects of the Policy

The key aspects of the Commission's policy can be summarised as follows. State Aid at regional airports may be permitted, but only provided that aid can be shown to be:

- related to the development of genuinely new routes;
- limited in time to three years (or five years in the case of the most remote regions);
- proportionate to the costs involved, which includes relating total aid to the number of passengers expected;
- not exclusive to the recipient of the aid.

In the case of the Charleroi decision, the agreements signed with Ryanair were found to contravene these rules because the aid was for an excessive period, namely 15 years; was exclusive to Ryanair; and, in the case of the aid provided by BSCA, was found to be excessive relative to the costs and risks to which the airport was exposed.

General Comments

The guideline's approach to identifying acceptable forms of State Aid is inevitably formulaic, but from the point of view of economic efficiency fails to address to key aspects of the circumstances surrounding the aid that will be crucial to the assessment of the economic impact that it has. In particular, the guidelines fail fully to address the issue of the competitive conditions surrounding the airport (and air services) in question and, in addition, underestimate the complexity of identifying the appropriate private investor test (or competitive benchmark) against which the arrangements in question should be judged.

Competitive Conditions

Clearly, by subsidising the initial start-up costs of certain services, market entry by an airline or airlines is encouraged that might not have occurred in the absence of the aid. But this situation only has material implications in terms of allocative efficiency, and hence a possible impact on 'trade between Member States' if the services in question are in competition with existing services at that airport or elsewhere.

The Guidelines capture this effect by focussing aid on new routes that are not currently served either by air or by high-speed rail. This definition has two shortcomings. First if we focus entirely on competition to provide transport services it is unclear what constitutes a 'new' route. Secondly the premise of the guidelines is that it is competition between airlines (or between airlines and trains) that may be impacted by the aid. This misses the important point that it may be competition between *airports* that is affected by the aid.

Route competition Competition between airline routes may not be as straightforward as identifying origin-destination pairs for which services are already offered. As I have already noted, the low-cost airline market has facilitated the development of a new demand for air travel, where the specific point of destination is no longer necessarily the only potential target of the traveller. Rather, different destinations are being chosen by passengers in competition with each other from broad classes of destination for specific sorts of holiday, such as city breaks, beach holidays, clubbing, ski trips etc.

With this in mind the competitive impact of supporting new routes at a regional airport may, in principle, be much wider than would be suggested by a simple analysis of origin-destination pairs.

Inter-airport competition In addition, the guidelines do not address the issue that airports themselves compete with each other to provide landing services to airlines. By altering the start-up costs that an airline experiences at a given airport, the aid received may affect the terms of competition between the beneficiary airport and its rivals.

Even if the routes in question are not in competition with existing routes, the choice of originating (or destination) airport has been distorted by the aid, which may have a knock-on impact on the prices and investment decisions at existing rivals. Thus proximity to existing airports and the extent of potential competition between them should be an important consideration. In the Charleroi case, the fact that Charleroi Airport itself is only 30 km from Brussels Airport is a relevant factor in the assessment of the competitive effects of the aid.

The Competitive Benchmark

I have argued above that competitive pricing at airports is a complex matter and will depend on the competitive conditions between airports and the existence of spare capacity or otherwise at the different available airports. As such the private investor test cannot rely on either short-run or long-run marginal (or incremental) cost as the appropriate competitive benchmark against which the arrangements between airport and airline should be compared.

Given the lumpiness of airport investment it is to be expected that at any point in time efficient airport pricing could be either significantly below or significantly above long-run incremental cost, depending on the point in the investment cycle and the extent of congestion at the airport and its rivals. In particular, at a very empty airport with little or no traffic it may be efficient to set charges that do little more than cover its short-run avoidable costs.

In addition, I have noted above that a number of further factors could be involved that would justify a divergence between marginal cost and price for aeronautical charges, including network effects between services at the same airport (or even between different airports) and the knock-on effects of aeronautical activities on the value of other commercial services provided at the airport. These factors would tend to cause a privately owned airport to discount aeronautical charges relative to a purely cost-based benchmark.

It is interesting to note that Ryanair argued strongly, in response to the Commission's findings, that the arrangements between itself and Charleroi were not unusual as compared to other arrangements that it had signed at other airports. Thus, in Ryanair's view at least, the arrangements at Charleroi represented normal commercial terms and did not need to be justified further by reference to support for regional development etc.

On grounds of spare capacity alone it is to be expected that an airport as empty as Charleroi would be prepared to offer very heavily discounted landing charges in order to encourage the use of its facilities. A profit-maximising operator would not, however, be expected to reduce landing charges below short-run marginal cost just to achieve better use of its assets.

However the existence of externalities and further network effects may also be very significant indeed in the context of a virtually empty airport, which may further justify reductions in landing charges, possibly below marginal cost.

The addition of services at Charleroi on a scale offered by Ryanair could, for instance, be expected to add significantly to the value of other ancillary services there which could justify significant reductions in charges. This would be especially true given the low base from which the airport was starting.

In such an 'extreme' case of an airline beginning operations on a major scale at an almost empty airport the externality effect of new services may also interact with the potential for bandwagon/

network effects to further increase the value that the carrier provides to the airport, and thus justify further reductions in landing charges. By delivering scale of operations which the airport had never previously enjoyed, it could be argued that Charleroi could reasonably have expected Ryanair's presence to stimulate the sort of dynamic effects described above: promoting improved transport links to surrounding towns and cities, and also services at the airport such as hotels, business centres and retail outlets. This in turn would make Charleroi a much more attractive option for other airlines, which could result in additional airlines and services being attracted to the airport.

Given this collection of factors it is necessary to question the applicability of a cost-based private investor test for State aid in the case of Ryanair's deal at Charleroi.

Furthermore, a related criticism of the Commission's approach with regard to the Charleroi decision concerns the differential way in which it treated landing charges levied by the region of Walloon and ground handling charges levied by BSCA. The Commission did not apply the private investor test to landing charges because it took the view that the region was acting as public regulator and so could set the charges at whatever level it chose. While I cannot comment on the legal status of this opinion, clearly from an economic efficiency point of view this distinction is meaningless. Economic distortion can be created by setting landing charges too low (or too high) in exactly the same way that it can be created from setting ground handling charges too low (or too high). The Commission is treating landing charges as a tax to be set at whatever level the regional authority sees fit because the charge is determined by the regional authority, not by the airport. But there is no economic justification for this distinction. Landing charges need to be set to cover the costs of the runways, taxi-ways, ATC facilities etc. over the life of these investments. The pricing of access to these facilities should not be exempted from the appropriate assessment of economic efficiency and distortion simply due to the administrative organisation of which body determines which element of the airport's charges.

Specific details In addition to the general problems of identifying the competitive benchmark, the Commission's Guidelines raise some additional specific questions regarding the rules the Commission proposes to apply to such cases. I deal with two such issues below: time limits on the provision of aid and the requirement that any terms are not exclusive to one operator.

Time Limits

The Commission's objective is to allow subsidy for the start-up of new routes, but to ensure that in the long run only routes that can be profitable in their own right should be supported. Hence the Commission considers that aid should not be made available for more than three years (or five years in particularly disadvantaged regions). Such limits are understandable, given the Commission's desire to avoid airlines being provided with long-term operational aid from governments. Nevertheless, fixed limits of this kind may be incompatible with the regional development objective and with the nature of commercial airport pricing.

Regional development objectives The logic of the Commission's position is that a region requiring assistance can generate enough traffic to sustain a commercial air service, requiring only some additional start-up costs in terms of additional marketing activities or greater seat price discounts to build the critical mass of passengers. But it is not obvious that this is correct.

If economic activity in the region is not strong enough to sustain profitable air services, then it is doubtful that this situation can be turned around over a three-year period. It is doubly doubtful that this situation can be addressed solely through the mechanism of providing a start-up subsidy to an air service. It is hard to imagine a commercial airline committing to a long-term project to provide air services to an area with low latent levels of demand on the basis of relatively short-term support for start-up costs. The risks of such an enterprise would seem to be excessive.

Hence if persistent economic deprivation is seen as the motivation for the aid, and air services are seen as a significant contributor to strengthening the regional economy, then it would seem necessary at least to contemplate support over a longer period than only three years, subject to evidence showing the benefits that such aid could bring.

The restriction placed on aid in this way would seem to have the effect of limiting its applicability to locations where the commercial viability of air services is marginal, in the sense that it is thought that sufficient demand exists or will exist in the near future, but the short-run costs of developing that demand might prevent the service from starting in the first place. Thus it would seem that by limiting the applicability of aid the rules may favour support to regions at intermediate levels of development, rather than the most disadvantaged ones.

Commercial airport pricing I have already outlined that it may be rational profit-maximising behaviour to set prices very low at very underutilised airports, because any contribution to overheads from attracting traffic is preferable to the operator to leaving airport assets lying idle.

But a situation of persistent spare capacity could exist over many years. Experience with Stansted Airport in the UK has shown that when it operated with excess capacity it set prices below the cost-recovery level it was permitted to charge under UK regulations. This situation persisted for considerably more than the three-year period envisaged by the Commission, and has only come to an end recently as capacity constraints begin to bite. The implication of the Commission's guidelines is that it could be very difficult for a publicly owned airport to commit to offering discounted landing charges over a period reflecting a realistic view of its capacity position. This could act as a disincentive for low-cost carriers to use these airports.

Exclusivity

At Charleroi the Commission took the view that the agreements between Ryanair, the airport and the region were unacceptable because they were exclusive to Ryanair. While recognising the possible distortion to competition that could arise from offering airport charges that discriminate between different carriers, it is not clear that the Commission's interpretation of exclusivity is one that would promote efficient airport pricing.

The key distinction should not be whether an airline is the only one that has negotiated a particular deal with the airport authority, but rather whether the terms of the deal are intrinsically reasonable and whether other airlines could have negotiated the same deal. Furthermore, the fact that a similar deal might not subsequently be available to a second carrier should not imply that the original deal is inefficient or discriminatory.

If an airline is willing to make a long-term commitment to an otherwise empty airport, this has enormous value to the airport operator, because it means that a significant proportion of the capacity of the airport is now being used that would otherwise have gone to waste. We should expect an efficient airport operator to reflect this value in a very competitive level of aeronautical charges. The greater the level of commitment the airline is willing to make (in terms of scale or time), the greater the value will be to the airport and the better the terms it will be prepared to offer. At the same time, the commitment on the part of the airline is costly and risky, especially at an airport where the long-term level of demand is unproven. Hence the airline will seek significant incentives in order to commit to any such arrangement.

Furthermore, if there are significant network effects between air services then an airline that acts as a first mover will be in a position to demand a share of the expected benefits that its entry will create for the airport.

So we see that an airline moving with scale into an empty airport creates real benefits for that airport and takes significant risks itself. Once it has done so, however, the position at that airport is altered. Subsequent entrants will not bring with them the same level of benefit for the airport, especially if entering on a lower scale. Furthermore, the risks of entry for subsequent airlines may be much reduced, because the original entrant has already helped to create a passenger-base for that airport and has established it as a viable place of operation. Thus it would be understandable for the terms offered by the airport to a first mover would be more generous than to later entrants.

Thus in the Charleroi case, what the Commission has identified as an exclusive agreement may in fact be a rational agreement between the airport and the first-moving airline. The same deal may not be available to other airlines because the same circumstances no longer exist. But such an outcome need not be inefficient. Indeed it may be essential if a relatively empty regional airport is to develop a critical mass of services.

Conclusions

The 'private investor' test is a central part of the Commission's evaluation of State Aid. This test relies on the ability of the authorities to identify the appropriate counterfactual, i.e. the 'competitive' level of prices that could be expected to prevail at an airport. If the authorities get this assessment wrong then it is likely that they may incorrectly identify the presence (or absence) of State aid in any particular case.

I have outlined here, however, that this assessment is complex and highly context sensitive. Airports with spare capacity may have incentives to set very low prices for very long periods of time, because that is the only way they can achieve any significant utilisation of their assets until demand has built up to the point where capacity can earn any return at all.

As a consequence it would be wrong to conclude that landing charges which make little or no contribution to capital costs fail the private investor test at a largely empty airport. Furthermore, while the Commission's approach allows for time-limited assistance of up to three years, this period is likely to be significantly too short as it is likely that any period of time for which an airport maintains substantial excess capacity could be much longer than this.

The existence of a complex web of externalities further complicates the successful application of the private investor test. Air services from one operator have major positive externality impacts on other operators, by bringing passengers to the airport (at hub airports), and on wider airport services, including retail, parking and hotels. Profit-maximising airports will discount landing charges below cost in order to reflect the value of these externalities. In the case of a very empty airport, the entry of a new carrier with significant scale may facilitate a step change in the overall level of commercial activities. Incorporating the value of this externality to the airport may invalidate the cost-principle as a useful benchmark in determining the presence of level of State aid implied by any specific set of airport charges.

References

Barbot, C. (2004) 'Low Cost Carriers, Secondary Airports and State Aid: An Economic Assessment of the Charleroi Effect', FEP Working papers No. 159, October 2004.

Competition Commission (2002), BAA plc.

Court of First Instance of the European Communities (2008), Case T 196/04 Ryanair Ltd v Commission of the European Communities, State aid – Agreements entered into by the Walloon Region and the Brussels South Charleroi airport with the airline Ryanair – Existence of an economic advantage – Application of the private investor in a market economy test, Luxembourg 17 December 2008 http://eur-lex.europa.eu/LexUriServ/LexUriServ.do?uri=CELEX:62004A0196:EN:HTML.

Crocioni, P. and Newton, C (2007), State Aid to European Airlines, A Critical Analysis of the Framework and its Application', in Lee, D. (ed.) *Advances in Airline Economics*, vol. 2, Amsterdam: Elsevier, 147–70.

European Commission (2004), Commission Decision of 12 February 2004 concerning advantages granted by the Walloon Region and Brussels South Airport Charleroi to the Airline Ryanair in Connection with Installation at Charleroi, *OJ* L 137/1-62 of 30 April.

European Commission (2005), Community Guidelines on Financing of Airports and Start-up Aid to Airlines Departing From Regional Airports.

Frontier Economics (2006), Economic Consideration of Extending the EU ETS to Include Aviation, March 2006, http://www.frontier-economics.com/_library/publications/frontier%20paper%20-%20economic%20consideration%20of%20extending%20the%20EU%20ETS%20to%20include%20aviation%20-%20mar06.pdf.

Frontier Economics (2007), The De-designation of Stansted Airport, October 2007, http://www.frontier-economics.com/_library/publications/ Frontier%20paper%20-%20de-designation%20of%20Stansted%20airport%20Oct%202007.pdf.

Gröteke, F. and Kerber, W. (2004),'The Case Of Ryanair – EU State Aid Policy On the Wrong Runway', Philipps-Universität Marburg Working paper No. 13/2004.

Nicolaides, P. and Bilal, S. (1999), 'An Appraisal of the State Aid Rules of the European Community – Do They Promote Efficiency?', *Journal of World Trade*, 33(2), 97–124.

Office of Fair Trading (2004), Public Subsidies, OFT750, November 2004.

Rohlfs, J.H. (2001), *Bandwagon Effects in High Technology Industries*, Cambridge MA: MIT Press.

Starkie, D. (2002), 'Airport Regulation and Competition', *Journal of Air Transport Management*, 8, 62–72.

Chapter 24

Competition for Airport Services – Ground Handling Services in Europe: Case Studies on Six Major European Hubs[1]

Cornelia Templin

Introduction

Ground handling activities describe the (often barely noticed) services between flights. They include elements such as check-in of passengers, handling of cargo, and transportation of luggage; but also catering, fuelling, fresh water and toilet services. These services are very important for the airlines' performance, and consequently for their competitiveness.

The liberalization of the airline industry during the last decades led to far-reaching changes concerning market structures, both horizontal and vertical – with strong impact also on the ground handling business. Potentials to differentiate airlines' products 'in the air' are low; their competitive advantage lies to a great extent on pre- and post-flight activities like convenient check-in procedures, punctuality, fast and efficient (transfer) processes, and short baggage delivery times. Hence, service quality on the ground is essential due to the increasing competition between airlines.

Furthermore, airline alliances and hubbing, as well as the emergence of low-cost carriers, change the needs of airlines. Multi-station contracts, complex transfer processes as well as 'light' handling services become more and more important aspects of ground handling contracts.

Additionally, the financial situation of airlines causes pressure on ground handling prices and more selective use of services. Worldwide crises with sudden cutbacks of flights and increasing security needs add to the challenges in the aviation industry that have direct impact on the ground handling industry.

Competing airlines and alliances reinforce competition between airports. But they also influence the competition for airport services – the pressure from the airlines was the reason for deregulating ground handling services at European airports during the 1990s. Airlines were exposed to full competition and did not want to rely on highly regulated suppliers any more. This ultimately led to the implementation of Council Directive 96/67/EC on access to the ground handling market at Community airports, introducing competition to ground handling markets in Europe.[2]

Today, many ground handling services such as landside cargo and passenger handling are generally completely open to competition, whereas airside baggage services and most services on the ramp are provided in different market structures throughout the world. In North America, for instance, ground handling markets are fully liberalized. Airlines dominate the markets, with several independent ground handlers offering handling services to the remaining airline

---

1 Excerpts of an analysis conducted as part of a doctoral thesis, see Templin (2007). All figures are from 2004 unless otherwise indicated.
2 See European Council (1996).

customers. In Asia, the markets are still quite regulated and airlines mainly provide handling. Normally only one additional independent handler serves as a competitor in a sort of 'regulated duopoly'. European airports, in contrast, have heterogeneous market structures. Following the EU deregulation, independent handlers entered and joined handling airlines and airport operators. Nevertheless, there is still limited competition among most of the European airports. At present, predominantly handling oligopolies exist, in addition to a few completely open markets and some handling duopolies.

This chapter focuses on competition among airports concerning airside handling services at European hubs.[3] After a brief introduction of the handling industry, as well as landmarks of deregulation in Europe, the effects of the deregulation on different stakeholders are described. Thereafter, the circumstances at six major European hubs are discussed, followed by concluding remarks and an outlook on future developments.

The Ground Handling Industry

Providers of ground handling services can be airlines, airports or independent handling companies. Airlines service their own aircrafts (self handling) and may also provide this service to other carriers (third party handling). Thus, airports and independent companies offer third party handling.

Airlines usually self handle their flights at their home base airports. To improve the utilization of resources, they also regularly offer their services to other airlines, commonly on a reciprocal basis between airlines at their respective hubs. Recently, some airlines have started to outsource their handling activities to concentrate on their core competencies and because of cost-oriented make or buy decisions.[4]

Airports provide third party handling at times due to historical reasons. For instance in Germany only airport authorities were allowed to provide handling services after World War II, supervised by the Allies. But now German airports too have started to cease their handling activities due to more favourable cost structures of competing independent third party handlers.[5]

Until recently, independent handlers were largely restricted to local markets. Meanwhile, their market share was increasing constantly, not only because of the opening of the ground handling markets in Europe but also because of substantial acquisition activities worldwide. A lot of independent handlers offer their services at several airports, either worldwide or in certain regions of the world.

The industry consists of local, regional and global ground handling companies. In addition to over 200 small, mostly local handlers worldwide there are also a few big and international companies that offer their services at a number of airports in various countries.[6] Global players serve up to 190 airports in 43 countries and have revenues up to US$1.5 million.[7] They usually have fairly small market volumes at these airports, with an estimated average of 20,000–30,000

---

3   Concentrating on baggage and ramp handling.
4   See KPMG (2002), p. 2 and Nana (2003), pp. 21–2. Alaska Airlines was one of the first airlines to outsource ramp operations at their hub in Seattle, with 140 daily flights: see Ground Handling International (2006), p. 31.
5   See Piper (1994), p. 51 and Nana (2003), pp. 21–2.
6   See Walker (2000), p. 7.
7   See Swissport (2008).

turnarounds per year.[8] From this perspective, Fraport is playing in a different league, as the biggest independent supplier at Frankfurt airport, serving 192,000 turnarounds in 2004.[9]

The world market of ground handling has a volume of about €32 billion.[10] Airlines still dominate the market (59 per cent); the remaining market is split between independent handlers (24 per cent) and airports (17 per cent). It is expected that the share of the independent handlers will increase in 2010 to about 45 per cent of the total volume,[11] leading to further consolidation throughout an industry that is still very fragmented. The eight biggest companies hold only 11 per cent of the total market share, see Figure 24.1.

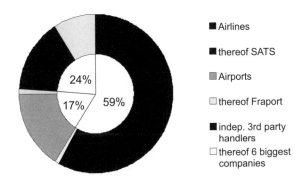

Figure 24.1 Major players in the ground handling industry[12]

Introducing Competition

After deregulating the European aviation industry at the beginning of the 1990s, airlines lobbied for competitive structures for their suppliers to have more influence on their own competitive situation, since the market structure for European ground handlers at that time was often monopolistic. Sometimes handling was provided by the flag carrier of the respective country, while at other airports the sole supplier was the airport company. Even if there were more suppliers active at an airport, market entry was restricted most of the time to only a few handlers. Neil Kinnock, EU Commissioner for Transport at that time, therefore stated on behalf of the airlines 'it is essential for them to be able to choose which handler they want, including handling for themselves, rather

---

8   An average volume to successfully operate at a hub is estimated at 20 to 30, sometimes even up to 55, turnarounds per day. This leads to 7,000–20,000 turnarounds per annum, depending on the operating costs and pricing structures at the relevant airport: see Templin (2007), p. 145.

9   See Swissport (2006), Fraport (2005), p. 24 and Templin (2005a), p. 1.

10 The evaluation of the total market volume varies largely. Experts estimate the volume at €32–35 billion (or about US$40 billion): see Swissport (2005), p. 5; Smith (2004), p. 6; and Nana (2003), p. 21. The World Trade Organization (WTO) calculates a volume of only €22 billion (approximately US$27 billion): see WTO (2005), p. 2.

11 See KPMG (2002), p. 2 and ACI Europe (2004), p. 16. According to Swissport the current market share for independent handlers is already 30 per cent, see Swissport (2005), p. 5.

12 See Conway (2005), p. 48. These are Servisair (former Penauille Servisair), Swissport, Fraport, Worldwide Flight Services, Aviance UK, Menzies, Aviapartner and SATS. Besides Fraport and SATS (a subsidiary of Singapore Airlines); the big suppliers are independent third party handlers.

than just having to accept the price and service of the monopoly handler. That is the situation now in most airports ...'.[13]

The main goals of the EU Directive 96/67/EC for deregulating ground handling services at European airports were to lower prices, increase quality and give airlines a better choice when selecting handlers. It set standards for all member states, which had to translate the Directive into national law. By the beginning of 1999, almost all member states had implemented the Directive and started opening up their markets at airports with more than 2 million passengers per year.[14]

However, the EU acknowledged the special circumstances airside services were in and provided for the possibility to restrict the number of competitors for ramp, baggage, fuel and oil handling, as well as some parts of freight and mail handling. Member states (not airports themselves) could ask for a limitation, which was only granted if special restrictions applied at the relevant airport (like lack of capacity or security problems). If the number of handlers remained limited, a tender process had to be carried out. The tender process was also described in the Directive; it was supposed to be non-discriminatory and licences could be issued for up to seven years.[15]

Impact of Deregulation

An empirical analysis of the competition at six major airports in Europe shows similarities as well as differences. The airports visited were London Heathrow (LHR), Paris Charles de Gaulle (CDG), Frankfurt (FRA), Amsterdam Schiphol (AMS), Madrid Barajas (MAD) and Rome Fiumicino (FCO). Qualitative semi-structured interviews had been carried out with several airport experts in the relevant markets to learn from their experience about market entry and competition on the ramp. The interview partners represented between 50 and 100 per cent of the ground handling community at the respective airports.[16]

All airports are hubs of their national carrier and the respective alliance. They handled volumes of between 28 and 67 m passengers and 150 to 260,000 aircraft turnarounds[17] in 2004. On the demand side, customer structures are very different. Focusing on alliances, CDG, AMS and FCO are SkyTeam hubs, LHR and MAD serve as oneworld hubs and FRA is the only Star Alliance hub – see Figure 24.3. This might be the reason why Lufthansa and Star Alliance have this dominant position at their hub in Frankfurt, with 72 per cent of the movements at this airport. Compared to the presence of the other alliances at their hubs – with oneworld having only 50 per cent at LHR and SkyTeam 61 per cent at CDG (and the other three hubs with a figure in between) – Star Alliance in Frankfurt is by far the most powerful customer group.[18] For the characteristics of the observed airports see Figure 24.2.

---

13   Kinnock (1996), p. 11.
14   See SH&E (2002), p. 40.
15   See European Council (1996). The Directive also allows keeping facilities of the central infrastructure in one hand if complexity, cost or environmental impact does not allow division or duplication: see ibid., Article 8.
16   See Templin (2007), p. 18. All experts were responsible for the sales of handling products.
17   Half the aircraft movements, a figure commonly used in ground handling.
18   Data analysis based on OAG (2005).

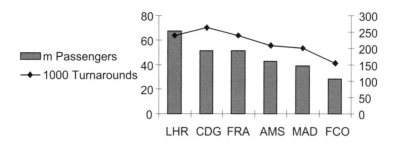

Figure 24.2 Overview of the six major European airports in the analysis[19]

On the supply side, all the airports had different situations concerning their ground handling markets.[20] Some airports had a completely deregulated market, for example LHR and AMS, while others had restricted the number of providers. Altogether, 24 ground handling companies offered their services at these six airports: see Figure 24.3. The numbers ranged from 11 airside ground handling companies at LHR to only two third party handlers at FRA and MAD. Three of the six airport companies offered ground handling themselves (at CDG, FCO and FRA), while the others left this field to airlines and third party handlers. At most airports, at least the home carrier and sometimes also other airlines provided self handling, usually combined with third party handling. Only at two airports (AMS and MAD) did some self handlers not offer additional third party handling, whereas at another airport (FRA) not even the home carrier used the possibility to self handle. The volume of third party handling in percentages ranged from 27 per cent in Madrid with four self handlers to 100 per cent in Frankfurt, with no self handlers.

Airlines providing self and/or third party handling

Air Canada	Air Europa	Air France	Alitalia Airport
American Airlines	British Airways	EAS	KLM
Martinair	Iberia	Spanair	United Services

Airports providing third party handling

ADP Handling	ADR Handling	Fraport

Independent handlers providing third party handling

Acciona Airport Services	Aviance UK	Aviapartner
Groupe Europe Handling	Ineuropa Handling	Menzies Aviation
Penauille Servisair	Plane Handling	Swissport

Figure 24.3 Ground handling companies at six major hubs in Europe[21]

---

19 See ACI (2005a) and ACI (2005b).
20 The analysis was carried out in 2004. In the meantime, the ground handling markets at some of these airports has changed dramatically. For instance, the airport operator in FCO no longer offers handling and the market is fully open. At MAD one additional licence was issued and one of the incumbent handlers did not receive a new licence.
21 See Templin (2005b), p. 11.

Impact on Airlines

As described previously, the airlines, as the ground handlers' customers, were the target group to benefit when the EU Directive was implemented. Between one and four new handlers entered the markets at the observed airports. Airlines had a larger variety of suppliers to choose from at lower costs. On completely open markets handling prices fell by as much as 50 per cent and at airports with limited competition prices dropped about 30 per cent.[22]

However, the development of quality levels after the implementation of the EU Directive was less transparent; it differed from airport to airport as well as from handler to handler. The pressure to reduce costs and the increase in interfaces and complexity, along with better customer orientation – with a wider variety of products to choose from – led to a mixed picture.[23]

Altogether, the market was more and more driven by demand. Airlines exploited their growing negotiating power, combined with more detailed service level agreements, to strengthen the commitments of handlers. The more dynamic a market was, the shorter was also the contracting period.[24] In markets with a small number of handlers, long-term contracts were generally rewarded with price reductions. In more competitive markets, airlines preferred to discuss contracts more often to take advantage of the competitive pressure and reduce their handling costs even further. The larger choice of handlers with an increased negotiating power could be an explanation of why airlines did not make much use of the new possibilities to self handle.

Impact on Passengers

As successful as the deregulation was for airlines, the direct impact on end users – the passengers – was small. Airport-related costs add up to 11–13 per cent of the airlines' costs. However, ground handling costs represent only 4–7 per cent of the total costs.[25] This suggests that price reductions between handlers and airlines did not cause substantial price reductions for passengers.

Concerning the quality of handling, the chances for passengers to notice a difference in handling services was higher than that of observing reduced costs. The time spent on check-in and baggage claim and the quality of service regarding the delivery of transfer baggage were perceptible. But the provided service level depended mainly on the airlines' strategy and had to be agreed between the airline and the handler. Changes of handling quality therefore depended less on the competitive situation of the handlers than on the airlines' objectives and the capabilities of the chosen handler.[26]

---

22   See SH&E (2002), pp. 21–3.
23   See SH&E (2002), pp. 23–4.
24   In deregulated markets contracting periods generally were one to three years; in partly deregulated markets the periods vary between two and five years: see Templin (2007), p. 173.
25   According to a study conducted at Cranfield University, station and ground handling costs represented about 17 per cent: see AEA (1998), attachment 2. More recent studies, conducted by IATA and AEA, estimate the airport-related costs at between 11 and 13 per cent: see IATA (2002), p. 52 and Le (2004), p. 12. Concerning ground handling the estimates vary according to the airlines' business model, at between 4 and 7 per cent: see Smith (2004), p. 11 and Schmitz (2004), p. 277.
26   Pressure on costs can lead to a more efficient use of resources, but also to leaner service levels, lower reserve capacity and the use of too little resources such as personnel and equipment to maintain a certain service level.

Incumbent Suppliers

The most serious change for the incumbent suppliers was the loss of market share, as shown in Figure 24.4. This decrease of absolute handling volumes, accompanied by significant price reductions, led to considerable revenue losses. In consequence, the experienced ground handlers had to adapt their cost structures and increase their flexibility and productivity. Furthermore, due to the increasing negotiating power of airlines, incumbents had to reconsider their customer relationship management and adjust their pricing as well as product strategies to offer competitive services.

The downside of increased competition was inefficient processes and inefficient use of resources. Benefits from economies of scale and scope were sometimes reduced because, initially, more personnel and equipment were needed to service the same amount of traffic. We observed that handlers usually did not reduce their workforce and equipment after a loss of market share.[27] Given that there were space constraints on the ramp of most airports, this also increased barriers to entry. In addition, more operating companies per airport led to an increase of interfaces and occasionally to more complex handling procedures and divided responsibilities. The more parties involved, the more coordination was necessary to ensure safe and seamless handling processes.[28]

Market Entrants

Market entrants on the other hand had the chance to start ground handling operations.[29] This was usually through the application for an operating licence in completely open markets, or the participation in a tender process in markets with a limited number of entrants. Additionally, they had the possibility to enter markets through the acquisition of active handlers. Technical entry barriers were fairly low in open markets, but even there ground handlers had to fulfil certain institutional criteria set by the airport before they could start operations.[30] Disputes in licensing new competitors occurred at several airports that had a limit on market entrants. Handlers that were not given a licence filed objections against the tender process or the final decision, which postponed market entries up to one year.[31]

Moreover, even when newcomers obtained a licence, most of the time their entry was accompanied by a slow gain of market share in the first years of operation.[32] New competitors had to build up a reputation first to win the customer's trust and to make them change their supplier. Regularly, airlines used the emergence of new competitors merely to increase their negotiating power to receive price reductions with their long-time handling partners.

However, success factors for new entrants were their competitive cost structures compared to the incumbent handlers. They also had lean and flexible organizational structures and short decision processes, as well as a very strong customer orientation. Very important for new handlers were also a good reputation and handling know-how that they obtained in their international operations. Entrants that were part of global handling companies used this as an additional competitive factor.

---

27   See Templin (2007), p. 83.
28   See Templin (2007), pp.81–2.
29   Additionally, airlines now have the possibility to start self handling.
30   For instance the presentation of statements concerning the financial situation and business plans, evidence for quality and security of operations – including insurance coverage – as well as experience and qualification of employees: see Templin (2007), p. 160.
31   See Templin (2007), p. 163.
32   See also Figure 24.8.

Their customers could compare service levels between airports at which they operated or even combine contracts for two or more stations into one-stop shopping arrangements, which offered networking advantages on top.

Within five years of operation, new competitors gained in all the six markets studied – up to one-third of the market volume in the case of LHR, and at FCO even two-thirds of the total traffic (see Figure 24.4).

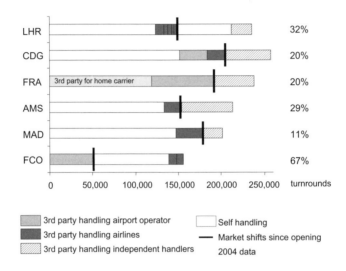

Figure 24.4   **Shifts in market share after deregulation (1999–2004)**[33]

Impact on Employees

Besides airlines as customers and handlers as suppliers, there were other stakeholders that were affected by the deregulation of ground handling markets. Since ground handling is a very labour-intensive industry,[34] the implementation of the EU Directive consequently had a significant impact on employees. Due to substantial shifts in market share, a transfer of personnel between ground handlers occurred at some airports. To lesson the negative impact on the workforce, the Directive envisaged measures to protect employment rights.[35] New employers had to take over the rights and especially the costs of the transferred workforce from incumbents. Implementing the Directive into national law, Spain, Italy, Germany integrated a rule to protect the workforce that allowed the transfer of personnel according to the shift of market volume. In Madrid and Rome this opportunity was extensively used by the former monopolists. In Frankfurt, the sole provider did not make use of this possibility.[36] Great Britain and France on the other hand had national laws that allowed shifts of workforce according to the transferred functions. These laws that were independent of the

---

33   See Templin (2007), p. 76.
34   Up to 80 per cent of the total handling costs are labour related, see SH&E (2002), p. 97.
35   See European Council (1996), Article 18.
36   The laws for transferring workforce in Germany as well as in Italy were judged to be illegal in the meantime – see European Court of Justice (2004) and European Court of Justice (2005); but the process is still practised in Spain.

provision in the Directive were applied both in London and Paris. Only in the Netherlands were transfers not allowed; companies had to adjust the volume of their workforce by themselves.[37]

In addition to the shift of mere workforce volumes, the pressure on costs led to significant changes for handlers and their employees. The later a handling company started operations at an airport, the lower the wages generally were. New entrants paid up to 30 per cent lower wages, and working times were longer and more variable than for employees working in incumbent companies.[38]

To meet the needs for flexible and less expensive workers, incumbents as well as new handlers relied to a greater extent on part-time and temporary workers. They also used more limited contracts to be able to facilitate the termination of employment relationships according to the shifting of handling volumes. Furthermore, there was a tendency to employ less-qualified staff or keep qualifications to a minimum to lower costs.[39]

For the employees this led to less attractive and sometimes unstable working conditions, to possible transfers to new entrants or even to unemployment. Due to the increasing competition, the workers therefore had to be seen as the stakeholders with the most disadvantages.

Impact on Airports

Airports as stakeholders were also impacted by the deregulation, but the effects were less significant than expected. In the beginning, most airport operators were afraid of decreasing quality concerning airport performance and security as a consequence of liberalization that in the end did not occur, or at least did not last very long. Although there were considerable space limitations, combined with a temporary decrease in safety due to the increase of personnel and equipment, the performance of the airports studied was not reduced.[40] One of the reasons for this smooth conversion to more competition in ground handling markets was certainly the gradual deregulation process the European commission implemented with the Directive, allowing the member states to limit the number of new entrants.

Competition on the Ground: A Look at the Airports in Detail

London Heathrow

London Heathrow (LHR) served as the role model for ground handling competition in Europe. As the biggest European airport, with about 67.3 million passengers in 2004, the airport was served by 11 airside ground handling companies.[41] Seven of these were the airlines or their subsidiaries – British Airways, American Airlines, Air Canada, Air France Services, Alitalia Servizi, KLM Ground Services and United Airlines. All of them provided self and third party handling. The four

---

37 See Templin (2007), p. 68. Substantial shifts in market share already led to significant dismissals in Amsterdam; see ibid., p. 87.
38 See Bender (2005), p. 8. In addition, new independent handling companies do not have such expensive benefits as airlines and airport handlers offer their employees.
39 See Templin (2007), pp. 88–9. Even though temporary workers sometimes have higher average wages than employed workers, the flexibility to take them up on demand makes the use of temporary workers attractive.
40 See Templin (2007), pp. 90–92.
41 The often mentioned 13 handlers include two cargo handlers, see SH&E (2002), p. 19.

remaining third party handlers were Aviance UK and its sister company Plane Handling, Penauille Servisair (now Servisair) and Menzies. The airport itself did not offer ground services. For the market presence of these handling companies in 2004 see Figure 24.5. However, none of these handlers offered their services at all of the (then) four terminals at Heathrow. The small airline handlers especially concentrated on the terminal where they were located. Only British Airways and the four third party handlers served customers at two or three terminals.[42]

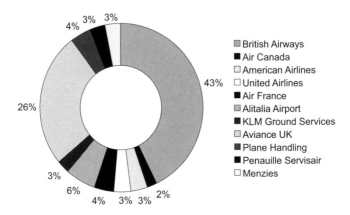

Figure 24.5 Market split at London Heathrow 2004 (turnarounds)[43]

Although there were quite a lot self handlers at LHR, their market volume reflected only about 53 per cent of the total market. The remaining handling volume of 47 per cent was third party handling of all 11 providers. The reason for this was the small market share of each of the six small airline handlers – their self handling accounts on average for 2 per cent of the total market share. British Airways, as the biggest self handler, was responsible for 43 per cent of the market share. For the third party handlers, Aviance UK was by far the biggest with approximately 26 per cent of the total market volume. The other providers each served between 1 and 4,5 per cent of the third party market, including British Airways with 2 per cent.[44]

The market at Heathrow could be described as very dynamic. Before the directive came into force, there were eight airlines providing self and third party handling. The airport authorities of Heathrow airport tried to avoid the complete opening of the market due to capacity problems and tried to use the directive to reduce the number of handlers to six. Several third party handlers objected to the demanded reduction; the regulating body, the Civil Aviation Authority (CAA), finally decided to completely open up the market.[45]

Three independent handlers entered the market right after the complete opening in 1999; the fourth started in 2001. Two of them entered the market by asking for operating permits, while the other two started by buying former self handlers. Shifts in market share occurred due to the strategic decision of British Airways to reduce third party handling and to concentrate on self

---

42 See Figure 24.12.
43 See Templin (2005b), p. 14.
44 See Templin (2007), pp. 35–38.
45 See CAA (1998a). London Gatwick was more successful with its claim; there the number of handlers is restricted to four, see CAA (1998b).

handling. Furthermore, some airlines gave up self handling and contracted with the new third party providers. In 2004, Swissport, as one of the first global handlers to enter the market at Heathrow, abruptly exited the market after five years of service.[46] Menzies, being familiar with the Heathrow market as a landside cargo handler, decided to step in and started airside ground handling in 2005.

Even though the market was completely open to all handlers that wanted to offer airside ground handling at Heathrow, there was an administrative entry barrier when it came to moving around the four terminals at the airport. If a handler wanted to provide services to an airline at a terminal it was not yet serving, it had to complete an approval process set up by the airport authority. This procedure was implemented by the British Airports Authority (BAA) after the decision of the CAA to limit the number of handlers due to capacity restrictions. The motivation was to protect active handlers and airlines operating at the respective terminals.[47] Therefore, a case study to evaluate the effects of an additional handler had to be conducted. First of all, the minimum handling volume to be shifted had to represent at least 5 per cent of the terminal volume. Moreover, additional infrastructure needs of the potential new handler (e. g. check-in counters, equipment parking or baggage belts) were weighed against the released infrastructure of the former handler. In the end, the airport authority had to decide if a handler was allowed to offer its services at the terminal.[48]

Paris Charles de Gaulle

Paris Charles de Gaulle (CDG) was the second largest airport in Europe, with 51.3 million passengers in 2004. After implementing the Directive, the market was partly opened and three new independent handlers were admitted. Altogether, five different handling companies offered their services to airlines. These were a 100 per cent subsidiary of Aéroports de Paris (ADP), Alyzia, and Air France as the two established incumbents. Penauille Servisair (now Servisair), Groupe Europe Handling and Swissport started operations as new entrants in 2000 and 2001. For handling volumes of suppliers in Paris see Figure 24.6. Contrary to London, no other airline besides home carrier Air France was performing self handling, which came to about 56 per cent of the total market volume. The third party volume was split between all five handlers, Air France providing approximately 9 per cent of the third party volume.[49]

Since the number of handlers was limited, the licences offered to handlers had to be tendered, and a selection process took place for each terminal. ADP used its airport privilege expressed in the Directive and was de facto selected.

But in Paris – as already experienced in London – airlines did not have a choice between all five handlers but between two or three, depending on the terminal they were located at. Only the airport operator ADP offered handling at all three terminals; Air France operated at Terminals 1 and 2. All three new independent handlers received licences only for one terminal each.[50] Additionally, there were temporary restrictions at Terminals 2 and 3 for third party handlers granted by the European Commission due to capacity constraints, which expired in 2000.[51]

---

46   See Swissport (2004).
47   See Templin (2007), p 161.
48   Usually, the requests are granted, but there has been at least one case where the transfer was not allowed by BAA, see Templin (2007), p. 161.
49   See Templin (2007), pp. 35–7.
50   See Figure 24.12.
51   See SH&E (2002), p. 12.

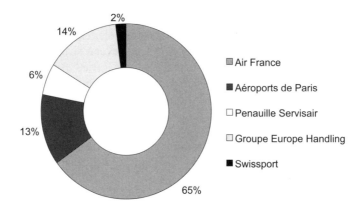

Figure 24.6 Market split at Paris Charles de Gaulle 2004 (turnarounds)[52]

The ground handling market at Charles de Gaulle was less dynamic than the one in London. Nevertheless, almost half of the market open to third parties (meaning not blocked due to self handling of Air France) was gained by new entrants since 2000.[53] ADP suffered most from market shifts, but Air France experienced a decline of customers as well.

An interesting characteristic of handling at CDG was the extensive use of subcontractors. These companies acted on behalf of the licensed handlers and did not contract directly with airlines. Handlers used subcontractors to lower their costs and to transfer parts of their operating risk. Most of the time, subcontractors offered their services to several handlers at the same time. As a result, far more than the above mentioned five handlers served the market in Paris, even if they were not allowed to contract these services to airlines and thus were only indirectly part of the competition.

Frankfurt

Around 51.1 million passengers travelled through Frankfurt airport (FRA) in 2004. Despite the large handling volume, no airline was performing self handling;[54] not even the home carrier, Lufthansa, took advantage of this option at its hub airport (with a share of about 63 per cent of the total market volume). Since the market had been restricted to only two third party handlers, the providers were the airport operator Fraport[55] as the incumbent and Acciona Airport Services as the new entrant in 2000. Fraport was – as ADP in Paris – de facto selected and Acciona won the tender process that took place in Frankfurt.

In the beginning, Acciona was allowed to offer its services only in Terminal 2 until the end of 2000. Like CDG, the airport FRA was granted an exemption by the EU Commission for Terminal 1 due to capacity restrictions.[56] Since then, all airlines had the choice between Fraport and Acciona. Until 2004, Acciona gained up to 20 per cent of the market share – see Figure 24.7.

---

52 See Templin (2005b), p. 14.
53 See Figure 24.4.
54 FedEx being an exemption since they are performing self handling for their freighters.
55 The ground handling department is a vertically integrated business unit within the airport operator.
56 See SH&E (2002), p. 12.

After a phase of testing the performance of the new competitor, the first major changes took place two years after market entry. Even after the second handler entered the market, the bulk of Lufthansa flights were still handled by Fraport.[57] However, several airline customers switched to the new competitor, including most members of the oneworld Alliance. The dynamics of the market can be seen from Figure 24.8, they are characteristically for all former monopolistic markets.[58]

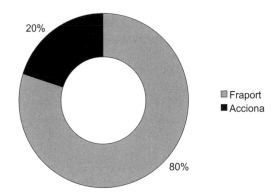

Figure 24.7   Market split at Frankfurt 2004 (turnarounds)[59]

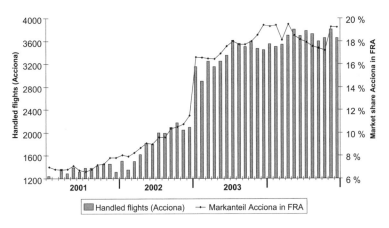

Figure 24.8   Market growth of Acciona at Frankfurt[60]

---

57 However, Lufthansa decided in 2002 to break up the monolith and had Acciona handle its regional branch CityLine, followed by the ramp handling of preferred customers in 2004. This showed that contracts were not as incontestable as generally assumed when the markets opened up. Nevertheless, Lufthansa and Fraport agreed on another five-year contract until 2010. During this period, the contracted handling volume can be seen as incontestable.
58 See Templin (2007), pp. 171–2.
59 See Templin (2005b), p. 14 and Templin (2007), pp. 35–6.
60 See Acciona (2006).

Amsterdam Schiphol

Amsterdam (AMS) – number four of the European airports in 2004, with 42.5 million passengers – was, like Heathrow, completely open to ground handlers. Five suppliers received operating permits for third party ramp handling, although only four companies used them. Martinair waived the right to handle other airlines' customers, although the airline was not interested in giving up self handling. Home carrier KLM was the most important handling company providing self (60 per cent) and third party handling (9 per cent). Therefore, self handling represented approximately 62 per cent of the market and all four third party handlers operating at Schiphol airport had more or less the same market share (see Figure 24.9).[61] As in Heathrow, the airport itself did not offer handling for third parties.

With the implementation of the directive, the market was completely opened at AMS. Before that, there were already three competitors offering third party handling. Aviapartner and Dutchport entered the market in 1999, acquiring operating permits. Two additional companies, Menzies and GlobeGround (now Servisair) started operations in 2000 by acquiring incumbent suppliers. In 2002, third party handler Dutchport ceased operations after only three years of service. Unlike at Heathrow airport, the parting handler was not replaced.

Every airline flying to AMS could choose between all of the four third party suppliers. Since all handlers were operating at critical mass, competition was very dynamic and the fight for market shares was the strongest compared to the other observed airports. Additional market entries were not expected.

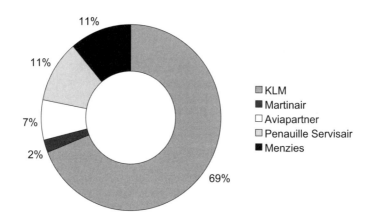

Figure 24.9 Market split at Amsterdam Schiphol 2004 (turnarounds)[62]

Madrid Barajas

About 38.7 million passengers travelled through Madrid (MAD) in 2004. This ground handling market was one of the first to be deregulated in Europe. Spain started its deregulation process a few years before the Directive came into force. This is why incumbent home carrier Iberia had to adapt to competition already in 1997 when Ineuropa, as the second third party handler, entered

---

61   See Templin (2007), pp. 35–7.
62   See Templin (2005b), p. 14.

the market in the course of a tender process. Up to then Iberia had provided self and third party handling as a monopolist at Spain's largest airport.

MAD was still a regulated market in 2004: only Iberia (16 per cent) and Ineuropa (11 per cent) were allowed to offer third party handling. All airlines could choose from these two handlers in all terminals; the airport operator Aena did not offer handling services. In addition, there were three self handlers without third party licences present at the airport besides Iberia (53 per cent) – see Figure 24.10. These airlines were handling themselves already before the opening of the market. Altogether, this led to the largest amount of self handling at the observed airports (73 per cent).[63]

All airports in Spain with more than 1 million passengers per year had to open up their markets to one third party handler in addition to the incumbent handler, Iberia, in 1997. Handlers were selected during tender processes throughout Spain organized by Aena. While a decrease in handling prices was already part of the selection criteria, the most obvious and more often reported effect of competition was the increase in quality of services at MAD.[64] Moreover, heavy capacity and space constraints obstructed operations due to the extra amount of equipment needed by the second third party handler and the three self handlers, combined with the quite limited space on the ramp.[65]

Ineuropa gained about 40 per cent of the third party market after 1997 but Iberia is still the leading supplier, providing handling for the remaining 60 per cent. In the beginning, market shares shifted gradually from the former monopolist and slowed down towards the termination of the licence. Initially, the handling licence for Ineuropa was issued for seven years. However, the subsequent tender process was postponed until 2006.[66]

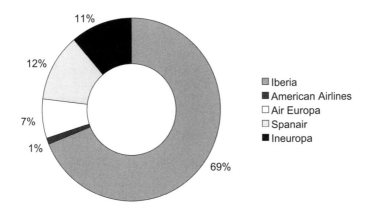

Figure 24.10 Market split at Madrid Barajas 2004 (turnarounds)[67]

---

63   See Templin (2007), pp. 35–6. One of the self handlers was interested in the second third party licence as the tender took place but was not selected.
64   See SH&E (2002), p. 24.
65   See Templin (2007), pp. 91–2.
66   See Templin (2007), p. 163.
67   See Templin (2005b), p. 14.

Rome Fiumicino

The ground handling market at Rome Fiumicino (FCO), with 28.1 million passengers, was restricted to three third party and three self handlers in 2004. The airport operators subsidiary ADR Handling, as the former sole provider, was accompanied by two airlines that provided self as well as third party handling. Home carrier Alitalia (handled by Alitalia Airport) and Air One as the regional Italian carrier (handled by EAS) decided to self handle and were later selected as the two additional third party handlers.[68] Until then, no handler independent of airport or airlines offered handling to third parties, whereas the remaining self handling licence was not used by any airline. Altogether, self handling of the two airlines added up to 56 per cent of the market; all other airlines arriving at FCO could choose between all three suppliers – see Figure 24.11 for their market shares.[69]

The opening up of the market was followed by major structural changes in the handling market at FCO. First, the formerly vertically integrated handling business unit of the airport operator was outsourced.[70] Within this new subsidiary, separate departments were formed right from the beginning that were later outsourced as 100 per cent subsidiaries of Alitalia and Air One to provide their self handling. The third party handlers were again selected through tender processes, where both new self handlers applied and in the end received their licences. The airport operator ADR Handling was, like Fraport in Frankfurt and ADP in Paris, de facto selected as third party handler.

With 33 per cent, ADR Handling handled the biggest third party volume in Rome. Both airline handlers Alitalia Airport and EAS served about 5 per cent each of the third party market. However, the major market shifts away from the airport operator occurred due to the airlines' decision of to self handle. Before that, they were the two biggest third party customers of ADR Handling at the time of the handling monopoly.

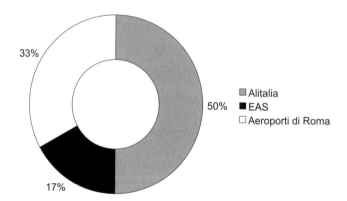

Figure 24.11 Market split at Rome Fiumicino 2004 (turnarounds)[71]

---

68   Alitalia Airport and EAS are respective 100 per cent subsidiaries.
69   See Templin (2007), pp. 35–7.
70   Initially, the subsidiary of the airport operator was owned by the airport (51 per cent) and the independent ground handling company Ogden, later Menzies (49 per cent). But the latter sold its shares back to the airport operator in 2003: see ADR Handling (2002), p. 8.
71   See Templin (2005b), p. 14.

Conclusions and Outlook

Altogether, the deregulation of the ground handling markets led to more competitive market structures at airports in Europe. The positive effect for airlines, besides decreasing costs and – at some airports – increasing quality levels, was the additional variety when choosing handlers. Even if an airline was satisfied with the current handler, the enhanced negotiating power helped to improve the airlines' position and to lower handling prices. As described in the last chapter, the mere number of handlers at an airport was not crucial for the assessment of the negotiating power since not all handlers could offer their services at all terminals at the respective airport. The number of handlers to choose from varied from two to six: for the distribution of handlers according to terminals in 2004, see Table 24.1.

For handlers, a few administrative entry barriers still exist at the observed hubs even after deregulation was implemented in Europe. Either market entry is regulated according to the Directive or local rules apply, as for instance in the case of Heathrow. Even Amsterdam has minimum standards new handlers have to comply with before starting their handling business in the completely deregulated market at Schiphol.

But despite these remaining entry barriers, ground handlers gain new opportunities to grow their businesses in deregulated markets. They have chances to develop international strategies and to offer their services at various airports within the EU and even beyond. This multi-station strategy offers new competitive advantages to other handlers and makes them adapt to the needs of airlines which increasingly ask for multi-station contracts.[72]

As for the market structures, self handlers generally stayed in the markets: two new self handlers emerged in Rome, whereas two airlines outsourced their handling and left the market at Heathrow. Additional competitors entered the markets at all observed airports. Moreover, two third party handlers left the open markets in London and Amsterdam, but no exits took place in the still limited markets of the other four airports.

Table 24.1   Third party competition per terminal[73]

Market	Terminal	Airport operator	Airline	Independent Handlers	Total no. of Handlers
LHR	1		3	3	6
	2		2	2	4
	3		2	3	5
	4		1	1	2
CDG	1	1	1	1	3
	2	1	1	1	3
	3	1	1		2
FRA	1, 2	1		1	2
AMS	1, 2, 3		1	3	4
MAD	1, 2, 3		1	1	2
FCO	A, B, C	1	2		3

---

72   Economies of scale and scope are believed to be negligible. Transaction cost savings may explain this trend, but this topic has to be researched further.
73   See Templin (2007), p. 212.

In a sustainable market, no additional supplier wants to enter the market because they do not see a chance to be successful. Looking at the six airports and the ground handling scene, the picture is as different as described above. The markets in AMS and LHR are fully liberalized and seem to be sustainable. Nevertheless, the leaving supplier at Heathrow was replaced, whereas a departing handler at Schiphol was not substituted.

At airports with limited access, sustainability still has to be proved. With a large market volume, only two third party handlers and no self handlers operating, the market at FRA seems to be very attractive for additional handlers from a competitive point of view. Problems concerning the size of the market exist at CDG and MAD as well as FCO. Paris might be interesting for new handlers but since it has limited the scope of the licences to certain terminals, there might not be enough handling volume for another handler if this artificial barrier to entry remains.[74] At Madrid and Rome, self handling companies considerably reduce the handling volume for third party handlers.

The EU is planning on further opening up the markets with a revision of the Directive. But the first proposal was postponed and further studies on the effects of the deregulation were assigned. The changes concentrate on increasing competition; some of the possible changes are:

- an increase in the minimum number of third party handlers at airports with more than 10 million passengers p.a.
    - four instead of today two third party handlers at airports with over 20 million passengers;
- a wider definition of self handling which would allow airlines to 'self handle' their alliance partners
    - decreasing market volume for third party handlers but a bigger chance to self handle due to the increased handling volume;
- legal unbundling of airports and airlines and their ground handling *business*
    - forced outsourcing of ground handling for vertically integrated airport and airline handling companies;
- an abolition of the airport privilege
    - airport operators offering third party handling would have to apply for a licence like the other third party handlers;
- an implementation of minimum quality standards for ground handlers
    - to protect performance standards;
- an implementation of social standards for workers[75]
    - to protect the workforce from excessive disadvantages due to competition;

Therefore the market structure will be in motion for the coming years, at least until all ground handling providers have adapted to the changes due to the deregulation and taken advantage of the chances offered by the continuing deregulation.

---

74 However the distance between the terminals and the efficiencies of a probably necessary split in operations have to be considered carefully.

75 See Templin (2007), pp. 207–8.

References

Acciona (2006): Traffic Development 2001–2004 (Flights): Market Share FRA, Acciona Airport Services, www.acciona-fra.de/pic/facts/trafic_dev_xl.gif (14 February 2006).

ACI (2005a): Passenger Traffic 2004 final, Airports Council International, www.aci.aero/cda/aci/display/main/aci_content.jsp?zn=aci&cp=1-5_9_2_ (13 December 2005).

ACI (2005b): Traffic Movements 2004 final, Airports Council International, www.aci.aero/cda/aci/display/main/aci_content.jsp?zn=aci&cp=1-5-54-57_9_2__ (13 December 2005).

ACI Europe (2004): Agreed service levels required, in: E-communiqué, September/October, issue 157, www.aci-europe.org (18 February 2006).

ADR Handling (2002): Report on Operations and Financial Statements as of December 31, 2001, Aeroporti di Roma Handling, ADR Group, www.adrhandling.it (14 February 2006).

AEA (1998): Benchmarking of Airport Charges, Information Package, February.

Bender, W. (2005): Mitarbeiter Dialog: Wir machen Fraport fit, Management Conference Fraport AG, 19 January 2005, unpublished.

CAA (1998a): Decision of the Authority under Regulation 10(1) of the Airports (Groundhandling) Regulations 1997 on an application by Heathrow Airport Ltd for a determination to limit the number of suppliers authorised to provide airside services to third parties, GH 2/98, Civil Aviation Authority, United Kingdom, www.caa.co.uk/erg/ergdocs/finaldecision.pdf (04 January 2006).

CAA (1998b): Decision of the Authority under Regulation 10(1) of the Airports (Groundhandling) Regulations 1997 on an application by Gatwick Airport Ltd for a determination to limit the number of suppliers authorised to provide airside services to third parties, GH 1/98, Civil Aviation Authority, United Kingdom, www.caa.co.uk/erg/ergdocs/galdecision98.pdf (04 January 2006).

Conway, P. (2005): Revolution delayed, in: *Airline Business*, January, pp. 46–8.

European Council (1996): Council Directive 96/67/EC of 15 October 1996 on access to the groundhandling market at Community airports, Official Journal L 272, 25 October 1996, pp. 36–45.

European Court of Justice (2004): Judgement of the Court of 9 December 2004 in Case C-460/02: Commission of the European Communities v Italian Republic (Air transport – Groundhandling – Directive 96/67/EC).

European Court of Justice (2005): Judgement of the Court of 14 July 2005 in Case C-386/03: Commission of the European Communities v Federal Republic of Germany (Failure of a Member State to fulfil obligations – Airports – Groundhandling – Directive 96/67/EC).

Fraport (2005): Fraport Ground Services – Master Presentation, Fraport AG, August, unpublished.

Ground Handling International (2006): The US and Canada – Outsourcing and economy have been highlighted factors over the last 12 months for the industry, in: *Ground Handling International*, 11, no. 1, February, pp. 30–34.

IATA (2002): Annual Report 2002, International Air Transport Association for the 58th Annual General Meeting, Shanghai, June.

Kinnock, N. (1996): The liberalisation of the European aviation industry, in: *European Business Journal*, 8, no. 4, pp. 8–13.

KPMG (2002): Handling pressure: A snapshot of the aviation ground handling industry, KPMG Corporate Finance, Spring.

Le T.M. (2004): Making the most of a marginal business – The 1996 EU Handling Directive: what next? Presentation at AEA, 6th Annual Ground Handling International Conference, 1 December 2004, Budapest.

Nana, J.-P. (2003): Assistance au sol: conjuguons nos talents, in: *Aéroports Magazine*, 335, January–February, pp. 20–26.

OAG (2005): Official Airline Guide, OAG Worldwide Limited, 6, no. 12, June.

Piper, H.P. (1994): Deregulierung im Flughafenbereich: Bodendienste der deutschen Verkehrsflughäfen, in: *Internationales Verkehrswesen*, 46, no. 1–2, pp. 51–2.

Schmitz, P. (2004): Liberalisierung der Bodenverkehrsdienste um jeden Preis?, in: *Internationales Verkehrswesen*, 56, no. 6, pp. 276–8.

SH&E (2002): Final Report: Study on the quality and efficiency of ground handling services at EU airports as a result of the implementation of Council Directive 96/67/EC, Report to European Commission, SH&E International Air Transport Consultancy.

Smith, C. (2004): A Valued Business? – A Financial Overview, Presentation by SH&E International Air Transport Consultancy, 6th Annual Ground Handling International Conference, 30 November 2004, Budapest.

Swissport (2004): Swissport UK Limited is placed in administration effective November 16th 2004, 16 November 2004, News Release, http://www.swissport.com/mediacenter/index_news.php?id=197&ref=archive (13 December 2005).

Swissport (2005): Profile 2005, Swissport, http://www.swissport.com/download/publications/profile_05.pdf (27 February 2006).

Swissport (2006): Facts & Figures: Key Figures for 2005, http://www.swissport.com/aboutus/facts.shtml (29 April 2006).

Swissport (2008): The Swissport Profile, http://www.swissport.com/corporate/index.php (16 April 2008).

Templin, C. (2005a): Deregulation of Ground Handling on six European Airports, http://www.garsonline.de/Downloads/050609/Templin_paper_new.pdf (29 April 2006).

Templin, C. (2005b): Deregulation of Ground Handling on six European Airports, Presentation at GARS Workshop, June 2005, http://www.garsonline.de/Downloads/050609/Templin_slides.pdf (08 May 2006).

Templin, C. (2007): Deregulierung der Bodenabfertigungsdienste an Flughäfen in Europa – Konsequenzen für Märkte und Wettbewerber, Dissertation, Cologne: Kölner Wissenschaftsverlag.

Walker, S. (2000): *The Liberalisation of Ground Handling Services: The Impact and Implementation of the European Directive*, Global Aviation Reports, London: SMi Publishing.

WTO (2005): Communication from Australia, Chile, the European Communities, New Zealand, Norway and Switzerland – Trade in Services to the Aviation Industry: A case for commitments under the GATS, World Trade Organization, Council for Trade in Services, TN/S/W/29, 16 February.

Chapter 25
Airport Competition: Market Dominance and Abuse

Peter Lewisch

Introduction

This chapter addresses the issue of airport competition from its negative side, that is, in terms of the absence of sufficient competition. This absence of competition may simply stem from the fact that airports enjoy a monopoly position over their own infrastructure. Whereas in a functioning market, market forces would control the behaviour of all market participants, these normally endogenously generated constraints do not exist in dominated markets. It is for this reason that general competition law aims to substitute these absent market constraints with rules that subject the dominant firm to a set of specific legal obligations. The respective provision of European competition law is enshrined in Article 82 EC-Treaty.[1] National competition law contains similar regulations for cases without cross-border concerns. These obligations do not demand specific positive behaviour, but prohibit any abuse of a dominant position.

The chapter provides an analytical approach to identify the conditions for stable market power of airports (bottleneck approach). Also discussed are the conditions under which this market power may erode. Questions of deregulation, both with respect to general aviation and to the provision of infrastructure for ground handling services, are reviewed. The chapter then discusses the twofold regulatory environment of general and sector-specific competition law. In conclusion, the paper analyses the core question of abuse and the pertinent European Court of Justice (ECJ) case law, both with respect to aviation and ancillary services (in particular ground handling services).

The Three-Tier Structure of Transport Markets and the Bottleneck Nature of Airports

The Three-Tier Structure of Transport Markets and Aviation

In transport markets there are several analytically distinguishable layers which build upon each other in order to ultimately allow the end customer to satisfy his/her demand for the transportation service

---

1 Article 82 EC-Treaty reads as follows: 'Any abuse by one or more undertakings of a dominant position within the common market or in a substantial part of it shall be prohibited as incompatible with the common market in so far as it may affect trade between Member States. Such abuse may, in particular, consist in: (a) directly or indirectly imposing unfair purchase or selling prices or other unfair trading conditions; (b) limiting production, markets or technical development to the prejudice of consumers; (c) applying dissimilar conditions to equivalent transactions with other trading parties, thereby placing them at a competitive disadvantage; (d) making the conclusion of contracts subject to acceptance by the other parties of supplementary obligations which, by their nature or according to commercial usage, have no connection with the subject of such contracts.' For a legal-doctrinal analysis of Article 82 see Lewisch (2007).

(see Knieps, 1996). Typically, this structure is threefold: infrastructure, capacity management, transport services (Table 25.1).[2]

The first layer is the provision of the earthbound infrastructure by the infrastructure holder. This infrastructure is used by the suppliers on the service level for the provision of transportation services (i.e., ports are used for ferry services and tracks are used for railway services). In the realm of aviation, airports are the providers of this earthbound infrastructure, which consists of the runways that allow planes to take off and to land, and to taxi to their parking positions.

The second layer, safety and capacity management, comprises heterogeneous services that complement the provision of the earthbound infrastructure. These services can sometimes be seen as an independent input for the provision of transport services. It is a question of the technological and institutional characteristics of the respective transport industry, whether and to what extent these services are provided, either together with the infrastructure access or separately. In the railway sector, the holder of the tracks takes care of the capacity management. In aviation, this second layer – highly complex in nature – is typically independent, provided by separate entities. It covers both the guidance of aircraft through the airspace (including their descent or ascent near the airport) and the allotment of the take-off and landing capacity.

The third layer consists of the transportation providers. Ultimately, the respective earthbound infrastructure and the respective capacity management are used by the providers of the transportation services to render these services to consumers. Rail companies run carriages on the tracks, ferry companies operate vessels for boating services and, in the aviation sector, airlines operate aircraft for flight services.

Airport Infrastructure as a Bottleneck Facility

Let us now turn to the market structure for each layer and the potential for self-sustaining competition (see again Knieps, 1996, at p. 90).

Such an analysis – quite unsurprisingly – shows that the market for the provision of air-traffic services is one where airlines do not enjoy systematic market power: Both potential and actual competition is viable and, as can be observed in many areas, fierce.[3] With regard to the second layer (safety and capacity management), competition *on* the market, i.e. simultaneous capacity management by various competing entities, is likely to cause confusion and is, hence, not desirable. What is feasible, however, is competition *for* the market in the sense of a tendering procedure for the provision of these services.

Table 25.1 Threefold structure of transport markets

Provision of transportation services	Vessels, planes, trucks, carriages
Safety and capacity management	
Provision of earthbound infrastructure	Harbours, airports, streets, tracks

---

2   Capacity management services, often provided by an independent middle layer, may, however, also be bundled into infrastructure services.

3   This is, of course, not to say that suppliers would not want to soften the effects of competition by cartel arrangements; in fact one would, depending on the characteristics of the market and the respective barriers to entry, assume some such behaviour.

Regarding the provision of earthbound infrastructure, the traditional analysis suggests the unfeasibility of actual or potential competition because of non-contestability.[4] Airport infrastructure is, hence, a bottleneck facility. The monopolistic control over these facilities grants stable market power (Table 25.2).

Table 25.2   Market power in threefold structure of transport market

Provision of transportation services	Potential for competition? Typically yes
Safety and capacity management	Potential for competition? Typically mixed
Provision of earthbound infrastructure	Potential for competition? Problem of non-contestability

One may rephrase the above analysis in terms of standard competition law terminology: The upstream market is the market for the provision of access to the airport infrastructure. This access is the indispensible input for the downstream service market. In this downstream market, the airlines provide transportation services to customers. Whereas the transportation market is typically competitive, the upstream market is not. The only provider of access to the runways is the airport itself. Viewed thus, the access market appears as a market in itself, with the infrastructure provider holding a dominant (namely, monopolistic) position.

European competition law reflects this position. In a series of cases the European Court of Justice (ECJ) has treated ports and airports as distinct markets for the product of access rights to the infrastructure.[5] Services on these markets are monopolistically provided by the infrastructure holder, which therefore enjoys a dominant position.

Potential for Competition between Airports

Airports as the Object of Choice

Let us now, on the basis of the bottleneck concept, explore the potential for competition between airports. At first glance, it seems as if the bottleneck characteristics of the earthbound infrastructure would eliminate any potential for competition. Airports seemingly derive stable market power from the very fact that there is no way to take off and land at a given airport other than by using its runways. How then can an airport possibly lose its dominant position? One can develop the answer along the following lines.

There is, obviously, no possibility of bypassing a specific airport's runways if access to that airport is at issue. However, an airport may lose, irrespectively of its continued bottleneck control over its runways, its monopoly position if choice options for the demand side (i.e., the airlines)

---

4   The holder of the earthbound infrastructure cannot be reasonably attacked because of a systematic divergence in the costs relevant to the decision. Potential competition will not work because of the asset-specificity of the necessary earthbound investment. The concept of monopolistically held, non-contestable infrastructure as a basis for stable market power has been developed by Knieps (1996, at p. 77).

5   For ports, see ECJ 10.12.1991, C-179/90 regarding the port of Genoa and ECJ 12.02.1998, C-163/96 regarding the port of La Spezia. For airports, see ECJ 29.03.2001, C-163/99; similar decisions have been rendered by the European Commission, see Commission Decision of 10.02.1999, OJ 1999 L 69, 24.

broaden. To the extent that the airlines may satisfy respective end-user demand by switching to some other airport, all feasible airports become possible candidates for take-off and landing. Each airport's bottleneck infrastructure is then in competition with other comparable infrastructures. Thus previously stable market power is eroded.

Demand Characteristics on the Downstream Market as the Decisive Factor

Whether the loss in market power will actually occur depends on the characteristics of the downstream transport market and the consumer interests (i.e., the nature of the demand on the downstream market).

The downstream market typically is heterogeneous. For business travel, time and location sensitivity is high; for leisure travel, relatively low. The demand by airlines translates to the demand by their end customers. The consequence of this view is that the holder of a similar runway may differ in its market position according to the various segments of demand, i.e., according to the market for which the airport provides the input. The monopolistic control over the similar runway may constitute market dominance in some cases (namely in those where time and location sensitivity is high) and not in others.

More technically, competition law sees transport markets as product markets characterized by a point-of-origin/point of departure (O-D) pairs approach.[6] This concept implies that, in the light of the consumer demand, the respective product market is separated into pairs of connections between certain end points. The provision of infrastructure-access to airports on the upstream market is then an input for the airlines as providers of flight services between points of origin and departure, respectively. The consumer demand for transport between two cities may or may not be satisfied by different modes of transport in competition with each other and/or by different airports. A German Court of Appeal (OLG Düsseldorf) held in a recent ruling that in the market for point-to-point business flights (with extremely high end-user demand for location and time of arrival/departure) an airport enjoys a monopoly position if the next available airport is located a distance of 35 km away.[7]

Along similar lines, the European Commission has, in a railway case (Comp 37.685 GVG/FS, Commission Decision of 27.08.2003), derived the problem of access to rail tracks from the end-user transport market to be served. It has, thus, distinguished between the upstream market for access to the track infrastructure (and the traction market) and the downstream market for rail passenger transport (which it viewed, in line with the O-D approach) as a set of separate markets for transport between certain end points (in the case at hand, between several German cities and Milan). Based on the finding that modes of transport were not interchangeable for these bundles of routes from a consumer viewpoint because of market characteristics,[8] the Commission argued that – in the absence of viable alternatives – only the holder of the respective train infrastructure in Italy (the Ferrovie dello Stato) was able to provide and assign train paths for the relevant geographical

---

6   The European Commission has developed this approach in air transport cases (i.e., Commission Decision 2002/746/EC of 5 July 2002 Austrian Airlines/Lufthansa (OJ L 242, 10.9.2002, p. 25) and extended it to railway transportation.

7   OLG Düsseldorf, ruling of 18.10.2006, VI-U(Kart)1/05, published in Wirtschaft und Wettbewerb 2007, p. 380.

8   Comp 37.685, item 60: 'In transport, under certain conditions passengers may consider air travel, high-speed rail travel, coach and car travel to be interchangeable modes of transport. This depends on the concrete characteristics of the service, for instance the travelling time'. And at item 66: 'Similarly, there are considerable quality differences between travelling by train and by aircraft.'

market. Therefore, the Commission approved a dominant position held by Ferrovie dello Stato and, ultimately, also an abuse of this dominant position. The spirit of this decision suggests that, in the case that either alternative modes of transportation (car, plane, coach) or alternative routes (i.e. via Austria or France) were feasible, Ferrovie dello Stato would not have enjoyed a dominant position over its infrastructure with respect to the satisfaction of the demand on the downstream end-user market.[9]

Let us now revert to aviation. The general point is that monopoly control over a certain bottleneck infrastructure ceases to purport stable market power when this infrastructure (the respective airport) becomes the object of choice for users. Whereas much of the air traffic is time- and location-sensitive (with limited scope for competition), some potential for competition exists.

Consider, as an initial example, the 1970s and 1980s, when flights from Europe to East Asia required a refuelling stop in the Middle East. Various airports (Muscat, Oman, and so on.) were potential candidates for these refuelling stops, without any of them holding an ideal location. With respect to long-haul traffic between Europe and the Far East none of these airports held a dominant position (Figure 25.1).

Of course, the situation would be different if point-to-point traffic rather than various route options was in question. In the former case, when it comes to serving the downstream consumer market of point-to-point transportation from any location (say, Cairo) to Muscat, Muscat airport is likely to hold a monopoly position in the upstream market.[10] It is, hence, the demand characteristics of the respective end-user market that determine the degree of market power in the upstream infrastructure market.

The same analysis applies to hub airports in Europe with respect to their potential to satisfy the demand for inter-continental traffic. In this respect, London-Heathrow is in competition with Paris, Zurich, Madrid, Milan, Frankfurt and Vienna. At all of these airports, airlines collect incoming flights from (Eastern) Europe in order to process them across the Atlantic (and vice versa). For example, the role of Paris airport is effectively constrained by the market presence of the other aforementioned airports. Given that these airports compete for the respective traffic, they lose dominance.

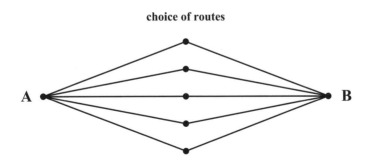

Figure 25.1 Competition between airports for fuelling stops

---

9   Such as would, for example, be the case with respect to long-distance transportation from Hamburg to Marseille, where access to the Italian railway network could be substituted by access to the Austrian, Swiss and French networks.

10   This position would depend on the availability of nearby airports and of different modes of transport (boat, taxi) as viable market alternatives.

But what about the conceivable objection that due to the specific characteristics of the aviation industry (namely the close ties of certain hubs to certain airlines), the potential for competition would not effectively constrain the dominant position held by the airports? If an airline can effectively operate only via its own home hub, the availability of other hub airports might not introduce competition. A closer look reveals that the ties between airlines and airports do not matter. What is necessary for competition to work is the choice option for the end-user between various airlines (independent of possible ties of those airlines to certain hub airports), not necessarily for the providers of transportation services between various infrastructures. Or in other words: if consumers can switch between various airlines (however linked to certain airports), the respective competition translates back to the infrastructure market.

To see this point in a broader context, imagine a small island with two ports (one long and narrow, the other broad and short). Imagine that vessel company 1 runs only boats of type 1 (fitting only port A), whereas vessel company 2 runs boats of type 2 (fitting only port B).[11] Both ferry lines serve the island from the same port on the mainland, with an identical schedule and the potential to expand output (in terms of a tighter schedule and increased passenger volume). Of course in the long run, each ferry-line may acquire a different type of boat in order to serve the other port. But even in the short run (that is, under static circumstances) neither port A nor port B enjoy stable market power as long as the customers are free to choose between the two ports.[12] If port A abuses its monopoly position, the respective deterioration of economic conditions would be passed on to end-consumers, who could and would move to ferry-line 2 and satisfy their demand there and vice versa.

As long as, on the downstream transport market, consumers consider services from their starting point to either port interchangeable, it does not matter whether the providers of transportation services have access to both ports. Whereas each port provides an indispensable input for the transport services of either ferry-line, neither port can exploit this position given the presence of the other port.

Consequences for Airport Competition

Airports enjoy monopolistic control over their infrastructure. Whether or not this control purports stable market power depends on the demand characteristics of the downstream transport service market. Generally, one would expect airport competition to be of limited scope. Hub-airport competition is real and functioning. Competition between regional airports regarding point-to-point traffic is less likely to be realistic, depending, however, on the density of airports[13] and their respective catchment areas.[14]

---

11   One could, of course, make a comparable example with airports (one airport, where only aircraft of type 1 can land, the other being accessible only for aircraft of type 2); though this example is not realistic.

12   Assuming that the costumers are indifferent as to services to either port.

13   There is some decision practice regarding merger control between regional airports. If these airports serve the same region, competition authorities usually assume a cautious approach. We shall revert to the issue of airport density later on in the text.

14   In the area of merger control, there are some cases concerning airport mergers/acquisitions. The British Monopolies and Merger Commission blocked a proposed Belfast airport merger in 1996; a proposal for the acquisition of Exeter airport by an airport group was abandoned in 2005 following a sceptical reaction from the Competition Commission. Neither case has led to a formal decision. On the other hand, there are a variety of merger cases not involving close geographical connections, where the European Commission declined serious doubts as to the compatibility with the common market. See Case IV/M.1035 of 22.12.1997 or

In principle, a similar approach is also adopted by the European Commission in its 'Community Guidelines on Financing of Airports and Start-Up Aid to Airlines Departing from Regional Airports' (2005/C-312/01). Item 11 of these Guidelines holds that 'competition scenarios are evaluated case by case, based on the markets in question'. As a rule of thumb, also in item 11, the Guidelines assume:[15] 'Major international hubs are competing with similar airports in all the transport markets concerned, with the level of competition depending on factors such as congestion and the existence of alternative transport, or in certain cases with large regional airports.'[16]

The Three-Tier Structure of Transportation Markets, Vertical Disintegration and Deregulation

Let us now pursue the three-tier analysis with respect to issues of deregulation. As regards deregulation and market liberalization, there are considerable differences between the various transportation sectors. The three-tier analysis allows us to explain these differences. It allows us also to compare the role of airports as infrastructure holders in aviation (as a provider of runways) to that as infrastructure providers for ground handling services.

Vertical Disintegration in Transport Markets and in Aviation

The three-tier structure provides the analytical foundation for the deregulation of transport markets (and other network economies). Railways may serve as an example. In earlier times, it was commonly held that the owner of the earthbound infrastructure would naturally also provide the capacity management and transportation services on the tracks; competition was considered infeasible because of the non-contestability of the earthbound infrastructure. The idea of vertical disintegration, namely to separate the layers of the three-tier transportation system and to examine their respective potential for competition, has opened up the way for deregulation. Such analysis has revealed that even if the infrastructure layer remained non-contestable, competition on the transport service level would be feasible under the condition that providers of transportation services enjoy equal access to the same, monopolistically held infrastructure.

Under such a scenario of deregulation, two key regulatory issues remain. On the one hand, the infrastructure holder still enjoys a monopoly position with respect to its infrastructure which may give rise to the problem of monopolistic overpricing. On the other hand, regulatory law has to guarantee discrimination-free access to the infrastructure.[17]

Comp/M.2927 of 20.09.2002. See also, regarding Berlin's airports, the lack of objections by the Commission in cases IV/M.1255 of 21.05.1999 and Comp/M.2262 of 05.02.2001.

15   In this respect, the guidelines refer to the 'Study on Competition between Airports and the Application of State Aid Rules' by Cranfield University, June 2002.

16   Item 11: 'Large regional airports may be competing not only with other large regional airports but also with the major Community hubs and land transport, especially if there is high-quality land access to the airport. ... Small airports do not generally compete with other airports except, in some cases, with neighbouring airports of a similar size whose markets overlap.'

17   In the railway sector, this problem is particularly acute because the provider of the infrastructure is itself – via its 'transportation subsidiary' – also a provider of transportation services. With respect to access to the infrastructure, the incumbent operator is in a position to be player and arbitrator in one person. Discrimination-free access to the infrastructure is, therefore, of paramount importance.

In the aviation sector, vertical disintegration is a phenomenon of the past. The separation of the various layers took place decades ago.[18] Typically, airports are not active in the transportation market. Since airports do not have any incentive to favour their own service unit over a competing airline, the problem of discriminatory access to the earthbound infrastructure is dramatically reduced. Moreover, access rights to the infrastructure are, at least under the current regime for large airports, not granted by the infrastructure holder as such, but by a slot-coordinator. This independent entity, the slot-coordinator, controls the allocation of access rights. What remains is the conceivable problem of monopolistic overpricing. This problem is dealt with by regulation, most notably by price-cap regulation (Figure 25.2).

Vertical Disintegration and Ground Handling Services

Let us now turn to the market for ground handling services. Ground handling services are typically provided on the airside part of the airport. These services can only be provided by agents that have access rights to the airport infrastructure. The airport is the monopolistic holder of the airside infrastructure, namely, the airfield.

One can, therefore, distinguish between the upstream market of access to the infrastructure and the downstream service market, on which ground handlers provide their services to airlines. Originally, both layers were monolithically provided by the airport as the infrastructure holder. Duplication of the bottleneck infrastructure of an airfield is unfeasible. It is, however, conceivable to introduce competition on the market for ground handling services by opening up access to the infrastructure. Provided that potential suppliers of ground handling services are granted access to the monopolistically held infrastructure, competition on the downstream market may thrive. And, in fact, this vertical disintegration and deregulation is a result of the law (namely the Ground Handling Directive 96/67/EC).

In light of the above, it becomes apparent that the situation is very similar to that of railway services. The airport as the infrastructure holder is also itself (typically by means of an affiliated company) provider of ground handling services. In its role as supplier on the upstream market, the airport, therefore, has an incentive to favour its own services. What results is a problem of discriminatory access that the law has had to take into account (Figure 25.3).

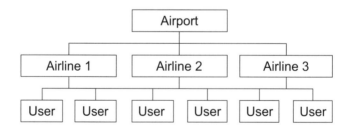

Figure 25.2 Threefold structure in aviation

---

18 Deregulation, hence, had the focus of opening up the nationality oriented characteristics of the transportation market.

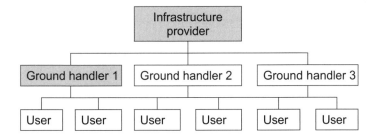

Figure 25.3 Threefold structure and vertical disintegration in ground handling

As a peculiarity of ground handling services, there is the problem of scarcity of space on airfields. Whereas, in principle hundreds of different rail companies may run their carriages on the same tracks, congestion is likely due to competition on the airfield. How then to allow limited competition there? The economist would advise using competition *for* the market as a partial substitute for competition *in* the market: if, due to space constraints, only one competitor can be allowed in the market, then the use of a tendering process allows for the selection of the fittest competitor in the market.

Market Dominance and Deregulation: On the Relation of Sector-Specific Regulation and General Competition Law

Sector-Specific versus General Competition Law

The deregulation of certain industries is typically the result of a deliberate policy choice. This choice is implemented by a set of regulatory rules, specifically tailored to open up the respective industry and to ensure a framework for workable competition (sector-specific competition law). The regulatory rules would typically oblige the holder of the monopolistic infrastructure to open up its facilities for competitors on the downstream market; it would enforce discrimination-free access to the infrastructure and also deal with pricing issues, namely the problem of monopolistic overpricing.

Saying that deregulation is effectuated by sector-specific competition law begs the question as to the relationship of the rules of general competition law and the rules for dominant companies. The answer is that these sets of rules are independent of each other, duplicating the constraints on the dominant company. On the one hand, the dominant company is subject to the detailed obligations of sector-specific competition law. On the other hand, the dominant company enjoys market dominance and must, therefore, refrain from abusing its dominant position. Furthermore, some aspects of market abusive behaviour obligate the holder of a bottleneck facility to open up its bottleneck facilities for potential newcomers in the downstream market: in this respect even the rules of general competition law may serve as a market opener.[19]

---

19   See the following section, 'Refusal to deal'.

Transitory Role of Sector-Specific Competition Law?

In some industries, sector-specific competition law is considered to play a transitory role designed to open up the monopoly and to allow competition to gain a foothold. To the extent that competition thrives, the scope of regulation may shrink and, ultimately, phase out. This phasing out of sector-specific regulation rests on certain assumptions regarding the source of this regulation, namely the monopolistic bottleneck. What is assumed is that the basis of the bottleneck may get eroded by new developments and market forces or, at least, that general competition law will suffice to discipline the incumbent operator. In other markets (for example, railways), where the bottleneck is stable and specific, a similar development is hardly conceivable: here, regulation of the monopolistic bottleneck is likely to be permanent.

There are two issues that merit attention here. First, the bottleneck nature of a specific (earthbound) infrastructure is not a static attribute of the respective facility, but the outcome of a certain given technological and factual environment, in which – due to the current cost structure – this infrastructure cannot be challenged by a competitor. It is, therefore, entirely conceivable that a facility which has a bottleneck infrastructure loses this attribute because of a shift in the respective cost structure. It is not the old infrastructure monopoly losing its control over what has established the bottleneck, but the impact of this control being eroded in the presence of new facilities.

Secondly, the market power of airports (with respect to their control over the airfield) is, in itself, stable and unlikely to be eroded by technological advancements. Still, supply shocks that open up new airport opportunities for airlines do exist, and their effects can be quite dramatic. Since the end of the Cold War, many military airports have been transformed into civil airports, tremendously increasing the opportunities for airlines. Depending on the specifics of the airport and its catchment area on the one hand, and the characteristics of consumer demand on the other hand, such a development may erode the bottleneck nature of a previously monopolistically held facility.[20]

Abuse of a Dominant Market Position

The Nature of Abusive Behaviour

General competition law does not prohibit the emergence or maintenance of a dominant position. The law aims to prohibit the abuse of a dominant position. Even though the respective law contains a general clause and also concrete examples of abusive behaviour, the interpretation is difficult. On the one hand, there still exists the concept of exploitative abuses, that is, of abuses by which the dominant firm uses its dominant position to extract monopoly rents from consumers. On the other hand, there is no doubt that the true economic problem with abuses lies in its potential for anti-competitive foreclosure. It is this second category that deserves attention.

Let us discuss two cases of possible abuse, namely refusal to deal and price discrimination.

---

20   The Military does not transform its airports into civil airports. The respective work is money intensive. Regional authorities do have budgets to perform these services and/or to offer attractive conditions for new (low-cost) airlines to assume services. What follows is the state-aid problem of regional airports. For a discussion of this problem, see the contribution by Kristoferitsch, Chapter 22 in this volume.

Refusal to Deal

The legal framework   Let us turn first to a category of cases that rank under the rubric of vertical abuse. These cases concern abusive behaviour on the upstream market that could possibly exclude access to the neighbouring downstream market for potential competitors.

The most prominent of these cases is refusal to deal. This category of abuse is heterogeneous. However, all sub-categories share the common feature: that a dominant position in the upstream market reduces, withholds or impedes access to a certain input that is required for the potential service provider on the downstream market to effectively compete. The incentives for the dominant firm to prevent access to the downstream market stem from the fact that this dominant firm has some vested economic interest in certain providers on the downstream market. Often the firm holding a dominant position on the upstream market is also, directly or via a subsidiary, present on the downstream market, and by impeding access for potential competitors, it favours its own economic position.

Refusal to deal cases can be divided into three categories:

- first, the termination of an existing business relationship by the dominant company without proper reason;
- second, the selective refusal to deal with one possible transaction partner despite the existence of comparable business relationships with other partners (a discrimination between transaction partners);
- third, refusal to deal in the sense of declining access to one's own (monopolistically held) facilities for potential competitors on the downstream market in order to maintain a monopoly position on this derived downstream (service) market.[21] In this category the abuse concept works as a market opener, as it forces the holder of a facility to open access to this facility for new entrants as well. It is clear that the conditions under which enforced access should be granted have to be carefully balanced. The case law of the Commission and the ECJ defines these conditions in the light of what is called 'The Essential Facilities Doctrine'. Under this concept, the monopolistic holder of a facility which is essential for providing services on the downstream market (as it cannot be duplicated physically or reasonably by any player on the downstream market) has to grant (compensated) access to this facility, provided that this access is feasible.[22]

Refusal to deal in aviation and non-aviation   Let us now explore the impact of the refusal to deal category for airports. Typically, airports hold a dominant position on the upstream market. Is it realistic to assume that an airport has incentives to impede access to its facility for downstream service providers? Going back to the three-tier structure, one immediately recognizes the difference between aviation and non-aviation (ground handling).

In the case of aviation, the airport is typically not a provider of transportation services on the downstream transportation market. The economic interest of the airport consists of selling its services to airlines and in expanding the transportation market volume, not in constraining or reducing it. Refusal to deal is, hence, unlikely to be an issue.

---

21   This refusal may be mantled by an actual offer with unfavourable business conditions (so-called constructive refusal).

22   Lewisch (2007) item 340. The leading case by the ECJ is the ruling of 28.05.1998, C-7/97 concerning Oscar Bronner. One sees the similarity between the bottleneck concept and the essentiality of a facility. A bottleneck infrastructure is also essential for competition on the downstream market.

In the realm of non-aviation (ground handling and other services) the airport is itself a provider of services on the downstream market. Thus, the airport has a clear incentive to impede access to its monopolistically held infrastructure for the sake of favouring its own service arm. A case decided by the Swiss Competition Commission of 18 September 2006 (Wirtschaft und Wettbewerb 2007, p. 432) addressed non-aviation outside the realm of ground handling, namely valet parking at Zurich airport. Zurich airport reorganized valet parking and terminated the existing contracts with two providers of valet parking services so that only regular parking in parking garages provided by the airport remained. The competition commission, first, assumed that Zurich airport held a dominant position on the relevant market for the 'provision of airport facilities for the off airport-parking of passengers' and, second, considered the termination of the respective contracts to constitute an abusive foreclosure.

A second case concerns the opening of a market by reliance on the prohibition of abusive behaviour. This second case, decided by the European Commission, concerned the airport at Frankfurt/Main (Commission Decision 98/190/EC, IV/34.801 of 14.01.1998). The airport authorities refused the request by a potential new ground handler to grant access to its airfield. This request occurred well before the enactment of the respective sector specific deregulation and was, therefore, solely decided on the basis of Article 82 EC-Treaty. The Commission assumed that the airport, being the sole provider of airport facilities at this location, enjoyed a dominant position on the upstream infrastructure market. As to the merits of the case, the Commission considered the airport's refusal to deal abusive, namely as an attempt to monopolize the market for airfield services.

Price Discrimination

Let us now turn to the issue of price discrimination. Again, the three-tier analysis of aviation provides a good framework. If the economic self-interest of airports consists in expanding their business, why would the airport as infrastructure holder selectively favour/disfavour certain service providers? The obvious answer is that an apparent self-interest for such discrimination would require some kind of linkage to the benefited service provider.

In the Portuguese airport case (judgement by the ECJ of 29.03.2001, C-163/99), a system of discounts on landing charges at Portuguese airports, namely both quantity discounts and a 50 per cent reduction for domestic as opposed to international flights, was at issue.[23] The reason for the Portuguese airports, as state managed entities, to offer these discounts was the simple fact that the airlines benefiting the most from these discounts[24] were national Portuguese airlines.[25] Such discriminatory treatment can be avoided if either the airports or the airlines, or both entities, are privatized.[26]

---

23   Because the airport administration was a government matter, this case did not directly concern an abuse case, but was a matter of state responsibility for failure to comply with the competition law rules (one of them being the prohibition of abusive behaviour).

24   Whereas the system of quantity discounts was formally open to all airlines, it selectively favoured those airlines flying the most at Portuguese airports, namely the national airlines. The same held true for the 50 per cent reduction for national flights.

25   In this sense, they were subsidiary companies of the State as the provider of the infrastructure.

26   In the Paris airport case (judgement of the ECJ of 24.10.2002, C-82/01 P), the managing company of the Paris airports was held to have violated the prohibition of abusive behaviour 'by using its dominant position ... to impose discriminatory commercial fees in the Paris airports ... on suppliers or users engaged

Whereas one would expect that cases of abusive price discrimination tend to disappear as the market develops and privatizations take place, this is not usually the case. The reason for the ambiguous situation regarding price discrimination and differing rebates is a legal one, namely the current process of modernization regarding Article 82 EC-Treaty. This modernization process, inspired by the European Commission, is not aimed at amending the text of Article 82, but at providing a new conceptual foundation that would allow for more consistent and focused application. Prior to this modernization phase, both the Commission and the ECJ have followed a very restrictive position regarding legitimate price discrimination by dominant companies, requiring some objective cost factor (i.e. differences in the production or the transaction) that would explain the different cost treatment. The more economics based approach is, in turn, less concerned with price discrimination as such, as long as it is not used for foreclosure purposes (i.e. to selectively favour one's own affiliate company). This more relaxed position grants ample opportunity for price discrimination according to market conditions (i.e. according to differences in the elasticity of demand, or for some variant of peak-load pricing or externally based pricing). It would be a positive development if the case law of the ECJ would follow this position with respect to transport markets in general and airports in particular.

Summary

Typically, airports hold a monopolistic position over their infrastructure. Potential for competition emerges to the extent that airports become the object of choice. The respective choice options depend on the characteristics of the downstream transport market, namely consumer preferences. Even though access to a certain airport's infrastructure remains indispensable to airlines when carrying out flight services at a specific location, airports lose market power if consumers' travel preferences allow them to switch between locations. A similar position is taken in the European Commission's 'Community Guidelines on Financing of Airports and Start-Up Aid to Airlines Departing from Regional Airports'. Whereas fuelling stops in the Middle East on flights from Europe to Asia provide an early example of such airport competition, inter-continental traffic at European hub airports is a current illustration of this.

The three-tier structure of transport markets serves as an ideal analytical tool to study and compare the role of airports as infrastructure providers in the realm of aviation and in ground handling. Different to aviation, airports regularly assume a twofold role as infrastructure provider and as service provider with respect to ground handling. The economic approach also provides a framework of reference with respect to the decisions of the ECJ and the European Commission, which are discussed in this chapter.

Sector-specific competition law and general competition law are two separate, and mutually independent, legal fields that both aim at constraining the economic behaviour of airports as the holders of a dominant bottleneck facility. With respect to possibly abusive behaviour, the three-tier structure of transport markets allows for an economic and legal evaluation of refusals to deal by airports both in aviation and in ground handling. An increasingly economically oriented position is also warranted regarding price discrimination, in particular rebate schemes: it is here, where not only cost differences, but differences in market conditions (i.e. characteristics on the demand side) should legitimate price differentiation.

---

in ground handling or self-handling activities relating to catering ... to the cleaning of aircraft and to the handling of cargo' (item 17).

References

Knieps, G., 1996 *Wettbewerb in Netzen*, J.C.B. Mohr (Paul Siebeck), Tübingen.
Knieps, G., 2005 *Wettbewerbsökonomie* (2nd edn), Springer Berlin, Heidelberg.
Lewisch, P., Artikel 82 – Kommentierung. In: Mayer, H. (ed.), *Kommentar zu EU- und EG-Vertrag, 78. Lieferung 2007*, Manz'sche Verlags- und Universitätsbuchhandlung, Vienna.
Wirtschaft und Wettbewerb, 2007, http://www.wuw-online.de.

Chapter 26
Airport Competition: A Perspective and Synthesis

Peter Forsyth

Introduction

Airports are no longer seen as necessarily locational or natural monopolies, and thus competition between them is seen as a distinct possibility. As the chapters in this book show, there are many aspects of competition between, and within, airports. Head-to-head competition across the product range is not common, though it does exist. However competition for different services and market segments does exist, and is sometimes intense. Thus market power and competition issues with airports are not matters of black and white – rather they are of varying shades of grey. This means that empirical and policy issues are both subtle and complex.

There are number of themes which arise in this book. They include:

- What is the extent of this competition, and how intense is it?
- What preconditions are there for this competition to come about, and how effective might it be?
- What are the most significant impediments to the development of airport competition, and how might they be addressed?
- What policy problems are developing from airport competition?

Each of these questions is discussed in this chapter. However, as with other areas of inquiry, there remain some unanswered questions – these are posed in the conclusion.

The Extent and Intensity of Competition

The chapters in this volume have documented a wide variety of ways in which airports compete. These include:

- competition across the board, for traffic in general;
- competition for specific types of traffic, for example low-cost carriers (LCCs) – see Lei et al. – or for cargo traffic (see Tretheway and Andriulaitis);
- competition in specific services, for example retail (see Kincaid and Tretheway; Graham), car parking and for facilities for services such as maintenance;
- competition between major airports as hubs;
- competition as bases for airline operations, especially LCC bases;
- competition between destinations, leading to indirect competition between airports at different destinations (see Kincaid and Tretheway), along with competition for services (see Morrell);

- competition within airports such as between terminals (see Reynolds-Feighan) and ground handling (see Templin).

This competition can take place over several variables, in particular:

- pricing and charging;
- quality of services;
- investment and available capacity – this affects the ability to compete in the long run, quality and congestion.

Documenting the extent of competition is quite straightforward. It is much more difficult to measure the intensity of competition. Is competition sufficiently strong to eliminate market power or significantly constrain its use? In some dimensions, this appears to be so. Some airports compete strongly against each other to be the bases for LCC operations – this is especially so for airports which only serve LCCs. Airports can be quite competitive with each other, as well as off-site providers, in retail. Numbers of sellers and their proximity to one another are not the only determinants of market outcomes. In addition, whether or not users such as airlines possess countervailing power (see Button) ,and the extent and nature of airline competition (see Pels and Verhoef), can affect outcomes.

From a policy perspective, the most critical issue concerns how strongly airports are competing for their core business – handling flight and passenger movements. If competition is strong, regulation is unnecessary and probably counterproductive. If competition is weak or non existent, the airports will have market power, and if they are privately owned and profit-oriented, or publicly owned without any pressure to reduce costs, they will make use of it by charging high prices. Regulation, explicit or light-handed, will be needed if it is desired to keep prices down. Evidence suggests that competition is sufficiently strong amongst the smaller UK airports to keep prices down and render regulation unnecessary (see Starkie). Competition among smaller airports, if allowed, might also be strong enough in other countries, such as Germany (see Strobach; Malina) or France, to leave airports unregulated. In some countries such as Greece, while there are several airports spread across a region, they tend to serve local markets and not compete (see Papatheodorou).

However there does not seem to be much evidence of strong competition between the airports of medium to large cities. In the cases of very large cities, such as Paris and London, with several airports, the potential for competition has been eliminated by common ownership (see Forsyth and Niemeier). Competition between the airports of different cities is not likely to be sufficiently strong as to eliminate the need for regulation, since most cities with major airports in Europe are sufficiently far apart for them to be weak substitutes for most of their traffic.

The Preconditions for Competition

Ultimately, for competition to be effective, users need to be free to choose between suppliers. In particular, it is essential for the ultimate users, mainly passengers but also freight shippers, to be able to choose. Different airports need to be able to provide essentially the same or highly substitutable product to the passenger – travel between a specific origin and specific destination. Airports which are close to one another may be regarded by passengers as good substitutes. Hubs can compete not just if airlines regard them as substitutes; it is also necessary that potential passengers regard alternative routings through different hubs as effective substitutes. Airports might compete not directly for passenger traffic but for airlines, as some airports do for LCC bases. This might not give rise to much competition between origin destination markets for passengers, though it can set up a degree of destination competition, which can lessen the market power of the airports at the competing destinations. Airports which are distant from one another can compete as cargo hubs if cargo airlines and forwarders regard different routings as good substitutes. Passengers may have some flexibility over where they make retail purchases, and airports can compete in this area. Airports can also compete in the related services that they or their partners offer to airlines, such as facilities for maintenance. However in all of these cases, the ultimate decision-maker, often though not always the passenger, must have the freedom to choose between the options. For this to be the case, several preconditions need to be fulfilled.

Proximity and Overlapping Catchments

The greatest scope for competition is where airports are close to one another and have overlapping catchments. This would happen if there are two or more airports in a city with capacity to handle more traffic. While one airport might be the preferred option for residents of some districts in the city, the closeness of the airports means that most residents or visitors to the city have a real choice of which airport they use. Another locational environment which is conducive to competition is where airports are spread fairly evenly over a densely populated wider region, as is the case in Britain and, to a lesser extent, Germany. For most passengers to a particular destination or a particular origin, there may be a choice of two or more airports.

One issue which is less well settled is where there is some, though limited, overlap between catchment areas of airports (see Starkie). If each airport is the sole effective supplier to, say, 85 per cent of the passengers in its catchment area, but there is an overlap for the remaining 15 per cent of passengers, how effective will competition be? If it is difficult for the airports to price discriminate and offer more favourable terms to the passengers who do have a choice, any price or quality competition will affect all passengers. However, would it be worthwhile for an airport to dilute its revenues from the majority of its customers in order to win a greater share of a minority?

The concept of overlapping catchments is not a precise one, and the delineation of catchment borders depends on what is assumed about the willingness of travellers to spend time coming to and going from airports. The strength of competition between airports depends on to what extent travellers regard them as substitutes. If travellers have a strong preference for short access times and costs, even airports which are quite close to one another will been poor substitutes and not compete. Alternatively, if travellers are willing to incur long travel times to and from airports, even distantly separated airports may be competitors. This will vary of course by the dominance of short- versus long-haul flights. Passengers travelling long-haul international flights might view airports as substitutes, while short-haul domestic passengers would not. This highlights the need to estimate travellers' willingness to travel to access airports, and how they trade off access times for other attributes of airports. To this end, the theory and evaluation of consumer choice is of particular relevance. This theory is discussed (see Gaudry; Hess) and empirical aspects of choice are also explored (see Hess and Polak; Strobach; Malina). The evidence suggests that sometimes travellers do have an effective choice of airport, but they may have a strong preference for low access times and general convenience. This would imply that competition between airports would not be strong. The passengers who use LCCs probably would not have as strong a preference for low access times, and they would be willing to travel further to save money. Hence airports catering for this market segment will compete more strongly.

Spare Capacity and the Ability to Handle Extra Demand

For airports to compete effectively, it is necessary that they have the capacity with which to handle any extra traffic they win (see Elliott). There is little point in reducing prices, or increasing quality at a cost, if the extra demand cannot be accommodated. For effective competition, all airports which are close enough to compete need to have spare capacity. If one airport has spare capacity and a nearby airport is full, the former can gain traffic from the latter, but not vice versa. The full airport has no incentive to reduce its charges or increase its quality if doing so is costly. The airport with spare capacity will have flexibility over its prices, but will lower these only to the extent that this adds to profits. It will take advantage of its rival's excess demand position, and will not be forced to reduce its prices down to cost.

Thus competition is observed where airports do have spare capacity. This is the case for most or all of the smaller airports in Britain. There are also examples of city airports which have spare capacity and which face competition from secondary airports, especially for the business of LCCs (see Forsyth). Sometimes there will be several airports in a large city which all have some spare capacity; however, typically in multi-airport cities, some airports are full and airports are all owned by the same entity, or both. In London, there will be little scope for the airports to compete if and when they are under separate ownership, in the short to medium term, because of the tightness of capacity constraints.

Independent Ownership

For airports to compete, they need to be under separate ownership. This is recognised in the proposals to force BAA to divest itself of one or two of its London airports (see Forsyth and Niemeier). As with other institutional aspects, the notion of ownership is becoming less clear-cut (see Wolf et al.). In particular, there are instances of airport alliances forming, such as that proposed between Amsterdam Schiphol and Aéroports de Paris. One motivation of alliances is to align the objectives of separate entities with one another, and to thus remove the scope for competition. Minority shareholdings may be used to strengthen the alignment of objectives and lessen competition. Mergers of independent airports will eliminate competition between them – often this is the primary objective of the merger.

Absence of Regulatory Hurdles to Competition

Regulation has the potential to affect the way in which competition works between airports. Airports are often regulated, especially when they are privately owned or operated. Regulation can be cost based or incentive based, as is the intention with regulation applied to the London airports. Regulation can involve the specification of prices, quality levels and investment programmes. If regulation sets these parameters, it overrides the ability of airports to compete on them. Regulation can also impact on the incentives to compete – if an airport is almost guaranteed a level of revenue no matter what it does, what incentive does it have to compete? A regulated city airport (for example, Hamburg) might be subjected to competition from an unregulated secondary airport (for example, Lübeck) – how it responds to this competition depends on how it is regulated. If and when the London airports are separated, it is proposed that they continue to be regulated. This poses real issues of how regulation will impact on competition and how the regulation can be designed to foster rather than hinder competition.

Problems in Achieving Competition

Difficulties in Entry

Entry and exit are key aspects of the competitive process – in many 'competitive' industries, there is a continuous flow of entry and exits. This is less so for infrastructure industries with large sunk investments, such as airports.

Entry is often, though not always, difficult to achieve with airports (see Müller-Rostin et al.). If an airport with market power is charging high prices and earning high profits, it is rarely easy for another airport to enter and compete the prices down. Airports are associated with major environmental costs, and require large amounts of land, and obtaining planning authority to build is typically difficult to obtain. When new airports are built in cities, delays in obtaining planning permission of over a decade are not uncommon. Cost conditions with airports are relevant, though complex (see Pels et al.). Setting up a new airport in a city from scratch involves a very large capacity commitment, much of which is sunk. This will be risky if the incumbent airport has spare capacity and a more convenient location. Across Europe, and elsewhere, there are very few examples of private investors setting up new airports. There have been some entries, mainly from former military airports which are converted to commercial use. With these airports, the main large indivisible expenditure has already been sunk, and the additional expenditure to enable the airport to handle passengers or cargo is not necessarily large. Thus there have been some entries in Britain and, interestingly, from publicly owned airports in Germany. The effects of airport size on unit cost are not clear-cut – there is some evidence of falls in operating costs per passenger as airports become larger, though this effect is not likely to be so strong as to make smaller airports inherently uncompetitive (especially given the importance of other factors such as access times and costs). Also it may be that larger airports are operating on a higher cost function due to indivisibilities, governance or x-inefficiencies.

In the long run, growth in demand and the need for new capacity can be consistent with the promotion of competition. When demand for an existing airport is expected to outstrip its capacity, governments sometimes give the go-ahead for a new airport. At this stage, they can choose to facilitate competition by requiring separate ownership of the new and existing airport. Alternatively they can limit completion by giving the existing airport the right to develop the new airport (as the UK government did by allowing BAA to develop Stansted – see Starkie, 2004). Typically, governments choose this latter approach.

Restrictions on Investment and Capacity Expansion

To be competitive, an airport must have spare capacity. Many airports, especially in Europe, face demand which is at, or in excess of, capacity for part or all of the day. This currently will stop them from competing for a particular segment of traffic should they wish to do so. In an unconstrained environment, a firm which is seeking to compete but which has inadequate capacity will invest to expand capacity. However, with airports, the typical situation is that the airport does not have discretion over significant capacity expansion, particularly runway expansion. Even where it is not restricted from investing by economic regulation, it will need permission to expand from local and central government planning authorities. Granted the environmental sensitivities associated with airports, this permission will be slow to obtain, and sometimes expansion is not permitted under any circumstances. Thus it is possible to have a group of independently owned airports with the scope to compete, but capacity limitations and the difficulties in expanding capacity will prevent them from doing so, even in the medium to long term. Even with separate ownership of the London airports, achieving a situation in which they have the capacity to compete will take several years to engineer. In a similar way restrictions such as noise regulations or night curfews hamper the ability to compete due to a lack of environmental capacity or capacity expansion.

Traffic Segregation and Airport Specialisation

A city might have two or more airports which have the scope to compete. However, planners often assign specific roles to different airports – one is to handle one type of traffic while the other is to handle a different type of traffic. Thus one airport might be reserved for domestic traffic and another mainly international traffic. In Europe some airports were reserved for scheduled traffic while charter traffic was required to use other airports – with the growth of LCCs and the decline of charter operations, this distinction is now breaking down. Such assignment of specific traffic to specific airports has the effect of limiting the scope for competition. If there are real operational or economic gains to be obtained by this segregation of traffic, they need to be weighed against the gains from competition. Such gains are probably questionable – segregation may be required for bureaucratic convenience, or it may be a means of protecting preferred users (by allowing preferred airlines to access the more convenient airport).

Investment and other Coordination under Competition

The argument that separation of ownership, and competition, will lead to inefficient lack of coordination between airports in investment and other aspects of operations is potentially one of the most telling arguments against competition. The most important aspect of coordination is investment coordination, though other aspects could be present. Until recently management of BAA has made very effective use of the argument – that of the lack of coordination in investment which would come about under separate ownership of the London airports – in opposing break-up.

The argument is that separate decision-making by independent airports would lead to an inefficient pattern of airport investments in a city or region. There could be either under- or oversupply of capacity, and what capacity is provided could be inefficiently located. In an oligopoly context, it is possible that investment in capacity would be inefficient; either though firms underinvesting to keep prices high, or some firms undertaking excessive investments to forestall competitors. Hence it may be more efficient to have single overall control of the investment programme, and the single airport owner would invest where needed, and when needed. Such a single owner would also be potentially better informed than independent owners because it would be well informed about all the options and all traveller preferences.

This argument is superficially appealing, though it does not fit very closely with the reality of airport investment decision-making. There are many problems with it. These include:

- The argument presupposes that the objective of the airport company is to maximise welfare – in reality a privately owned airport company will not seek this objective, and it will perhaps seek to maximise profit; though regulation, which it is very likely to be subject to, will distort its objectives and it may well seek inefficient outcomes, e.g. though maximising size or similar objectives.
- A profit-maximising firm would have a strong incentive to invest excessively so as to forestall entry by competitors.

- If the airport firm is a regulated monopoly, it may be under little pressure to perform or invest efficiently. For example, it may use profitable investments to cross-subsidise inefficient and unprofitable investments.
- Much investment in airports is very long term and, under competition, airports will be aware of their rivals' plans and will have a strong incentive to take them into account.
- Most major investments at airports are subject to detailed scrutiny by planning authorities and governments, and airports do not have the freedom to simply invest as they wish. Governments can, and typically do, produce overall plans for airport investments in a city or region. Coordination of investments can be imposed by governments through various mechanisms such as regulation or veto powers.
- Similar situations, with sunk costs and long lead times, exist in other infrastructure industries, such as electricity, and there is little evidence that separated ownership leads to less well coordinated investments than separated ownership.

Empirical information on the benefits from coordination is not readily available. In the context of the proposed separation of the London airports, BAA has claimed that separation would result in a one-off cost of £106 million, along with an annual cost from loss of economics of scale of £100m (Competition Commission, 2008, Appendix 4). In addition it claims non-quantified costs from loss of coordination of investment. The Competition Commission considered that the costs of separation would be much lower than this, though neither side has provided a rigorous or thorough assessment.

Competition Policy Issues

Recognising that there is some scope for competition amongst airports gives rise to a number of policy issues. Some of these are:

- whether competition can be an adequate substitute for regulation;
- whether, at the time of privatisation, groups of airports should be sold together or separately;
- whether mergers or alliances between airports should be permitted or prohibited;
- whether subsidies to regional airports should be permitted.

Competition as a Substitute for Regulation

If airports possess market power, they can be regulated. However, it is well recognised that regulation has its problems. Incentives to keep costs down, provide the right quality and invest will be weakened if airports are regulated. There is a danger that incentive regulation will revert to much more cost-based regulation, with implications for performance. Competition, if it is feasible, is a better way of keeping prices and costs down than regulation.

Competition amongst airports is not sufficiently strong to enable a general recommendation that regulation should be dispensed with. Nonetheless, there are situations where competition is sufficiently strong and regulation is unnecessary.

An example of this is with the small to medium-sized airports in the UK. Many of these airports are privately owned, and they are not subject to price regulation. They compete with each other, and there is little evidence of the presence or abuse of market power (see Starkie). It could be that their behaviour is being affected by an implicit threat of regulation – this could be so with BAA. Even some large airports which are or have recently been regulated, including Manchester and Stansted, have not been setting as high price levels as they are permitted to do under the price caps.

The presence of competition may also impact on how tight regulation needs to be. There will be situations where competition will be present, but it will be constrained and sometimes weak; though it may be sufficiently strong to lessen the opportunity for serious use of market power. In such situations, it may not be necessary to implement full regulation – light-handed regulation may be sufficient. Light-handed regulation may not keep prices exactly at cost, but it will give the airports more flexibility with which to operate. It will give the airports more scope to negotiate with airlines over investment and with users over the quality of service to be provided. When there are two or three airports in a city competing, albeit in a constrained manner, light-handed regulation may be preferable to either no regulation or full regulation. This could be the case for London, or other cities which choose to separate the ownership of large city airports.

Privatisation and Separation

In a number of cases, governments own groups of airports, some of which might be able to compete with one another. If these governments decide to privatise their airports, they can choose to sell them separately or as a group. In some cases, airports are distant from one another, and there is little scope for competition between them – in such situations, it does not matter much whether they are separated or not. In Europe, because distances between airports are often small, there is the possibility of stimulating competition by selling airports separately. In the UK, the London airports were sold as a group, though now they are likely to be broken up. It is possible that the argument for separation and competition is not strong in some circumstances – airports may have little spare capacity, and may not have much freedom to invest to change this. In addition, there are arguments against separation. In particular, it has been argued that the coordination of investment is better under common ownership. This argument has not yet been tested empirically – would separate airports overinvest in an attempt to gain market share? In the case of investment programmes, city airports are very constrained by regulators and planning authorities and the scope for uncoordinated and excessive investment is limited. Separation of the London airports should provide some evidence on competition between busy city airports and on investment coordination between them, though this evidence will be slow in coming and also case specific.

Mergers, Alliances and Anti Competitive Behaviour

With mergers and alliances being proposed between airports, competition authorities will have to determine their attitude to them. Already, the competition authorities of Austria and the Slovak Republic have had to address the issues raised by the takeover of Bratislava airport by Vienna – which was proposed, though which now has been abandoned – in the light of opposition from Slovak authorities (see Wolf et al.). Clearly, with alliances between airports which are widely separated, there will few problems with lessening of competition (though two airports which are far apart can compete as hubs, as for example Schiphol and Charles de Gaulle). Competition authorities will have to assess the scope for competition between the airports, and assess the reduction in welfare when airports which could compete either merge or form an alliance. They will have to set this cost against any benefits that the merger or alliance might deliver. An emerging test will be that posed by the proposed alliance between Amsterdam Schiphol and Aéroports de Paris airports. The proponents claim large benefits from the alliance, though the benefits mentioned are somewhat vague. These two airports may be sufficiently distant from one another to have only a small overlap in their catchment areas, but they can compete as hubs.

Competition authorities may not just be concerned about structural issues such as mergers. As with other industries in which market power exists, there is the issue of anti-competitive behaviour, such as refusal to deal (see Lewisch). There is also the problem of airports affecting competition in downstream airline markets. General competition law may be sufficient to address problems which might arise, and airport-specific measures might not be necessary.

Subsidy Policies

Subsidies to airports will affect the competitive balance between them, and will very often result in an inefficient distribution of traffic between the airports (see Elliott; Kristoferitsch; Morrell on subsidy issues). The European Union has been concerned about the effects of subsidies on airline competition, and has sought to ensure that subsidies at an airport are available to all airlines serving it. However, such subsidies can distort competition between airports. It is often argued that airports stimulate economic activity in a region, are 'jobs machines' and can be used to attract tourists (through low airport charges attracting LCCs). In light of this, it is not unreasonable that regional governments will seek to attract business to their regions. While this makes sense for the region considered on its own, it does not make sense for nations or groups of regions, since the traffic won by one region comes, at least to a large extent, from other regions. The gain by one region is offset by losses faced by others.

The impacts of subsidies will depend on the circumstances at specific groups of airports. If one airport wins traffic from another which has ample spare capacity, subsidies will distort the pattern of traffic (see Forsyth). On the other hand, if a subsidy given to a secondary airport with ample capacity helps win traffic from a congested airport, it may lessen the overall cost of delays if this congestion is not efficiently handled at the congested airport. Even if the airport has efficiently priced its congestion, through use of a slot system, it is possible that there may be an advantage in shifting traffic from it because road access to the busy airport is subject to congestion which is not efficiently priced.

Another policy problem is that it may be difficult to determine whether a subsidy is present and, if so, how large it is. Straightforward cash subsidies may be easy to identify. However, an airport may have been set up with infrastructure facilities, such as a runway, which it paid very little for (as is often the case with former military airports) – is this a subsidy if the facilities had no alternative uses? Sometimes secondary airports claim that, in the past, the major airports have been the recipients of substantial subsidies. This may be so, but does this imply that they continue to have an advantage because of past support?

Concluding Remarks – Some Unanswered Questions

Any survey such as this book necessarily raises a number of unanswered questions. Some of these go to the heart of how competition between airports will work, and others have key policy implications, such as whether the gains from breaking up airport groups to promote competition will be offset by losses from lack of coordination. Some of the questions which emerge are:

- If airports are permitted to compete, how strong will competition between them be? Airports are separated in space, and location determines how convenient they are for travellers and

shippers. How low do access costs and times have to be for users to regard different airports as good substitutes? As noted, the answer to this depends on the nature of the traffic, with airports being better substitutes for one another for price-conscious leisure traffic than for business traffic.
- If airports which have catchment areas that overlap, does the presence of this overlap enhance competition strongly or not? Airports will compete for traffic in the overlapping area, but does the existence of this area lead to stronger competition in parts of the catchment area which do not overlap?
- Can the UK experience of effective competition between small to medium-sized airports be replicated in other countries? Conditions for competition in the UK are favourable, with a large number of airports spread over a densely populated region.
- How can competition and regulation be reconciled? Airports might be sold to separate owners and have the scope to compete, but competition might not be strong enough to dispense with the need for regulation. Regulation will have a strong influence on the way competition works, and it will be important to design regulation in a way such that at least it does not hinder competition, and, preferably, enhances competition. The solution to this poses a secondary question – if competition is imperfect but regulation is costly, when does it make sense to remove regulation?
- What are the gains from single ownership and coordination of airports that are close together; and how large are they compared to the gains from competition? Governments and competition authorities are having to determine whether single firms can own close-by airports with the potential to compete. Strong claims have been made for the benefits of coordination, but how valid are these?
- How large are the gains from alliances and common ownership of airports that are not adjacent, but which may compete in some market segments, for example for LCC traffic or as hubs? Airports do compete as hubs, but how strong is this form of competition, and is it strong enough to keep prices down? If alliances are formed between airports with no potential to compete, there are no competition policy issues. If competition is reduced in some market segments, competition authorities will need to determine whether the gains from such alliances outweigh any anti-competitive effects.

References

Competition Commission (UK) (2008) *BAA Airports Market Investigation: Provisional Decision on Remedies*, Competition Commission, London, December.

Starkie (2004) 'Testing the Regulatory Model: The Expansion of Stansted Airport', *Fiscal Studies*, 25, 389–413.

Airport Competition: Some Key References

Introduction and Overview

Air Transport Group, Cranfield University (2002), *Study on Competition between Airports and the Application of State Aid Rules, Final Report, Vol. 1*, September European Commission, Directorate-General Energy and Transport, Directorate F-Air Transport.

Barrett, S. D. (2000), 'Airport competition in the deregulated European aviation market', *Journal of Air Transport Management* 6, 13–27.

Dresner M., Windle R. and Yao Y. (2002), 'Airport barriers to entry in the US', *Journal of Transport Economics and Policy* 36, 389–405.

Fewings, R. (1998), Provision of European airport infrastructure, *Avmark Aviation Economist*, July, pp. 18–20.

Forsyth, P. (2006), 'Airport Competition an: Regulatory Issues and Policy Implications', in: Lee, D. (ed.), *Advances in Airline Economics: Competition Policy and Antitrust*, Amsterdam, Elsevier, pp. 347–68.

Starkie, D. and Thompson, D. (1985), 'Privatising London's Airports', Institute for Fiscal Studies, Report 16, London.

Travellers Choice and Airport Competition

Blackstone, E., Buck, A. J. and Hakim, S. (2006), 'Determinants of airport choice in a multi-airport region', *Atlantic Economic Journal* 34, 313–26.

Dresner, M. and Windle, R. (1995), 'Airport choice in multiple-airport region', *Journal of Transportation Engineering* 121, 332–7.

Harvey, G. (1987), 'Airport choice in a multiple airport region', *Transportation Research A* 21, 439–49.

Mandel, B. (1999), Measuring Competition –Approaches for (De)Regulated Markets, in: Pfähler, W., Niemeier, H.-M. and Mayer, O., *Airports and Air Traffic, Regulation, Privatisation and Competition*, Frankfurt, Peter Lang, pp. 71–92.

Pels, E., Nijkamp, P. and Rietveld, P. (2003), 'Access to and competition between airports: a case study for the San Francisco Bay area', *Transportation Research A* 37(1), 71–83.

Airport Strategies: How Airports Compete

Dresner M., Lin, J. and Windle, R. (1996), 'The impact of low-cost carriers on airport and route competition', *Journal of Transport Economics and Policy* 30, 309–28.

Jarach, D. (2005), *Airport Marketing – Strategies to Cope with the New Millennium Environment*, Aldershot, Ashgate.

Pels, E., Nijkamp, P. and Rietveld, P. (2000), 'Airport and airline competition for passengers departing from a large metropolitan area', *Journal of Urban Economics* 48, 29–45.

Vowles, T. M. (2001), The 'Southwest Effect' in multi-airport regions, *Journal of Air Transport Management* 7, 251–8.

Warnock-Smith, D. and Potter, A. (2005), 'An exploratory study into airport choice factors for European low cost airlines', *Journal of Air Transport Management* 11, 388–92.

Airport Competition for Freight

Gardiner, J., Ison, S. and Humphreys, I. (2005), 'Factors influencing cargo airlines' choice of airport: an international survey', *Journal of Air Transport Management* 11, 393–9.

Ohashi, H., Kim, T.; Oum, T. H. and Yu, C. (2005), 'Choice of air cargo transshipment airport: an application to air cargo traffic to/from Northeast Asia', *Journal of Air Transport Management* 11, 149–59.

Policy Issues

ACI Europe (1999), 'European airports: a competitive industry', Policy paper submitted by the Airport Council International (ACI) Europe Policy Committee.

Barbot, C. (2006), 'Low-cost airlines, secondary airports, and state aid: an economic assessment of the Ryanair–Charleroi Airport agreement', *Journal of Air Transport Management* 12, 197–203.

Competition Commission (2008), BAA market investigation provisional findings report, August, http://www.competition-commission.org.uk.

European Commission (2004), The Commission's decision on Charleroi airport promotes the activities of low-cost airlines and regional development. IP/04/157, Brussels, European Commission.

European Commission (2005), Community guidelines on financing of airports and start-up aid to airlines departing from regional airports. Communication from the Commission, Official Journal C 312, 9 December.

Starkie, D. (2002), 'Airport regulation and competition', *Journal of Air Transport Management* 8, 63–72.

Airport Index

Amsterdam Schiphol 41, 67, 100, 107–108, 122–123, 130, 141–143, 277, 345, 396, 406–409, 430, 434
Athens 277–290

Barcelona 67, 142, 266
Basel 269–275
Berlin Schönefeld 252, 266
Berlin Tegel 107, 253, 266
Bratislava 347–351
Brussels 28, 68–69, 99, 142–147

Charle de Gaulle Paris 17, 19, 28, 33, 41, 67, 72, 121–123, 141–145, 277, 396, 401–404, 408
Charleroi 15–23, 68–69, 93, 121, 278–279, 312, 365–376, 379–390

Dublin 353–363

Edinburgh 312, 293–294, 299, 321–323

Frankfurt 67, 95, 107, 122, 130, 141–147, 240, 249–252, 269–273, 277, 287, 395–397, 404–405, 408
Frankfurt Hahn 12, 16, 36–38, 68–69, 97, 142, 240, 278, 287, 365–366

Glasgow 69, 100, 107, 293–294, 299, 312, 321–323

Hamburg 12–13, 29, 36, 69, 107, 115–116, 121, 142, 241, 246, 252, 266

London Gatwick 11, 19, 23, 28, 39, 67, 108, 121, 134, 142, 185, 293–294, 298–304, 312, 321–336
London Heathrow 11–12, 15–16, 28, 39, 41, 67, 96–97, 103, 106–108, 121, 134, 141–142, 185, 277, 286, 292–294, 298, 302–304, 321–336, 396, 401–403, 406–410
London Luton 15–19, 32, 39, 68–69, 121, 134, 185, 278, 293–295, 299–305, 312, 322–323, 332
London Stansted 11–24, 28, 32, 68–69, 93, 121, 134, 142, 185, 279, 287, 293–295, 298–304, 312, 321–336, 371, 389, 431, 434
Lübeck 12–13, 8, 69, 77, 121, 134, 241, 252

Madrid 67, 142, 396–400, 406–407
Manchester 11–23, 39, 65, 69, 108, 142, 293–303, 311–312, 323
Münich 15–16, 36–38, 67, 107, 142, 252, 269–274

Orly Paris 67, 121, 142, 266

Prague 41, 266

Robin Hood Doncaster 39–42, 292–303
Rome 15, 67–69, 408–410

Strasbourg Airport 19–22, 269
Stuttgart 269–276

Vienna 28–29, 96, 107, 142, 347–351

Index

Abuse of dominant market position 413–415, 422
Aeronautical revenue 38, 311–317
Air cargo 137–147, 343
Air travel behaviour 151–176
Airline alliances 65–73, 339–351
Airline choice 55, 177–195, 198–207
Airport
    access mode choice 153–195, 262–267
    alliances 89–109, 339–353, 430–436
    brands 134, 343–344
    charges 89–92
    economic regulation 11–17, 27–32, 77–85, 91–93, 291–292, 307–308, 321–336, 431
    gateway competition 137–147
    ownership 17, 33–39, 291–295, 301, 307, 339–351, 430–436
    pricing 47–57, 60, 66, 78–86, 322–324, 334, 399, 428
    quality 108–117, 262–274, 321–335, 396–410, 428–430
    regional development 19, 28–29, 287, 311, 438
    retailing 92, 100, 120, 125, 307, 427–429
    subsidies 18–19, 32, 42, 82–86, 366–376, 379–390, 433–435
    support for new routes 18, 22, 381, 385–388
Aer Rianta 353–360
Altmark Trans-case 368–376
Attraction to passengers 261–275

Bilateral monopoly 61–74
Bottleneck – concept 413–425
Branding 92, 134, 344
BAA Plc. 294–309, 321–336, 361–362, 430–436

Catchment Area 261–275, 91–101, 429–436
Charleroi Airport decision 20–21, 369–376, 379–390
Choice modelling 151–230, 429
Civil Aviation Authority (UK) 292–306, 323–336, 361–362

Commission for Aviation Regulation (Ireland) 354–362
Community Guidelines 366–377
Competitive advantage 89–101, 198
Concentration 277–288
Congestion 47–55, 81, 321–328, 428–435
Convenience Factors 261–268
Cost function 47–55, 103–118, 43
Cost recovery 78–84, 382–389
Countervailing power 59–74, 250–253, 297, 428
Cournot 29–30, 47–55

Deregulation of ground handling 396–401, 428
Discrete choice 151–174, 220
Double marginalization 53–55, 341–342
Dublin Airport Authority 354–361

Economies of scale 31–42, 54, 77–84, 90–97, 117, 302, 340–341
Economies of scope 31–35, 291, 306, 340–351
Employees 399–401
Empty core 60
Essential facilities doctrine 423
EU directive of ground handling 393–400
EU rules 17–18
European Commission 365–374
Externalities 381–390

Five forces framework 89–91
Fixed costs 33–34, 42, 49–52, 117, 132, 291–302
Flight Frequency 179–187, 263–267
Freight forwarders 140–146, 429

Galbraith, J. K. 60–73
Game theory 61
Generation-Distribution 209–230
Gini Index 281
Gravity model 207–230
Greece 277–287
Ground Handling 393–410, 428

Herfindahl-Hirschman Index (HHI) 281–284
Horizontal integration 97, 341, 349–351
Hub 393–410
Hub-and-spoke systems 65

Intermodal hub 11–13
Internalization 48–57
Investment in airports 31–34, 295–307, 321–336, 383–388, 428–434
Ireland 353–364

Landing fee discounts 18–23
Leisure travel 152–160, 278–287, 315
Logit
   Mixed 173, 181–190
   Nested 206–225
London 266, 291–308, 321–336, 396–409, 428–434
Low Cost Carriers (LCC) 64–74, 77–84, 89–100, 278, 296–300, 311–317, 365–376, 427

Marketing 93–99, 119–134, 145, 341–347
Market entry and exit 27–43, 66, 431
Market power 27–36, 47–57, 59–73, 239–243, 250–253, 321–332, 340–342, 366, 414–418, 425, 427–435
Market shares in ground handling 406–408
Mode split 203–214, 230
Monopoly
   Locational 32, 47–48, 85, 341, 427–429
   Natural 31–33, 59–74, 77, 104, 307, 427–429
Monopsony 62–73

Negotiating power 296–297, 398–399, 409
Network effects 381–390
Networks 54–55
Non-aeronautical revenue 14, 38, 92, 311–314

Oligopoly 30, 48, 62, 335, 341
Operating aid 370–376
Operational costs 103–117
Ownership 59–63, 291–307, 321–336, 339–351, 428–436

Panel Data Analysis 313–314
Passengers 109–117, 119–139, 393–410
Path choice 197–215, 228–230
Price discrimination 61, 71–72, 79–80, 86, 325–328, 422–425

Principle of private investor 368–374
Privatisation 89–101, 291–307, 322, 344–350, 433–434
Product differentiation 94–96, 139, 302
Public-Private Partnership (PPP) 285–287
Public Service Obligation (PSO) 278–280

Ramp services 393–407
Refusal to deal 423–424
Regional Airports 78, 86, 277–279, 286, 291–307, 311–317, 433–435
Regulation
   Airport 27–28, 36, 59–73, 104, 291–292, 354, 360, 428–430
   price cap 29, 81, 321–336, 434
   single-till 322–326, 354
Road feeder services 141–143
Ryanair 13–14, 20–24, 278–279, 285, 297–300, 355–357, 365–376

Secondary airports 68, 77–86, 430, 435–436
Sector-specific competition law 413, 421–425
Services of general economic interest 365–376, 379–380
Slots 47–51, 60–62
Start-up aid 365–377
State aid 17–23, 365–376, 379–390
Secondary airports
   Competitive impact 34–35, 77–86, 427, 430
   Strategy 93–99, 427, 430

Tourism 277–287
Traffic accounting matrix (TAM) 197–203, 229
Translog function 103–116
Transport markets 413–425
Terminal competition 357–362

United Kingdom 291–308, 311–317, 321–323, 333–336, 428–436

Value chain 61–63, 73
Variable costs 104–118
Vertical disintegration 419–421
Vertical foreclosure 422–425
Vertical integration 98–101

X – inefficiency 59, 73, 326–328